W9-CCI-604

A professor of zoology at the University of Nebraska, **Paul A. Johnsgard** has been intensely interested in upland game birds and waterfowl for most of his life. With the aid of National Science Foundation research grants and a Guggenheim fellowship, he has traveled North America from the Bering coast of Alaska to the tropical forests of Yucatán and southern Mexico, and from Puget Sound to Newfoundland, seeking out and studying virtually every native species of waterfowl and upland game.

Dr. Johnsgard has published the results of his studies in more than forty technical papers and journal articles, and in three previous books: *Handbook of Waterfowl Behavior* (1965), *Animal Behavior* (1967), and *Waterfowl: Their Biology and Natural History* (1968) *(see below)*.

Recognizing the need for a comprehensive survey of the biology of grouse and quails, and the opportunities for research provided by these birds, Dr. Johnsgard devoted his entire research time from 1968 through 1971 to the present project. To obtain distributional and ecological information on the lesser known quail species he made two extensive trips to Mexico, and several trips to the western states and Canada to obtain firsthand information on all of the North American grouse species.

Grouse and Quails of North America

Grouse and Quails of North America

Paul A. Johnsgard

UNIVERSITY OF NEBRASKA · LINCOLN

Publishers on the Plains
UNP

International Standard Book Number 0-8032-0810-3
Library of Congress Catalog Card Number 77-181596

Most recent printing shown by first digit below:
2 3 4 5 6 7 8 9 10

Manufactured in the United States of America

To my children—Jay, Scott, Ann, and Karin—
in the sincere hope that they and their
children will be able to enjoy these
birds as much as I have

Table of Contents

List of Illustrations

Plates

FIGURES

Tables

Preface

NEXT to the waterfowl, upland game birds have always occupied a special place in my heart. Some of my earliest memories are of riding along dusty North Dakota roads in the mid-thirties on pheasant and prairie chicken hunts with my father, long before I was able to carry a gun myself. My recognition of upland game as something other than exciting targets began during a 1952 tour of North Dakota game refuges while collecting waterfowl breeding records for an undergraduate special project. On one chilly May morning Merrill Hammond, biologist of the Lower Souris National Wildlife Refuge, drove me to a sharp-tailed grouse display ground. I watched the "dancing" of the grouse with fascination but, in retrospect, in virtual ignorance. At that time, terms like *sign stimuli, fixed action patterns,* and *isolating mechanisms* were foreign to me, and I was inclined to view the birds' behavior as a wonder of nature rather than as an intricately beautiful example of natural selection.

My next few years were spent intensively watching waterfowl, and through them I gradually gained insight into the significance of social behavior patterns in avian reproduction. Often while watching mallards displaying I would think back on the morning I watched sharp-tailed grouse, and ponder the parallels and differences between the lek displays of grouse and the social pair–forming displays of ducks. It wasn't until almost ten years later, in 1962, that I had an opportunity to renew my memories of grouse display. Then, in southeastern Nebraska, I spent an unforgettable April morning in the midst of a prairie chicken booming ground, and I became an immediate addict to grouse watching. As an ethologist, I could finally understand the evolutionary significance of these fantastic behavior patterns, and appreciate the marvelous opportunities that the grouse provided for behavioral studies under natural conditions.

Unfortunately, grouse do not readily adapt to captivity, nor can their social behavior patterns be studied to advantage in such situations, thus I made no attempt to establish a captive flock. Instead, I decided that the New World quails provided a great potential for experimental behavioral studies and taxonomic research that had been largely overlooked by other investigators. Besides being relatively easy to keep and to breed in captivity, they exhibit a complex vocal repertoire that may readily be subjected to acoustical analysis. Further, the prior records of hybridization under natural conditions and in captivity suggested studies not only of possible genetic interest but also of potential taxonomic significance. Finally, the quails' ecological and behavioral adaptations provided such a striking contrast to those of their relatively close relatives, the grouse, that a comparison of the two and an evaluation of the possible reasons for these strong differences appeared warranted.

My plans for a comparative study of New World quails first took form in the fall of 1966, and were greatly facilitated by a National Science Foundation research grant (GB–7666X) awarded in the spring of 1967. This grant allowed me two summers of field work in Mexico during 1969 and 1970, where I traveled over ten thousand miles by car, establishing distributional limits and obtaining live specimens of various species of Mexican quails. In the winter of 1968–69 I first decided that a book-length summary of grouse and quail biology was worth undertaking, and during the academic year 1969–70 I began to actively collect references and wrote the first drafts of the early subject-heading chapters. I did not begin writing species accounts until the academic year 1970–71, during which I was granted a leave of absence by the University of Nebraska Research Council. Their financial assistance, and that provided by a Guggenheim Foundation fellowship during the spring and summer of 1971, allowed the completion of the manuscript.

The writing of a book on an assemblage such as the grouse or quails is greatly facilitated by the enormous body of technical literature that results from their importance as game birds. A useful work by Charles Crispens, Jr., *Quails and Partridges of North America: A Bibliography*, was published in 1960 and includes over two thousand references. No comparable bibliography exists for the North American grouse, but the Fish and Wildlife Service's *Wildlife Review* has abstracted over six hundred works published between 1935 and 1970 dealing with North American grouse species. Of these, 40 percent were concerned with the ruffed grouse, 18 percent with prairie chickens, about 10 percent each with sage grouse, blue grouse, sharp-tailed grouse, and ptarmigans, and 3 percent with the spruce grouse. During the same period nearly eight hundred publications on North American quail species were abstracted, of which approximately

80 percent dealt with the bobwhite, 10 percent with California quail, 6 percent with Gambel quail, and the remaining 4 percent concerned scaled, mountain, and harlequin quails. Far too many research studies on both grouse and quail have also been hidden in game agency reports that never are formally published and thus are, in effect, buried without benefit of epitaph. Except in a few necessary cases, where such information has been presented that was not otherwise available on a species, these sources were not used in this book. A recent index summary of published Pittman-Robertson research (Tait, 1968) provides a useful literature guide.

Far more interesting than digging through library stacks to obtain information have been my opportunities to see under natural conditions most of the species included in the book. Of the twenty-five included species, I have observed in life all of the nine species of grouse, both of the introduced partridges, and all but two of the fourteen species of quails. These birds have been observed in such diverse areas as the arctic tundra near Hooper Bay, Alaska, the lowland rain forests of Chiapas, and the Sonoran desert of Arizona. For example, during three memorable weeks in 1970 I waded in hip-deep snow along Trail Ridge of Rocky Mountain National Park while photographing white-tailed ptarmigan, climbed the humid and misty cloud forests of Hidalgo in search of bearded tree quail, and sweltered under a blistering Acapulco sun while trapping barred quail with mist nets. These great diversities in their ecology are one of the attractions of the grouse and quails; virtually every major community type in North America has been successfully occupied by one or more species of the group. As a result, every state and province in the United States and Canada supports at least one species of grouse or quail that may be legally hunted.

Partly because they were written at different times, the two major sections of the book have slightly differing outlooks. The first nine chapters, which are generally comparative in nature, are written in a somewhat formal, technical fashion. The individual species accounts were written with the thought in mind that not only professional biologists but also hunters and bird watchers will perhaps be reading them, and some attempt has been made to make them less formidable than the earlier chapters. Purists may object to this dual philosophy. Yet, in looking back on my own development as a biologist, it was the sections on habits, life history, or life story in the classic ornithological references that first captured my attention, and only much later did technical aspects of ornithology appeal to me. Thus, it is hoped that the people who obtain this book to read the species accounts will perhaps take an occasional look at the earlier chapters, and that the theoretical ecologist or evolutionist will also admit that his data must be based on actual living birds that possess both esthetic and scientific beauty.

A word of explanation about the basis for inclusion of species might be in order. All native species of grouse and quail occurring north of the Mexico-Guatemala border are included in the species accounts. By extending the geographic coverage to Panama, it would have been necessary to include *Colinus cristatus* (or "*leucopogon*"), *Rhynchortyx cinctus*, and four additional species of *Odontophorus*. Virtually nothing is known of the ecology or reproductive biology of any of these species, thus their inclusion in this book would have no great value. On the other hand it was decided to include both the gray partridge and the chukar partridge, since these species are well established in North America and considerable research on their biology has been carried out. In addition, they provide an interesting comparison with the true New World quails in terms of their ecology and behavior. In contrast, the ring-necked pheasant was purposely excluded; it has been well described in several monographs and is apparently not as closely related to the native quails as are the two introduced partridges.

Although I have been actively involved in research on the grouse and quails for four years, I must honestly say that very little in the present book represents new and original information. Nearly all of the findings reported are those of others, and the most that I can claim is credit for bringing them together in a single volume. Lest the reader believe that little research is left to be done on North American grouse or quails, he need only read the accounts of such species as the elegant quail, the harlequin quail, or the Mexican tree quails. Even for such intensively studied species as the bobwhite and ruffed grouse much more research might be done; I hope one of the virtues of this book will be to point out some of the great gaps or weaknesses in our knowledge. When initiating my research on grouse and quail after so many years of studying waterfowl, I felt as if I were embarking on an uncharted ocean. Since then I have discovered no new continents or even any major islands, and at most have simply confirmed or remeasured the depths already plumbed by others. Yet, inasmuch as any new voyage is an exciting one, I hope that others will see fit to follow me.

No voyage of any length is normally undertaken alone, and I must here express my great appreciation to the persons and agencies that assisted me. Foremost among the agencies that have assisted me are the National Science Foundation, the J. S. Guggenheim Foundation, and the Research Council of the University of Nebraska, all of which provided financial support for this study. Other institutions that have provided data, lent specimens, or allowed me to utilize their collections, are the American Museum of Natural History, the United States National Museum, the Chicago Field Museum of Natural History, the University of California

Museum of Vertebrate Zoology, the Los Angeles County Museum, the Denver Museum of Natural History, the James Ford Bell Museum of Natural History in Minneapolis, and the Chicago Zoological Park. The Cornell University Laboratory of Ornithology very kindly allowed me to reproduce a previously unpublished painting by L.A. Fuertes, and in addition provided copies of several sound recordings. Nearly all of the United States and Canadian game and wildlife agencies provided me with information about hunting seasons and, in many cases, data on estimated upland game harvests. The Secretaría de Agricultura y Ganadería of Mexico and its director general, Dr. R. H. Corzo, facilitated my Mexican field work and provided the necessary permits for collecting specimens.

Among the individuals who have personally assisted me I am particularly indebted to C. G. ("Bud") Pritchard, who painstakingly prepared five of the color paintings included in the book, and whose meticulous attention to the smallest details of feather and soft-part characteristics unfortunately cannot be adequately reproduced by the printing process. Likewise, on short notice John O'Neill set aside his other obligations to produce two stunning paintings of Mexican quail species that testify both to his great artistic abilities and to his personal familiarity with these tropical forest birds. Without the splendid paintings by these artists the book would have much less value and attractiveness. Charles Hjelte of the Colorado Department of Natural Resources very kindly allowed me to reproduce three excellent paintings done for that department by Dexter Landau, for which I am most grateful.

Other persons who personally helped me are too numerous to mention individually, but I cannot neglect Andrew Prieto or Edmund Sallee, who accompanied me on my Mexican trips, or Clait Braun, James Inder, and John Lewis, who assisted me with my field work in the United States and Canada. Dr. Starker Leopold gave me valuable advice and information; were it not for the groundwork provided by his research in Mexico my own work there would have been much more difficult and time-consuming. Many persons provided photographs, and although not all of them could be used, I wish to extend my thanks to David Allen, George Allen, Clait Braun, Edward Brigham, Glenn Chambers, Don Domenick, Kenneth Fink, Sean Furniss, Harvey Gunderson, C. G. Hampson, Joseph Jehl, K. C. Lint, Stewart MacDonald, M. Martinelli, Alan Nelson, Raphael Payne, Bruce Porter, C. W. Schwartz, Roger Sharpe, Charles Shick, Robert Starr, and Mary Tremaine. In particular, I appreciate Ken Fink's generous donation of his outstanding collection of grouse and quail photographs for my use.

Dr. Ingemar Hjorth very kindly allowed me the use of two of his published illustrations, for which I am very grateful. Many persons assisted me by allowing me to observe or photograph birds in their collections, providing me with valuable specimens, or supplying me with information. Among these are F. E. Strange, William Huey, William Lemburg, and Glen Smart.

The use of the facilities of the Department of Zoology of the University of Nebraska has been of great benefit to me, and I must acknowledge the work of several of my graduate students, especially Daniel Hatch and Calvin Cink, in caring for birds and in maintaining incubation and rearing records. I owe a special debt of gratitude to Viki Peterson and Mrs. Janette Olander, who as departmental secretaries often neglected more pressing duties to type or retype a section of the manuscript without the slightest hint of complaint.

Finally, and most importantly, I must thank my wife, Lois, for patiently enduring too many summers alone, and for lovingly accepting too little gratitude in return.

Introduction

NEARLY all of the gallinaceous birds that are native to North America are included in two taxonomic groups, the grouse-like species of the subfamily Tetraoninae, and the quail-like species of the subfamily Odontophorinae. The former represent a temperate and subarctic group of about sixteen species which collectively have a widespread distribution in the Northern Hemisphere, and over half of which are found in North America. The latter group is a strictly Western Hemisphere assemblage that collectively includes about thirty species, almost half of which occur north of the Mexico-Guatemala border. Most of the remaining quails are tropical forest birds of northern and western South America about which very little is known. Thus, evidence suggests that North America was originally doubly colonized by early gallinaceous stock; from the south by basically tropical-forest-adapted birds that have evolved into the present array of quail species, and from the north by relatively arctic-adapted forms that have given rise to the present species of ptarmigans and grouse. Convergent evolution of these two separate but related stocks has since allowed much of North America to become inhabited by birds having similar ecological adaptations and in some cases overlapping distributions.

Within each of the two ancestral groups, evolutionary radiation has developed an interesting spectrum of anatomical variations, ecological adaptations, and behavioral specializations. These latter two aspects—adaptational niche variations associated with habitat differences, and behavioral variations associated with maximal reproductive efficiencies under varied climates, habitats and contacts with associated species—are the primary subjects of this book. Anatomical and physiological considerations will be given some attention in the early chapters, but the primary focus will be on the living bird in its natural environment.

In the species accounts, the summaries of the ranges are in general derived from *The American Ornithologists Union Check-list of North American Birds* (1957), modified as necessary to take recent changes and new information into account. This will be referred to as the "*A.O.U. Check-list*." Likewise, the ranges of the strictly Mexican species are generally based on the *Distributional Check-list of the Birds of Mexico* (1950) by Friedmann, Griscom, and Moore (referred to as the "*Check-list of the Birds of Mexico*"). In cases where subspecies have been described since the publication of these books, they are listed but are identified as not yet verified. In a very few instances, subspecies described earlier but not recognized by the A.O.U. by 1957 have been recognized here. Also, contrary to current A.O.U. practice, most of the accepted subspecies have been given vernacular English names. However, such subspecies have normally been designated by simply adding a descriptive term to the species' vernacular name, so that confusion in species identification may be avoided. This usage of special vernacular names was felt desirable in view of the rather broad species concept employed in this book and the proposed merging of certain forms that have usually been recognized as separate species. In a few instances this has forced a deviation from vernacular names of American species as used by the *A.O.U. Check-list*. I have avoided possessives in English vernacular names, using for example Gambel quail rather than Gambel's quail. For strictly Mexican species I have in general followed the vernacular terminology used by A. S. Leopold in *Wildlife of Mexico: The Game Birds and Mammals*. Measurements indicated for each species were largely derived from those appearing in *The Birds of North and Middle America*, part 10, by R. Ridgway and H. Friedmann. Unless otherwise indicated, measurements for the folded wing represent unflattened wings, and tail measurements are from the tip of the tail to the point of insertion.

Part I

Comparative Biology

1

Evolution and Taxonomy

EVOLUTIONARY HISTORY

THE modern array of grouse-, quail-, and partridge-like species occurring in North America is the result of three processes: evolution and speciation within this continent, range expansion or immigration from Central America and Eurasia, and recent introductions by man. The last category accounts for the presence in North America of the chukar and gray partridges, which are both natives of Europe or southern Asia and typical representatives of the quail-like and partridge-like forms that have extensively colonized those land masses. It is still necessary to account for the presence of the nine or so species of grouse-like forms that are native to this continent, as well as the fourteen or fifteen species of New World quails that occur north of the Guatemala-Mexico border. In general, the evidence clearly indicates that the New World quails had their center of evolutionary history and speciation in tropical America, whereas the grouse are a strictly Northern Hemisphere group that perhaps originated in North America but which now occur throughout both this continent and Eurasia and at present represent about an equal number of species in each of the two hemispheres. North America therefore has provided the common ecological conditions to which two distinctly different groups of gallinaceous birds have become independently adapted and have undergone somewhat convergent evolutionary trends.

The evolutionary history of grouse- and quail-like birds on this continent is a long one, going back to at least Oligocene times, from which an indeterminate quail-like fossil is known, an addition to a unique fossil quail genus *(Nanortyx)* (Tordoff, 1951). Perhaps *Paleophasianus* from the Eocene represents the earliest grouse-like fossil (Holman, 1961), although it is more probably a species of limpkin (Cracraft, 1968). Other known North American fossil species are summarized in table 1. According to Larry

TABLE 1

FOSSIL QUAILS AND GROUSE FROM NORTH AMERICA*

	Quails	*Grouse*
Lower Oligocene	*Nanortyx inexpectatus* Weigel	
Lower Miocene	*Miortyx aldeni* Howard	*Palaealectoris incertus* Wetmore
Middle Miocene	*Cyrtonyx cooki* Wetmore; *Miortyx teres* A. H. Miller	*Tympanuchus stirtoni* A. H. Miller
Upper Miocene		*Archaeophasianus roberti* (Stone); *Archaeophasianus mioceanus* (Shufeldt)
Middle Pliocene	*Lophortyx shotwellii* Brodkorb	
Upper Pliocene	*Colinus hibbardi* Wetmore	
Lower Pleistocene		*Tympanuchus lulli* Shufeldt
Middle Pleistocene	*Colinus suilium* Brodkorb; *Neortyx peninsularis* Holman	*Palaeotetrix gilli* Shufeldt†; *Dendragapus nanus* (Shufeldt)‡; *Dendragapus lucasi* (Shufeldt); *Tympanuchus ceres* (Shufeldt) (also upper Pleistocene)
Total fossil genera	3	3
Total modern genera	3	2
Total fossil species	8	9
Neospecies from archeological sites	6	7

*Based on Holman, 1961, Brodkorb, 1964, and Howard, 1966

†*Dendragapus gilli* according to Jehl, 1969

‡Not separable from *D. lucasi* according to Jehl, 1969

Martin,* the Oligocene and Miocene forms share a number of common characteristics and in general are cracid-like. On this basis it seems a reasonable assumption that both groups may have been derived from cracid-like ancestors during mid-Tertiary times.

The present array of grouse and quail indigenous to America north of Guatemala includes nine species of grouse (ten if *Tympanuchus pallidocinctus* is recognized) and fifteen species of quails (fourteen if *Cyrtonyx ocellatus* is not recognized), as shown in table 2. Evidence that North America may be regarded as the evolutionary center of the grouse includes the fact that it has more total genera and more endemic genera than does Eurasia, although the differences are slight. In contrast, Central and South America exhibit the largest total species number of species as well as the largest number of endemic quail species (nearly all of which are in the large genus *Odontophorus*), whereas North America exhibits the largest number of genera and endemic genera. Since the apparently most primitive genera (*Dendrortyx* and *Odontophorus*) are of Mexican or more southerly distribution, it seems apparent that the center of origin of this group must be regarded as Middle American.

TABLE 2

Distribution of Extant Species of Grouse and New World Quails

	Central and South America	*North America*	*Eurasia*	*Total*
Grouse†				
Total genera	—	5	4	6
Endemic genera	—	2	1	—
Total species	—	9	9	16
Endemic species	—	7	7	—
Quails				
Total genera	6	8	—	9
Endemic genera	1	3	—	—
Total species	20	14	—	30
Endemic species	15	10	—	—

†Based partly on Short, 1967; *T. pallidocinctus* not recognized by him.

It is difficult to determine which of the extant genera of grouse is most like the ancestral grouse types. Short (1967) argues the *Dendragapus* includes the species that possess a greater number of primitive features than do the species of any other extant genus. However, he also mentions two species

* Larry Martin, 1971: personal communication.

of *Bonasa*, two of *Lagopus*, and one of *Tympanuchus* that exhibit presumably ancestral traits, leaving only the genus *Centrocercus* as a relatively specialized genus. I am inclined to regard *Centrocercus* and *Tympanuchus* as the most highly specialized of the extant genera; both of them presumably evolved independently from forest-dwelling forms as arid habitats expanded during the late Tertiary times. I would similarly favor regarding the Holarctic genera *Dendragapus* and *Lagopus* as being nearest the ancestral types in general morphology, with the tundra-dwelling adaptations of *Lagopus* representing a more recent development than the forest-habitat adaptations of *Dendragapus*. The Holarctic genus *Bonasa* and the Old World genus *Tetrao* can then be considered somewhat more specialized offshoots of ancestral *Dendragapus-Lagopus* stock which have remained adapted to temperate forest habitats. These ideas are summarized in figure 1, which provides a suggested evolutionary tree for the extant grouse genera. This diagram seemingly differs considerably from that proposed by Short (1967), but actually represents an only slightly different way of emphasizing what are essentially very similar ideas. Our suggested sequences of genera are identical except for the position of *Centrocercus*, which I believe should be listed adjacent to *Dendragapus* to emphasize better its independent origin from *Tympanuchus*.

Similarly, the extant species and genera of New World quails can be grouped by their relatively primitive or specialized characteristics. There can be little question that the arboreal and long-tailed forms in the genus *Dendrortyx* exhibit a large number of generalized traits, and must therefore be regarded as nearest the hypothesized ancestral quail type. Holman (1961, 1964) reported that this genus exhibits numerous skeletal characteristics suggestive of those found in less advanced gallinaceous families, and, in addition, is the most aberrant extant genus of the group. Second only to *Dendrortyx* in generalized characteristics is the large and similarly forest-adapted but more ground-dwelling genus *Odontophorus*, which shares several primitive traits with *Dendrortyx*. Both genera also exhibit distribution patterns that center in Middle America or northern South America, the presumed area of evolutionary origin of the group.

From this central cluster of forms, it is relatively easy to derive, on zoogeographical, anatomical, and ecological grounds, two independent evolutionary lines in the New World quails. One such line leads in a generally northerly and more xeric-adapted direction, and presumably gave sequential rise to *Philortyx*, *Oreortyx*, *Callipepla*, and *Colinus* (which also moved south), as suggested in the accompanying evolutionary tree (fig. 1). The genus *Philortyx* is clearly transitional in its morphology and other characteristics between the suggested ancestral quail and these specialized

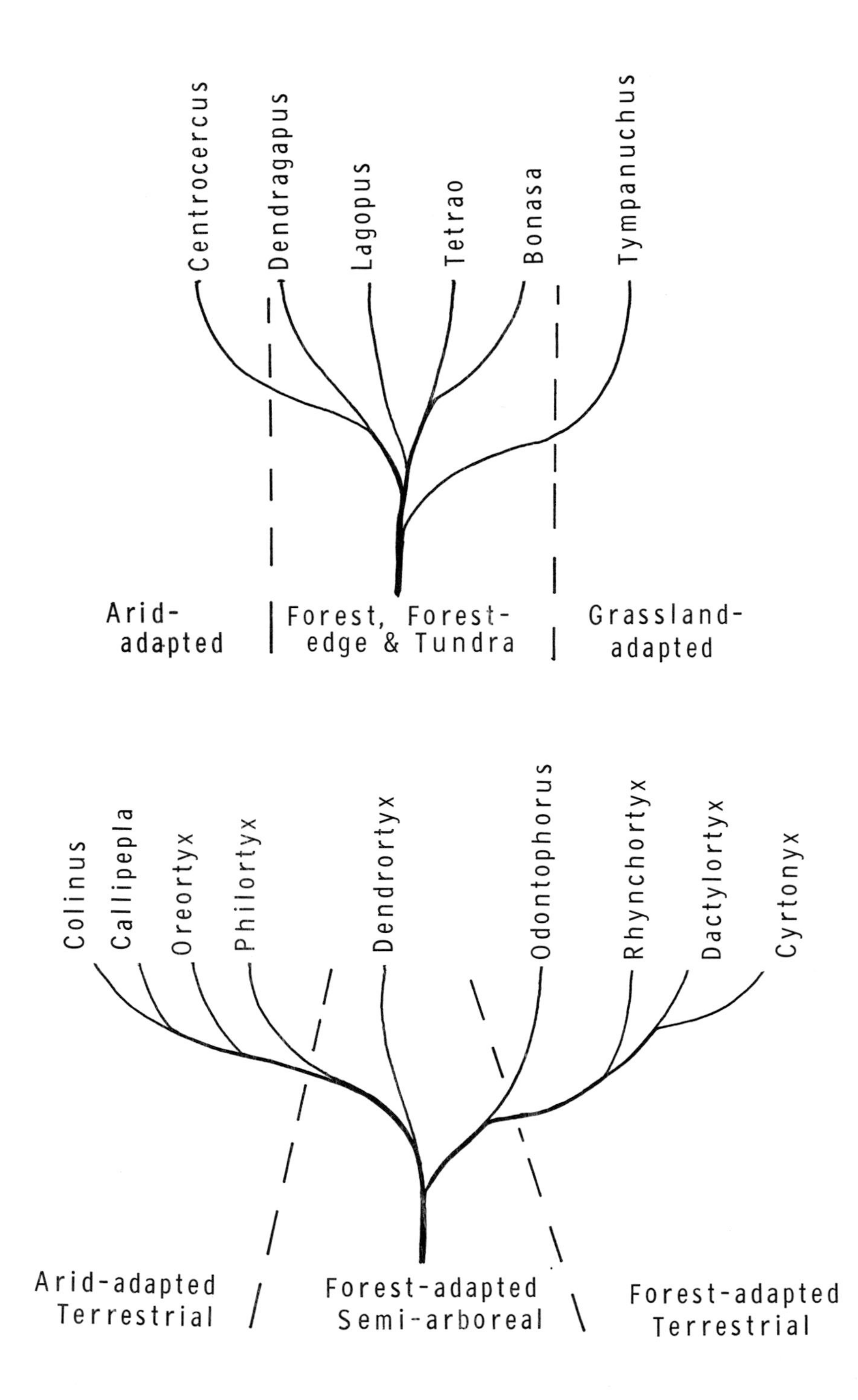

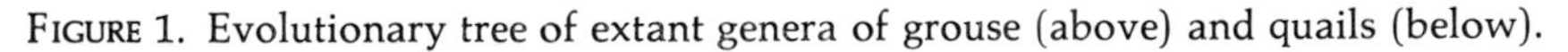
FIGURE 1. Evolutionary tree of extant genera of grouse (above) and quails (below).

and crested types, which are mostly seed-eating forms associated with open and often arid habitats.

From the *Odontophorus* nucleus, it is likewise fairly easy to derive the remaining three genera, *Dactylortyx*, *Cyrtonyx*, and *Rhynchortyx*. These are mostly Middle American forest dwellers that are in two cases relatively more specialized for digging for bulbs, rootlets, and tubers than for seed-eating. The long-legged and weak-toed *Rhynchortyx* differs in this regard, but nonetheless exhibits distinct skeletal similarities to *Cyrtonyx* and *Dactylortyx* (Holman, 1961).

GENERAL TAXONOMIC SEQUENCE AND HIGHER CATEGORIES

Until fairly recently, the traditional American treatment of the grouse has been to designate them as a distinct family, Tetraonidae, although the 1886 *A.O.U. Check-list* also included the New World quails in this family. Familial recognition of both the Tetraonidae and the Odontophoridae occurred with the third edition of the *A.O.U. Check-list* in 1910 and in the case of the grouse has persisted ever since. Other major authorities who have given a corresponding ranking to the grouse include Peters (1934), Ridgway and Friedmann (1946), Wetmore (1960), and Hudson et al. (1966). But recently a number of other writers have urged a reclassification of the group as a subfamily (Tetraoninae) of the Phasianidae. Some of the authors who have supported this view include Delacour (1951), Mayr and Amadon (1951), Sibley (1960), Brodkorb (1964), Holman (1964), Streseman (1966), Short (1967), and others. Hudson et al. (1966) admit that their basis for retaining familial status for the grouse is rather weak; it apparently stems in part from the fact that the grouse genera they studied were obviously much more closely related to one another than they were to any other genera. This would not seem to be sufficiently strong reason to maintain the family, in my view, nor would the obviously adaptive feathered condition of the tarsus and nostrils and the pectinate toes seem to justify such separation.

The level of separation of the New World quails is somewhat more difficult because of problems of separating real phyletic affinities from convergent similarities between this group and the Old World partridges and quails. Apart from occasional familial separation (Odontophoridae), as used for example in the 1910 edition of the *A.O.U. Check-list*, the group has generally been included in a subfamily of the Phasianidae. This was the procedure followed by Peters (1934), Ridgway and Friedmann (1946), Mayr and Amadon (1951), Sibley (1960), Brodkorb (1964), Holman (1964, but not 1961), Hudson, Lanzillotti, and Edward (1959), Hudson et al. (1966), Short (1967), and others. In these cases the Old World quails either were regarded as a separate subfamily, Perdicinae (Ridgway and Friedmann, 1946), or were more commonly included in the large subfamily Phasianinae (e.g., Peters, 1934; Sibley, 1960; Holman, 1964; Brodkorb, 1964; Short, 1967). A tribal (Odontophorini) recognition of the New World quails within the subfamily Phasianinae was advocated by Delacour (1961), while Streseman (1966) suggested closer affinities with the Old World quails by listing the New World species as a tribe of the subfamily Perdicinae. This question of relative closeness of relationship to the Old World quails and partridges seems to be the most important criterion in deciding whether the New World quails should be given subfamilial rank or simply listed as a tribe of the Phasianinae. On the basis of chromosomal studies, Jensen (1967) concluded that the New World quails are probably not as closely related to *Coturnix* and Old World partridges as they are to *Phasianus*. Hudson, Lanzillotti, and Edward (1959) and Hudson et al. (1966) reported a considerable number of similarities between New World quails and various Old World forms, particularly *Alectoris*, and seemed uncertain whether subfamilial separation was warranted. Arnheim and Wilson (1967) provide biochemical data suggesting close relationships between representatives of the New World quails and the Old World partridges and quails. Holman's (1961, 1964) evidence on skeletal anatomy, including some fourteen criteria, provides the strongest support for maintaining subfamilial separation and is the primary basis for the classification followed here. It would also seem desirable to distinguish taxonomically the true pheasants and their relatives (as recognized by Delacour, 1951) from the remaining Old World quails, partridges, and francolins, which may perhaps be best achieved by tribal separation, although several genera (*Ptilopachus, Ophrysia, Galloperdix,* and *Bambusicola*) provide intermediate characteristics.

Finally, it has been urged by several recent writers (e.g., Sibley, 1960; Brodkorb, 1964; Hudson et al., 1966; Streseman, 1966; and Short, 1967) that the turkeys and guinea fowl should probably be given no more than subfamilial recognition, but that the hoatzin (*Opisthocomus*) only very doubtfully belongs in the order Galliformes (Hudson et al., 1966). The summary of galliform classification shown in table 3 takes these recommendations into account.

GENERIC AND SPECIES LIMITS

As with many groups of birds that have been subjected to sexual selection and selection for reproductive isolation in a polygamous or promiscuous

TABLE 3

SUMMARY OF SUGGESTED GALLIFORMES CLASSIFICATION

ORDER GALLIFORMES

- Superfamily Cracoidea
  - Family Megapodidae—megapodes or mound builders (10 spp.)
  - Family Cracidae—chachalacas, guans, and curassows (38 spp.)
- Superfamily Phasanoidea
  - Family Phasianidae—pheasant-like birds (199 spp.)
    - Subfamily Meleagridinae—turkeys (2 spp.)
    - Subfamily Tetraoninae—grouse and ptarmigans (16 spp.)
    - Subfamily Odontophorinae—New World quails (30 spp.)
    - Subfamily Phasianinae—Old World pheasants (144 spp.)
      - Tribe Perdicini—Old World partridges, francolins, and quails (95 spp.)
      - Tribe Phasianini—pheasants, peafowl and jungle fowl (49 spp.)
    - Subfamily Numidinae—guinea fowl (7 spp.)

mating system (Sibley, 1957), the classification of the grouse has been confused by a plethora of generic names having little if any phylogenetic significance. Fortunately, Short (1967) has reviewed this situation from the viewpoint of both Eurasian and North American forms and has effectively stated the case in favor of elimination of several unnecessary generic names. Among the North American forms, these include the genera *Canachites (=Dendragapus)* and *Pedioecetes (=Tympanuchus).* At the species level, the American Ornithologists Union (1957) has already seen fit to merge *Dendragapus franklinii* with *D. canadensis,* and *D. fuliginosus* with *D. obscurus,* as essentially allopatric populations that are best regarded as subspecies.

The only remaining question relative to the grouse is that posed by the "lesser" form of prairie chicken, *Tympanuchus pallidocinctus,* which is still recognized as specifically distinct by the *A.O.U. Check-list.* Short (1967) summarized the evidence favoring the view that this population should likewise be regarded as only racially distinct from *T. cupido* and questioned the evidence presented by Jones (1964a) supporting species separation. More recently, Sharpe (1968) has also contributed his views, which in general are in agreement with those of Jones. The question is one that is impossible to provide with a clear-cut answer, and the conclusion one reaches reflects in large measure one's personal philosophy about the primary function of the species category. No additional evidence on the question has been gathered in this study, but *T. pallidocinctus* will not be given the space or attention that has been accorded the better-defined species.

Among the quails, problems of generic recognition are limited to relatively few instances. Most authorities (Peters, 1934; Ridgway and Friedmann, 1946; *A.O.U. Check-list*, 1957) recognize the genus *Lophortyx* as distinct from *Callipepla*. An adequate anatomical separation of these two genera has yet to be made, and the biological and anatomical validity of distinguishing them has been recently questioned by Sibley (1960), Holman (1961), Phillips, Marshall, and Monson (1964), Hudson et al. (1966), and others. Delacour (1961, 1962) synonymized both these two genera and *Oreortyx* and *Philortyx* as well but failed to provide adequate reasons for this procedure. I have suggested (1970), as has Holman (1961), that *Colinus* is clearly so closely related to the *Callipepla-Lophortyx* complex that it too is a highly questionable genus. Yet, since such lumping of *Colinus* with these other forms would tend to obscure the close relationships of the three bobwhite species with one another, I have refrained from doing so in this book. It is of some interest that the crested forms of bobwhite were once generically distinguished *("Eupsychortyx")* from the noncrested ones *(Colinus)* in a manner analogous to the separation of *Callipepla* from *Lophortyx* largely on the basis of crest condition.

At the species level, the primary problem concerns the possible justification for recognizing *Cyrtonyx ocellatus* as distinct from *C. montezumae*. This case, like that of the lesser prairie chicken, involves an allopatric population which is clearly a result of fairly recent separation. The biology of *ocellatus* is as yet unstudied, but until it can be proved to the contrary, it would seem most probable that the form should be regarded as a highly distinctive race of *montezumae*. In deference to tradition, however, it is listed separately in this book, although no individual account of its biology will be included.

Similarly, Mayr and Short (1970) have suggested that the Yucatán population of bobwhites *(Colinus nigrogularis)* is probably conspecific with *C. virginianus*. The question is complicated by the presence of a series of highly variable populations of *Colinus* extending from Guatemala all the way to northern Brazil. These have usually been regarded as consisting of two species *(C. cristatus* and *C. leucopogon)*, although as many as three species were recognized by Todd (1920). Monroe (1968) has argued for the lumping of these population groups into the single species *C. cristatus*, which thus exhibits as much plasticity in plumage variation in Middle and South America as does *C. virginianus* in Mexico and the United States. I am at present uncertain whether *nigrogularis* is phylogenetically closer to the *cristatus* group or to *virginianus*, and Holman (1961) reported that in its skeletal anatomy *nigrogularis* exhibits a generally intermediate condition (resembling *virginianus* in four of twelve characters, *leucopogon* in two characters, and being unique in six characters). Cink (1971) reported stronger vocal similarities between *nigrogularis* and *virginianus* than between *nigrogularis* and *cristatus*. A possible extreme solution would be to consider the entire complex of allopatric populations as a single species, but such a position cannot be justified on the basis of current knowledge, and representatives of the extreme types (*virginianus* and *cristatus*) are known to differ considerably in downy plumage, egg coloration, and nearly all vocalizations other than the male "*bob-white*" notes.

On the basis of these considerations, a list of the species included in this book is shown in table 4. Rather than being listed in taxonomic sequence, they have been organized according to zoogeography and the major plant community types with which they are most closely associated. A detailed identification of habitat preferences and range of ecological distributions is not possible in such a tabular comparison, but the individual species accounts in the second section of this book will provide a more accurate analysis of habitat characteristics of each species. What is of interest here is the large number of tropical and arid-temperate community types that have been colonized by the New World quails, and the corresponding habitat segregation in arctic and temperate community types of the North American grouse. Only in the case of the greater prairie chicken and the bobwhite is any ecological overlap indicated in the table, and certainly these two species also exhibit marked niche differences. The general geographic distribution of these vegetational communities is illustrated in figure 2, which has been derived from various sources. With a few exceptions, this map illustrates the distribution of potential climax vegetational types rather than successional or disturbance conditions.

An abbreviated systematic synopsis of the species included in this book follows, with subspecies excluded since they are listed under the individual species accounts:

Family Phasianidae: pheasant-like birds

Subfamily Tetraoninae: grouse and ptarmigans

Genus *Dendragapus* Elliot 1864

(Subgenus *Dendragapus*)

1. *D. obscurus* (Say) 1823: blue grouse

(Subgenus *Canachites* Stejneger 1885)

2. *D. canadensis* (Linnaeus) 1758: spruce grouse

Genus *Centrocercus* Swainson 1831

1. *C. urophasianus* (Bonaparte) 1828: sage grouse

Genus *Lagopus* Brisson 1760

1. *L. lagopus* (Linnaeus) 1758: willow ptarmigan
2. *L. mutus* (Montin) 1776: rock ptarmigan
3. *L. leucurus* (Richardson) 1831: white-tailed ptarmigan

Genus *Bonasa* Stephens 1819

1. *B. umbellus* (Linnaeus) 1776: ruffed grouse

Genus *Tympanuchus* Gloger 1842

1. *T. cupido* (Linnaeus) 1758: pinnated grouse
2. *T. phasianellus* (Linnaeus) 1758: sharp-tailed grouse

Subfamily Odontophorinae

Genus *Dendrortyx* Gould 1844

1. *D. macroura* (Jardine & Selby) 1828: long-tailed tree quail
2. *D. barbatus* Gould 1844: bearded tree quail
3. *D. leucophrys* Gould 1844: buffy-crowned tree quail

Genus *Philortyx* Gould 1844

1. *P. fasciatus* (Gould) 1844: barred quail

Genus *Oreortyx* Baird 1858

1. *O. pictus* (Douglas) 1829: mountain quail

Genus *Callipepla* Wagler 1832

(Subgenus *Callipepla*)

1. *C. squamata* (Vigors) 1830: scaled quail

(Subgenus *Lophortyx* Bonaparte 1838)

2. *C. douglasii* (Vigors) 1829: elegant quail
3. *C. gambelii* (Gambel) 1843: Gambel quail
4. *C. californica* (Shaw) 1789: California quail

Genus *Colinus* Goldfuss 1820

1. *C. virginianus* (Linnaeus) 1758: bobwhite
2. *C. nigrogularis* (Gould) 1843: black-throated bobwhite

Genus *Odontophorus* Vieillot 1816

1. *O. guttatus* (Gould) 1838: spotted wood quail

Genus *Dactylortyx* Ogilvie-Grant 1893

1. *D. thoracicus* (Gambel) 1848: singing quail

Genus *Cyrtonyx* Gould 1844

1. *C. montezumae* (Vigors) 1830: harlequin quail
2. *C. ocellatus* (Gould) 1836: ocellated quail

Subfamily Phasianinae: Old World pheasants, partridges, francolins, and quails

Tribe Perdicini: Old World partridges, francolins, and quails

Genus *Perdix* Brisson 1760

1. *P. perdix* (Linnaeus) 1758: gray partridge

Genus *Alectoris* Kaup 1829

1. *A. chukar* (Gray) 1830: chukar partridge

TABLE 4

Ecological Distribution of North American Grouse and Quails

Vegetation or region	*Representative quail*	*Representative grouse*
Tundra		
Alpine		White-tailed ptarmigan
High arctic		Rock ptarmigan
Low arctic		Willow ptarmigan
Coniferous forest		
Western montane		Blue grouse
Northern boreal		Spruce grouse
Hardwood; hardwood-coniferous		
Northern deciduous		Ruffed grouse
Evergreen chaparral	Mountain quail	
Grassland; grassland-forest		
Shortgrass; Brushland		Sharp-tailed grouse
Tallgrass-forest ecotone	Bobwhite	Greater prairie chicken
California grassland	California quail	
Shortgrass-desert ecotone		Lesser prairie chicken
Desert scrub		
Sage; sage grassland		Sage grouse
Sonoran scrub desert	Gambel quail	
Chihuahuan scrub desert	Scaled quail	
Tropical deciduous forest		
Northern Mexico	Elegant quail	
Central Mexico	Barred quail	
Yucatan Peninsula	Black-throated bobwhite	
Pine-oak forest		
Northern Mexico	Harlequin quail	
Southern Mexico	Ocellated quail	
Tropical evergreen forest	Singing quail	
Lowland rain forest	Spotted wood quail	
Cloud forest		
Western Mexico	Long-tailed tree quail	
Eastern Mexico	Bearded tree quail	
Southern Mexico	Buffy-crowned tree quail	

FIGURE 2. Distribution of major natural vegetation communities in North America.

2

Physical Characteristics

ALL of the grouse, quails, and introduced partridges of North America share a number of anatomical traits which provide the basis for their common classification within the order Galliformes. Among these are the facts that they all have fowl-like beaks and four toes. In all the North American species the hind toe is elevated and quite short, thus is ill-adapted for perching. There are always ten primaries, thirteen to twenty-one secondaries, and twelve to twenty-two tail feathers (rectrices). Aftershafts on the contour feathers are well developed, especially in the grouse, and true down feathers are infrequent. A large crop is present, and is associated with the largely granivorous (seed-eating) behavior of most quails, and the more generally herbivorous (leaf-eating) diets of grouse. The egg colors range from pastel or earth tones (buff, cream, olive, etc.) to white, with darker spotting prevalent among those species having nonwhite eggs. The nest is built on the ground, and incubation is by the females alone or occasionally by both sexes (some quails and partridges). The young are down-covered and precocial and are usually able to fly short distances in less than two weeks. They are cared for by the female (most grouse) or by both parents (some ptarmigans, all quails). A number of external structural characteristics typical of grouse, quails and partridges are shown in figures 3 and 4.

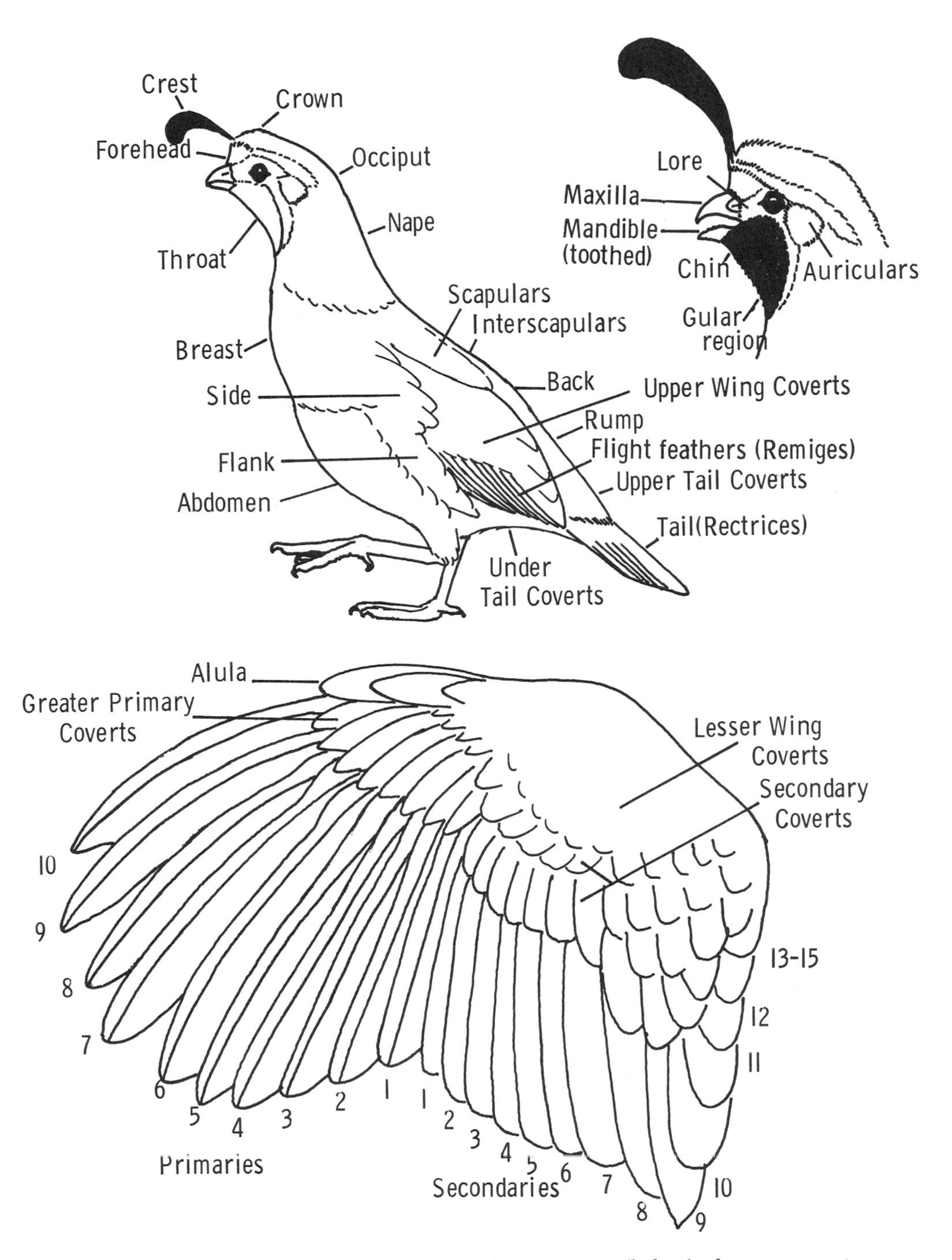

FIGURE 3. Body regions and feather areas (above) and wing regions (below) of a representative quail, with number sequence of the remiges indicated.

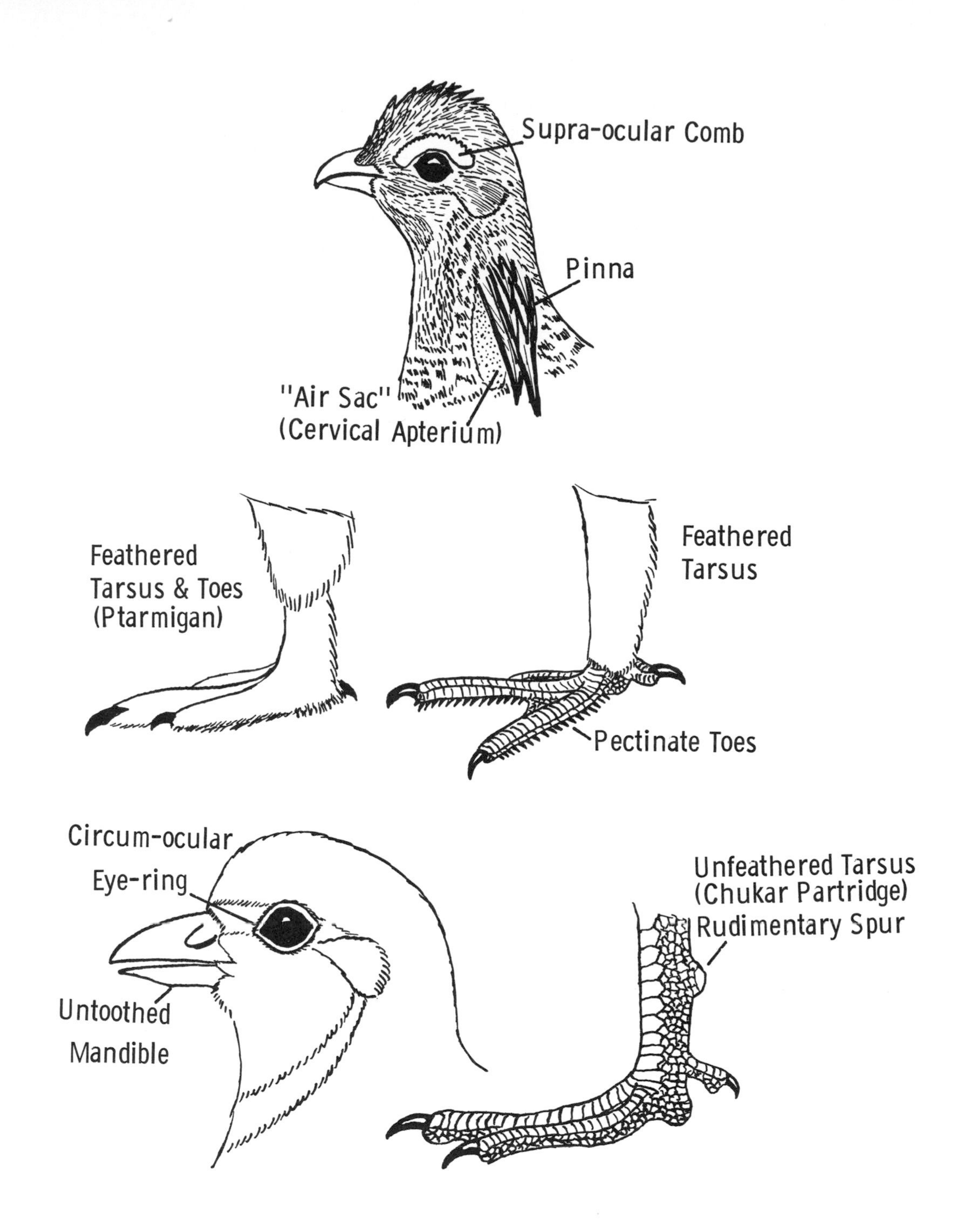

FIGURE 4. Structural characteristics typical of certain grouse species (above) and Old World partridges (below).

Additionally, the grouse may be characterized by the fact that they have feathered nostrils and feathering on the legs that usually extends to the base of the toes. Among ptarmigans this feathering extends to the tips of the toes in winter.

In the other grouse species, the toes have marginal comb-like membranes (pectinations) present in winter. Males of several species of grouse have large unfeathered areas (apteria) at the sides or front of the neck, which can be exposed and enlarged by the inflation of the esophagus. The skin associated with these "air sacs" may be variously colored, or the feathers around the area may be specialized in shape or color, but the true air sac system associated with the lungs is not directly connected to these structures. A bare area of skin (eye-comb) is usually present above the eyes in mature males also. Grouse are not normally highly gregarious, but during fall and winter some species that migrate considerable distances may form large flocks. Grouse are usually polygamous or promiscuous, but the ptarmigan are relatively monogamous. At least sixteen secondaries and twenty-two rectrices are present, but in some species (ptarmigans) the central pair closely resembles the upper tail coverts, while in others (sage grouse) some upper tail coverts may easily be confused with rectrices.

The New World quails can be distinguished from the grouse and their Old World relatives by the fact that they are relatively small (usually under twelve inches), the nostrils are unfeathered, and the edge of the lower mandible is slightly serrated or toothed (inconspicuous in some species). There are thirteen to sixteen secondaries, ten to fourteen rectrices, and the tarsus is not feathered nor is a spur present in males. The bill is very short and stout, and the toes and claws are well developed in many species for digging. They are all monogamous, and the male normally remains to help rear the young. The quails are usually highly gregarious, and occur in coveys at all times except during nesting.

The two successfully introduced Old World partridge species differ from the New World quails in that they lack serrations on the cutting edge of the lower mandible and have fourteen to eighteen rectrices, and males sometimes exhibit slight spurs on the legs. These two species are monogamous; further, males may occasionally participate in incubation and often help rear the young, although opinions differ on these points.

ADULT WEIGHTS

Weight characteristics of adults are of some interest, since they not only provide the hunter with an indication of the trophy values of his game but are also important in the consideration of anatomical adaptations to the environment. Thus, body weights in relation to climatic conditions encountered by the species, heart weights in relationship to total body weights (Hartman, 1955), or body weights in relation to egg weights (Lack, 1968) are all significant relationships and provide useful indices of ecological and physiological adaptations. A summary of reported adult weights is therefore provided (tables 5 and 6) as they have been reported in the literature. In general, the selected references represent the largest sample sizes available and do not take into account the possibilities of geographic or seasonal variations in weights, as are known to occur often. Representative studies on the geographic or seasonal variations in adult weights include those of Gullion and Gullion (1961), Stoddard (1931), Boag (1965), Zwickel, Buss, and Brigham (1966), and Bump et al. (1947).

EGG CHARACTERISTICS

The coloration, markings, and other physical characteristics of bird eggs have particular ecological interest. To some extent the physical characteristics of eggs might be expected to be the result of evolutionary relationships, but the requirements for concealment under the existing ecological conditions are probably of primary significance in the interpretation of egg coloration and patterning characteristics. In table 7 an abbreviated summary of the physical characteristics of the eggs of North American grouse and quails is presented. Known or estimated incubation periods are also indicated, and it may be seen that in all known cases these range from twenty-one to twenty-seven days. There is no apparent relationship between egg size and incubation period; the only clear example of ecological specialization in the entire group is the unusually short (twenty-one- to twenty-two-day) incubation period of the ptarmigans. The longest known incubation periods for any New World quail are those of two tropical species, the bearded tree quail, which we have recently found to have a twenty-eight- to thirty-day incubation period, and the spot-winged wood quail (*Odontophorus capueira*), with an approximate twenty-six- to twenty-eight-day period (Flieg, 1970).

It is also of interest to compare the egg size to the size of the adult female. This is perhaps most easily done by determining the ratio of the fresh egg's weight to that of the female (Lack, 1968). Average weights of fresh eggs for all the species concerned are not available, but is is possible to calculate the volume of an egg quite accurately when its linear measurements are known. Stonehouse (1966) suggests a convenient formula for calculating volume as follows:

$$\text{Volume (cc)} = .512 \times \text{length (mm)} \times \text{diameter (mm)}^2$$

TABLE 5

Adult Weights of North American Grouse

Species	*Mean or Range of Means*	*Maximum Weight*	*References*
Sage grouse			
Male	2010–2835 gm (71–100 oz.)*	3175 gm (112 oz.)	Patterson, 1952
Female	1142–1531 gm (40–54 oz.)*	1531 gm (54 oz.)	Patterson, 1952
Blue grouse			
Male	1150–1275 gm (14–45 oz.)*	1425 gm (50 oz.)	Boag, 1965§
Female	850–900 gm (30–32 oz.)*	1250 gm (44 oz.)	Boag, 1965§
Spruce grouse			
Male	501 gm (17.7 oz.) (14 birds)	630 gm (22 oz.)	Stoneberg, 1967
Female	450–548 gm (16–19 oz.)*	606 gm (21 oz.)	Stoneberg, 1967
Willow ptarmigan			
Male	535–696 gm (19–25 oz.)*	804 gm (28 oz.)	Parmelee, Stephens, and Schmidt, 1967
Female	525–652 gm (19–23 oz.)*	749 gm (26 oz.)	Irving, 1960
Rock ptarmigan			
Male	466–536 gm (16–19 oz.)*	575 gm (21 oz.)	Irving, 1960
Female	427–515 gm (15–18 oz.)*	550 gm (20 oz.)	Johnston, 1963
White-tailed ptarmigan			
Male	323 gm (11.4 oz.) (24 birds)	430 gm (15.2 oz.)	Johnson & Lockner, 1968
Female	329 gm (11.5 oz.) (14 birds)	490 gm (17.5 oz.)	G. Rogers (*in litt.*)
Ruffed grouse			
Male	604–654 gm (21.5–23.3 oz.)*	770 gm (27 oz.)	Nelson & Martin, 1953†
Female	500–586 gm (17.9–20.9 oz.)*	679 gm (24 oz.)	Bump et al. 1947
Greater prairie chicken			
Male	992 gm (35 oz.) (22 birds)	1361 gm (48 oz.)	Nelson & Martin, 1953†
Female	770 gm (29 oz.) (16 birds)	1020 gm (36 oz.)	Nelson & Martin, 1953†
Attwater prairie chicken			
Male	938 gm (33.1 oz.) (10 birds)	1135 gm (40 oz.)	Lehmann, 1941
Female	731 gm (25.7 oz.) (6 birds)	785 gm (28 oz.)	Lehmann, 1941
Lesser prairie chicken			
Male	780 gm (27.6 oz.) (20 birds)	893 gm (31.5 oz.)	Lehmann, 1941
Female	722 gm (25.5 oz.) (5 birds)	779 gm (27.5 oz.)	Lehmann, 1941‡
Sharp-tailed grouse			
Male	951 gm (33 oz.) (236 birds)	1087 gm (43 oz.)	Nelson & Martin, 1953†
Female	815 gm (29 oz.) (247 birds)	997 gm (37 oz.)	Nelson & Martin, 1953†

*Mean weights of these species vary considerably with season and/or locality.
†Reported as fractions of pounds by authors.
‡Reported as pounds and ounces by authors.
§Reported in graphic form, points interpolated.

TABLE 6

Adult Weights of Quails and Partridges

Species	*Sample Size*	*Mean Weight*	*Maximum Weight*	*References*
Long-tailed tree quail				
Male	4	433 gm (15.3 oz.)	467 gm (16.5 oz.)	Warner, 1959
Female	3	390 gm (13.8 oz.)	446 gm (15.7 oz.)	M.V.Z. data‡
Mountain quail				
Male	30	235 gm (8.2 oz.)	292 gm (10.3 oz.)	
Female	24	230 gm (8.2 oz.)	284 gm (10.0 oz.)	Miller & Stebbins, 1964
Barred quail				
Male	7	130 gm (4.6 oz.)	139 gm (4.9 oz.)	M.V.Z. data‡
Female	6	126 gm (4.4 oz.)	148 gm (5.2 oz.)	
Elegant quail				
Male	15	175 gm (6.2 oz.)	207 gm (7.3 oz.)	M.V.Z. data‡
Female	11	169 gm (6.0 oz.)	188 gm (6.6 oz.)	
Gambel quail				
Male	390	161 gm (5.7 oz.)	187 gm (6.6 oz.)	Campbell & Lee, 1953
Female	337	156 gm (5.6 oz.)	192 gm (6.7 oz.)	
California quail				
Male	418	176 gm (6.2 oz.)	206 gm (7.3 oz.)	Nelson & Martin, 1953*
Female	272	162 gm (6.0 oz.)	206 gm (7.3 oz.)	
Scaled quail				
Male	143	191 gm (6.7 oz.)	234 gm (8.2 oz.)	Campbell & Lee, 1953
Female	132	177 gm (6.2 oz.)	218 gm (7.7 oz.)	Nelson & Martin, 1953*
Bobwhite (eastern U.S.)				
Male	899	173 gm (6.1 oz.)	255 gm (9.0 oz.)	Nelson & Martin, 1953*
Female	692	170 gm (6.0 oz.)	240 gm (8.5 oz.)	
Black-throated bobwhite				
Male	3	137 gm (4.8 oz.)	146 gm (5.1 oz.)	Klaas, 1968
Female	3	139 gm (4.9 oz.)	152 gm (5.4 oz.)	Berrett, 1963
Spotted wood quail				
Male	16	300 gm (10.6 oz.)	358 gm (12.6 oz.)	Van Tyne, 1935; Hartman, 1955;
Female	5	288 gm (10.2 oz.)	316 gm (11.2 oz.)	Paynter, 1957
Singing quail				
Male	12	212 gm (7.5 oz.)	266 gm (9.4 oz.)	Warner & Harrell, 1957
Female	3	189 gm (6.7 oz.)	206 gm (7.3 oz.)	
Harlequin quail				
Male	45	195 gm (6.9 oz.)	224 gm (7.9 oz.)	Leopold & McCabe, 1957
Female	22	176 gm (6.2 oz.)	200 gm (7.1 oz.)	
Gray partridge				
Male	87	396 gm (14 oz.)	454 gm (16 oz.)	Nelson & Martin, 1953*
Female	57	379 gm (13.7 oz.)	432 gm (15.3 oz.)	
Chukar partridge (Turkish race)				
Male	44	557 gm (19.6 oz.)	631 gm (22.3 oz.)	Bohl, 1957†
Female	50	444 gm (15.7 oz.)	520 gm (18.5 oz.)	
Chukar partridge (Indian race)				
Male	20	614 gm (21.7 oz.)	722 gm (25.5 oz.)	Christenson, 1954†
Female	(both sexes combined)	501 gm (17.7 oz.)	545 gm (19.2 oz.)	

*Reported as fractions of pounds by authors.
†Reported as pounds and ounces by authors.
‡Museum of Vertebrate Zoology, University of California.

TABLE 7
Egg Characteristics and Incubation Periods

Species	*Spotting*	*Basic Color*	*Dimensions (mm)*	*Incubation (days)*	*References (for incubation)*
Sage grouse	Moderate	Buffy green or brown	55 x 38	25–27	Patterson, 1952
Blue grouse	Moderate	Buff or pale brown	48.5 x 35	24–25	Godfrey, 1966
Spruce grouse	Moderate	Buff or pale rust	43 x 31	21	Pendergast and Boag, 1971
Willow ptarmigan	Heavy	White to pale brown	43 x 31	21–22	Westerkov, 1956 Jenkins et al., 1963
Rock ptarmigan	Heavy	White to pale brown	42 x 30	21	Godfrey, 1966
White-tailed ptarmigan	Moderate	White to reddish buff	43 x 29.5	22–23	Braun, 1969
Ruffed grouse	Slight or none	Buffy white to cream	38.5 x 30	24	Bump et al., 1947
Greater prairie chicken	Slight or none	White to olive buff	43 x 32.5	24–25	McEwen et al., 1969
Lesser prairie chicken	Slight or none	White to buff	42 x 32.5	25–26	Coats, 1955
Sharp-tailed grouse	Slight	Fawn to chocolate or olive	43 x 32	24–25	McEwen et al., 1969
Bearded tree quail	None	Dirty white	46.6 x 31	28–30	this study
Barred quail	None	White	30 x 23.7	22–23	F. Strange, pers. comm.
Mountain quail	None	Pale buff to cream	34.5 x 26.5	24–25	F. Strange, pers. comm.
Scaled quail	Slight	Pale buff to cream	32.5 x 25	22–23	various studies
California quail	Moderate	Pale buff to cream	32 x 25	22–23	various studies
Gambel quail	Moderate	Pale buff to white	31.5 x 24	22	various studies
Elegant quail	None	White	34 x 24	22	this study
Bobwhite	None	White	31 x 25	22–23	Stoddard, 1931
Black-throated bobwhite	None	White to buff	30.5 x 23	24	this study
Spotted wood quail	Slight	Creamy white	40 x 29	?	Wetmore, 1965
Singing quail	None	White & yellow	31 x 25	?	Warner & Harrell, 1957
Harlequin quail	None	White	32 x 24	24–25	F. Strange, pers. comm.
Gray partridge	None	Pale olive	35 x 27	24–25	McCabe & Hawkins, 1946
Chukar partridge	Moderate	Pale brown or creamy	45 x 31	22–24	various studies

Assuming that the fresh egg has an average specific gravity of 1.08 (Barth, 1953), the preceding formula can be modified as follows:

$$\text{Weight (gm)} = .552 \times \text{length (mm)} \times \text{diameter (mm)}^2$$

Using this formula, estimated fresh weights of eggs were calculated from the linear measurements presented in table 7 and are summarized in table 8. In addition, a calculated total estimated clutch weight, based on reported average clutch sizes (see table 12), is indicated as an index to the relative physiological drain on the female in laying an entire clutch. It may be seen that a female's average clutch may represent as little as 20–25 percent of her own weight, as in spruce grouse and ptarmigan, to as much as 90 percent of her weight in certain quail species. Since some of these quail species are persistent renesters, it would seem that such a large investment of energy in a clutch is not detrimental as long as sufficient food is available. Captive bobwhites and other quail regularly lay over one hundred eggs per year (up to three hundred recorded) and may lay as many as five hundred in a lifetime (Kulenkamp and Coleman, 1968), clearly indicating their high capacity for channeling food energy into egg production.

TABLE 8

RELATIONSHIP OF ADULT FEMALE WEIGHT TO ESTIMATED EGG AND CLUTCH WEIGHTS

	Est. Egg Weight (gm)	*Percentage of Female Weight*	*Average Clutch Size*	*Percentage of Female Weight*
Sage grouse	44	3.4	7.4	25.2
Blue grouse	33	3.6	6.2	22.4
Spruce grouse	23	4.2	5.8	24.4
Willow ptarmigan	23	3.3	7.1	23.1
Rock ptarmigan	21	4.1	7.0	28.7
White-tailed ptarmigan	21	6.4	5.2	33.3
Ruffed grouse	19	3.8	11.5	43.7
Greater prairie chicken	24	3.1	12.0	37.2
Lesser prairie chicken	24	3.3	10.7	35.3
Sharp-tailed grouse	24	2.9	12.1	35.1
Mountain quail	13	4.7	10.0	47.0
Scaled quail	11	6.2	12.7	78.7
California quail	11	6.7	13.7	91.8
Gambel quail	10	6.4	12.3	78.7
Bobwhite	11	6.4	14.4	92.2
Harlequin quail	10	5.7	11.1	63.3
Gray partridge	14	3.7	16.4	60.7
Chukar partridge	24	5.4	15.5	83.7

FEATHERS AND OTHER EXTERNAL ADAPTATIONS

As in nearly all birds, the contour feathers of grouse and quail are arranged in definite tracts, or pterylae, which do not differ much among the included species. The general arrangement of these tracts is shown in figure 5. At the edges of these tracts "half-down" or semiplume feathers regularly occur, and true down feathers sometimes occur on the neck and wings. There are usually numerous long and nearly hairlike filoplumes scattered among the contour feathers; these become especially conspicuous in adult male sage grouse when they are erected during display.

The general arrangement of the feather tracts is very similar in grouse and quails. The major differences to be noted are that in quails the dorsal feather tract has only a small apterium and is nearly continuous with the upper cervical tract, whereas in grouse these tracts are well separated, forming a large dorsal apterium. In quail species the lower cervical tract is also forked more anteriorly on the throat than is true of grouse (Clark, 1899). McCabe and Hawkins (1946) provide a description of the feather tracts of the gray partridge, which more closely resembles the New World quails in both these regards.

The number of primaries is the same (ten) throughout the group, but their relative lengths differ somewhat. Clark (1899) reports that in the New World quails (at least the United States genera) the longest primary is the sixth. In the North American grouse the sixth, seventh, and eighth are of about uniform length, followed by five and nine, four and ten, and finally three, two, and one. According to Clark (1899), the number of secondaries is fourteen in harlequin and scaled quail, fourteen to fifteen in bobwhites, fifteen to sixteen in the "*Lophortyx*" species, and sixteen in the mountain quail. Ohmart (1967) reports only fourteen true secondaries in both the scaled quail and the three "*Lophortyx*" species. Among the grouse they vary from fifteen to sixteen in the ruffed grouse, seventeen in the spruce grouse, eighteen in the blue, sharp-tailed, and pinnated grouse, eighteen to nineteen in ptarmigans, and twenty-one in sage grouse (Clark, 1899). In most of these species the secondaries grade gradually into the scapulars and proximal coverts and thus become very difficult to count accurately. The arrangement of the wing feathers is shown in figure 3.

Many of the New World species of quail bear elaborate crests that may be similar or different in the sexes. These insert in a distinctive arrangement on the crown. In the mountain quail this crest is made up of two feathers, while in "*Lophortyx*" six to nine are present. Although the scaled quail lacks such a distinctive crest, it too has an arrangement of ten crest feathers similar to that found in the typically crested species (Ohmart, 1967).

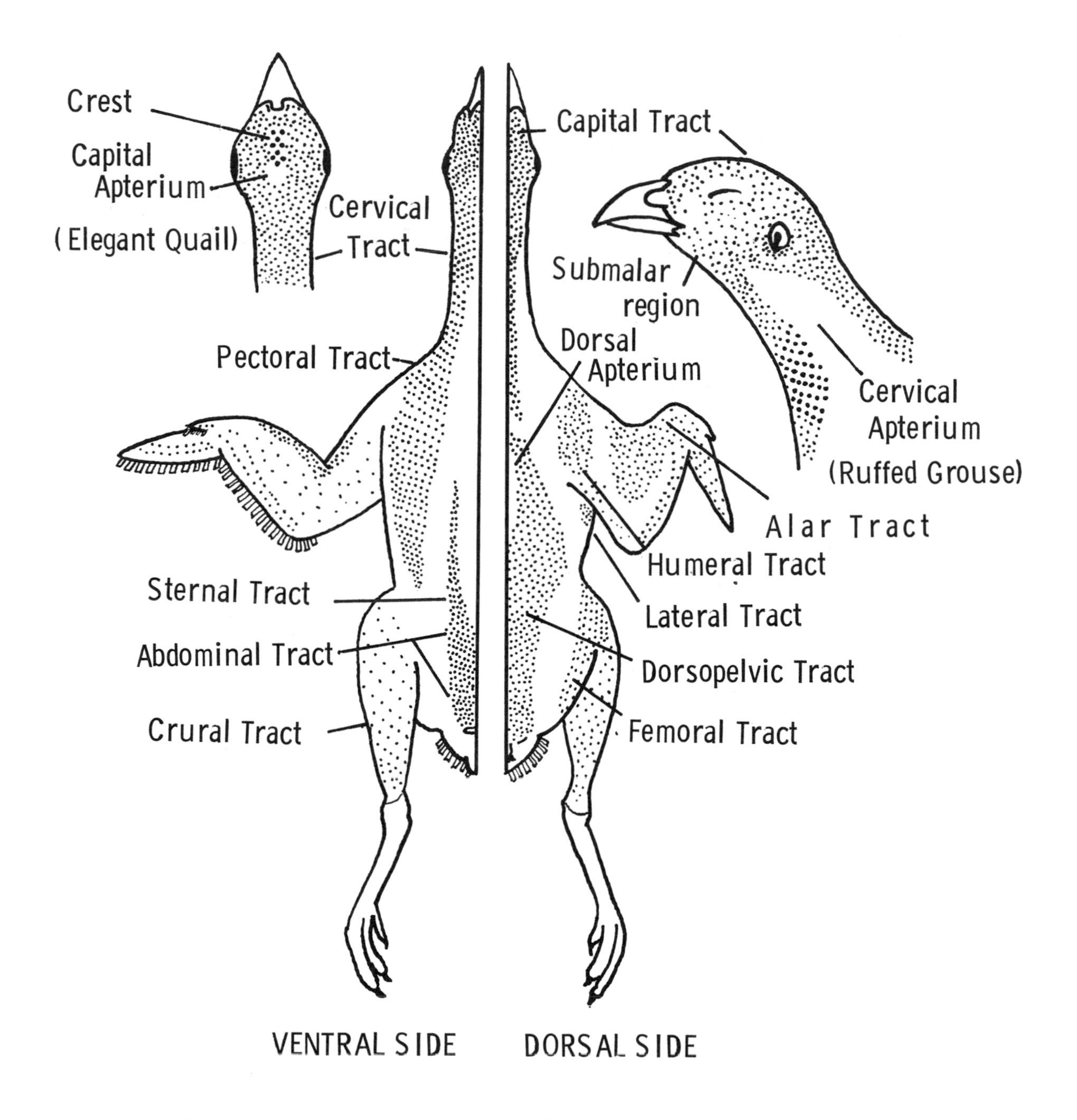

FIGURE 5. Feather-tracts of grouse and quails (primarily after Ohmart, 1967, and Clark, 1899).

None of the grouse species possess such elaborate crests, but several have special tracts of feathers on the neck or have unfeathered areas in this region. In the ruffed grouse, the special "ruff" feathers are borne on the lateral branches of the lower cervical tract, and there is no marked apterium between the lower and upper cervical tracts. However, the dozen or so feathers making up the pinnae of the pinnated grouse are similarly borne on each side of the upper cervical tract, below which is a large apterium (Clark, 1899). In the greater prairie chicken this apterium is yellowish in color, presumably because of subcutaneous fat, whereas in the lesser prairie chicken it is more reddish. The sharp-tailed grouse has a similar apterium which appears reddish to violet when expanded by esophageal inflation, but this species lacks specialization of the feathers above and below. The sage grouse lacks lateral neck spaces, but there is a large and somewhat oval apterium on each side of the neck, located quite low and somewhat frontally. These spaces are about 45 by 25 millimeters in older males, and about 25 by 13 millimeters in females (Brooks, 1930). The bare skin is olive gray, but appears yellowish when expanded during display. The lower and laterally adjacent breast feathers of male sage grouse are curiously bristly, which was once thought to be a result of wear, until Brooks (1930) discovered that newly grown feathers have the same appearance. They evidently produce the rasping or squeaking sound made when the foreparts of the wings are brushed over the lower breast during display (Lumsden, 1968).

Although the blue grouse lacks such specialized feathers on the neck, males do expose rounded areas of the neck during "hooting," which are emphasized by the whitish bases of the surrounding neck feathers. The exposed skin in these areas varies from a condition (in the interior races) of being thin, flesh-colored, and changing to purplish red when expanded to (in the coastal races) being highly thickened, gelatinous, and corrugated, and of a deep yellow color. These conditions presumably result from subcutaneous fat deposits, which are less evident during the nonbreeding periods (Brooks, 1926).

As has been mentioned, the sharp-tailed grouse lacks specialized neck feathers associated with display, but Lumsden (1965) has found that the tail feathers are unusually developed in this species and are related to the tail-rattling noises made during display. The rectrices in males are very stout basally but taper rapidly. Ventrally the shaft projects in two keels, but dorsally the shaft is rounded and projects only slightly. The outer webs of the vanes are stiff and curve sharply downward, and the inner webs are also thickened. Each clicking sound is produced by lateral feather movements, during which the inner web catches on the ventrally projecting shaft of the inwardly adjacent feather web, and after some resistance the two disengage, producing a click. Simultaneously the curved outer webs brush over the dorsal surface of the next outwardly adjacent feather, producing a scraping sound. Additional nonvocal noise in males of these species may be produced by foot stamping, and to a lesser degree the same can be said for the pinnated grouse. In the greater, Attwater, and lesser prairie chickens tail-spreading or tail-clicking noises that are taxon-typical occur during display (Sharpe, 1968).

VOCAL APPARATUS AND SYRINGEAL SOUND PRODUCTION

In addition to the nonvocal means of sound production such as the feather-scraping, foot-stamping, and wing-clapping sounds made either during flight as in *Dendragapus* (Wing, 1946) or while on the ground as in ruffed grouse, sound can also be produced by internal means. These include sound production by the syrinx in conjunction with the inflation and deflation of the esophageal "air sacs" of various grouse, and by the syrinx alone in all species of grouse and quails.

The anatomy of the syrinx of grouse and quail is relatively simple, and is quite similar to that of the domestic fowl, as described by Myers (1917) and Gross (1964). A diagram of the syringeal anatomy of the domestic fowl and a representative species of grouse is shown in figure 6.

The syrinx of gallinaceous birds is tracheo-bronchial in location; that is, it occurs at the junction of the trachea and the paired bronchi. The syrinx consists of a variable number of partially fused tracheal rings, collectively called the tympanum. The bony structure that is located at the junction of the trachea and the two bronchi is the pessulus, which provides important support for the two pairs of tympaniform membranes. One such pair consists of the external tympaniform membranes that are located between the fused tympanum-pessulus complex and the first pair of bronchial rings. A second pair of internal tympaniform membranes are situated medially between the pessulus and the second pair of bronchial rings. The tension on these membranes can probably be increased either by stretching the neck or by pulling the trachea forward through the action of the *tracheolateralis* muscles. The tension can also be reduced by contracting the *sternotrachealis* muscles, which insert anteriorly to the syrinx on the sides of the trachea. When these latter muscles are in a normal state of tension the internal and external tympaniform membranes are held well apart and air can pass unimpeded between them. When the muscles are contracted, however, the membranes are brought closer together and air resistance builds up pressure in

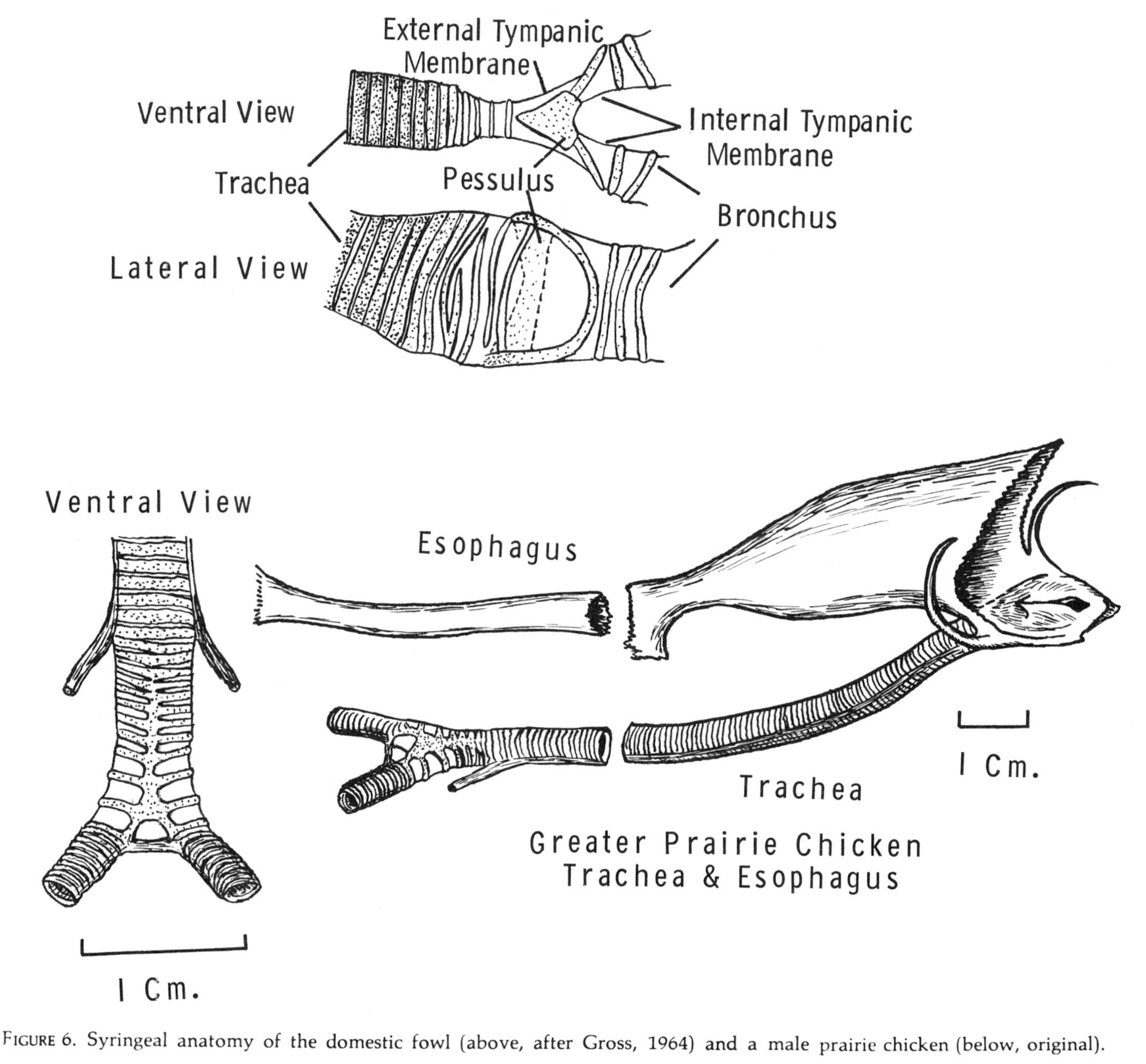

FIGURE 6. Syringeal anatomy of the domestic fowl (above, after Gross, 1964) and a male prairie chicken (below, original).

the bronchi, lungs, and air sacs. As air then passes outward between the membranes they are drawn more closely together and are set into vibration (the Bernoulli effect) thus producing sound (Gross, 1964).

In domestic fowl at least, the frequency (pitch) of the induced vibrations and associated sound production can be increased by stretching the tympaniform membranes and thus increasing membrane tension (Harris, Gross, and Robeson, 1968). Simply changing the length of the tracheal tube within the limits imposed by anatomy evidently has little effect on the fundamental frequency established by the surface dimensions and tension of the tympaniform membranes. Since smaller species have tympaniform membranes narrower in width it is not surprising that the fundamental frequencies of their calls average somewhat higher than the corresponding calls of larger relatives (Sutherland and McChesney, 1965). The fundamental frequency of vocalizations in similar-sized species having essentially identical syrinxes is thus regulated by the tension of the tympaniform membranes, which vibrate at a rate that is proportionate to the square root of their tension (Harris, Gross, and Robeson, 1968).

In species such as the turkey which have an interbronchial ligament posterior to the tympaniform membranes, posterior movement of the trachea and syrinx caused by contraction of the *sternotrachealis* muscles cannot force the bronchi back very far and instead alters the shape of the syrinx. Specifically, the internal and external tympaniform membranes are pushed closer together and the latter are stretched, thus increasing the fundamental frequency (Gross, 1968).

It is also clear that two different fundamental frequencies are sometimes simultaneously produced by one bird. This "internal duetting" is theoretically explainable by assuming that each side of the syrinx can operate independently of the other (Greenewalt, 1968), or perhaps there is a simultaneous activation of the internal and external tympaniform membranes under differential tension.

Few if any of the vocal sounds produced by grouse and quails are pure tones, rather, in addition to a basic or fundamental frequency that is generated by the vibration of the tympaniform membrane, there are usually also a considerable number of higher overtones or harmonics, which are progressive multiples of the fundamental frequency. These harmonics are of varying loudness, or amplitude, since they are differentially amplified or dampened by the resonating characteristics of the tracheal tube and pharynx. The acoustical effect of the trachea, oral cavity, and beak is thus to tune the bird's vocalizations to a resonant frequency which serves to sharpen the pitch and, perhaps, to reduce the number of harmonics (Harris, Gross, and Robeson, 1968).

There is no direct relationship between the fundamental frequency of a vocalization (which is regulated by the vibrations of the tympaniform membranes) and the resonant frequency, which is determined by physical characteristics such as the length of the tracheal tube and its associated resonating structures. The resulting sound is therefore a composite of these two independently determined acoustic characteristics. Although as an individual animal matures, the growth of its syrinx and trachea results in a concomitant lowering of both the fundamental frequency and the resonant frequency, those two variables can also have contrasting effects. For example, during "head-throw" calls the increased tension on the tympaniform membranes causes an increase in the fundamental frequency, while the stretching of the tracheal tube results in a lowering of the resonant frequency.

By means of a simple formula, the expected resonant frequency and its associated harmonics can readily be calculated for a tracheal tube of any length. Harris, Gross, and Robeson (1968), for example, compared such calculated frequencies with the observed frequencies that they generated by using differing lengths of an excised trachea and syrinx from a domestic fowl. They concluded that the trachea and bronchi combine acoustically to form a single resonant tube, and the formula they used indicates that they assumed that the combined structures represent a closed-tube acoustical system. However, Sutherland and McChesney (1965) made somewhat similar calculations for calls recorded from live individuals of two species of geese, and concluded that the vocal apparatus had resonance characteristics more closely related to those of an open tube than those of a closed tube. Thus, in an open-tube sound system only the odd-numbered harmonics above the resonant frequency should be expressed, whereas in a closed-tube system both the even-numbered and odd-numbered harmonics will be amplified. In figure 7 the calculated resonant frequencies and expected harmonics are shown for open-tube and closed-tube tracheal tubes ranging in length from five to twenty centimeters, and for frequencies up to eight thousand Hz. A comparison of these curves with the harmonic patterns produced by quail and grouse species (see Sonagrams in figures 18 to 20) will illustrate the point that an open-tube acoustic system appears to be present in grouse and quail vocalizations. We may conclude, therefore, that the fundamental frequencies of these birds' vocalizations result from the vibration rates of the tympaniform membranes but that the relative amplitudes of the fundamental frequencies as well as their associated harmonics are differentially amplified or dampened according to the resonance characteristics of the tracheal tube and pharynx.

In male grouse of those species that inflate their esophageal "air sacs" during sound production, additional complexities arise. This vocal process

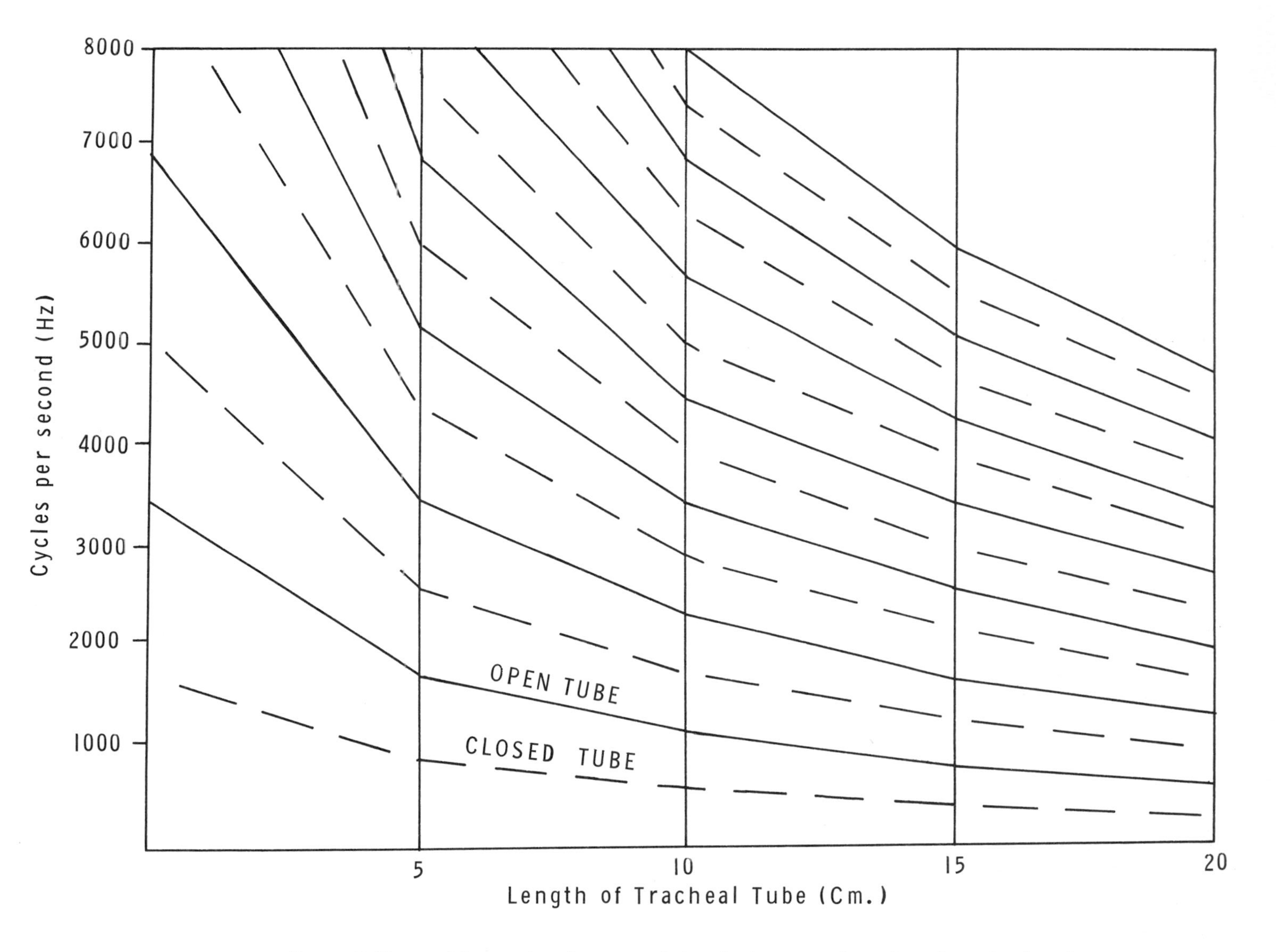

FIGURE 7. Expected harmonics of open- and closed-tube resonating tubes of varying lengths.

has been studied by Gross (1928) for the pinnated grouse, and presumably the same principle applies to the other species. When this species "booms," the beak is closed, the tongue is raised upward against the roof of the mouth, and the internal nares become blocked. The glottis thus opens directly in front of the esophagus, and the latter fills with air passing out of the trachea. The expanded anterior end of the esophagus then becomes part of the resonating structure, and the total length and volume of the sound chamber is considerably increased. This combination of the trachea and esophagus is acoustically similar to that of a cylindrical tube and an associated expansible chamber. The resonant frequency of such a combination of tube and cavity is inversely proportional to the volume of the cavity (Harris, Gross, and Robeson, 1968). This clearly accounts for the low fundamental frequency characteristics of such calls (under two hundred Hz.). Besides having the obvious visual signal value associated with the inflation of the unfeathered neck region, these low frequency sounds have considerably greater carrying power than do high frequency sounds of the same amplitude. Alfred Gross (in Bent, 1932) has mentioned that the booming sounds of the heath hen sounded softer than did the bird's more typical calls, yet carried considerably further. The ecological value of booming is thus clearly apparent.

OTHER ANATOMICAL AND PHYSIOLOGICAL ADAPTATIONS

In common with many other gallinaceous birds, the grouse and quails possess a blind sac on the dorsal wall of the cloaca, which is called the bursa of Fabricius (figure 8). In younger birds this typically opens directly into the cloaca, while in sexually mature birds it regresses in size and may completely disappear. The bursa does not always open into the cloaca and instead may be occluded by a thin membrane, so its presence cannot in all cases be detected by probing. The function of the bursa is now known to be that of antibody production (Warner and Szenberg, 1964), and its removal or inactivation interferes with immunological processes in the animal. Since the relative size and activity of the bursa decreases with age, this structure has been used as a supplementary means of estimating age in gallinaceous birds. Gower (1939) indicates its usefulness through the first November of a bird's life in determining the ages of ruffed, sharp-tailed, pinnated, and spruce grouse, as well as the gray partridge. It also has some limited value in estimating ages of California quail (Lewin, 1963), but the age-related differences in upper primary covert coloration in this group are obviously much more convenient.

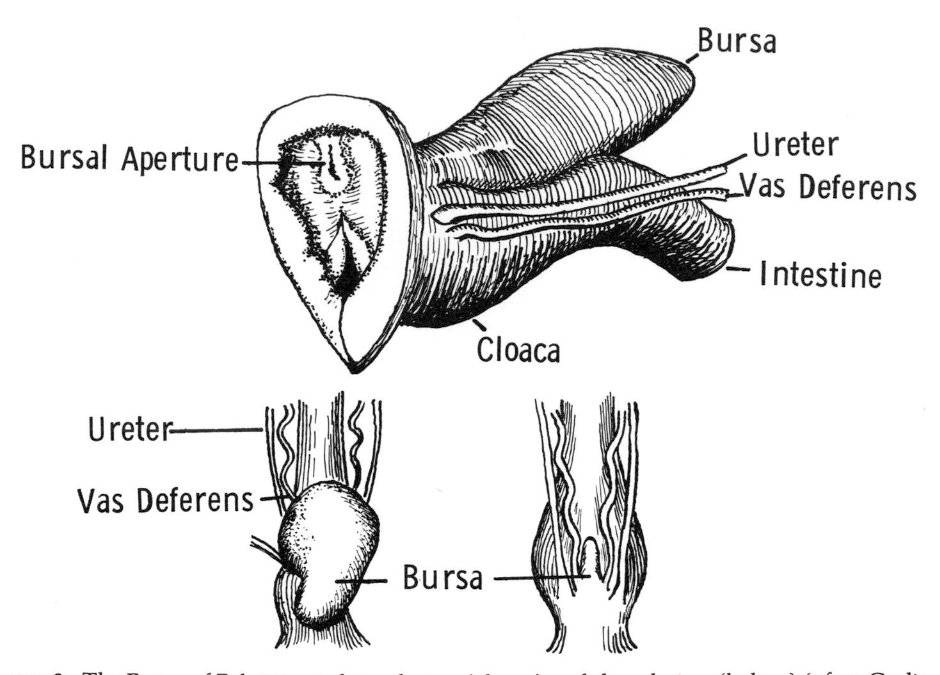

FIGURE 8. The Bursa of Fabricius in lateral view (above) and dorsal view (below) (after Godin, 1960).

In most or all of the Galliformes, another outpocketing of the digestive tract occurs at the junction of the small and large intestines. Here a pair of caeca occur, which vary greatly in length among different species but are particularly long in the grouse. The function of these caeca is apparently primarily to provide a place for the bacterial breakdown on cellulose and similar fibrous materials that cannot be handled by the digestive enzymes produced by the bird. Leopold (1953) surveyed the relative development of these caeca in grouse and quail species, and found that most North American grouse with adult weights of about five hundred grams have caeca averaging about forty-four centimeters in length, with the sage grouse having the longest caeca (sixty-eight to seventy-eight centimeters) of all species studied. By comparison, although gray and chukar partridges also weigh nearly five hundred grams, their caeca lengths averaged about seventeen centimeters. Adult quail, ranging in weight from about one hundred and seventy to two hundred and fifty grams, had caeca lengths of from ten to seventeen centimeters. Additionally, grouse, which are generally herbivorous, exhibited somewhat longer total intestine lengths than did quail, which are largely granivorous. These findings support the idea that the adaptive function of the caeca in grouse species is to provide for bacterial decomposition of cellulose.

Another interesting physiological difference between grouse and quails is in their water requirements. Regrettably little is known of this situation in grouse, but probably only the sage grouse might be expected to exhibit physiological adaptations allowing minimal water requirements. Edminster (1954) states that young and old sage grouse can go many days without water, but that birds will travel considerable distances to water and that good populations occur only where a supply of water is available. With regard to the ruffed grouse, Bump et al. (1947) report that three of six adult birds died in less than four days when deprived of both food and water, but twelve birds all survived a period of nine days when given water but no food.

In the case of quail, it is known that at least the bobwhite, Gambel, and California quails can survive indefinitely without water if succulent food is available, but when fed only dried seeds they gradually lose weight. Minimum actual water needs for these birds when fed on a dry diet are about 1.6 milliliters of water for the Gambel quail, 1.9 milliliters for the California quail, and 4.5 milliliters for the bobwhite, representing from about 1.1 to 2.5 percent of the adult weight (McNabb, 1969). All these species absorb both water and salt in the intestine when forced to drink excessively salty water, but the Gambel quail is able to produce the most concentrated urine, while the bobwhite is least able to concentrate its urine. Thus, the water requirements of these species are in direct relationship to their relative ecological distributions.

Another physiological adaptation for desert living that Gambel and California quail exhibit is their ability to tolerate short-term body temperature increases of up to four degrees Centigrade above normal levels (Bartholomew and Dawson, 1958). The relatively small volume–to–surface area ratios of these quails also favor their survival in situations where body heat loss must be sufficient to avoid overheating (Miller and Stebbins, 1964). In contrast, the insulating values of ptarmigan plumage appear to be among the highest of any bird species yet studied (Sturkie, 1965). Apparently this high capacity for insulation in arctic species results from the barbules at the tips of the contour feathers being unusually soft and having extended processes that cling to adjacent feathers when erected, thus trapping air (Irving, 1960). The feathered legs and toes of ptarmigan probably serve both as snowshoes and insulation.

Quantitative data are still lacking, but it appears that grouse may have considerably larger hearts relative to their body size than do quails. In general, smaller birds have relatively larger hearts than do larger ones (Hartman, 1955). Yet, Johnson and Lockner (1968) report that the three species of ptarmigans have heart sizes ranging from 0.87 to 1.85 percent of the body weight, whereas Hartman (1955) reports that in two genera of quails of smaller average sizes these ratios are only 0.34 to 0.39 percent. It is quite possible that the relatively large hearts of ptarmigans are related to their migratory movements (Irving, 1960); among the three ptarmigan species the relative heart size is not correlated with altitudinal distribution (Johnson and Lockner, 1968).

One of the interesting and reproductively significant ways in which the New World quails differ from grouse is in the tendency of male quail to take over the incubation and brooding of an abandoned nest or a group of young, sometimes allowing a female to produce a second brood. Broody behavior in these birds is associated with the development of a "brood patch," an extensively defeathered area which first forms in the area of the lateral apteria but which eventually includes much of the ventral body surface (Jones, 1969a). Unlike some passerines that have been studied, the brood patch does not develop in females before egg-laying, but rather forms during early stages of incubation. Incubation behavior evidently produces a strong release of prolactin through visual or tactile stimulation provided by the eggs, which in synergism with gonadal steroid hormones stimulates the defeathering that results in the formation of the incubation patch (Jones, 1969a). Most male quail do not form incubation patches in spite of their high prolactin levels during testis regression, but will do so if visually stimulated by the presence of an abandoned nest. In both sexes of the California quail it has been found that prolactin alone will not produce defeathering, but rather this defeathering results from synergism with estrogen, progesterone, or testosterone. In contrast to many passerines, in which only females form a brood patch in response to the synergistic effects of estrogen and prolactin, or to phalaropes, in which only males form a brood patch in response to the combined effects of testosterone and prolactin, the quails appear to provide an intermediate physiological situation that is clearly of adaptive value in this group.

3

Molts and Plumages

AN understanding of the molts and plumages of the quails and grouse is of great importance to the applied biologist, for they provide clues that are valuable for determining age and sex of individual birds without resorting to internal examination. They thus offer a means of analyzing wild populations as to sex and age composition, which are basic indices to past and potential reproductive performances and probable mortality rates. Additionally, molts and plumages are generally species-specific traits, which have resulted from pressures of natural selection over a long period of time in a particular habitat and climate. The ecology of the species is of major importance in this regard; species occurring in more northerly regions may undergo their molts more rapidly than those in southerly ones or, as in the case of the willow ptarmigan, certain races may even lack particular plumages that occur in populations existing in other areas having different climates.

From the time they hatch, all grouse and quails exhibit a series of specific plumages, separated by equally definite molts, that are comparable in nearly all species. The only known exception to this occurs in the genus *Lagopus*, which is unique in having an extra molt, and thus a supplementary plumage, intercalated between its summer and winter plumages. This special case will be dealt with as required; the following summary will thus serve to provide the basic sequences and terminology that describe the molts and plumages found in the North American grouse and quails.

NATAL PLUMAGE

All galliform birds hatch covered with a dense coat of down that serves to insulate the body and to provide camouflage for the precocial young, which typically leave the nest shortly after hatching. This natal plumage is generally extremely similar among related species and, because of the lack of known selective pressures for rapid divergence in downy patterns during speciation, often provides more valuable clues to evolutionary relationships than do adult plumage patterns.

POSTNATAL MOLT AND JUVENAL PLUMAGE

Virtually at the time of hatching, or at least within the first week of life, the first indications of the juvenal plumage become apparent through the emergence of the secondary and inner primary feathers and the rectrices (tail feathers). The two outermost juvenal primaries and the innermost of the juvenal secondaries appear later than those situated near the middle of the wing. All native galliform species have ten primaries, which are counted outwardly from the most proximal one, while the number of secondary feathers is somewhat greater and varies among species, with the innermost secondaries sometimes designated as "tertiaries" (although like typical secondaries they insert on the ulna rather than on the humerus). The secondaries are counted inwardly, from the feather nearest the first of the primaries (which insert on the bones of the hand). The third secondary is typically the first to emerge, followed in sequence by the progressively more proximal ones, while the two outermost ones often emerge at about the same time as those near the proximal end. At the same time that the primaries and secondaries are growing, all the upper greater coverts begin growth. The upper coverts for the two outermost primaries actually begin to grow before their associated primaries and possibly serve as functional substitutes for these flight feathers, which are typically delayed in development. Correlated with this, the ninth upper primary covert of the juvenal plumage is often notably more pointed and larger than are the adjoining coverts.

The juvenal remiges (primaries and secondaries) and rectrices (tail feathers) are scarcely fully grown before they begin to be pushed out by the remiges and rectrices of the next plumage, but during the short time they are present the rest of the body is being transformed from a down-covered one to one covered with contour feathers. This transformation is called the *postnatal molt*, and is a complete molt. The feathers which replace the natal down are called *juvenal* feathers, and the associated age category is called the *juvenile* stage.

As the juvenal remiges and rectrices are appearing in the manner described above, other juvenal feathers begin to emerge on both sides of the breast and backwards toward the flanks. Shortly, juvenal feathers also appear on the crown, base of the neck, scapular region, and upper legs, spreading toward the back. The greater and lesser upper wing coverts are fully grown before their associated remiges, and are followed by the median coverts. These upper coverts appear in advance of the lower coverts (Dwight, 1900). Before all the juvenal feathers have appeared throughout the head region, the first signs of the next (post juvenal) molt will be evident in the loss of the inner juvenal primaries and the emergence of new (first-winter) primaries in their places. This occurs as early as eighteen days after hatching in willow ptarmigan and blue grouse and occurs within the first month of life in most or all species. The juvenal primaries are molted outwardly at roughly five-day intervals for the inner ones, and at increasing intervals for the outer ones (table 9). The two outer juvenal primaries (numbers nine and ten) will have just completed their growth shortly before the eighth juvenal primary is dropped. Except in rare instances these two outer juvenal primaries are never normally molted in the species under consideration here. In the ring-necked pheasant this does occur, but it is not typical of the introduced Old World partridges (*Perdix* and *Alectoris*). An important difference between the New World quails and the grouse occurs in association with the postjuvenal molt of the eight inner primaries. In the grouse (as well as in *Perdix* and *Alectoris*) the associated juvenal greater upper primary coverts are also molted in the postjuvenal molt, whereas in all the New World quails so far studied the juvenal greater upper wing coverts of these primaries are not molted but rather are held through the winter and spring until they are molted in the annual (postnuptial) molt (van Rossem, 1925; Petrides, 1942). Since these feathers are marked with more buffy or lighter tips than are the upper primary coverts of adult birds, this difference provides an alternate and more reliable method of determining age in New World quails than the examination of their outer two primary feathers for signs of wear and fading.

The juvenal secondaries, as well as all the juvenal rectrices, are also rapidly lost at about the time that the juvenal primaries are being shed. The juvenal rectrices may be dropped almost simultaneously, as in the bobwhite and rock ptarmigan (Watson, 1962c; Salomonsen, 1939), molted from the lateral follicles toward the middle ones (centripetally), as in the blue grouse (Smith and Buss, 1963), or molted from the central follicles outwardly (centrifugally), as in the California and scaled quails (Raitt, 1961; Ohmart, 1967). The gray partridge also has an imperfectly centrifugal postjuvenal molt of the rectrices (McCabe and Hawkins, 1946). The juvenal

TABLE 9

Average Age (in Days) of Start and Completion of Growth of First-winter Primary Feathers in Representative Grouse and Quails

Species	Primary number (counting from inside); A = starts growth; B = grown								Authority
	1	2	3	4	5	6	7	8	
Grouse	A/B	A/B	A/B	A/B	A/B	A/B	A/B	A/B	
Sage grouse									
males	24/35	29/42+	34/49+	40/56+	47/77+	59/84+	74/105+	102/140+	Pyrah, 1963
females	24/42	27/42	33/49+	38/56+	45/63+	54/77+	66/91+	91/126+	
Blue grouse	17/38	23/44	26/50	35/59	41/68	47/80	59/95	71/113	Zwickel & Lance, 1966
Willow ptarmigan	18/—	25/—	30/—	35/—	40/—	46/—	53/—	65/91	Westerskov, 1956
Ruffed grouse									
New York	14/45	20/49	27/63	35/68	42/77	49/83	61/98	74/119	Bump et al., 1947
Ohio	23/46	27/54	34/54	40/68	47/73	56/88	68/102	85/124	Davis, 1968
Greater prairie chicken	28/56	35/56	41/64	49/77	55/84	60/84+	70/—	82/126+	Baker, 1953
Quails									
Scaled quail	26/50	30/60	35/65	39/75	45/85	57/93	64/115	107/150	Ohmart, 1967
California quail	29/55	32/62	38/70	46/80	52/90	62/108	72/121	100/141	Raitt, 1961
Bobwhite	26–30/ 54–58	33–37/ 56–60	40–44/ 60–64	44–50/ 70–75	52–58/ 81–89	58–63/ 99–107	69–77/ 120–28	97–105/ 146–54	Petrides & Nestler, 1943; Rosene, 1969
Harlequin quail			42/—	49/—	56/—	77/—	98/—	119/ 133–35	Leopold & McCabe, 1957
Partridges									
Gray partridge	24/—	27– 31/—	33– 38/—	39– 45/—	47– 52/—	55– 59/—	67– 73/—	86–87/ 115–25	Petrides, 1951
Red-legged partridge	29/—	34/—	41/—	49/—	58/—	70/—	86/—	105/130	Petrides, 1951

secondaries are lost in about the same sequence as they emerged; starting from the third, the molt proceeds inwardly, with the two outermost secondaries dropping as the more proximal secondaries are being lost.

Body feathers of the juvenal plumage are surprisingly similar in both the grouse and quail groups. Typically white or pale buffy shaft-streaks are conspicuous, especially on the upper parts; these often expand near the tip of the feather to form distinctive hammer-shaped markings. Apparently only the ptarmigans lack these distinctive juvenile markings. Usually the sexes are nearly identical in this plumage.

POSTJUVENAL MOLT AND FIRST-WINTER PLUMAGE

The postjuvenal molt (or "prebasic," according to Humphrey and Parkes) gradually replaces the juvenal body feathers with the more distinctly species-specific feathers of the first-winter (or "basic") plumage. The postjuvenal molt is virtually complete in all the species considered here, involving all the body feathers and all the flight feathers with the exception of the two outermost (primaries nine and ten) and their coverts. Additionally, the upper greater (and possibly other) coverts of the more proximal primaries are retained in the New World quails, as earlier mentioned.

Because the outer two juvenal primaries are retained during the post-juvenal molt, they will normally be carried by the bird until its next complete molt, which occurs after the next breeding season. This is the case at least in the species considered here, although Petrides (1942) reports that in the chachalaca (*Ortalis vetula*) these primaries are retained only until February or March, when a complete wing and tail molt occurs. Because of their relatively long persistence, the outer two primaries are usually subjected to considerable fading and wear; they thus provide a basic method of age estimation, especially in grouse, which do not retain distinctive juvenal upper primary coverts. Limitations to their value in determining age come from two sources; possible difficulties in estimating their wear relative to that of the more proximal primaries and from occasional aberrations in wing molt. This latter problem may result from a precocious molting of one or both of the outer juvenal primaries during the first fall (as regularly occurs in pheasants) or from an abnormally arrested molt in which the juvenal remiges are retained longer than normally. Several examples of each of these aberrant variations have been reported, and are important to note because of their obvious implications in the accuracy of age estimation techniques (table 10).

The gradual loss of the eight inner juvenal primaries, and their replacement by primaries of the first-winter plumage, provides an excellent method of estimating the ages of young grouse and quail between about three and fifteen or twenty weeks of age, by which time the last of these primaries will have completed their growth. Growth rates of representative species of all the United States genera of grouse and quails have been studied and for these an estimation of age is possible by determining the extent of primary replacement and growth during this period (table 9). Undoubtedly there may be some population variations in growth rates of these species, and hand-reared birds may develop at somewhat different rates from wild ones, but the availability of such ageing criteria is extremely valuable for back-dating probable hatching periods based on the examination of young birds.

First-winter secondaries replace juvenal secondaries at the same time this process occurs in the primaries or slightly later, and by the time the last of the juvenal secondaries have been shed the young bird will be well into its acquisition of the *first-winter* plumage. By the time the bird is four or five months old it should have completed growth of all its first-winter flight feathers and lost all its juvenal feathers other than those few wing feathers that are carried through the winter. With the loss of its juvenal body feathers the bird can be classified as an *immature* rather than a juvenile. Except in the ptarmigans, no further molt will occur until at least the following spring among the species considered here. However, the three ptarmigan species present a special case, for which an additional plumage stage and molt cycle must be mentioned.

SUPPLEMENTARY POSTJUVENAL MOLT AND SUPPLEMENTARY PLUMAGE

In at least the North American species of ptarmigans, a special plumage situation exists that must be mentioned here. One unique fact is that when the two outer juvenal primaries emerge (at two or three weeks of age) they do not resemble the other brownish primaries, but rather have the white vanes typical of the first-winter primary. Indeed, Salomonsen (1939) considers them to represent first-winter rather than juvenal primaries, but to do so is to accept the view that a major evolutionary difference between the primary molt of ptarmigans and all other grouse exists, and it seems more reasonable to believe that the coloration of the two outer juvenal primaries has only been adaptively modified in the genus *Lagopus* in relation to ecological requirements for concealing coloration. The postjuvenal molt of the body feathers of young ptarmigans likewise begins unusually early, at about four or five weeks of age, and the first feathers of preliminary winter plumage begin to appear. These initially consist of vermiculated or mottled feathers rather than pure white ones. Some juvenal feathers are retained for a time, including ones on the throat, breast, and hindneck. As this postjuvenal molt is being completed, a second stage of molt ("supplementary postjuvenal molt") begins, which replaces the last of the juvenal body feathers with pure white feathers and which also replaces the grayish or brownish feathers grown during the earlier stages of the postjuvenal molt with new white feathers. The body plumage held during the first winter thus includes both some of those feathers acquired during the preliminary postjuvenal molt, such as those on the abdomen, the under tail coverts, under wing coverts, legs, and toes, as well as others acquired during the later or supplementary postjuvenal molt, all of which are white (Salomonsen, 1939).

PRENUPTIAL MOLT AND FIRST NUPTIAL PLUMAGE

In most grouse and quail relatively little and possibly no additional molting occurs after the assumption of the first-winter plumage and the first breeding season. Dwight (1900) reported that in the genera *Colinus*, *Callipepla*, and *Cyrtonyx*, and possibly also in *Oreortyx*, there is a restricted renewal of feathers in the face and throat regions of these quail prior to the onset of breeding. The occurrence of such a prenuptial molt in the New World quail has been questioned by later investigators (Raitt, 1961; Raitt and Ohmart, 1966) but observations on hybrid quail support its existence (Johnsgard, 1970). Dwight reported a correspondingly restricted chin and head molt in species of the genera *Tympanuchus* and *Bonasa* and possibly but not definitely in species of *Dendragapus*. A fairly extensive prenuptial ("prealternate") molt was reported by Watson (1962c) in Cuban bobwhites. He also mentions that whereas the head is the last site to complete the postjuvenal molt, it is the first to begin the prenuptial molt in this species.

In the ptarmigan species there can be no question about the occurrence of a prenuptial molt ("pre-alternate" according to the classification of Humphrey and Parkes, 1959) and a distinctive nuptial (or "alternate") plumage. The extent of this molt may vary with age, sex, and latitude, but at this time the males first become markedly different from females. The male willow ptarmigan thus assumes its characteristic rusty brown upperparts, while male rock ptarmigan acquire vermiculated grayish feathering, and females of both become decidedly barred in appearance. The molt of the female may proceed somewhat more rapidly and be more extensive than in the male. However, at least in the rock ptarmigan, both sexes retain through the summer at least some portions of the preceding winter plumage, including feathers of their legs, toes, under wing coverts, and some upper wing coverts (Salomonsen, 1939).

POSTNUPTIAL MOLT AND SECOND WINTER PLUMAGE

Except in those species such as sage grouse in which sexual maturity may not be attained the first year, the bird will normally have attempted to breed while still in its first nuptial plumage. The timing of the following postnuptial molt is generally associated with endocrine changes related to changes in gonadal activity. In any case, it is typical for all the species considered here to begin a complete body molt in late summer, with the males generally somewhat in advance of the females. At this time the primaries will begin to be molted in outward sequence from the first through the tenth, the secondaries will be dropped starting with the outermost ones and proceeding proximally, and the rectrices will begin a gradual or rapid molt. The adult tail molt of grouse, like that of most pheasants (Beebe, 1926), is generally centripetal, as reported by Bendell (1955b) for blue grouse, and by Bergerud, Peters, and McGrath (1963) for willow ptarmigan, but may be virtually simultaneous, as indicated by Salomonsen (1939) for the rock ptarmigan and Stoneberg (1967) for the spruce grouse. In the bobwhite (Watson, 1962c) and probably in most or all other New World quails, the adult tail molt is centrifugal, providing an apparent basic difference in the molting sequences of grouse and quail. This may not be universal however; Baker (1953) mentions a greater prairie chicken specimen that was undergoing an apparent centrifugal tail molt.

At the same time the wing and tail feathers are being molted, the body feathers are being renewed, approximately in the same order that they originally grew in during the postjuvenal molt. Except in the ptarmigan, all of the feathers that grow in during the postnuptial molt will be carried through the following winter and represent the second winter plumage. It is of interest to note that only at this time will the last traces of the juvenal plumage be lost—namely, the two outermost primaries, their coverts and, in the case of the New World quails, the upper greater coverts of the other primaries as well. There are also a few cases of arrested molt known in which these outer primaries are not dropped but are carried through a second winter, as noted in table 10.

TABLE 10

REPORTED ABERRATIONS IN PRIMARY AND SECONDARY MOLT OF GROUSE AND QUAILS

I. Examples of Precocious Molt

A. Molt of ninth juvenal primary in first autumn, or of both ninth and tenth juvenal primaries

1. Chukar partridge:	Reported in six of eighteen early-hatched birds by Smith (1961).
2. Interior bobwhite:	Reported in two Wisconsin specimens by Thompson & Kabat (1950).
3. Florida bobwhite:	Loveless (1958) reported that 33.4 percent of 138 south Florida birds molted beyond the eighth primary, with 30.4 percent molting their ninth, and 3.0 percent both their ninth and tenth. Further, 5.1 percent of the birds molted some of the upper primary greater coverts, starting distally, and two males molted all but one of these coverts. Precocial primary and covert molt was also noted in South Carolina bobwhites by Rosene (1969).
4. Cuban bobwhite:	Reported in eight of eighty-one specimens by Watson (1962c).

II. Examples of Arrested Molt

A. Retention of one or more secondaries abnormally long

1. Chukar partridge:	One female retained two juvenal secondaries through the first autumn (Watson, 1962c).
2. Blue grouse:	Two adult males retained all but two secondaries for at least thirteen months and through the postnuptial molt (Bendell, 1955b).

B. Retention of primaries abnormally long

1. Scaled quail:	One specimen retained tenth juvenal primary through second autumn (Wallmo, 1956a).
2. Bobwhite:	Late-hatched birds frequently retain seventh juvenal primary (as well as ninth and tenth) through first autumn (Thompson and Kabat, 1950). Adults may retain ninth or tenth primaries through postnuptial molt and at least until midwinter (Rosene, 1969).
3. Blue grouse:	Two adults retained their ninth and tenth primaries through the postnuptial molt (Bendell, 1955b).
4. White-tailed ptarmigan:	Retention of juvenal primaries through a second autumn reported (C. Brown, quoted in Ellison, 1968a).

SUPPLEMENTARY POSTNUPTIAL MOLT AND SUPPLEMENTARY WINTER PLUMAGE

As noted, the ptarmigans differ from the other grouse in the postnuptial molt sequence, and they exhibit an early or preliminary postnuptial fall molt in adults that corresponds to the early postjuvenal molt of young birds. In this mixed white and grayish plumage adult male rock ptarmigan closely resemble females, and both can hardly be differentiated from immature birds (although the old birds will be replacing their two outer primaries at this time). This stage is referred to as the preliminary second winter plumage. A few body feathers will still be retained at this time from the summer plumage, including (in males) some greater wing coverts or tertiaries and some mantle or hindneck feathers. Females retain many lower breast or flank feathers, some inner median and greater coverts, some tertiaries, and some scattered upper breast, throat, and mantle feathers. These summer feathers, plus the grayish fall feathers just acquired and including some of the upper parts, some flank feathers, the tertiaries, and some upper wing coverts, are now quickly replaced with white feathers by a special supplementary postnuptial molt (Salomonsen, 1939). Observations by Höst (1942) on captive willow ptarmigan clearly indicate the importance of photoperiod not only in regulating the timing of molt in willow ptarmigan but also in influencing the pigment characteristics of the new feathers. Höst found that by exposing birds in winter plumage to artificially long photoperiods starting in November, he could induce the precocious assumption of the spring nuptial plumage and even stimulated a female to lay a clutch of eggs in December and January. One of the males that had acquired a nuptial plumage at the beginning of February was then exposed to a seven-hour photoperiod, upon which it molted directly back into a white winter plumage without passing through an intervening fall plumage. However, five birds that had their daylight reduced in August passed through a short fall plumage before assuming their winter plumage.

SECOND NUPTIAL PLUMAGE

The second nuptial plumage is acquired in the same manner as the first nuptial plumage, and later plumages and their intervening molts are repetitions of the earlier ones. Once the juvenal outer primaries have been lost in late summer, it is generally almost impossible to recognize birds in their second fall of life from older age categories.

A summary (figures 9 and 10) of the foregoing information with respect

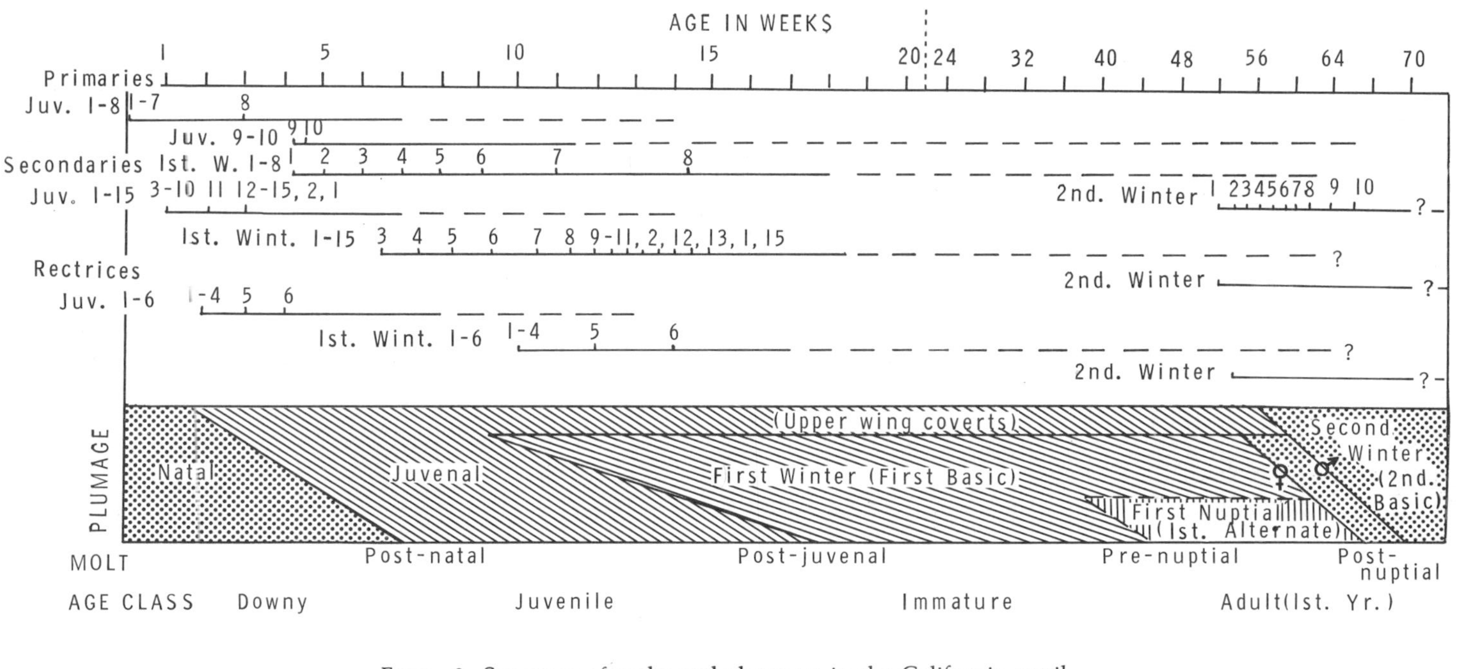

FIGURE 9. Sequence of molts and plumages in the California quail.

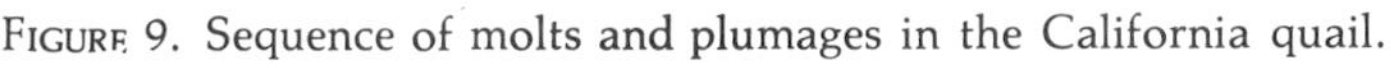

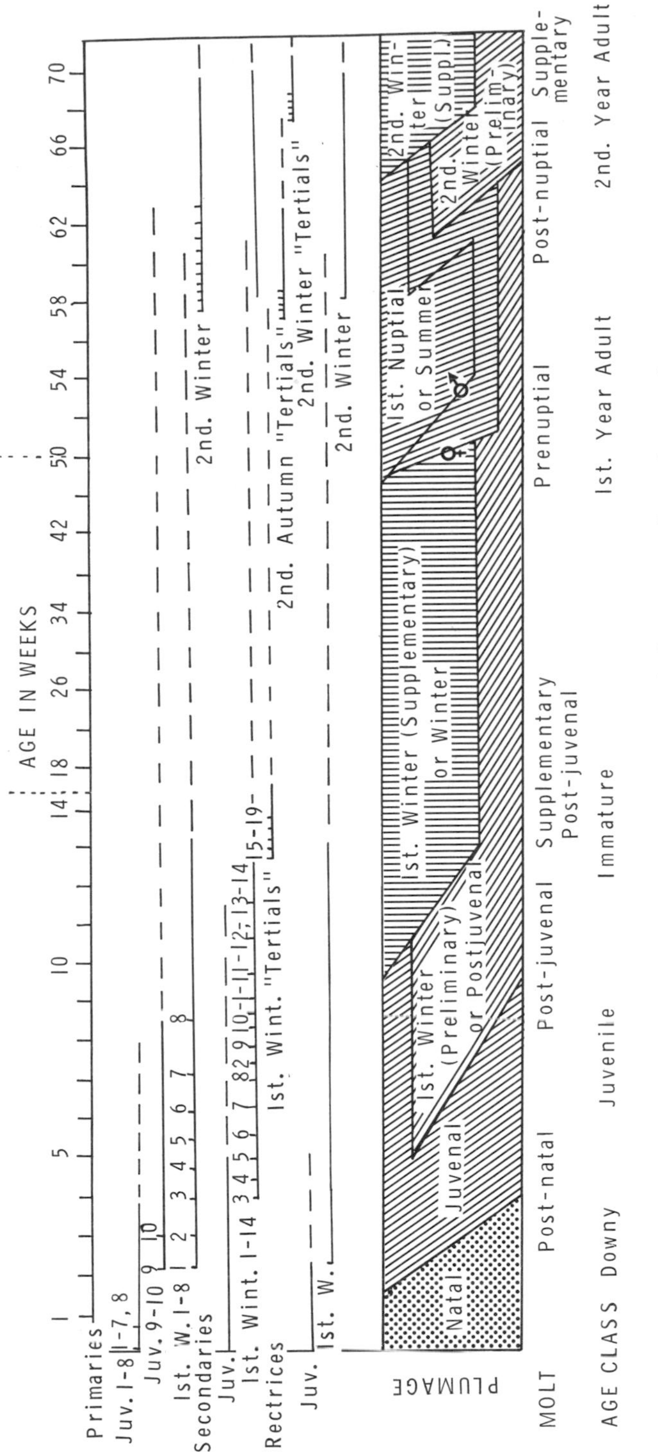

FIGURE 10. Sequence of molts and plumages in the rock ptarmigan.

to a representative species of quail (based mainly on information on the California quail by Raitt, 1961) and the rock ptarmigan (based mostly on data provided by Salomonsen, 1939). The relatively greater complexity of the ptarmigan plumages and the compression of the natal and juvenal plumages into the minimum possible time spans are apparent in these diagrams.

ENDOCRINE AND ENVIRONMENTAL CONTROLS OF MOLTING

Timing of the postnatal and postjuvenal molts can probably be regarded as age-dependent, progressing as rapidly as food supplies and general bodily development allow. Additionally, in the ptarmigan, ecological requirements for coloration changes related to the seasons may place special demands on molt timing in young birds. Watson (1962c) found that Cuban bobwhites apparently have more prolonged postjuvenal and prenuptial molts than do Florida populations of the same species, which he suggests might be related to differences in midwinter temperatures of the two regions. Watson suggests that, although gonadal hormones may help to regulate molt timing by their temporary inhibiting effects, thyroid activity is probably responsible for the initiation of molting and that breeding and molting are not under the same hormonal control. Raitt and Ohmart (1966) likewise suggest that there is no directly causative relationship between the regression of gonadal activity and the onset of the postnuptial molt they observed in Gambel quail.

Although the thyroid hormone is most commonly implicated in molt initiation, recent evidence (summarized by Sturkie, 1965, and by Lofts and Murton, 1968) indicates that molting may be relatively independent of thyroid activity, or at least the increased metabolic activity associated with molting may not indicate direct thyroid control of the latter. Juhn and Harris (1955, 1968) found that injected progesterone can initiate molt in adult female domestic fowl, and that prolactin stimulates molt in capons when given alone or in conjunction with progesterone. Shaffner (1955) and Adams (1956) also reported on the molt-stimulating effects of progesterone. Jones (1969b) found that progesterone injection alone did not stimulate defeathering associated with incubation patch development in the California quail, but that this hormone in conjunction with prolactin has such effects.

Although molt might be initiated by progesterone alone in various year-round breeders, this effect evidently does not occur among seasonally breeding birds (Kobayashi, 1958). In seasonally breeding forms there may instead be a synergistic relationship between progesterone and thyroxin relative to molt control, since Kobayashi found that thyroidectomy had the effect of inhibiting molt induction through progesterone treatment in such species.

In addition to direct endocrine controls, external factors such as photoperiod changes may be additional regulators of molt, as suggested by Höst's (1942) early experiments with willow ptarmigan. Lofts and Murton (1968) have reviewed the evidence on this point and have confirmed that at least some north-temperate photoperiodic species of birds require a postnuptial exposure to reduced photoperiod not only to regain their photosensitivity relative to reproduction but also for the normal temporal completion of their molt.

4

Hybridization

THE study of hybridization between species, under either natural or artificial conditions, provides information of value for a variety of reasons. In general, it may be expected that the incidence of crossbreeding between populations existing under natural conditions will be related to their nearness of relationship, and information of taxonomic interest may be obtained from such study. Furthermore, the relative survival and fertility of the resulting hybrids should provide an indication of the degree of genetic difference between the parental types, and thus genetic information may be available through experimental hybridization studies. Hybrids provide favorable material for studying the chromosomal numbers and configurations among related species, and when they are fertile the degree of phenotypic variation in second or backcross generations may be used to estimate genetic differences controlling specific traits. Finally, the presence or absence of natural hybridization between closely related forms occurring in the same habitats may provide a clue as to the degree of niche overlap and interspecies competition for habitat resources. Therefore, if the basis for periodic or local hybridization between two forms that normally do not hybridize can be established, the ecological differences that normally prevent hybridization may possibly be deduced.

For various reasons, the grouse and quails of North America exhibit a rather surprisingly high tendency to hybridize, even among species belonging to seemingly different genera. Peterle (1951) reviewed the cases of intergeneric hybrids reported in gallinaceous birds, and Cockrum (1952) provided a more complete survey of hybridization in North American birds. Sibley (1957) commented on the taxonomic significance of hybridization in grouse, and a similar review of the significance of hybridization in the New World quails is available (Johnsgard, 1970). For a complete listing of all known hybrids of gallinaceous birds, including those reported from Europe and Asia, the summary by Gray (1958) may be consulted.

GROUSE HYBRIDS

Virtually all known cases of hybridization among the North American grouse species have involved naturally occurring hybrids. This is largely a reflection of the difficulties of keeping and breeding grouse in captivity. The only case of hybridization among North American grouse under captive conditions known to me is the production of several hybrids (including reciprocal crosses) between greater and lesser prairie chickens in 1969 and 1970 by William Lemburg of Cairo, Nebraska.* He has also attempted, without success, to obtain backcross hybrids from a wild-caught female greater prairie chicken × sharp-tail grouse mated to males of both of these species.

All of the North American genera of grouse (as recognized here) have been involved in intergeneric hybridization except for *Bonasa* and *Centrocercus.*[1] In addition, intrageneric hybridization has occurred in *Tympanuchus, Dendragapus,* and probably also in *Lagopus.*

Intrageneric Hybrids

Hybridization within the genus *Lagopus* has still not been certainly proved, but would seem highly probable on the basis of the extensive area of geographic contact between the willow and rock ptarmigan. Gray (1958) summarized references to British specimens of possible hybrids between these two species but questioned their authenticity. Todd (1963) mentioned one specimen from Labrador that he examined, which he thought might be an abnormally colored willow ptarmigan or possibly a hybrid. Harper (1953) described a subadult male ptarmigan collected in Keewatin that had a bill depth of 8.5 millimeters (vs. 7.75 maximum for his series of rock, and 9–10.5

*William Lemburg, 1970: personal communication.

1. A record of hybridization between the sage grouse and the sharp-tailed grouse has recently been published (*Wilson Bulletin* 73:491–93).

for willow ptarmigan). Its upper tail coverts were longer than are typical of rock ptarmigan, but its weight and wing measurements were less than are typical of rock ptarmigan. Some black feathers were present at the base of the bill and behind the eye. Harper concluded that it must be a hybrid or a highly aberrant willow ptarmigan.

A single reported specimen representing hybridization between the blue grouse and spruce grouse has been reported (Jollie, 1955). This bird was obtained in Idaho where, although the ruffed grouse is common, both of the parental species are evidently rare. These two species overlap extensively in their ranges from western Montana through Idaho and Washington, and north to the Yukon Territory and apparently occupy generally similar habitats through much of this range.

The other genus of grouse that has been involved in intrageneric hybridization is *Tympanuchus*, and in this case there is no question that hybridization between the sharp-tailed grouse and the greater prairie chicken has occurred repeatedly under natural conditions. In a recent summary, Johnsgard and Wood (1968) pointed out that natural hybridization has been reported in every state and province where natural contact between these species has occurred. These include four Canadian provinces from Ontario to Alberta and the Dakotas, Colorado, Nebraska, Iowa, Minnesota, Wisconsin, and Michigan. The highest known incidence of hybridization so far reported is on Manitoulin Island, Ontario, where the two species have recently come into contact and between 5 and 25 percent of the total grouse population may be of hybrid origin. The complete spectrum of plumage patterns exhibited by such birds would indicate a clear capacity to produce second-generation or backcross offspring, but so far little information is available on the relative reproductive success of hybrids as compared to the parental types. The few observations made so far (Lumsden, 1965; personal observations) suggest that hybrids are usually able to occupy only peripheral territories on display grounds that are dominated by pures of either species, and are probably at a considerable reproductive disadvantage in spite of their apparent fertility.

Lagopus × *Dendragapus* Hybrids

At least three specimens of natural hybrids between willow ptarmigan and spruce grouse have been reported so far (Lumsden, 1969). These two species overlap extensively in their breeding ranges in eastern Labrador, northern Ontario, the Northwest Territories, Yukon Territory, British Columbia, and Alaska, but are ecologically isolated during the breeding season. Lumsden noted that in the area where two of the hybrids occurred, the Hudson Bay region of Ontario, spruce stands near rivers are in close proximity to heath and lichen communities. The last of the three reported hybrids came from York Factory, Manitoba, which is also near Hudson Bay and presumably represents similar habitat. Lumsden suggested that the measurements of one of the two preserved hybrid skins would suggest that hybrid vigor has influenced its size, but no information is available as to the possible fertility of this cross.

Dendragapus × Tympanuchus Hybrids

Only a single specimen representing this cross has so far been reported. Brooks (1907) described an apparent blue grouse times sharp-tailed grouse hybrid taken at Osoyoos, British Columbia. In spite of a seemingly substantial overlap in the breeding ranges of these two species, extending from the Yukon southeast through parts of British Columbia, Washington, Idaho, western Montana, Utah, and western Colorado, it appears that ecological differences in breeding habitats rarely would allow for possible interbreeding. Additionally, these two genera do not appear to be especially closely related.

Grouse × Pheasant Hybrids

In spite of the fact that the ring-necked pheasant is regularly placed in a separate subfamily from the grouse, many instances of apparent hybridization between these two groups have been reported. In North America several apparently authentic hybrids between the blue grouse and the ring-necked pheasant have been described, and one was captured and kept alive for several years (Gray, 1958; Hudson, 1955). There was also a case of probable hybridization between the ruffed grouse and the ring-necked pheasant in New York (Bump et al., 1947). In Europe, several reported hybrids between pheasants and ptarmigan have been reported (Gray, 1958), in spite of marked habitat differences exhibited by these species.

QUAIL HYBRIDS

Intrageneric Hybrids

All known natural hybrids among the New World quails involve the genus *Callipepla*, as recognized in this book.

The range of the scaled quail overlaps fairly extensively with that of the Gambel quail, primarily in New Mexico but also in western Texas along the Rio Grande, southeastern Arizona, and adjacent Mexico. Introductions of the scaled quail into central Washington have also resulted in a small amount of contact with California quail, and two hybrid specimens have been reported from that area (Jewett et al., 1953). Shore-Baily bred and reared a number of such hybrids, finding them to be fertile, and I have also reared fourteen first-generation hybrids of this cross (plate 114). Although I have not produced any second-generation offspring, several backcrosses to the California quail have been reared to maturity.

Wild hybrids between the scaled quail and the Gambel quail have been known to occur for some time; apparently the earliest published record is that of Bailey (1928), who described a hybrid shot in Grant County, New Mexico. This hybrid was the basis for a painting by Louis A. Fuertes, now in the collection of the Laboratory of Ornithology, Ithaca, New York. The laboratory very kindly gave me permission to reproduce this painting (plate 97), on the back of which is the following inscription:

An interesting half-breed Quail. Probably Gambels × Scaled. This bird was sent me by Mr. W. E. Watson of Pinos Altos, having been killed by him from a flock of Gambels Quail November 26, 1916 on Whiskey Creek a few miles from Pinos Altos, Grant County, New Mexico. It is a male, and brief description is:—Crest somewhat shorter than male Gambels and feathers not clubbed. Chin and throat, crest, and middle of belly patch rich chestnut. Thus chestnut taking the place of black in Gambels. Forehead light gray, hind head chestnut. Fore-breast color of Gambels but showing the shelled edgings and black shaft lines of Scaled in a lesser degree.

I have just sent the skin to Mr. L. A. Fuertes and he comments thereon, in part, as follows:—"This is the second instance I have known of wild hybridization of American partridges of different genera. The other was a male hybrid Mountain Plumed Quail and California Valley Quail. I painted the bird for Loomis but both the specimen and the picture were burned in the San Francisco fire."

R. T. Kellogg
Silver City, New Mexico
August 11, 1927

More recently, Phillips, Marshall, and Monson (1964) reported wild hybrids of this combination from various localities in southeastern Arizona, and Hubbard (1966) described an apparent backcross hybrid that had also been collected in Grant County. A hybrid was recently collected in Otero County by New Mexico game personnel. Captive hybrids of this cross have been obtained on several occasions. These were originally thought to be sterile, but second-generation embryos were brought nearly to the point of hatching at the Arizona–Sonora Desert Museum, and William S. Huey informed me that he reared about twenty apparent backcross hybrids. I have not been able to produce any second-generation or backcross hybrids from the three males and one female of this cross that have been present in my laboratory.

The Gambel quail also exhibits a very limited degree of natural contact with the California quail in southern California, and wild hybrids between these species have been reported from there (Miller and Stebbins, 1964). One male hybrid representing this cross that is in my laboratory has exhibited sexual activity but has not fathered any backcross offspring in either direction. It would nevertheless seem highly likely that this hybrid combination will prove to be fertile.

Two other intrageneric hybrid combinations have been reported in this genus. One is the cross between the California quail and the elegant quail, represented by hybrids reared in the London zoo, and the other is a cross between the scaled quail and elegant quail that was also produced under captive conditions (Banks and Walker, 1964).

Colinus × *Callipepla* Hybrids

Natural contact between the bobwhite and the scaled quail exists over a fairly broad zone extending from northern Mexico through west-central Texas, the Oklahoma panhandle, possibly extreme southwestern Kansas, and southeastern Colorado. Wild hybrids have been reported from three counties in Texas, and probable hybrids have also been seen in Oklahoma. Captive hybrids have been produced on a variety of occasions, including a considerable number that have been reared in my laboratory (plate 117). The females of this cross produce abnormally small eggs, which usually have exhibited no embryonic development. Most attempted matings with males of the parental species have proven unsuccessful but a few backcrosses have hatched.

Natural contact between the bobwhite and the other *Callipepla* species is virtually nonexistent, but introductions of the bobwhite into Washington, Oregon, and Idaho have resulted in some possible recent contact with the California quail. Furthermore, introductions of the bobwhite and California quail into Utah produced a relatively short-lived period of contact and resulted in the only known case of naturally occurring hybridization between these species (Aiken, 1930). This cross has also been obtained in captivity, and the hybrids evidently exhibit a very limited degree of hybrid fertility. A pair of such hybrids raised at the San Joaquin Game Bird Farms of Reedley, California, produced over one hundred eggs, of which four second-generation birds were successfully reared (plate 116). These hybrids were maintained in my laboratory for about two years, and the females produced uniformly small eggs that exhibited little or no embryonic development. Attempts to backcross the males to both parental species were also unsuccessful.

Although no wild hybrids between the bobwhite and Gambel quail are known, I have reared two hybrids representing this cross to maturity (Johnsgard, 1970). These two, a male and female, established a firm pair bond and exhibited normal sexual behavior (Prososki, 1970), but the female's eggs were somewhat smaller than normal and all of them were either infertile or exhibited early embryonic death. At least ten skins representing captive-raised hybrids between the elegant quail and the bobwhite, which is a hybrid combination previously unreported in the literature, are in the J. S. Ligon Collection at the University of New Mexico. Some of these specimens were the result of hybridization of the elegant quail with the masked bobwhite, while others involved one of the white-throated races, and the differences in body as well as throat pigmentation have been clearly transmitted to the hybrids, particularly the males.

Callipepla × Oreortyx Hybrids

The area of geographic overlap between the California quail and mountain quail is considerable, and includes much of California, Oregon, and Washington. The earliest record of hybridization between these species is that of Peck (1911), who described a specimen taken in Harney County, Oregon, which is now in the Museum of Vertebrate Zoology, Berkeley, California. A second specimen of unknown origin was painted by L. A. Fuertes (Peterle, 1951). I have reared to adulthood a male of this cross (male mountain × female California) but have been unable to obtain fertile eggs with attempted pairings with female California quail. There are no other reported cases of natural or artificial hybridization involving the mountain quail, although one unidentified skin in the J. S. Ligon Collection quite obviously is a mountain × California quail specimen.

Callipepla × Philortyx Hybrids

The barred quail and scaled quail exhibit no geographic overlap in ranges, and the barred quail is in natural contact only with the bobwhite and perhaps also with the elegant quail. Since the barred quail has only rarely been maintained in captivity, it is surprising that any hybrids at all have been produced. However, at the Centro de Investigaciones Basicas, Campo Agricola Experimental, Progreso, Guerrero, a variety of Mexican quail species are being raised for study and release. A single wild-caught male barred quail has been present for several years, and has been kept with a group of scaled quail. In 1969 it mated with a female scaled quail, as a result of which twelve hybrid offspring were reared (plate 111). When I saw the hybrids in June of 1970 all twelve (predominantly females) were still alive, but no eggs had been produced. The birds did not appear to be paired, nor had they exhibited any sexual behavior, according to the station manager, Sr. Alvaro Aragon.

Laboratory Hybridization of Quail

For the past several years I have been attempting to produce a variety of intergeneric and intrageneric hybrids of New World quails, for behavioral studies as well as for the genetic and evolutionary information that such hybridization might be able to provide (Johnsgard, 1970, 1971). Although the methods and some of the results have already been reported, an updated summary of hybrid fertility and hatching success is presented here (table 11). It may be seen that individuals representing eight different hybrid combinations (three of which are intergeneric on the basis of nomenclature used in this book) have hatched from pairing representing ten possible combinations. As was previously reported (Johnsgard, 1970), one of these intergeneric combinations had previously been unreported in the literature, and to my knowledge the crested bobwhite × bobwhite hybrid combination also is previously unreported. Not noted in the table is the recent hatching of two hybrids between the bobwhite and the black-throated bobwhite from twenty-one incubated eggs.

So far, only four hybrid combinations have been produced beyond the first generation. One of these is a backcross resulting from the mating of a male scaled quail with a female Gambel × scaled quail, from which eight offspring were hatched, although none was raised to maturity. The second successful backcross has been one produced by mating a male California quail with a female California × scaled quail hybrid. A total of twenty-two individuals representing this combination have been hatched, most of which were reared to their adult plumage. One F_1 hybrid pair, resulting from a mating of a male bobwhite with a female crested bobwhite × bobwhite has produced fourteen offspring. Attempts to produce F_2 Gambel × scaled quail and F_2 California × scaled quail have thus far failed, although the females of both these crosses lay normal-sized eggs and the males are obviously sexually active. In contrast, only abnormally small eggs have been produced by female hybrids between bobwhite and Gambel quail and between bobwhite and scaled quail. The small eggs laid by these females have been infertile or have usually undergone only limited embryonic development.

Attempts to produce F_2 bobwhite × scaled quail hybrids have proven fruitless, and backcross attempts in both directions have been made. To date only twenty backcross individuals have survived to hatching, and none has lived beyond two weeks after hatching. Interestingly, hybrid females appear to be relatively more fertile than are hybrid males, judging from our limited data. In all three cases where backcross pairing has produced either living offspring or survival until late embryonic stages, the maternal parent was the hybrid and the father was one of the parental species. In cases where the male parent was a hybrid and the female was a pure of one of the parental species, all of the eggs have proven to be infertile or have at most died early in embryonic development.

TABLE 11

FERTILITY AND HATCHABILITY OF HYBRID QUAIL EGGS

	Total Eggs	*Infertile Eggs*	*Embryonic Death*	*Total Hatched*
F_1 hybrid pairings				
Bobwhite x scaled (B x S)	197	92	52	53
Scaled x bobwhite (S x B)	338	303	35	0
Crested bobwhite x bobwhite (CB x B)	107	81	20	6
California x scaled (C x S)	47	21	7	19
Bobwhite x Gambel (B x G)	17	0	4	13
Scaled x Gambel (S x G)	28	17	4	7
Gambel x California (G x C)	13	11	2	0
Gambel x scaled (C x S)	9	6	2	1
Scaled x elegant (S x E)	5	3	1	1
Mountain x California (M x C)	34	27	4	3
Totals	795	561 (70.6%)	131 (16.5%)	103 (12.9%)
F_2 hybrid pairings				
F_1 BG x F_1 BG	16	6	10	0
F_1 GS x F_1 GS	33	26	7	0
F_1 CS x F_1 CS	266	264	2	0
F_1 BS x F_1 BS	370	370	0	0
F_1 BS x F_1 CS	9	9	0	0
Totals	694	675 (97.2%)	19 (2.8%)	0
Backcross hybrid pairings				
BB x F_1 CBB	22	5	3	14
F_1 BG x G	28	28	0	0
S x F_1 GS	32	11	13	8
F_1 CG x G	32	19	13	0
F_1 BS x B	146	146	0	0
F_1 BS x S	106	106	0	0
B x F_1 BS	249	165	64	20
S x F_1 BS	108	40	0	0
C x F_1 CS	66	12	32	22
Totals	789	600 (76%)	125 (15.8%)	64 (8.1%)

Since the hybrids mentioned here involve species having remarkable differences in head patterning and crest condition, it is of interest to consider the inheritance of these traits in the hybrids. On the basis of male hybrid specimens obtained in my laboratory, those I have seen in museums, and one literature description, it has been possible to produce a diagram indicating the male head plumages of eight hybrid combinations occurring among six different species of *Colinus*, *Callipepla*, and *Oreortyx* (figure 11). The diagram illustrates quite clearly the remarkable plasticity in facial patterning and probable visual signal characters (sign stimuli) that can be achieved by the addition or subtraction of feather pigments and the modification of crest shape and length. The genetic basis of both the feather pigmentation and the crest condition is as yet unknown, but it seems probable that these are under fairly simple genetic control. Certainly the variations in crest lengths and head pigmentation that are apparent in adjacent populations of the numerous Central and South American races of *Colinus cristatus* would suggest that this is the case. If so, it would seem that the evolution of distinctive male visual signaling devices under the influence of natural selection might occur quite rapidly and result in quite widely divergent appearances in the heads of fairly closely related species. The differences in the appearance of the head of the scaled quail from the closely related and geographically overlapping Gambel quail might provide a case in point. Here, a combination of differences in crest length, crest color, and throat color provides for a completely different head appearance, in spite of the fact that the colors of the back, flanks, wings, and tail are distinctly similar in these two species.

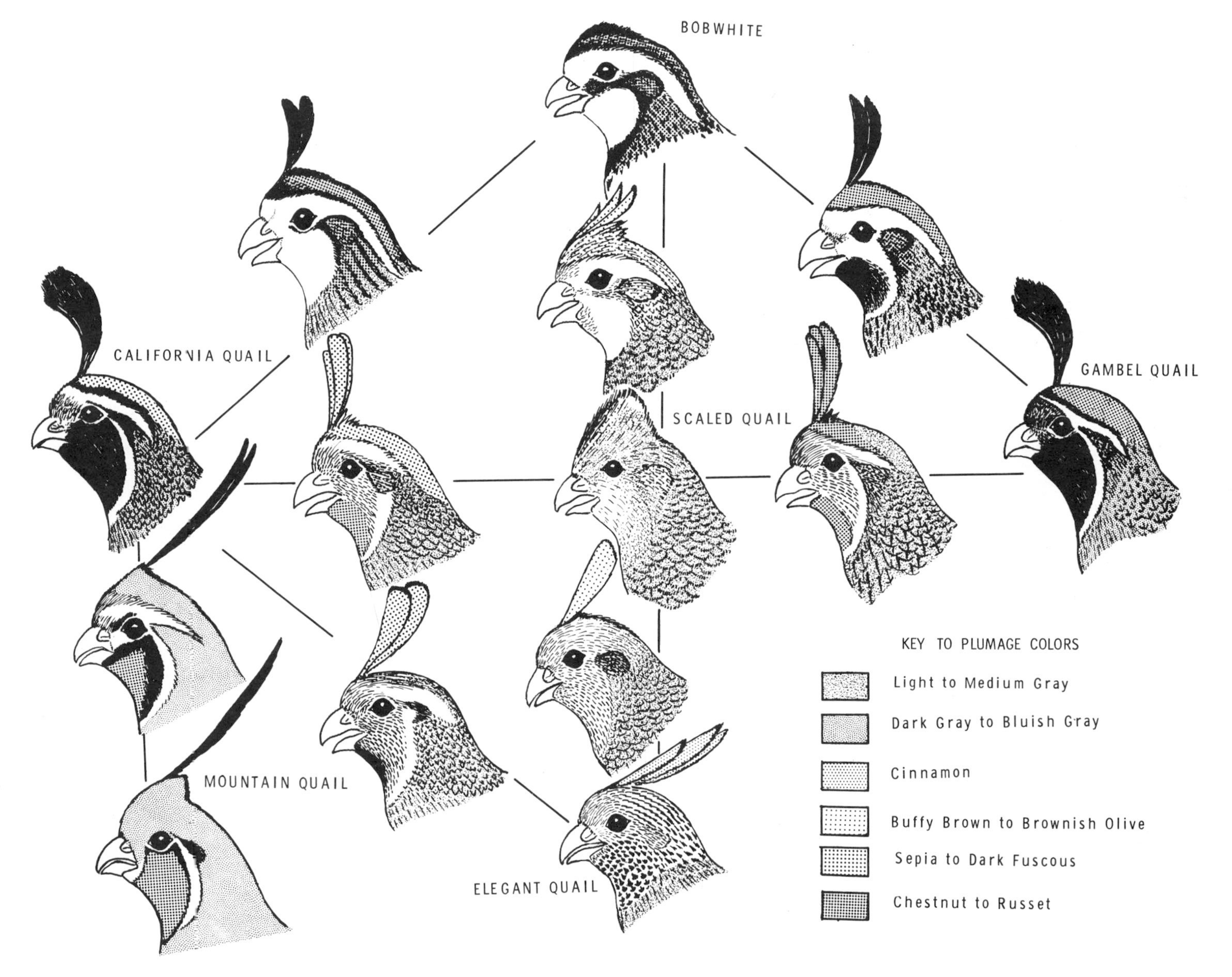

FIGURE 11. Head plumage patterns of hybrid quails, compared with parental species (from Johnsgard, 1971).

5

Reproductive Biology

THE reproductive potential of animal species is a compound result of numerous behavioral and physiological characteristics, most of which can be considered species-typical. These include such things as the time required to attain reproductive maturity, the number of nesting or renesting attempts per year once maturity is attained, the number of eggs laid per breeding attempt, and the number of years adults may remain reproductively active. These traits place an upper limit on the reproductive potential of a species, which is never actually attained. Rather, the actual rate of increase will only approach the reproductive potential, being limited by such things as the incidence of nonbreeding; the mortality rates of adults; decreased hatching success resulting from infertility, predation, or nest abandonment; relative rearing success; incidence of renesting and clutch sizes of renests; and similar factors that affect the reproductive efficiency. The relative involvement of the male in protecting the nest or the young may also influence hatching or rearing success. Among those species in which the male does not participate in nesting behavior, the relative degree of monogamy, polygamy, or promiscuity may strongly influence the reproductive ecology and population genetics of the species. Although many of these considerations will be treated under the accounts of the individual species, a general comparison of the grouse and quail groups as a whole are worth considering here, to see if any general trends can be detected.

AGE OF SEXUAL MATURITY AND INCIDENCE OF NONBREEDING

In the absence of evidence to the contrary, it must be assumed that all native quail species mature their first year. This is indicated by the apparent absence of nonbreeding females during favorable years under natural conditions, known regular breeding by females still carrying juvenal outer primaries, and consistent breeding under captive conditions of birds less than a year old. Bobwhites reared in captivity usually attain sexual maturity at between 139 and 185 days under lighted conditions (Baldini, Roberts, and Kirkpatrick, 1952), and scaled quail have laid fertile eggs in our laboratory within 160 days after hatching. We have also regularly obtained breeding from yearlings of all other quail species we have maintained in captivity.

Among the grouse, the situation may be different for at least some species. Bump et al. (1947) reported that nonbreeding by wild female ruffed grouse varied from none in most years to over 25 percent in some years. Weeden (1965b) found no indications of female nonbreeding in wild rock ptarmigan, although Maher (1959) found some evidence of nonbreeding in wild willow ptarmigan. Stanton (1958) reported that 25 percent of yearling female sage grouse failed to produce eggs, and Bendell and Elliot (1967) found that 25 percent of thirty-eight yearling female blue grouse were nonbreeders, compared with 4 percent of sixty-nine adult females. Yearling male blue grouse are nonterritorial according to these authors. Yet in this species, as in several other grouse, the highly promiscuous mating system allows for the achievement of effective fertilization of all females by a relatively small proportion of fully mature males.

Extensive nonbreeding during unfavorable years is apparently much more prevalent among quails than grouse, at least among the more northerly species of quails. Mountain quail may not nest at all in very dry years (Leopold, 1959). The same applies to scaled quail; precipitation occurring during the current spring and summer seems to be the most important influence on this species (Campbell, 1968). Little or no rainfall during the preceding winter and spring reduces the over-all nesting success of the California quail (Hungerford, 1964). Similarly in the chukar partridge extensive nonbreeding may occur in unusually dry years (Christensen, 1954), and the same may apply to bobwhites (Lehmann, 1946).

NUMBER OF NESTING OR RENESTING ATTEMPTS PER YEAR

No known instances of double-brooding have been reported for any North American grouse, and, indeed, known examples of renesting when nests are lost after incubation has begun are hard to find. Among the white-tailed ptarmigan Choate (1963) reports one definite renest; and the late clutches number only three or four eggs. Weeden (1965b) reported only one known case of renesting in rock ptarmigan, but noted that 3 percent of 228 nests and broods were late-hatching. Jenkins, Watson, and Miller (1963) mention that among Scottish red grouse definite renesting occurs in some years, and the clutch sizes of second nesting attempts are sometimes smaller than in first ones. They noted that five of seven marked birds laid again after their eggs were taken. Patterson (1949) estimated that a small incidence of renesting probably occurs in sage grouse, and Crunden (1959) subsequently reported one definite case. Stoneberg (1967) found no indication of renesting in the spruce grouse, and so far only two definite cases of renesting in the blue grouse have been reported (Zwickel and Lance, 1965). Renesting by ruffed grouse is apparently infrequent (Bump et al., 1947), with probably less than 25 percent of the unsuccessful females attempting to renest (Edminster, 1947). Ammann (1957) reported that no more than 10 percent of young sharp-tailed grouse hatched in Michigan could have resulted from renesting. Nests of the greater and lesser prairie chickens show a decline in clutch size toward the end of the nesting season (Hamerstrom, 1939; Baker, 1953; Copelin, 1963), suggesting a certain incidence of renesting, but until recently only in the Attwater prairie chicken had any verified cases been reported (Lehmann, 1941). However, Robel et al. (1970) found that three of fourteen radio-tracked greater prairie chicken females renested, one of them making two renesting attempts.

In contrast, the quail as a group show a greater tendency toward double-brooding and renesting, perhaps because of their monogamy and generally more southerly breeding distributions. Leopold (1959) reports that one or two renesting attempts may be made by mountain quail, but very early accounts suggesting that two broods of this species or of scaled quail are sometimes reared are yet to be verified. Evidence favoring double-brooding is strongest for the California and Gambel quails. McMillan (1964) reported that in favorable years up to 75 percent of the early broods of California quail are reared by males while the females renest. McLean (1930) reported one definite second brood in this species. Edminster (1954) states that there may be up to two renesting attempts, and Raitt (1960) stated that a few late broods hatched in August indicate probable renesting behavior. In the Gambel quail renesting attempts are reportedly common until mid-August (Gorsuch, 1934) or even early September (Raitt and Ohmart, 1966), and possible extensive double-brooding during a favorable year has been reported by Gullion (1956a), who believed that the earlier birds may be either cared for by males or left in the care of older birds of the year. Stanford (1953) reported that three captive pairs of bobwhites raised two broods, with the male taking over the first in each case.

Renesting in the gray partridge is highly probable. McCabe and Hawkins (1946) indicated that the average clutch size of probable renests is 9 eggs, or considerably under the clutch size of early nests. Mackie and Buechner (1963) suggested that in chukar partridges renesting until early July is probable in Washington state; in Turkey renesting occurs if the first nest is broken up early in incubation (Bump, 1951). The role of the male in the chukar partridge is still uncertain; in this species and the related redlegged and Barbary partridges the male evidently sometimes incubates the first nest while the female lays a second clutch (Goodwin, 1953). Watson (1962a) suggests that perhaps the male raises a brood when the population is low. Observations in the United States sometimes suggest that males may play no role in incubation and instead gather in flocks (Alcorn and Richardson, 1951; Bohl, 1957). Other studies indicate that males may be seen with about 10 percent of the females and broods (Mackie and Buechner, 1963) or may accompany broods fairly often (Galbreath and Moreland, 1953).

PARTICIPATION OF THE MALE IN INCUBATION AND DEFENDING THE BROOD

Since the availability of the male influences the likelihood of successful renesting and allows for possible double-brooding, a summary of male participation in breeding is of some interest. Among the grouse, no cases of male incubation have been reported. However, the male willow ptarmigan actively defends the nest and brood (Dixon, 1927; Conover, 1926; Watson and Jenkins, 1964). In the rock ptarmigan the male rarely stays with the brooding female and does not defend the brood (Weeden, 1965b) or if present may desert the brood when they can fly or even earlier (Bannerman, 1963). However, some instances of active brood defense have been seen by MacDonald (1970). In the white-tailed ptarmigan the male plays no part in the incubation or care of young (Choate, 1960).

Association of the male with the nest and brood is well established for most of the New World quails and introduced partridges. In a few species the male regularly assists in incubation or may occasionally assume the entire incubation duties. This has been reported in bobwhites (Stoddard,

1931), scaled quail (Schemnitz, 1961), and harlequin quail (Willard, in Bent, 1932). Males may also assume incubation duties if the female dies, as has been noted in bobwhites (Stoddard, 1931), Gambel quail (Gorsuch, 1934), and California quail (Emlen, 1939; Price, 1938). Some possible examples of a male's incubating the first nest so that the female may begin another were mentioned earlier for California quail, Gambel quail, and chukar partridge. In the gray partridge the male may possibly assist in incubation (Hart, 1943) and will typically remain with and defend the brood (McCabe and Hawkins, 1946).

Males of most New World quail species, whether or not they have actually assisted in incubation, will normally remain with the brood and defend it. Males are regularly seen attending females and broods of scaled quail (Schemnitz, 1961), mountain quail (Dawson, 1923; Bent, 1932), Gambel quail (Gorsuch, 1934), bobwhites (Stoddard, 1931), and California quail (Genelly, 1955; Emlen, 1939), in the last of which even broodless males may guard the young. Little information on this behavior is available for the tropical forest-dwelling species, but Skutch (1947) indicated that in the marbled wood quail (*Odontophorus gujanensis*) males participate in brood care. This also seems to apply to the harlequin quail (Leopold and McCabe, 1957), and to the singing quail (LeFebvre and LeFebvre, 1958).

CLUTCH SIZES AND EGG-LAYING RATES

The rate at which egg-laying in birds occurs presumably depends on how rapidly follicles can be ovulated and associated albumen can be secreted by the female, and for the species under consideration here this generally averages slightly more than one day per egg. Some estimates for various grouse species are 1.1 days per egg for rock ptarmigan (Westerskov, 1956), 1.3 days per egg for sage grouse (Patterson, 1952), and 1.5 days per egg for ruffed grouse (Edminster, 1947). Corresponding figures for quails and introduced partridges include 1.1 days per egg for bobwhite (Stoddard, 1931), 1.1 days per egg for gray partridge (McCabe and Hawkins, 1946), 1.3 days per egg for chukar partridge (Mackie and Buechner, 1963), and 1.4 days per egg for California quail (Genelley, 1955). Thus, in general, a clutch perhaps takes a few days longer to complete than there are eggs laid.

Clutch size data are difficult to be confident about, for not only do these figures tend to be influenced by the generally smaller clutches that are laid late in the season by renesting females but also there may be considerable geographic variation in the average sizes of first clutches in various parts of the range. Thus, clutch size figures for the gray partridge in England differ considerably from those in North America, and data for the white-tailed ptarmigan from Montana are quite different from observations made in Alaska. Nonetheless, since information on average clutch sizes is of such basic importance in the calculation of reproductive potentials of these species, a summary of published information on clutch sizes is provided (table 12). Among the grouse the smallest average clutch sizes occur among the ptarmigan and the coniferous-forest-dwelling species, while the ruffed grouse and the prairie- and grassland-dwelling species of *Tympanuchus* have clutch sizes of about a dozen eggs. Interestingly, the sage grouse falls closer to the species of *Dendragapus* in its average clutch size (and also in the appearance of its eggs) than it does to the prairie grouse. Clutch sizes among the quail species appear to be generally high, although the limited information on tropical-forest-dwelling genera such as *Dendrortyx*, *Odontophorus*, and *Dactylortyx* suggests that these species may have quite small average clutch sizes. Among the genera *Colinus* and *Callipepla*, the combined weight of the eggs in an average clutch often nearly reaches that of the female (table 8), thus these quails expend a relatively greater amount of energy in completing a clutch than do any of the grouse.

EGG HATCHABILITY AND HATCHING SUCCESS

All available evidence from field studies indicates that the incidence of infertility and embryonic death is probably so low among wild populations as to be almost insignificant. The most extensive observations available for any grouse species are those of Bump et al. (1947), which include data from over five thousand ruffed grouse eggs, while Stoddard (1931) provides information for the bobwhite on nearly three thousand eggs from nests found in the wild. These and other studies indicate that in general more than 90 percent of the eggs laid under these conditions are fertile and capable of hatching (table 13). The actual percentage of eggs which hatch, however, is invariably less, ranging from about 90 percent to as little as 15 or 20 percent, depending on the rate of nest desertion and predation. Substantial brood mortality usually occurs during the first month or so, further reducing reproductive success (table 14).

TABLE 12

Reported Clutch Sizes under Natural Conditions

Species	*Normal Range*	*Mean Clutch Size*	*References*
Sage grouse	7–13	7.39 (154 nests)	Patterson, 1952
Blue grouse	6–12	6.3 (51 nests)	Zwickel & Bendell, 1967
Spruce grouse	7–10	5.8 (39 nests)	Tufts, 1961
Willow ptarmigan	2–15	7.1 (Scotland, 395 nests)	Jenkins, Watson, & Miller, 1963
		10.2 (Newfoundland, 106 nests)	Bergerud, 1970b
Rock ptarmigan	3–11	7.0 (Alaska, 101 nests)	Weeden, 1965b
		6.6 (Scotland, 148 nests)	Watson, 1965
White-tailed ptarmigan	3–9	5.2 (11 nests)	Choate, 1963
Ruffed grouse	6–15	11.5 (1473 nests)	Bump et al., 1947
Sharp-tailed grouse	5–17	12.1 (36 nests)	Hamerstrom, 1939
Greater prairie chicken	5–17	12.0 (66 nests)	Hamerstrom, 1939
Lesser prairie chicken	6–13	10.7 (7 nests)	Copelin, 1963
Long-tailed tree quail	3–6	. . . . . .	Rowley, 1966
Mountain quail	6–15	10.0 (11 nests)	*P. R. Quart.*†
Scaled quail	5–22	12.7 (39 nests)	Schemnitz, 1961
Elegant quail	8–12	. . . . . .	Leopold, 1959
Gambel quail	6–19	12.3 (40 nests)*	Gorsuch, 1934
California quail	9–17	13.7 (16 nests)	Lewin, 1963
Bobwhite	7–28	14.4 (394 nests)	Stoddard, 1931
Harlequin quail	6–16	11.1 (24 nests)	Leopold & McCabe, 1957
Gray partridge	9–20	16.4 (470 nests)	McCabe & Hawkins, 1946
Chukar partridge	14–19	15.5 (4 nests)	Mackie & Buechner, 1963

*Calculated, excluding four obviously incomplete clutches.
†*Pittman-Robertson Quarterly* 8 (1948):10.

TABLE 13

Egg Hatchability and Hatching Success under Natural Conditions

Species	*Hatchability of Eggs*	*Percentage of Nests Hatching*	*References*
Sage grouse	. . . . . .	42.2% of 533 nests	Hickey, 1955
Blue grouse	ca. 98% of eggs in 36 nests*	75% of 36 nests	Bendell, 1955a
Willow ptarmigan (red		69% of 232 nests	Hickey, 1955
grouse)	84% of 2,464 eggs*	80.3% of 395 nests*	Jenkins, Watson & Miller, 1963
Rock ptarmigan	90% of 147 eggs (Scotland)	. . . . . .	Watson, 1965
	94% of 393 eggs (Alaska)	65% of 86 nests	Weeden, 1965a
White-tailed ptarmigan	. . . . . .	70% of 11 nests	Choate, 1963
Ruffed grouse	95.6% of 5,392 eggs (1st nests)*		Bump et al., 1947
	92% of 480 eggs (2nd nests)*	61.4% of 1,431 nests	
Sharp-tailed grouse	88.2% of 136 eggs*	40% of 176 nests	Ammann, 1957
Greater prairie chicken	90.9% of 343 eggs*	46% of 165 nests	Ammann, 1957
Mountain quail	95.8% of 82 eggs	57% of 14 nests	*P.R. Quart.*†
Scaled quail	90% of eggs in 6 nests	14.3% of 42 nests	Schemnitz, 1961
California quail	. . . . . .	24.8% of 83 nests	Glading, 1938b
Gambel quail	. . . . . .	24% of 44 nests	Gorsuch, 1934
Bobwhite	86% of 2,874 eggs	36% of 602 nests	Stoddard, 1931
Gray partridge	84.5% of 1,838 eggs	32% of 435 nests	McCabe & Hawkins, 1946
Chukar partridge	. . . . . .	25% of 16 nests	Harper, Harry, & Bailey, 1958

*Calculated from data presented by authors
†*Pittman-Robertson Quarterly* 8 (1948):10.

TABLE 14

ESTIMATES OF EARLY BROOD MORTALITY UNDER NATURAL CONDITIONS

Species	*Mortality Estimates*	*References*
Sage grouse	From 32 to 54% less reported in three studies	Hickey, 1955
	Average brood size reduced from 5.56 in June to 2.33 by August (48% brood loss)	Keller (in Rogers, 1964)
Blue grouse	Estimated 67% brood mortality by August	Bendell, 1955a
Willow ptarmigan (red grouse)	Average 52% of young from successful nests reared to August (48% brood mortality)	Jenkins, Watson, & Miller, 1963
Rock ptarmigan	Average 20.2% brood loss among 208 broods by late July	Weeden, 1965a
	Average brood size reduced to 3.6 young at 10-12 weeks	Watson, 1965
White-tailed ptarmigan	Approximate 33.1% brood loss among 41 broods in 1st 8 weeks	Choate, 1963
Ruffed grouse	Average brood mortality averaged from 60.9% (11 yr. avg.) to 63.2% (13 yr. avg.) in two areas	Bump et al., 1947
Sharp-tailed grouse	Average brood size reduced from 8.7 to 4.6 young (47% loss)	Hart, Lee, & Low, 1952
Greater prairie chicken	Average brood size reduced from 8.0 to 6.6 young (17.5% loss)	Baker, 1953
	Brood mortality of 46%	Yeatter, 1943
Attwater prairie chicken	Approximate 50% mortality in 1st month; 12% later	Lehmann, 1941
Mountain quail	Approximate 30% (range 0-55%) brood loss over 3 years	Edminster, 1954
California quail	Approximately 45-60% brood mortality by fall	Edminster, 1954
Gambel quail	Average 48% brood loss (range 42-51%) over three years	Edminster, 1954
Bobwhite	Approximately 25-40% brood loss in 16 weeks	Edminster, 1954
	Brood mortality 28.6% in 1st 8 weeks	Klimstra, 1950b
Gray partridge	Average brood size reduced from 12 to 8 by September (33% loss)	Yeatter, 1935

THE EVOLUTIONARY SIGNIFICANCE OF CLUTCH SIZE VARIATIONS

The question of the adaptive significance of the considerable variations in average clutch sizes for the species under consideration here (from about five to sixteen eggs) has recently been discussed by Lack (1968). He concluded that average clutch size in these species is generally inversely related to egg size; that is, species that have relatively small clutches typically lay relatively large eggs. The apparent advantage, for species with precocial young, of producing large eggs is that the young can be hatched at a relatively advanced and less vulnerable stage and can begin feeding for themselves and soon become independent of the parent. In this group, therefore, natural selection has seemingly compromised between allowing the largest clutch size that can be produced by the energy reserves of the female while retaining an adequate egg size that will allow the young to be hatched at a stage sufficiently advanced to favor their survival.

Assuming that natural selection fixes a relatively inflexible optimum egg size for each species (which can conveniently be estimated as the weight of the egg in proportion to the adult female's weight), the physiological drain on a laying female may thus be regarded as this constant multiplied by the average clutch size. It should also be noted that among all birds, smaller species tend to lay relatively larger eggs than do larger ones, apparently reflecting the minimal investment of energy needed to produce a viable egg. Lack (1968) believes that average clutch size in the gallinaceous birds must therefore be limited either by the number of eggs that the incubating bird can effectively cover, which he rejects, or by the average food reserves of the female as modified by the relative egg size. He suggests that the latter explanation best accounts for the variations in clutch sizes to be found in this group.

Lack makes a number of additional observations about clutch sizes in the pheasant-like birds. First, he notes that clutch sizes tend to be smaller in southern than in more northerly latitudes among related species; thus tropical forms are more likely to have smaller average clutches than are related species of the same size breeding in temperate or arctic regions. Second, Lack detected no clear correlation between clutch size and habitat of the species or the pair-bond characteristics of the species. He noted that only a weak positive correlation exists between egg size and incubation period, but did not consider other possible influences on incubation periods existing in this group, such as the length of the breeding season.

As may be noted in table 8, there is only a weak inverse relationship between the average weight of the egg in proportion to that of the female and the average clutch size in the species under consideration here. This trend is perhaps clearest in the grouse, of which the spruce grouse, rock ptarmigan, and white-tailed ptarmigan tend to have small average clutches and fairly large relative egg sizes, whereas the ruffed grouse, sharp-tailed grouse, and two prairie chicken forms have large clutches and smaller relative egg sizes. It is of interest, however, that the three ptarmigan species lay eggs of nearly the same size and that their average clutch sizes are nearly the same although they have markedly different adult weights. One would have expected that the willow ptarmigan might have a considerably larger average clutch size than the white-tailed ptarmigan.

The anticipated inverse relationship between egg size and clutch size breaks down completely in the New World quails; indeed, a positive relationship between these factors would seem to exist in this group, with the mountain quail and harlequin quail representing a small clutch–small egg condition and the California quail and bobwhite representing an opposite large clutch–large egg situation. The quail group as a whole, which on the average are smaller in body size than the grouse, rather surprisingly not only have relatively larger eggs, as might be expected from their average body sizes, but also have considerably larger average clutch sizes than do the North American grouse. This trend is clearly counter to the suggestion that egg size and clutch size characteristics are inversely related in these species.

If no strong case can be made for food reserves of the female as a major factor possibly limiting clutch size, alternate or supplementary factors must be considered. One possibility, that the clutch size is limited by the number of eggs that the adult can effectively incubate, is unpromising inasmuch as the large-bodied grouse typically produce smaller clutches than do most of the much smaller quail. It might be noted, however, that the grouse must cover their eggs more effectively, since they are mostly cool-temperate to subarctic breeders, whereas the breeding distributions of quails are more southerly and their eggs are less likely to be chilled during incubation. It seems unlikely that a ptarmigan could effectively incubate a dozen or more eggs, and each day that is invested in producing another egg not only reduces the time available for incubation and rearing of the young but also exposes the untended nest to possible predation that much longer.

If indeed the length of the breeding season is significant, and if the danger of chilling the eggs increases when the clutch size exceeds a number related to the size of the adult in proportion to the egg, then average clutch sizes should increase as breeding distributions are arranged from arctic or alpine areas to warmer ones, rather than the opposite as Lack has suggested. It is difficult to pick representative figures on frost-free periods for the habitats of the species in question, but it might be argued that among the grouse the species might be arranged in a northerly, or alpine, to southerly, or warm-temperate, series as follows: White-tailed ptarmigan, rock ptarmigan, willow ptarmigan, spruce grouse, blue grouse, ruffed grouse, sharp-tailed grouse, sage grouse, pinnated grouse. Except for the sage grouse, which commonly breeds in parts of Utah, Nevada, and Wyoming that have frost-free seasons of one hundred days or less, this series closely agrees with a progressively increasing average clutch size. It is unfortunate that clutch size data from different populations of widely ranging species, for example, from Alaskan compared with midwestern races of ruffed grouse and sharp-tailed grouse, are not available to show if any intraspecific north-south trends can be detected in the average clutch sizes of these forms.

Since nearly all of the species of quail breed sufficiently far south that the length of the breeding season is probably not a significant factor affecting their clutch sizes, it would seem that some other factor, such as food reserves or predation effects, might play a role. Provided that adequate food is available, it is quite evident, from studies of captive quail, that females can continue to lay eggs at approximate day-and-a-half intervals almost indefinitely. Instead, the factors limiting clutch sizes in these species might perhaps be the maximum number of eggs that the adult can effectively incubate or the increasing dangers of losing the entire clutch to predators during every day that the nest is left untended during the egg-laying period. Thus, an average clutch of from ten to fifteen eggs may require about twenty days to complete, and with each passing day the possibility of their discovery by predators is increased. Lack has dismissed the possibility that predation can effectively limit clutch sizes in birds, pointing out that for it to be fully effective the predation rate must exceed the rate of laying, or approximate nearly one egg per day. Yet, since predators usually destroy entire clutches or at least often cause desertion of the nest, they may become equally effective whenever the daily likelihood of predation exceeds the inverse of the then existing clutch size. As clutch size increases, fixed daily predation levels therefore become increasingly effective as a potential limiting factor, especially for species that are relatively defenseless or do not attempt to guard the nest prior to the start of incubation.

In figure 12 are presented the calculated effects of various daily predation levels on species that lay one egg per day, assuming a constant daily predation rate during the egg-laying period causing destruction or desertion of the entire clutch. For species that average a two-day interval between eggs, the indicated effects would be doubled (thus a 5 percent daily predation rate would have the effect of the 10 percent rate shown in the figure). The diagram demonstrates that species suffering a 20 percent daily predation level (20 percent of all initiated nests being destroyed each day) cannot effectively increase their clutch size after the third day of laying, and selection would thus favor the evolution of a clutch size of only three or four eggs. Similarly, those species exposed to a 10 percent daily predation loss cannot increase their effective clutch size beyond the eighth day. Species having a predation level of 5 percent per day can increase their effective clutch size only through the fourteenth to eighteenth day of laying, after which it levels off at eight eggs. Predation levels of less than 2 percent per day during egg laying are probably ineffective in keeping clutch sizes below the physiological limits of the female or the maximum number that can effectively be incubated, at least among species that lay an average of one egg per day.

Almost no field data on preincubation predation levels are available, but the high over-all incidence of quail nest losses through predation suggests that such losses may often reach significant levels. Since the completion of a clutch may require about twenty days, and incubation another twenty-one to twenty-four days, it follows that nearly half of all predation losses might be expected to occur before the start of incubation even if predation rates are not appreciably higher during the preincubation period. Edminster (1954) summarized field data from bobwhites and California quail indicating that some 60 to 80 percent of their nests are normally lost because of desertion or actual predation; if half of these losses occurred during the egg-laying period, it is clear that they might average at least 2 percent per day.

Stoddard (1931) reported that 37 percent of 602 bobwhite nests were destroyed by natural enemies and that 52 of the 65 nests lost to skunks were broken up before incubation started. Bump et al. (1947) found that 38.6 percent of 1,431 ruffed grouse nests were broken up, 89 percent of the disruption attributable to predators. Six studies summarized by Gill (1966) provide nest destruction estimates on 503 sage grouse nests, which averaged 47.7 percent losses (with a range of 26 to 76 percent). Recently, Ricklefs (1969) has calculated daily natural nest mortality rates for a number of North American game birds from data summarized by Hickey (1955) (see chapter 6). These calculated nest mortality rates for fifteen studies averaged 2.96 percent per day (with a range of 1.55 to 4.66 percent), which admittedly represents a minimal estimate, since the estimates are based on the entire nesting period (egg-laying plus incubation), whereas most nests are not found until the nesting period is partly over. If, in addition, it is true that in galliforms the mortality rates from predation are higher before incubation begins than afterwards, it is clear that such preincubation predation rates might have a significant role in influencing clutch size.

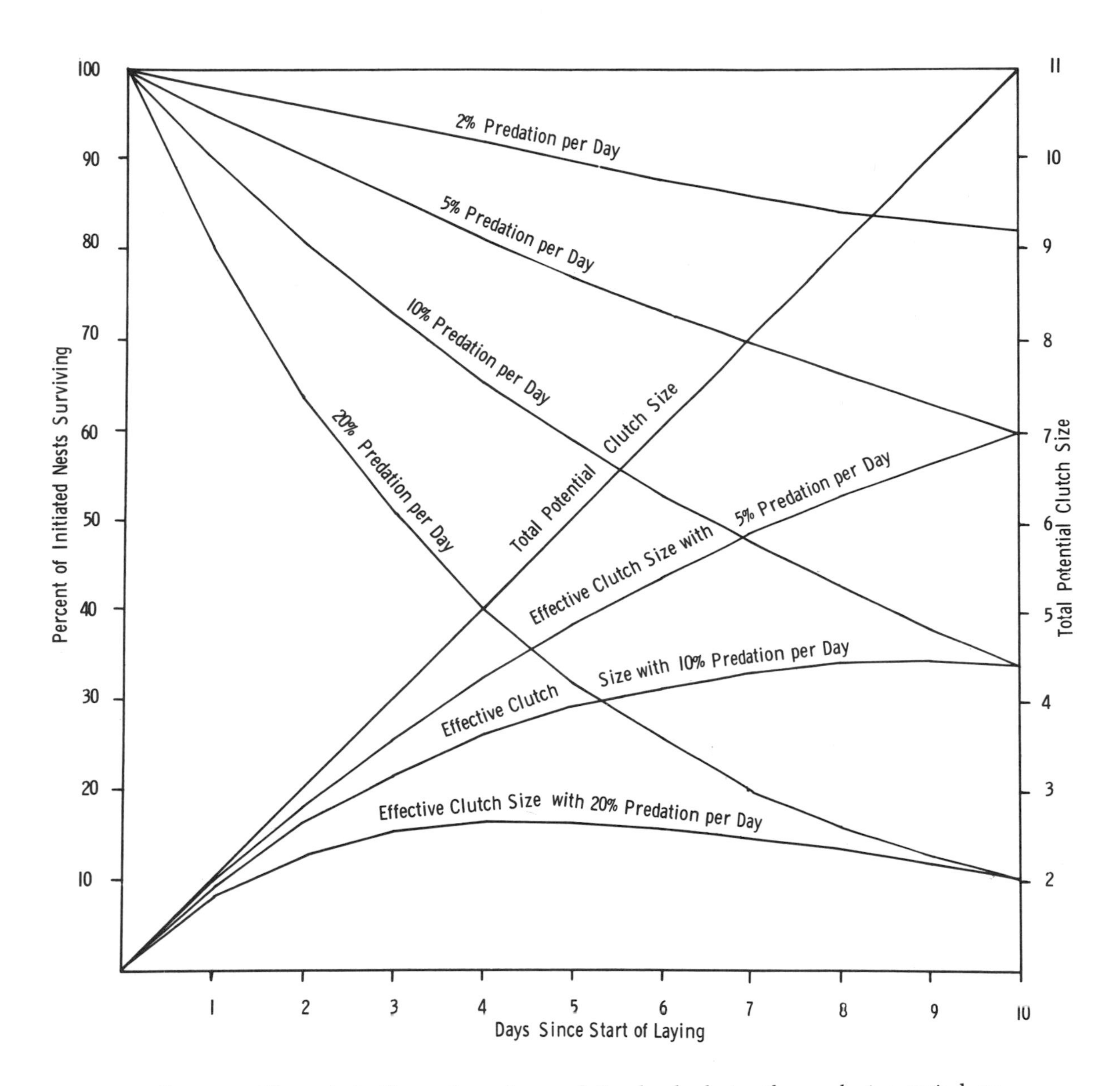

FIGURE 12. Theoretical effects of varying predation levels during the egg-laying period on effective clutch sizes, assuming an egg-laying rate of one per day and predation of the entire available clutch.

In summary, it would seem that available food reserves of the female probably play a subordinate role in limiting clutch sizes among grouse and quails and are probably important only among species that lay eggs so large or lay them so frequently that the female is unable to balance food intake against the physiological drain on her energy reserves. Otherwise the remarkably large clutches of quails and their persistent renesting behavior could not be accounted for. Among grouse, it is suggested that the need to complete a clutch rapidly and to lay no more eggs than can effectively be warmed by the female represents a significant factor in limiting clutch sizes of arctic- or alpine-breeding species, and is progressively less important for the more temperate-breeding forms. Limiting factors affecting clutch sizes of temperate-breeding species of grouse and quail might be related to the number of eggs that an adult can effectively incubate and to the predation levels during the relatively long egg-laying period, both of which would tend to allow fairly large rather than relatively small clutch sizes. It should finally be noted that the few tropical-forest-dwelling species of quail which have so far been studied appear to have quite small clutch sizes, suggesting that other limiting factors may play important roles under such ecological conditions. These factors might include relative food availability and predation rates, since Ricklefs's studies (1969) indicate that daily nest mortality rates of open-nesting passerine birds are higher in the humid tropical regions than in arctic, temperate, or arid-tropical areas; and ground-nesting quails might be similarly affected.

6

Population Ecology and Dynamics

LIKE other animals, grouse and quail exist as natural populations dependent upon particular habitats and vary in population density between the absolute minimum populations that have permitted past survival to fairly dense populations that may approach or even temporarily exceed the *carrying capacity* of the habitat. Each species may also have an upper limit on the density of the population, or a *saturation point,* which is independent of the carrying capacity of the habitat but is determined by social adaptations. Within the population as a whole, individual birds or coveys may have *home ranges,* geographical areas to which their movements are limited and within which they spend their entire lives. Part of the home range may be defended by individuals so that conspecifics of the same sex are excluded for part or all of the year; such areas of localized social dominance and conspecific exclusion are called *territories.* Among species lacking discrete territories and in which the social unit is the covey or flock rather than the pair or family, dominance hierarchies, or *peck orders,* may serve to integrate activities in the flock. These behavioral adaptations and habitat relationships play important roles in population ecology, and will be considered in detail in the individual species accounts. However, a preliminary survey may help to provide generalizations that will be useful to keep in mind when considering individual species.

Natural populations, whatever their densities, have definable structures in terms of the individuals that make up the population unit. Thus, their sex composition, as defined by *sex ratios*, and age composition, as similarly defined by *age ratios*, provide important information on the proportion of the total population that are potential breeders. The fall age ratio, readily determined by the number of immature birds appearing in hunters' kills, also provides the best information available to the field biologist about the success of the immediately past breeding season.

A final important characteristic of natural populations is the rate at which population recycling occurs, which in turn depends upon the *mortality* and *survival* rates characteristic of it. Mortality and survival are opposite sides of the same coin; as mortality rates increase, average survival probabilities decrease and *life expectancy* (or mean longevity) consequently decreases. Mortality rates can thus be used to determine a statistical measure of life expectancy among individuals of a population, and these data are of basic significance to the field biologist. Regardless of the actual mortality rate, all animals in a population must eventually die; the length of time required for a virtual 100 percent turnover of a population age-class is called the *turnover rate*. This figure corresponds to the maximum possible longevity that may be attained by 1 percent or less of the individuals in that population.

POPULATION DENSITIES

Since virtually all the species of concern here are game birds, information on estimated population densities may be found scattered widely through the technical literature. However, these figures are often not completely comparable; different techniques of census may give different results for the same species, to say nothing of their effects on different species, and the same population may have year-to-year fluctuations that must be taken into account. In addition, census data for some species (such as strongly territorial or lek-forming grouse) are most readily obtained during spring, while fall or winter data may be more readily obtained for species that form coveys and are most conspicuous at that time. Further, some census figures are calculated on the basis of territorial males per unit area, while others consider both sexes. Since the sex ratios of adult populations often differ considerably from a 50:50 ratio, it may be impossible to make the data exactly comparable.

Surprisingly little information is available on minimum tolerable population sizes in the grouse and quail, as Hickey pointed out (1955). These may vary considerably among various species; solitary species such as ruffed grouse and spruce grouse can perhaps tolerate quite low population densities, whereas highly social species such as quail and socially displaying grouse may have definite minimum thresholds of survival imposed by the physiological stress or inadequate behavioral stimulation of sparse populations. In general, however, the reproductive potential of most grouse and quail species is so great that populations which are drastically reduced by some means have the biological potential for rapid recovery as long as the habitat conditions are favorable. Rather marked population fluctuations are in fact quite common among certain grouse, particularly the arctic populations of ptarmigans and the more temperate populations of ruffed grouse, sharp-tailed grouse, and greater prairie chicken. Estimates of average population density for these species, at least in areas where major fluctuations are prevalent, must necessarily take these variations into account. The existence and possible causes of these periodic population fluctuations are much too complex and controversial to be considered here, and several review papers (such as Hickey, 1955) have dealt with the problem.

It seems evident that, whereas populations may exist over a wide range of densities at the lower limits, upper population densities of a species may have a definite limit. To some degree this is ultimately a habitat-imposed limit, the limiting factors being available food, nesting sites, winter cover, predation, and other density-dependent variables. In addition, territorial size may establish a maximum density, where the habitat might otherwise be capable of supporting a larger number of birds. Even in the absence of actual territorial boundaries the level of intraspecific fighting among reproductively active individuals may force mutual avoidance, causing a maximum spreading out of the population over the available habitat. To the extent that maximum population densities are the result of such species-typical behavioral traits rather than habitat variations, they should be fairly constant for a species in different parts of the species' range. If, on the other hand, maximum densities are primarily a reflection of the differential carrying capacities of the various habitats occupied by a species, they are likely to vary considerably between areas and in the same area from year to year.

In spite of difficulties, for the reasons mentioned earlier, in finding comparable data, it is of interest to compare estimated population densities of the species concerned here. These are in general late winter, spring, or adult breeding population figures (table 15).

Not unexpectedly, quail population densities are in general considerably greater than those of grouse, perhaps reflecting both their smaller sizes and thus lowered food requirements and the far greater sociality typical of these birds. It is generally true that quail densities average at least four times

TABLE 15

Some Reported Population Densities in Favorable Habitats (Expressed in Acres per Bird)

Sage grouse:	51 acres per male on strutting grounds in spring, Wyoming	Patterson, 1952
	13–21 acres per bird during fall in best habitats, Colorado	Rogers, 1964
Blue grouse:	9 acres per adult male, summer average, British Columbia	Fowle, 1960
	2.3–7.7 acres per male on summer range, British Columbia	Bendell & Elliott, 1967
	2.5 acres per female; 1.3 acres per male, British Columbia	Bendell, 1955a
Spruce grouse:	128 acres per territorial male, Montana	Stoneberg, 1967
	64–90 acres per male (30% of males territorial), Alaska	Ellison, 1968b
Willow ptarmigan (red grouse):	3.2–12.3 acres per male in spring, Alaska	Weeden, 1965b
	4.5–9.0 acres per pair in spring, Scotland	Jenkins, Watson, & Miller, 1963
Rock ptarmigan:	56–109 acres per male, spring, Alaska	Weeden, 1965b
	4.9–24.7 acres per territorial pair (peak year), Scotland	Watson, 1965
	19.8–74 acres per territorial pair (low year), Scotland	Watson, 1965
White-tailed ptarmigan:	12.8–42 acres per adult in summer, Montana	Choate, 1963
Ruffed grouse:	8–38 acres per adult during breeding season, New York	Edminster, 1954
	13.5–30 acres per adult in spring, New York	Bump et al., 1947
	3.4 acres per adult in spring (based on nests), Michigan	Palmer, 1954
Sharp-tailed grouse:	45 acres per bird in spring, Michigan	Ammann (in Edminster, 1954)
	16–25.6 acres per bird in late summer, Saskatchewan	Symington & Harper, 1957
Greater prairie chicken:	10–42.7 acres per bird (summary of 4 studies)	Trippensee, 1948
Lesser prairie chicken:	17–38 acres per adult male in spring, Oklahoma	Davison, 1940
Mountain quail:	2 acres per bird maximum spring density, California	Edminster, 1954
Barred & elegant quails:	Under 1 acre per bird locally, Mexico	Leopold, 1959
Scaled quail:	10.1 acres per bird in winter, Texas	Wallmo, 1956b
	0.84 acres per bird in winter, Oklahoma	Schemnitz, 1961

TABLE 15—(continued)

California quail:	1.7–3.9 acres per bird in late winter, California	Glading, 1941
	0.91 acres per bird in winter, California	Emlen, 1939
Gambel quail:	1.6 acres per bird in late winter, Nevada	Gullion, 1962
Bobwhite:	4–20 acres per bird in spring, good range (various states)	Edminster, 1954
Singing quail:	31 acres per pair, Tamaulipas, Mexico	Warner & Harrell, 1957
Harlequin quail:	21–23 acres per bird in summer, Chihauhua, Mexico	Leopold & McCabe, 1957
	27 acres per pair or family unit, Arizona	Brown, 1969b
Gray partridge:	3.5–5.3 acres per bird in winter, North Dakota	Hammond, 1941
	14–29.4 acres per bird in spring, Washington	Yocom, 1943

greater than do those of grouse, and certainly they show a greater degree of "clumping," because of reduced territorial tendencies and covey-forming behavior. Only the lek-forming grouse species exhibit a corresponding tendency toward contagious distribution patterns, which are related to the males' fidelity to vicinities of their display grounds even when these are not actively being defended. Quail populations also do not regularly exhibit the major oscillations of population density characteristic of some grouse, in spite of the fact that their reproductive potential is extremely high and rapid population increases are thus possible.

FLOCKING AND COVEY BEHAVIOR

Among the grouse, perhaps the best-known examples of flocking and covey formation are to be found among sharp-tailed grouse and pinnated grouse during late fall and winter. These migratory movements, often involving large flocks, were once conspicuous in such midwestern states as Minnesota, Iowa, and Missouri (Bent, 1932). Hamerstrom and Hamerstrom (1951) describe late fall "packs" of sharp-tailed and pinnated grouse that often numbered in the hundreds, sometimes as many as four hundred birds. Similar fall packs of spruce grouse once occurred, and migratory flocks of willow ptarmigan numbering in the thousands have been noted (Bent, 1932). Likewise, rock ptarmigan congregate in relatively large flocks during their seasonal movements to and from their breeding grounds.

In contrast, quail are to be found in coveys at all times other than during the breeding season, and even then nonbreeders may gather in coveys. No doubt for quail the covey represents the most efficient social unit for survival of these relatively defenseless birds, and its formation is facilitated by the reduced territorial tendencies and monogamous pair-bonding behavior of quail. Covey roosting may also be an important means of heat retention during winter. In bobwhites, for example, winter coveys usually consist of about ten to fifteen birds, the most efficient number for retaining heat in circular roosting. The maximum covey sizes of some species is probably a simple reflection of the over-all population density as well as the time of year, but there is a clear tendency for some quail species to form larger coveys than others. Large coveys are especially frequent in southwestern species such as the California quail and scaled quail, as the accompanying summary shows (table 16).

HOME RANGES AND TERRITORIES

Most quails and grouse are fairly mobile, but relatively few undertake true migrations. Vertical migrations are known to occur in such mountain-dwelling species as mountain quail, white-tailed ptarmigan, and blue grouse, and in the last-named species the winter range is actually at a higher altitude than is the summer range. The arctic-breeding rock and willow ptarmigans perform definite seasonal migrations in some areas (Bent, 1932), and Hamerstrom and Hamerstrom (1949, 1951) have summarized data on seasonal movements of the sharp-tailed grouse and pinnated grouse. The home ranges of these fairly mobile species must be the largest of any of the grouse, but detailed data are still lacking. Hamerstrom and Hamerstrom (1951) reported that band returns indicated sharp-tailed grouse movements of up to twenty-one miles, but most returns were obtained within three miles of the point of banding. A few transplanted sharp-tails were also found to have moved more than twenty miles before being shot. Fewer recoveries were obtained for the pinnated grouse, which is apparently the more mobile of the two species. Two banded greater prairie chickens moved as far as twenty-nine miles, and one moved approximately one hundred miles (Hamerstrom and Hamerstrom, 1949). Robel et al. (1970) used radio tracking to determine that greater prairie chicken ranges varied from under two hundred acres in late summer to more than five hundred acres during fall and spring, with adult males having maximum monthly ranges of more than twelve hundred acres during March.

Home range data for the other species of grouse are equally difficult to

TABLE 16

SOME REPORTED COVEY SIZES OF QUAILS AND PARTRIDGES

Species	Covey size	Source
Mountain quail:	Average of 21 coveys, 9.1 birds, range 3–20	Miller & Stebbins, 1964
Barred quail:	Average of 18 coveys, 12 birds, range 5 to 20 or 25	Leopold, 1959
Scaled quail:	Average of 325 coveys, 31.2 birds, range 4–150	Schemnitz, 1964
	Average of 56 coveys, 19.3 birds	Hoffman, 1965
Elegant quail:	Coveys range from 6 to 20 birds	Leopold, 1959
California quail:	Average of 4 winter coveys, 34.8 birds	Sumner, 1935
	Coveys usually 25–60 but up to 500–600	Leopold, 1959
Gambel quail:	Average of 40 coveys, 12.5 birds, range 3–40	Gullion, 1962
	Coveys usually 20–50 birds, but sometimes hundreds	Leopold, 1959
Bobwhite:	Average of 112 winter coveys, 13.8 birds, up to 28	Stoddard, 1931
	Average of 2,815 winter coveys, 14.3 birds, range 6–25	Rosene, 1969
Black-throated bobwhite:	Usually 7–15 birds in covey	Leopold, 1959
Spotted wood quail:	From 5 to 10 birds in covey	Leopold, 1959
	From 6 to 20 birds in covey	Alvarez del Toro (cited in Leopold, 1959)
Harlequin quail:	Average of 62 fall and winter coveys, 7.6 birds, range 3–14	Leopold & McCabe, 1957
Chukar partridge:	From 10 to 40 or more birds in covey	Leopold, 1959

obtain, partly because of difficulties in distinguishing home ranges (occupied but not defended areas) from areas of territorial defense in these species. Males of the forest-dwelling grouse may occupy a fairly large home range and establish territorial limits only where they encounter other males, so that possibly no firm distinction between home ranges and territories may be made (MacDonald, 1968). In the spruce grouse, males may occupy home ranges of 10 to 15 acres, or occasionally as little as three acres (Stoneberg, 1967), but both Stoneberg and MacDonald (1968) found that males spend most of their time within a small portion of their home range. Ellison (1968b) reported that territorial adult males remained on areas of 5 to 9 acres in early May, where display occurred and within which territorial behavior was seen. All adult males but only some yearlings held territories, and the latter's territories ranged in size up to 21 acres. Other nonterritorial immatures occupied "activity centers" of 6 to 16 acres in size, but sometimes moved more than a mile away from these centers. Nondisplaying or nonterritorial immature males have also been noted in ruffed grouse, blue grouse, and sage grouse. In late May and June the territorial males developed larger home ranges of up to 61 acres, and the nonterritorial birds wandered over areas of from 270 to 556 acres (Ellison, 1968b).

In the ecologically similar blue grouse, territory sizes appear to average somewhat smaller. Boag (1966) and Mussehl (1960) estimated territory size in this species to be from 1 to 2 acres, and Blackford (1963) provides diagrams indicating that eight territories averaged about 5 acres in size. Bendell and Elliott (1967) reported that territories were about 1.5 acres when blue grouse populations were high and from 5 to 11 acres when populations were low. About 30 percent of the males on the breeding range consisted of nonterritorial and wandering immature males. With regard to the forest-dwelling ruffed grouse, Marshall (1965) stated that one male remained within a 10-acre area during April and May, while Eng (1959) pointed out that males usually stayed within one hundred feet of their drumming logs during this period.

In the case of the open-country ptarmigans, several studies on breeding distribution patterns have been done. Weeden (1959) estimated that the territories of willow ptarmigan may range from 3.5 to 7 acres, and the data of Jenkins, Watson, and Miller (1963) suggest that breeding densities of red grouse in Scotland may allow territories of approximately this size, since from sixteen to forty males occupied territories on a 138-acres study area over a four-year period. Similarly, Watson (1965) reported that populations of rock ptarmigan in favored habitats might have territories of 1.2 to 3.5 hectares (3 to 8.1 acres). Schmidt (1969) indicated that the average territory of white-tailed ptarmigan in Colorado is from 16 to 47 acres (with smaller "areas of maximum use"), while Choate (1963) indicated that in Montana this species' territories average about forty by one hundred yards, or slightly less than an acre.

Territories of the lek-forming grouse are the smallest of any of the species concerned here. Dalke et al. (1960) indicated that in the sage grouse the master cocks had a territory forty feet or less in diameter (or 0.03 acre). Lumsden (1965) indicated that the central territories of sharp-tailed grouse were approximately fifteen by twenty-five feet (or 0.01 acre), while peripheral ones were larger. Robel (1965) indicated that territories of male greater prairie chickens varied from 23.6 to 106.5 square meters (or 0.006 to 0.026 acres), and Copelin (1963) stated that territories of the lesser prairie chicken were only about twelve to fifteen feet in diameter (or 0.002 to 0.004 acres).

Among the quail species, useful application of the principle of territoriality is very limited. Calling or singing by males, at least in the species well studied, denotes the presence of unmated but sexually active males rather than a breeding pair. Thus, in bobwhites, whistling males are simply surplus males (Stoddard, 1931; Bennitt, 1951). The territories of male bobwhites are at most ephemeral and mobile; the female's calls attract sexually active males, whose whistles serve as an advertisement of their presence (Robinson, 1957). The same probably applies to the scaled quail (Schemnitz, 1964). Similarly, in the California quail unmated males establish "crowing territories" near established pairs (Emlen, 1939; Genelly, 1955). Genelly reports that the crowing territories of the excess males may be spaced only about twenty or more feet apart and are as close to established pairs as the latter will allow. Neither California quail nor bobwhites actively defend their nesting sites, and most of the male-to-male fighting involves defense of the mate (Genelly, 1955). In the Gambel quail, pairs gradually form in the winter coveys; the coveys break up as pairs leave and as the unmated males become mutually intolerant and begin to establish individual crowing territories (Raitt and Ohmart, 1966). Estimated winter home range sizes are indicated in table 17 for representative quails. Evidence indicates that the size of these home ranges may vary considerably in different regions and habitats but that they probably average about twenty-five acres in favorable habitats.

The concept of typical territoriality with regard to the gray partridge and the chukar partridge is also of limited application. McCabe and Hawkins (1946) reported that the coveys of gray partridge remain intact until just before nesting. Blank and Ash (1956) report that neither *Perdix* nor *Alectoris* exhibits true territoriality. In the gray partridge establishment of a covey territory is the nearest thing to territorial behavior; covey composition is highly stable in this species. Pairing occurs before the selection of a nesting area, as is also true in New World quails, thus there is no correlation between the selection of mates and the establishment of a nesting area (Blank and Ash, 1956). Mackie and Buechner (1963) agree that typical territoriality is also absent in the chukar partridge. Males repel other males from their mates, thus the female, rather than a geographically defined area, is the object of defense. However, the rally call of mated males may serve to disperse the breeding population in this species (Williams and Stokes, 1965), and population dispersion is thought to be a basic function of avian territoriality.

TABLE 17

REPORTED HOME RANGES OF SOME NEW WORLD QUAILS

Mountain quail:	Nesting pairs occupied from 5 to 50 acres, California	*P.R. Quart.**
Scaled quail:	Winter covey home ranges averaged 52.3 acres, Oklahoma	Schemnitz, 1961
	Winter covey home ranges averaged 360 acres, Texas	Wallmo, 1956b
Gambel quail:	Winter covey home ranges averaged 20 acres, Nevada	Gullion, 1956b
California quail:	Winter covey home ranges averaged 26 acres, California	Emlen, 1939
Bobwhite:	Winter covey home ranges averaged 24 acres, Missouri	Murphy & Baskett, 1952
	Winter covey home ranges averaged 24 acres, Texas	Lehmann, 1946
	Winter covey home ranges (1,154 coveys) averaged 13.2 acres and ranged from 4 to 77 acres	Rosene, 1969

**Pittman-Robertson Quarterly* 11 (1951):10.

SEX RATIOS AND AGE RATIOS

The importance of obtaining data about the sex and age composition of game bird populations can scarcely be exaggerated. Such data are generally easy to obtain for the species under consideration here, since reliable techniques for determining sex and age are available for most species. Sex ratio data may provide useful indications of a species' relative reproductive efficiency. For example, adult (or "tertiary") sex ratios in strictly monogamous species such as most quails should clearly be as near 1:1 as possible in order to achieve efficient reproduction, whereas in highly promiscuous or polygamous species a sex ratio strongly favoring females probably represents the most efficient reproductive structure for the population. Nearly all the available data for grouse and quails (except sage grouse and blue grouse) indicate that sex ratios diverge from nearly equal numbers of the sexes at hatching to ratios favoring males in the adult population (table 18). A slight excess of males in renesting species such as

TABLE 18

Some Reported Sex Ratios
(Expressed as Percentage of Males in Population)

	Age Class	*Percentage Males*	*Sample Size*	*References*
Sage grouse	Immatures	45.3	2,693	Patterson, 1952*
	Adults	29.6	1,964	Patterson, 1952*
	Mixed ages	40.0	7,355	Rogers, 1964
Blue grouse	Immatures	50.0	. . . .	Boag, 1966
	Adults & subadults	40.0	. . . .	Boag, 1966
Spruce grouse	Immatures	48.3	766	Lumsden & Weeden, 1963*
	Adults	55.3	423	Lumsden & Weeden, 1963*
Willow ptarmigan (red grouse)	Adults	55.9	2,211	Jenkins, Watson, & Miller, 1963*
Rock ptarmigan	Adults	58.5	1,545	Watson, 1965*
Ruffed grouse	Immatures	51.2	17,577	Dorney, 1963*
	Adults	54.6	5,365	Dorney, 1963*
Sharp-tailed grouse	Immatures	56.0	2,108	Ammann, 1957
	Adults	60.0	889	Ammann, 1957
Greater prairie chicken	Immatures	54.9	306	Baker, 1953
	Adults	54.6	298	Baker, 1953
Lesser prairie chicken	Immatures	53.0	491	Lee, 1950
	Adults	47.0	532	Lee, 1950
Scaled quail	Young adults (1st 18 mo.)	47.4	213	Campbell & Lee, 1956
	Old adults (over 18 mo.)	58.9	141	Campbell & Lee, 1956
California quail	Immatures	50.8	6,335	Francis, 1970*
	Adults	57.3	4,347	Francis, 1970*
Gambel quail	Immatures	49.3	333	Raitt & Ohmart, 1968
	Adults	57.8	154	Raitt & Ohmart, 1968
	Young adults (1st 18 mo.)	51.4	215	Campbell & Lee, 1956
	Old adults (over 18 mo.)	55.8	525	Campbell & Lee, 1956
Bobwhite	Immatures	50.5	34,989	Bennitt, 1951
	Adults	59.0	7,521	Bennitt, 1951
Harlequin quail	Mixed (museum sample)	63.0	502	Leopold & McCabe, 1957
Gray partridge	Adults	58.0	115	McCabe & Hawkins, 1946
	Mixed	51.0	14,167	Johnson, 1964*
Chukar partridge	Mixed	50.0	116	Harper, 1958

*Calculated from data presented by authors.

TABLE 19

Some Reported Fall and Winter Age Ratios (Expressed as Percentage of Immatures in Population)

	Percentage Immature	*Sample Size*	*References*
Sage grouse	57.8%	4,657	Patterson, 1952
	51.4%	7,355	Rogers, 1964
Blue grouse	65% (late summer)	....	Boag, 1966
	57–65%	....	Hoffman et al. (cited in Bendell, 1955a)
Spruce grouse	64.4%	1,189	Lumsden & Weeden, 1963*
Willow ptarmigan	72%	5,266	Bergerud, 1970b
Rock ptarmigan	73–77%	....	Cited in Choate, 1963
White-tailed ptarmigan	33–47%	....	Choate, 1963
Ruffed grouse	77%	22,942	Dorney, 1963*
Sharp-tailed grouse	70%	3,926	Ammann, 1957
	63.5%	16,283	Johnson, 1964*
Greater prairie chicken	50.2%	604	Baker, 1953
Lesser prairie chicken	53.2%	932	Lee, 1950
Mountain quail	48%	198	Leopold, 1939†
Scaled quail	74%	1,219	Schemnitz, 1961
California quail	63.3%	5,603	Emlen, 1940*
	59.3%	10,682	Francis, 1970*
Gambel quail	76%	352	Raitt & Ohmart, 1968*
Bobwhite	82.3%	51,178	Bennitt, 1951
	82%	1,546	Marsden & Baskett, 1958
Harlequin quail	61%	57	Leopold & McCabe, 1957†
Gray partridge	79.5%	14,167	Johnson, 1964*
Chukar partridge	87–89.5%	....	Johnson, 1960

*Calculated from author's data.

†Based on museum skin samples taken at various times of year.

most quails may not be undesirable, inasmuch as it may assure that sexually active males will be available to fertilize renesting females whose mates have already reached a postreproductive condition. On the other hand, males of polygamous or promiscuous species may be selectively harvested without significantly reducing the reproductive potential of the population. Among such species in which only a single sex is hunted, prehunting and posthunting sex ratio changes provide a valuable means of calculating population sizes (Davis, in Mosby, 1963).

The acquisition of age ratio data is at least as important to biologists as a knowledge of sex ratios in wild populations. Hickey (1955) reviewed the history of age ratio studies and their application for wildlife biologists. He also summarized the then available data for age ratios of gallinaceous birds. In table 19 additional age ratio data are summarized, which for the most part have been chosen to supplement rather than to duplicate those figures provided by Hickey.

Age ratio data have two immediate applications. One such application is that they provide a means of estimating survival rates for relatively short-lived species, without the necessity of marking birds individually and obtaining recapture or recovery data. Marsden and Baskett (1958) used the technique of assuming that the percentage of immature birds in the fall hunting sample represented an estimate of the annual mortality rate of adults, and indeed these estimates are generally in close agreement with mortality estimates based on data from banded birds as summarized by Hickey (1955).

The second and more generally applicable use of age ratios is to supplement the evidence obtained from nesting and brood counts about the relative success of the past breeding season. By comparing the number of immature birds in the fall population with that of adults (or adult females, as is done by some investigators), an estimate of breeding productivity is possible. Thus, a ratio of 50 percent immatures to 50 percent adults in the fall kill sample would suggest a breeding season productivity of 100 percent, while a ratio of 75 percent immatures to 25 percent adults would provide a productivity factor of 300 percent. The ultimate limit on such productivity factors is determined by the average clutch size of the species, and the difference between the actual productivity ratio and the potential one (assuming an equal sex ratio in adults) might provide an estimate of the reproductive efficiency of the population. For example, a quail species with an average clutch size of twelve could attain a fall population of 86 percent immatures if conditions were ideal. A figure in excess of this would suggest that double-brooding might have occurred, or that an error in estimate resulted from differential sampling vulnerability of the two age classes.

Reported age ratio data for as many species of grouse and quail as possible are summarized in table 19. It should be apparent that such data are likely to vary considerably in different years or under different ecological conditions. Nevertheless, such data provide sample figures for interspecies and intraspecies comparisons and for illustrating the theoretical relationship just mentioned between clutch size and potential productivity. When tertiary sex ratio data are available, the possibility of inserting a correction factor based on the percentage of adult females in the breeding population is of course desirable.

MORTALITY AND SURVIVAL RATES

It has been emphasized that populations of animals can vary in density, in spatial distribution patterns (territoriality favors dispersion, sociality favors clumping), and in sex and age composition. Not only can the population be analyzed for immature and adult components but the adults themselves have age composition characteristics, with the relative frequency of the various age classes depending on the rate at which the animals die. It is possible to gather such mortality information only by marking individuals (preferably while still young enough to determine their exact age at the time of marking), releasing them, and resampling the population at later times to determine how long the marked individuals survive. A review by Farner (1955) provides the theoretical concepts and practical methods that are required in the performance of such investigations with birds, and it is beyond the scope of this short review to mention them here. A few ideas, however, are so basic to the understanding of this aspect of population dynamics that they must be considered individually.

The relative rate at which individuals in a population die is usually expressed as an *annual mortality rate* (M), which is the ratio of those individuals dying during a year to the number that were alive at the beginning of the twelve-month period, whatever its starting point. The *annual survival rate* (S) is the opposite ratio: the proportion of the animals still surviving at the end of a twelve-month period to those that were alive at its start. Thus, $S+M=1.0$, or $S=1.0-M$. Some examples of estimated survival rates appear in table 20. The total population may be subdivided into different *age classes* according to the year in which each individual was hatched. The population thus consists of varying numbers of one-year-olds, two-year-olds, etc. For the species under consideration here, all the individuals in a single age class will probably have actual ages

TABLE 20

SOME REPORTED ANNUAL ADULT SURVIVAL RATES

	Survival Rate(s)	*Reference*
Blue grouse		
Males	73%	Zwickel, 1966
Females	62%	Zwickel, 1966
Adults	72%	Bendell & Elliott, 1967
Yearlings	73%	Bendell & Elliott, 1967
Willow ptarmigan		
Norwegian race	22–23%	Hagen (cited in Hickey, 1955)
Scottish race	29–33%	Jenkins, Watson, & Miller, 1963
Newfoundland race	28%	Bergerud, 1970b
Ruffed grouse		
Adult males	47%	Gullion & Marshall, 1968
Greater prairie chicken		
Both sexes	28–38%	Hamerstrom & Hamerstrom, 1949
Sharp-tailed grouse		
Both sexes	40%	Ammann, 1957
California quail		
Immatures	26.7%	Raitt & Genelly, 1964
Adults	31.6%	Raitt & Genelly, 1964
Gambel quail		
Both sexes	28–40%	Sowls, 1960
Bobwhite		
Both sexes	22%	Marsden & Baskett, 1958
Gray partridge	20%	Westerskov (cited in Hickey, 1955)

within two or three months of one another, depending on the length of the breeding season. Each breeding season thus generates a new *cohort* of birds that have hatched during the same year and constitute a single age class. The length of time required for an entire cohort of hatched young to be essentially eliminated from the population is referred to as the *turnover period* or *turnover rate*. This is perhaps properly estimated on the basis of time required for 100 percent of the age class to be reduced to 1 percent of the original cohort, but practice varies in this regard (Hickey, 1955). The means proposed by Petrides (1949) for calculating an expected turnover rate is based on the assumption that the mortality rate is constant for all ages. It is therefore convenient to define the initial cohort as, for example, the birds alive at the start of the first October following hatching to avoid the problems of the higher mortality rates usually associated with the first few months of life. Obviously, turnover periods having a starting point consisting of 100 percent of the immatures surviving to fall will be longer than those based on a cohort of newly hatched young. Even shorter would be turnover rates based on 100 percent of the potential young, in the form of the total eggs laid. Although this last basis for defining a cohort is rarely if ever used in practice, it has one theoretical advantage. That is, by starting with the eggs laid rather than with some later stage, it is possible to introduce differential rates of prehatching, juvenile, and adult mortality rates in the construction of a *survivorship curve*, which not only provides a more realistic view of population diminution, but also introduces the possibility of calculating the rate of egg replacement potential in the adult age classes of the resulting survivorship series. This must be based on average clutch size estimates, knowledge of possible nonbreeding rates in younger age classes, and tertiary sex ratio information, but it provides a useful means of estimating the population regeneration potential of species having varying mortality rates of eggs, juveniles, and adults. Some examples of such calculations are presented in figures 13 to 15.

One of the most useful statistics that can be derived on the basis of known and constant mortality rates is an estimate of further *life expectancy* as of a prescribed initial date or age. Thus, a life expectancy figure may be defined as of the date of hatching, the date of fledging, or some later chosen time. In general, it is perhaps best designated for birds as the earliest age at which juvenile mortality rates have decreased to the point where they become virtually identical with adult mortality rates. This may be as early as the first September or October after hatching or possibly even a year later. In any case, the further life expectancy for any age class is in effect the length of time required to reduce the number of surviving individuals of that age class by 50 percent. The expectation of further life is thus an estimated *mean after lifetime*, or a mean longevity as of a selected initial date. Farner (1955) has suggested that an estimate of a mean after lifetime can conveniently be calculated, by using the following formula, if the mean annual mortality rate is known and if the mortality rate of the included age classes do not differ significantly from the over-all mean mortality rate:

$$\text{Mean after lifetime} = \frac{0.4343}{\text{Log}_{10}\ \text{S}}$$

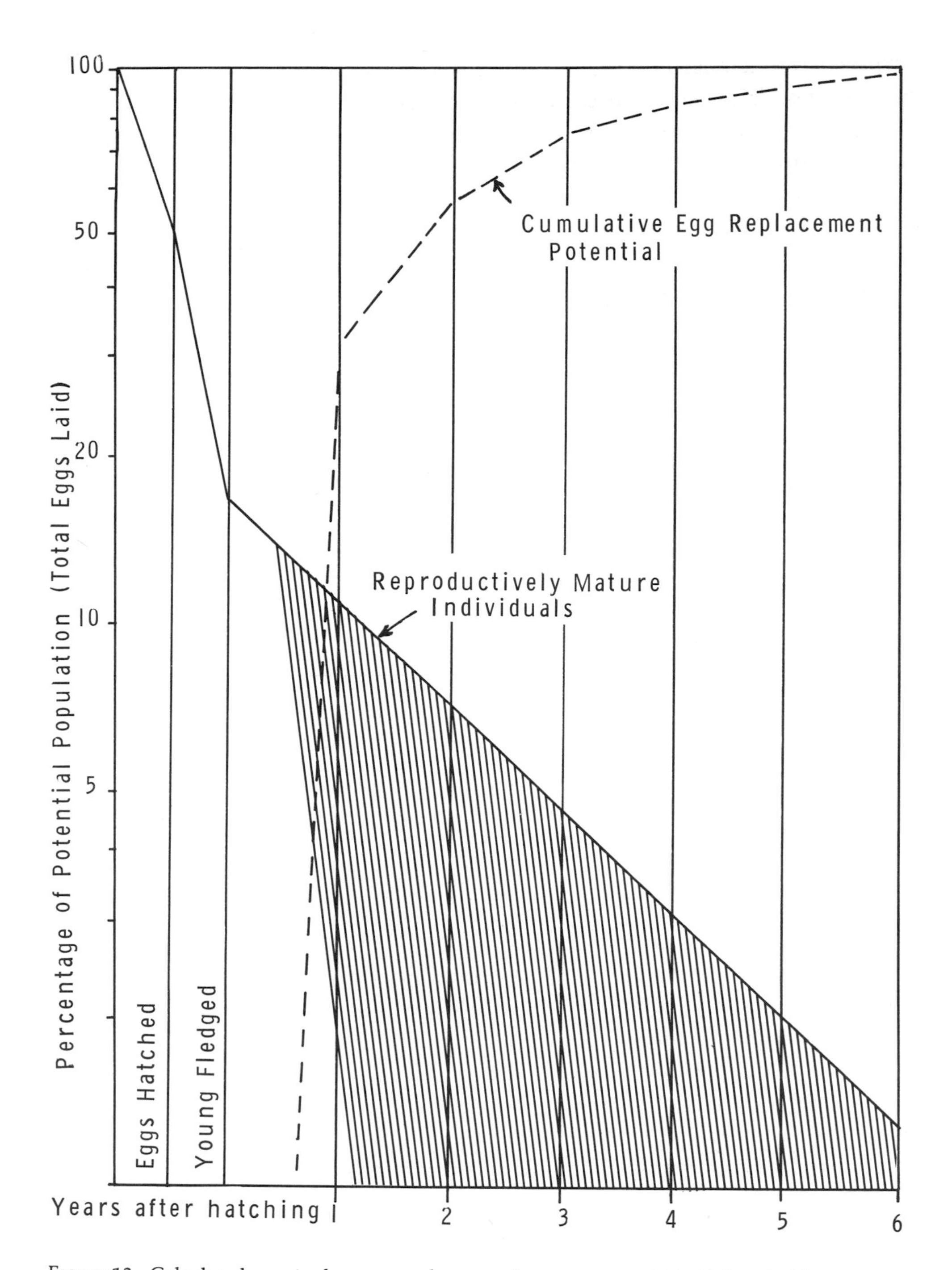

FIGURE 13. Calculated survival curve and egg replacement potential of female blue grouse based on field data. (Assumptions are of a 50% hatching and 33% rearing success, 62% annual survival after first fall, and an average clutch size of 6.2 eggs, with 25% of the first-year females nonbreeders.)

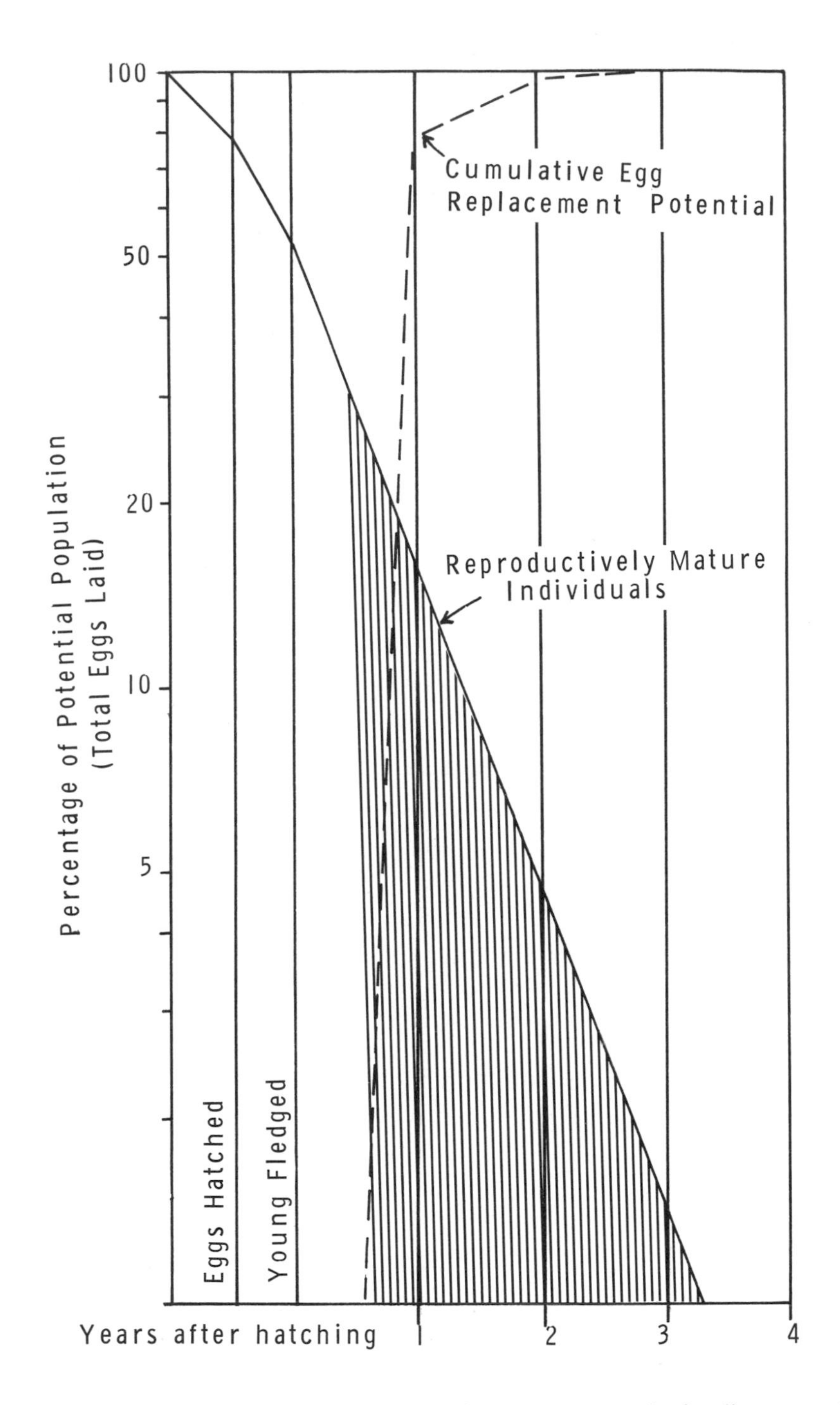

FIGURE 14. Calculated survival curve and egg replacement potential of willow ptarmigan. (Assumptions are of a 77% hatching and 33% rearing success, a 44% annual survival rate of both sexes after first fall, and an average clutch size of 7.1 eggs.)

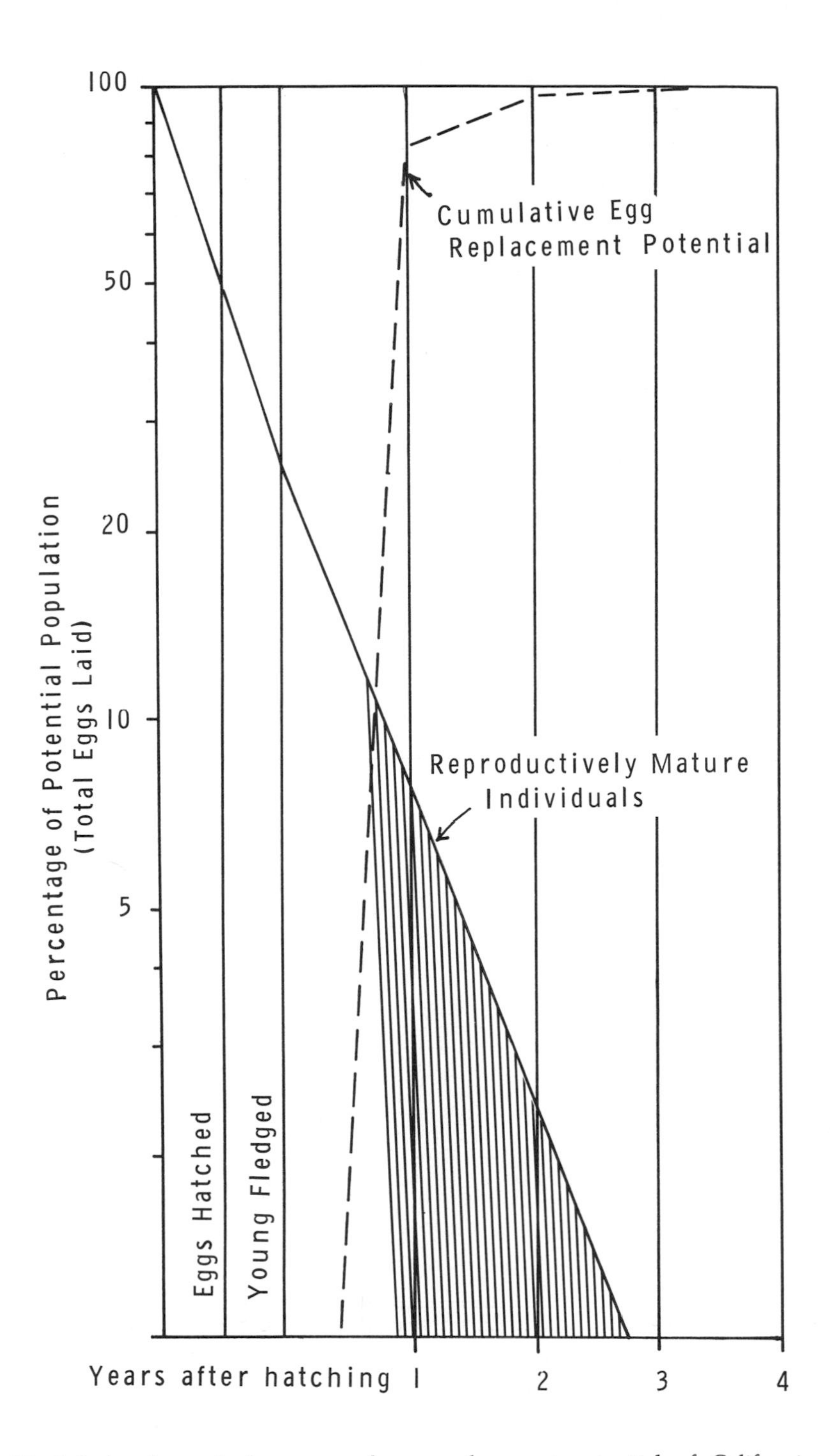

FIGURE 15. Calculated survival curve and egg replacement potential of California quail. (Assumptions are of a 50% hatching and 50% rearing success, a 42% annual survival of both sexes after the first fall, and an average clutch size of 14 eggs.)

If the selected initial date from which a mean after lifetime is calculated is chosen as some point following hatching rather than hatching itself, then of course the estimated mean after lifetime is not the same as the average life span. Rather, the average life span (or mean total longevity) will be somewhat less than the sum of the mean after lifetime estimate and the interval between hatching and the initially selected date, with the difference dependent on the higher mortality rates between hatching and the initially selected date. It might be noted that Lack (1966) has provided a convenient formula for computing further life expectancies in years by the following method, in which M equals the annual mortality rate:

$$\frac{2-M}{2M}$$

Recently, a valuable contribution by Ricklefs (1969) has concentrated on the significance of mortality rates of eggs and young, and he has provided a ready method of estimating short-term (weekly, daily, etc.) mortality rates for these important stages in the life cycle. He found that such mortality rates can be calculated by the equation:

$$m = \frac{-(\text{Log}_e \text{ P})}{t}$$

where m equals the mortality rate per unit of time (t) and P equals the proportion of nests or individuals surviving the total period considered, again assuming that mortality rates are constant throughout the entire period. As noted in the previous chapter, daily nest mortality rates are generally between 2 and 4 percent, whereas chick mortality rates are considerably lower (Ricklefs, 1969).

An equally useful formula is that proposed by Petrides (1949) for estimating the turnover period, this term being defined as the time required to reduce an original age-class cohort of 100 percent to its virtual elimination from the population. Such an effective end-point might be 5 percent, 1 percent, or 0.1 percent, depending on one's views. Petrides reported that the turnover period can readily be calculated by the following formula, again assuming that the mortality rate of different age classes does not vary significantly from the over-all annual mortality rate:

$$\text{Turnover period (years)} = \frac{\text{Log}_{10} \text{ of surviving fraction of cohort}}{\text{Log}_{10} \text{ S}} + 1$$

If 1 percent is chosen as the surviving fraction of the cohort that represents the virtual elimination of an age class from the population, then the formula can be restated simply as:

$$\frac{-2.0}{\text{Log}_{10} \text{ S}} + 1$$

In table 21 are presented some calculated mean after lifetimes (usually after the first fall of life) and estimated turnover periods among various species for which annual mortality estimates have been reported. In some cases these estimates of mean after lifetimes differ slightly from those reported by the original authors, the variations being the result of different techniques or assumptions, but in general the estimates are very close to those published earlier for these species.

Such calculated turnover periods should provide at least a general estimate of potential natural longevity, as represented by the oldest age class that might be encountered in natural populations. Potential natural longevity is likely to be less than potential longevity under ideal conditions, such as optimum conditions of captivity. In table 22 are presented some reported estimates of mean after lifetimes and records of unusual longevity for wild or captive individuals. It would seem that four or five years represents close to the potential natural longevity of most grouse and quail species, but available mortality rates of a few species (especially blue grouse and white-tailed ptarmigan) indicate that it might be considerably longer than this.

TABLE 21

SOME LONGEVITY ESTIMATES, BASED ON REPORTED SURVIVAL RATES

	Survival Rate(s)	*Mean Longevity after 1st Fall**	*Maximum Longevity and Turnover Period* †	*References*
Blue grouse				
Males	72.5%	3.1 yr.	15.3 yr.	Zwickel, 1966
Females	62.0%	2.09 yr.	10.6 yr.	
Willow ptarmigan				
Both sexes	30.0%	10 mo.	4.8 yr.	Jenkins, Watson, & Miller, 1963
White-tailed ptarmigan				
1st yr.	37.0%	0.99 yr.	Overall (57.9% S) 9.4 yr.	Choate, 1963
After 1st yr.	71.0%	2.92 yr.	After 1st yr. 14.4 yr.	
Ruffed grouse				
Males (after 1st winter)	47.0%	1.25 yr.	6.76 yr.	Gullion & Marshall, 1968
Sharp-tailed grouse	40.0%	1.10 yr.	6.0 yr.	Ammann, 1957
Greater prairie chicken	28.38%	11 mo.-1.1 yr.	9.4 mo.-1.03 yr.	Hamerstrom & Hamerstrom, 1949
California quail	28.8%	9.6 mo.	4.7 yr.	Raitt & Genelly, 1964
Bobwhite	22.0%	7.9 mo.	4.0 yr.	Marsden & Baskett, 1958
Gray partridge	20.0%	7.4 mo.	3.9 yr.	Westerskov, (cited in Hickey, 1955)

*Method of Farner (1955:409).
†Method of Petrides (1949), using 1% of original cohort as end-point.

TABLE 22

Some Longevity Estimates and Maximum Longevity Records

Sage grouse:	One banded female was recovered 7 years after banding. Returns of marked birds returning to strutting grounds one year later varied from 5 to 21 percent over 3 years. Dalke et al., 1963
Willow ptarmigan (red grouse):	Seven birds at least 4 years old were recovered from 12,050 banded. Estimated mean longevity of about 1 year from August following hatching. Jenkins, Watson, & Miller, 1963
White-tailed ptarmigan:	Twelve of 36 females and 16 of 31 males lived at least 5 years. Estimated mean longevity after first summer, 3.02 years; estimated maximum longevity of 13–15 years. Choate, 1963
Ruffed grouse:	Maximum known survival of 94 months by 1 of 978 marked birds. Mean life-span of 8.56 months for immature females; 8.63 months for immature males. Mean life-span of birds banded as adults was 25.3 months for males, 23.7 months for females. Gullion and Marshall, 1968
Sharp-tailed grouse:	One bird at least 7.5 years old from 93 banded birds. Mean longevity after full growth estimated from 1.51 years (females) to 1.61 years (males). Ammann, 1957
Greater prairie chicken:	Two birds, out of 597 banded, recovered in fourth year after banding. Hamerstrom & Hamerstrom, 1949
California quail:	One male banded as an adult was recaptured when at least 80 months old. Mean life expectancy after September following hatching is 9.7 months. Raitt and Genelly, 1964
Gambel quail:	Four out of 121 birds trapped as adults were alive 4 years later, and 10 out of 321 birds trapped as juveniles were alive 4 years later. Sowls, 1960
Bobwhite:	One out of 1,156 banded bobwhites was recovered in its fifth year. Estimated life expectancy after October following hatching is 8.5 months. Marsden and Baskett, 1958. In captivity known to live at least 8 years, still fertile at four to five years of age. Stoddard, 1931. One report of a captive individual surviving 9 years. Judd, 1905a

7

Social Behavior and Vocalizations

ONE of the most complex and fascinating aspects of grouse and quail biology is their social behavior, particularly that related to reproduction. Natural selection in the quail group has seemingly favored the retention of a monogamous mating system with the associated advantages of maintaining the pair bond through the breeding season. This system allows the male to participate in the protection of the nest, possibly participate in incubation, and later care for the brood. It also provides the possibility, if not the frequent actuality, that the male might undertake the entire incubation or rearing of the first brood, while the female is freed to lay a second clutch and rear a second brood in a single breeding season.

In addition, within the quails may be seen a breakdown of typical avian territorial behavior patterns, probably resulting from the greater survival value of ecological adaptations favoring sociality in these birds. Not only do these fairly vocal species benefit from their mutual alarm signals by remaining together but also their small size and catholic feeding behavior reduce the likelihood that the optimal breeding densities will exceed the carrying capacities of the habitat.

By contrast, in the grouse there is a clear indication that selective pressures have favored the retention of strong territorial behavior, and there is a direct relationship between a male's capacities to establish and maintain a favorable territory and his ability to reproduce successfully. This territoriality perhaps results mainly from the wide variation among males in their aggressiveness and reproductive vigor but also from the possibility that in these species the carrying capacity of the habitat in relation to the population density may be more significant for the species' survival than are the advantages of sociality. Thus, territorial behavior among males is conspicuous in all the grouse species.

EVOLUTIONARY TRENDS IN GROUSE SOCIAL BEHAVIOR

The size of the male's territory and the length of time during which it is defended vary considerably among grouse. From one possible extreme, that of defending a fairly large territory throughout the breeding season, within which a single female not only nests but she and her brood are also defended by the male, one may trace the progressive development of a reduced territorial size that is defended only until after fertilization of females has been completed and neither do the females nest within the territorial boundaries nor are they or their broods defended by the males. This trend toward the evolution of a polygamous or promiscuous mating system is associated with many parallel evolutionary trends. There is an increased pressure on males for enhancing their attraction value to females; thus a tendency exists for more elaborate or more conspicuous sexual signal systems among males. Since they no longer must remain near the female and the nest, pressures for protective coloration are countered by those of sexual selection, and increased behavioral and plumage dimorphism is to be expected.

Conspicuousness in male sexual displays can be enhanced not only by increase in body size and the exhibition of elaborate visual and acoustical signals in an individual male but also by multiplying such effects through the aggregation of several males. These counter pressures—those favoring the maintenance of definite and maximum territorial areas as a factor of reproductive success and those favoring the aggregation of several displaying males in a limited area to increase the likelihood of female attraction and reduce the danger of predators to individual males—have led directly to the evolution of arena behavior in several grouse species. This unlikely form of male communal display, in which individual male territories are closely adjacent, are relatively small, and serve only as mating stations, can evolve only under certain circumstances. First, the males must be totally freed from defending areas large enough for the females to nest within and also from defending the female during incubation and brooding. Next, the reproductive efficiency of a group of males must be greater than that of single males, either because of their greater attraction to females or because the assembled males are relatively safer from predators than are solitarily displaying ones. Further, to assure assortative mating there must be enough individual variation among the males in aggressiveness that territorial size or location is directly related to breeding success; these variations are perhaps most likely among species that require two or more years to attain full reproductive development. In addition, if male display aggregations are to develop it must be advantageous for the less successful males to associate with the more successful ones. It may be argued that such early experience increases the male's chances of holding a larger or more centrally located territory that will be more reproductively efficient later in its lifetime. Peripheral males participating in arena displays may be regarded as apprentices which reproductively benefit more from such experience than they would from establishing independent and solitary territories.

Since arena displays among grouse might logically be expected to evolve more readily in open-country habitats than in heavily forested ones, open-country and polygamous species are preadapted for the evolution of arena behavior. It seems quite probable that the arena behavior of sage grouse evolved independently from that of the prairie grouse (*Tympanuchus*), and the corresponding behavior of the European black grouse (*Tetrao tetrix*) may also have evolved independently. This last species is actually a woodland edge form, but its arena displays occur in open heaths. The communal leks of the black grouse were the earliest of the arena displays of grouse studied, and the term *lek* is now generally applied to arena behavior of all grouse. Koivisto (1965) suggested that *display ground* be used to describe the general topographic location in which social display is performed, *arena* be used to indicate the specific area (the collective territories), and *lek* be more broadly applied to both the birds and their arena. Similarly, the term *lekking* can be used to indicate the general process of communal male display in grouse.

To illustrate how arena behavior may have gradually evolved from more typical territorial behavior, a series of representative grouse specimens may be mentioned that provide reference points along this behavioral spectrum.

Of all the grouse, the willow ptarmigan's actions come closest to the presumed ancestral (or most generalized) type of reproductive social behavior. In this species fairly large territories are established by the male in fall (at least in nonmigratory populations). These individual territories are largest for the most aggressive males, and many young or inexperienced males may be unable to establish territories, especially in dense populations. The female is attracted to a displaying male, and a firm pair bond is formed. Sometimes males form a pair bond with two females and may breed with both. Territorial displays and defense continue after the pair bond is established, but such activities are diminished during the nesting season. At that time the male defends the female and nest and after hatching remains with the female and brood. After the brood is reared the territorial boundaries are again established.

In the rock ptarmigan and also in the white-tailed ptarmigan, the pair bond is established in the spring. At least in the rock ptarmigan, two or three females may sometimes be associated with a single territorial male, and Choate (1960) found some indications of polygamy or promiscuity in the white-tailed ptarmigan. The male continues to defend the territory while the female is incubating, although with reduced intensity, and the territory is abandoned about the time of hatching. The female and young may remain in the male's territory but are only infrequently accompanied by him, and he usually takes no part in defending the young. In the rock ptarmigan the male reestablishes his territory in the fall, while in the white-tailed ptarmigan this evidently does not occur until spring (Watson, 1965; Choate, 1963).

In the monogamous European hazel grouse (*Bonasa bonasia*), the male reportedly establishes his territory in the fall, with those in optimum habitats being the most successful in attracting females. A male usually remains on his territory, defending both it and the female during incubation and brooding periods, but only atypically performs distraction displays or utters warning calls to the female (Pynnönen, 1954). Some observers have, nonetheless, reported seeing males attending broods with females.

In the blue grouse exists a clearly intermediate stage between the one extreme of a monogamous or nearly monogamous pair bond associated with the establishment of a territory large enough to support the rearing of a brood and the other extreme of complete promiscuity and territorial defense limited to an area serving to attract females and provide a mating station. Other North American species that fall into this general category are the ruffed grouse and the spruce grouse, but the blue grouse will serve as an example.

Because of its winter migration, the blue grouse males probably first establish territories in spring. Although these areas may cover several acres, hooting is limited to particular places within the territorial boundaries. The home ranges occupied by females associated with territorial males may overlap the boundaries of several male territories. The typical mating system of blue grouse may thus be considered polygamous or promiscuous (Bendell, 1955c; Bendell and Elliott, 1967), but in local populations at least some birds may form strong pair bonds that persist until after the young hatch (Blackford, 1958, 1963). The location of the female's nest is not associated with the male's hooting sites, and the male does not defend the nest or brood. In general, male hooting sites are well separated and their territories are not contiguous, but in a few cases apparently communal male displays involving four or more males have been observed (Blackford, 1958, 1963). Males remain on their territories until their late summer migration, well after active territorial defense ceases.

The forest-dwelling capercaillie (*Tetrao urogallus*) of Europe provides a slightly more advanced stage in the evolution of communal displays, judging from such reports as those of Lumsden (1961b). He studied an arena with three territories (varying from three hundred to one thousand square yards in area) that did not have contiguous boundaries but were separated by twenty to forty yards. Four nonterritorial males visited the arena, all of which were apparently yearlings; they performed partial sexual displays and sometimes threatened one another but were ignored by the territorial cocks, between whose territories they moved at will. Up to nine females visited the display ground at one time, and of thirteen copulations seen, twelve were performed by a single male. Dement'ev and Gladkov (1967) found that sixty-six display grounds contained 630 males, collectively averaging 9.5 males per display ground (individual averages ranging from 2 to 12 males). However, Hjorth (1970) does not consider the capercaillie to be a lek-forming species.

In the related black grouse, the seasonal maximum number of males occupying a display ground averages about nine and ranges from three to twenty-six, the strongest one or two of which ("first-class") occupy relatively central territories (Koivisto, 1965). The territories of this species are nearly contiguous and range in size from one hundred to four hundred square meters (Kruijt and Hogan, 1967). Koivisto (1965) estimated that territories in this species may range from two to two hundred square meters, with no significant differences in the sizes of territories of first-class and second-class males. Immature males, which make up about one-third of the population, are either nonterritorial and are not tolerated by territorial males, or they occupy small and peripheral territories ("third-class" males). Koivisto believed that the primary survival value of these immature birds for the group is their tendency to warn the actively displaying males of the presence of danger. He found that there is a direct relationship between age and hierarchical position in the arena, the first-class males being mature birds that are the most fit for reproduction and also are the most successful in attracting females. Of forty-seven copulations observed by him, 56 percent were performed by first-class males. The value to the species of such assortative mating and the relative protection first-class males gained from the presence of the other categories of males appeared to Koivisto to be the primary evolutionary advantages of communal male display.

Among the North American grouse, corresponding arena behavior occurs in the pinnated grouse, sharp-tailed grouse, and sage grouse. In both the pinnated grouse and the sharp-tailed grouse, the average number of male birds occupying display grounds in general equals or exceeds the number reported for the black grouse. Copelin (1963) indicates that in the display grounds he studied the number of male lesser prairie chickens ranged from 1 to 43, and active grounds averaged 13.7 males over an eleven-year period. Robel's greater prairie chicken study area (1967) had from 17 to 25 resident males present in a three-year period. He found (1966) that 10 marked territorial males defended areas of from 164 to 1,069 square meters (averaging 518 square meters), and that the 2 males defending the largest territories in two years of study accounted for 72.5 percent of fifty-four observed copulations.

Numbers of male sharp-tailed grouse present on display grounds vary considerably with population density in Nebraska; leks of both this species and pinnated grouse average approximately 10 males, but sometimes exceed 20 and occasionally reach 40 or more. Hart, Lee, and Low (1952) reported that up to 100 male sharp-tailed grouse were observed on display grounds in Utah, but the average on twenty-nine grounds was 12.2 males. Evans (1961) confirmed that females select the most dominant males for matings, and Lumsden (1965) reported that on a display ground he studied one male accounted for 76 percent of the seventeen attempted or completed copulations seen. Scott (1950) concluded that the social organization of sharp-tailed grouse is more highly developed than that of the pinnated grouse but is not as complex as that of the sage grouse.

The sage grouse provides the final stage in this evolutionary sequence; it exhibits a higher degree of size dimorphism than any other species of North American grouse (adult weight ratio of females to males being 1:1.6–1.9), the display areas have a larger average number of participating males, and the central territories are among the smallest of any grouse species. Scott (1942) was the first to recognize the hierarchical nature of the territorial distribution pattern and to describe first-rank or master cocks, which were responsible for 74 percent of the 174 copulations that he observed. Dalke et al. (1960) reported that the territories held by master cocks were often forty feet or less in diameter, and Lumsden (1965) showed the territorial distribution of 19 males that exhibited an average distance from the nearest neighbor of about forty feet. In Colorado, 407 counts of strutting grounds indicated an average maximum number of 27.1 males present (Rogers, 1964). Patterson (1952) provided figures indicating that 8,479 males were counted over a three-year period on Wyoming display grounds, averaging about 70 males per display ground. Patterson reported one ground containing 400 males, and Scott's observations (1942) were made on a ground of similar size. Lumsden (1968) found that individual birds may have strutting areas that overlap those of other males, and that although entire groups of males may move about somewhat, the relative positions of the males remain the same. Furthermore, large sage grouse leks may have several centers of social dominance, and Lumsden suggests that these should be called conjunct leks. He believes that yearling males are not tolerated by old males in the center of the lek but can move about fairly freely near the edges of the arena. They probably do not normally establish territories until their second year, when they may become "attendant" males with territorial status. The remarkably large size and complex social hierarchy of sage grouse leks, as well as their extrordinarily complicated strutting performances, would seem to qualify this species as representing the ultimate stage in evolutionary trends discernible through the entire group. Since sage grouse are ecologically isolated from all other grouse species and are known to have hybridized only once, it would seem that these complex behavioral adaptations are the result of intraspecific selective pressures rather than the need for reproductive isolation from related forms.

A possible index of the intensity of sexual selection in promoting sexual differences in behavior and morphology of the sage grouse was indicated earlier as weight differences between adult males and females which approach ratios of 1:2. Corresponding ratios can readily be calculated for the other grouse species from table 5 in chapter 2. For the essentially monogamous ptarmigan species these female-to-male weight ratios range from about 1:1 to 1:1.09. For the blue grouse, spruce grouse, and ruffed grouse they range from 1:1.1 to 1:1.33, and in the prairie grouse they range from 1:1.14 to 1:1.31. These data would suggest that the intensity of sexual selection insofar as it might affect weight differences in the sexes is about the same in the lek-forming prairie grouse as in the non-lek-forming but polygamous or promiscuous forest-dwelling species. Data presented by Dement'ev and Gladkov (1967) indicate corresponding weight ratios for the black grouse of from 1:1.27 to 1:1.38, and for capercaillie the estimated ratio is 1:2.28, even higher than in sage grouse. Berndt and Meise (1962) report the adult weight ratio of females to males in the capercaillie to be from 1:2.08 to 1:2.25. This species and a closely related one are by considerable measure the largest of the grouse, and the ecological implications of both total body size and sexual differences in body size of these two species are still obscure.

Nonvocal Acoustical Signals in Grouse

The feather specializations found in the sharp-tailed grouse that are related to tail-rattling have been mentioned in chapter 2; it might also be mentioned that similar tail-rattling occurs in male sage grouse, that tail-clicking noises are made by pinnated grouse, and that a tail-swishing display occurs in Franklin spruce grouse, involving both alternate and simultaneous spreading of the rectrices (MacDonald, 1968). Likewise, foot-stamping sounds are made by males of many species; these are perhaps most apparent in the sharp-tailed grouse, but also occur in pinnated grouse, willow ptarmigan ("rapid stamping" of Watson and Jenkins, 1964), and probably other species.

A more interesting kind of nonvocal sexual signal used by male grouse is the drumming and clapping sounds made by various species, which apparently represent variably specialized or ritualized territorial flights. A rapid survey of the grouse with respect to such variations is instructive.

The territorial display flights of male ptarmigans may serve as a starting point from which the increasingly specialized variations of the other species may be derived. In the red grouse (willow ptarmigan), Jenkins and Watson (1964) report that the bird (either sex) "flies steeply upwards for about ten meters, sails for less than a second, and then gradually descends with rapidly beating wings, fanned tail, and extended head and neck. On landing, its primaries often touch the ground, and it then stands high with drooping wing, bobbing its body and fanning its tail in and out." Calling occurs during the ascent, descent, and after landing, with the loudness of the call and length of the flight varying with the bird's relative dominance.

Schmidt (1969) described the "scream flight" display of white-tailed ptarmigan, and Choate (1960) reported once seeing a male white-tailed ptarmigan fly upward in a nearly vertical flight, hovering, screaming, and gliding down in a single spiral, then landing with another scream about thirty-five feet from the starting point. This kind of flight was reported by Bent (1932) for the rock ptarmigan, in which the male flies upward thirty or forty feet, then floats downward on stiff wings until he is near the ground when he checks his descent and may sail up again, calling loudly. MacDonald (1970) has recently described this display of rock ptarmigan in considerable detail.

In the eastern Canadian and Alaskan forms of spruce grouse an apparently corresponding aerial display occurs as the male flies steeply downward out of a tree being used as a display perch, stops his descent about four to eight feet above the ground, and then descends rapidly with strongly beating wings (Lumsden, 1961a; Ellison, 1968b). In the Franklin spruce grouse males fly vertically and slowly up to a perch with whirring wings. They may then rush forward along the branch and spread the wings and tail, make three or four drum-like wing beats while standing upright, or perform an aerial wing-clap display (MacDonald, 1968). In this display the bird takes flight and at some point pauses in mid-air with a deep wing-stroke, following which he sharply strikes the wings together above the back and drops downward to the ground, with a second wing-clap following landing.

Short (1967) noted that males of Franklin spruce grouse have outer primaries that are more indented and more closely approach those of the Siberian spruce grouse (*Dendragapus falcipennis*) than they do those of the eastern race *canadensis;* thus it is probable that similar whirring or wing-clapping sounds are made during aerial displays in the Siberian species.

Corresponding drumming flight behavior is found in the blue grouse (Wing, 1946). Bendell and Elliott (1967) report that a "flutter flight" occurs in both sexes of the sooty blue grouse (*fuliginosus*) but that the noise produced is a ripping sound and apparently is not so elaborate as in the interior populations such as *richardsonii* and *pallidus.* Blackford (1958, 1963) reports that individuals (both sexes) of the former race perform a wing-flutter (or flutter-jump) display some eight or ten inches off the ground. Males perform more extensive drumming flights; they may also exhibit a fairly sharp whipping of the wings on alighting in a tree, and sometimes produce a wing-clap, consisting of a single loud wing note, presumably made in the same manner as by Franklin spruce grouse. In typical drumming flights the male jumps from his display perch, flies strongly upwards with whirring wings, and returns after a horseshoe-shaped flight course to a point near where he started (Blackford, 1963). Aerial rotations during display flights may also occur (Wing, 1946; Blackford, 1958).

The well-known drumming display of ruffed grouse would appear to be an exaggerated version of the drumming movements of the Franklin spruce grouse or a ritualized drumming flight in which the male has substituted wing-beating movements for the actual flight. No actual flight displays are known to occur in this species, but the related hazel grouse (*Bonasa bonasia*) exhibits both wing-flapping displays and actual display flights with associated calling (Pynnönen, 1954; Schenkel, 1958). Male vocalizations in these two species are limited: hissing sounds are made by the ruffed grouse, while whistling notes are produced by the hazel hen. The typical flutter-jump display, in which males make short, nearly vertical flights with strongly beating wings and sometimes with associated vocalizations, would appear to be an alternate evolutionary modification of the territorial song flights of ptarmigan. Typical flutter-jump displays occur in the prairie grouse and black grouse (Hamerstrom and Hamerstrom, 1960), as well as in the capercaillie (Lumsden, 1961b). Flutter-jumps of capercaillie, which have loud wing noises, are performed without associated vocalizations. Male sharp-tailed grouse only rarely utter calls at the start of these flights, which nonetheless are conspicuous in their open-country habitat. In the pinnated grouse calls may be uttered before, during, or after the display, and the black grouse utters hissing sounds during flutter-jumping. The sage grouse completely lacks a flutter-jump display, judging from all recent observations.

In summary, it would appear that the visually and acoustically conspicuous territorial flights of ptarmigans have, in the forest-dwelling grouse, been replaced by drumming, fluttering, or whirring flights; wing-clapping noises; and sedentary wing-drumming displays (table 23). In most of the lekking grouse they have been restricted to short and often quiet flutter-jumps, which are visually conspicuous in these open-country birds but are limited in length to the typically small territories.

As a final point, these aerial displays occur in both sexes of ptarmigan, are more common and better developed in males than in females of *Dendragapus* species, and are performed only by males in the lek-forming species of grouse. Ultimately, in the heavy-bodied sage grouse with its closely packed leks, the flutter-jump display has been lost altogether. Lumsden (1968) has suggested that the rotary wing movements made during strutting may represent the last vestigial remnants of the sage grouse's flutter-jump display.

The summary of major male social signals of grouse (table 23) may be compared with figure 16, which illustrates representative display postures of six grouse species, although it should be emphasized that these postures are not homologous in all cases. Rather, the drawings illustrate species-specific plumage characteristics that probably provide significant visual signals during display.

For additional comparison, table 24 provides a corresponding summary of male plumage features, postures and calls of representative New World quail species, which are also believed to provide species-specific signals in this group. Details on the acoustical and possible motivational variations in the calls listed and their apparent functions may be found in the individual species accounts, and the summary here is intended only as a general comparison with the grouse signals summarized in table 23. Corresponding postures assumed by male quails and partridges during the performance of some of these displays are illustrated in figure 17, which likewise are further explained in individual species accounts.

TABLE 23

SUMMARY OF MAJOR MALE SOCIAL SIGNALS IN REPRESENTATIVE GROUSE SPECIES

	Major Male Display Features		*Major Male Acoustical Signals*		*Major Male Display Postures and Movements*		
	"Air Sacs"	*Eye-comb*	*Vocal*	*Non vocal*	*Aerial*	*Strutting**	*Other Displays*
Sage grouse[1]	Yellowish	Yellow	*Wa-um-poo* Grunting	Wing-rustling Tail-rattling Air sac "plop"	None	Tail fanned equally	Shoulder spot
Blue grouse[2]	Yellow to reddish	Yellow to reddish orange	Hooting *Oop* call	Wing-clapping Wing-drumming	Drumming flight Flutter-jump	Tail fanned, tilted strongly	Short run with head low
Spruce grouse[3]	None	Red	Hooting Snoring	Wing-clapping Wing-drumming	Drumming flight	Tail fanned, "swished" laterally	Head-jerk with squatting Foot-tramping
Willow ptarmigan (red grouse)[4]	None	Red	Hissing *Kohwayo/Kohway/Korow/Ko Kok/Ka* etc.	Rapid stamping (audible?)	Flight song	Tail fanned, tilted strongly	Waltzing (circling) Rapid foot-stamping Bowing Walking in line Crouching with head-wagging
Ruffed grouse[5]	None	Orange (small)	Hissing	Wing-drumming	None	Tail fanned, tilted slightly	Short run Rotary head-shake
Pinnated grouse[6]	Yellow to red	Yellow	Booming or Gobbling, Cackling *Pwoik*, etc.	Tail-snapping Foot-stamping	Flutter-jump	Tail spread, snapped shut	Shoulder spot Circling Nuptial bow Running parallel
Sharp-tailed grouse[7]	Purplish to red	Yellow	Cooing Cackling *Lock-a-Lock*	Tail-rattling	Flutter-jump	Tail slightly spread, shaken rapidly Wings spread	Shoulder spot Circling Nuptial bow & posing Running parallel Foot-stamping (dancing)

* "Strutting" refers to high-intensity ground display; tail-cocking and wing-drooping present in all species

1. Based on Lumsden, 1968
2. Based on Brooks, 1926, and others
3. Based on Lumsden, 1961a, and MacDonald, 1968
4. Based on Watson and Jenkins, 1964
5. Based on Bump et al., 1947, and others
6. Based on Sharpe, 1968, and Hamerstrom and Hamerstrom, 1960
7. Based on Lumsden, 1965

TABLE 24

Summary of Major Male Social Signals in Representative Quail Species

	Major Male Display Features		*Major Male Display Postures*		*Major Male Sexual and Agonistic Calls*				
	Throat	*Crest*	*Frontal*	*Lateral*	*High intensity threat*	*Low intensity threat*	*Tidbitting*	*Separation*	*Advertising*
Mountain quail[1]	Chestnut	Straight, narrow	Wings partly spread	Poorly developed(?)	. . .	. . .	. . .	*Kow* (repeated)	*Plu-ark*
Scaled quail[2]	Buff	Straight, bushy	Wings drooped Head raised Crest up or down	Flanks spread Crest raised	Head-throw call	. . .	. . .	*Pay-cos*	*Whock*
Gambel quail[3]	Black	Recurved "Teardrop" black	Wings drooped Head raised Crest up or down	Flanks spread Body-shake Wing-flap Ground-pecking	*Meah* (with chin-lifting) *Wit-wut-whrr*	*Wit-wut*	*cu-cu*	*Ka-KAA-ka-ka*	*Kaa*
California quail[4]	Black	Recurved "Teardrop" Black	Wings drooped Head raised Crest up or down	Flanks spread	*Squill* (with head-throw)	*Wip-wip*	*tu-tu*	*Ca-ca'-caw*	*Cow*
Bobwhite[5]	White & black	None	Wings spread Head low	Flanks spread	*Hao-po-weih* (Caterwaul)	*Hoy* *Hoy-poo* *Squee*	*tu-tu*	*Koilee*	*Bob-white*

1. Based on previously unpublished studies
2. Based on previously unpublished studies
3. Based in part on Ellis & Stokes, 1966
4. Based in part on Williams, 1969
5. Based in part on Stokes, 1967

FIGURE 16. Male display postures of representative grouse, including (A) booming by greater prairie chicken, (B) dancing of sharp-tailed grouse, (C) strutting of sage grouse, (D) hooting of blue grouse, (E) strutting of ruffed grouse, and (F) strutting of spruce grouse. (From *Animal Behavior*, Wm. C. Brown Co.)

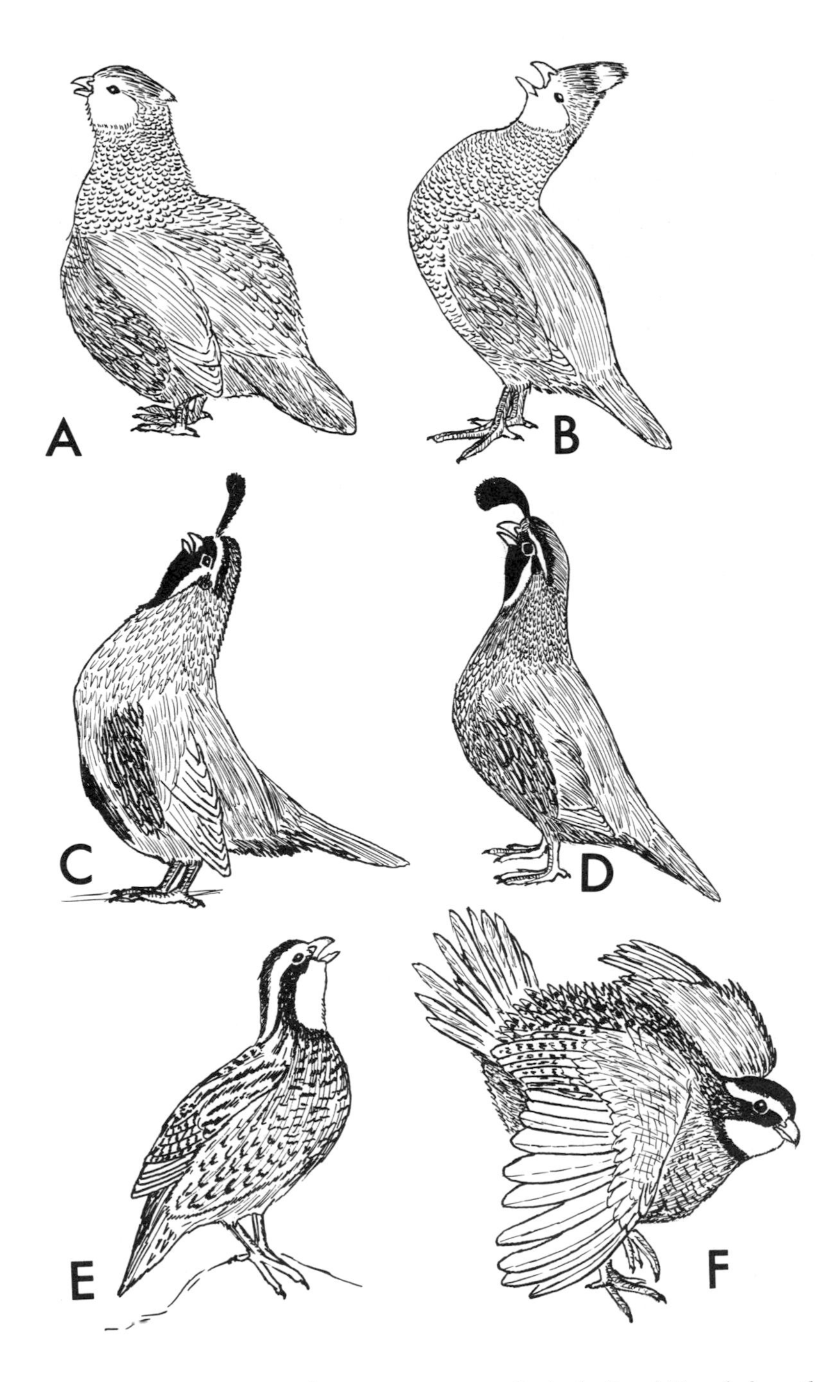

FIGURE 17. Male display postures of representative quails, including (A) scaled quail uttering *pay-cos* call, (B) scaled quail uttering head-throw call, (C) Gambel quail uttering *meah* call, (D) California quail uttering *squill* call, (E) bobwhite uttering *Bob-white,* and (F) bobwhite performing forward threat display. (Original, based on photographs.)

EVOLUTIONARY TRENDS IN VOCALIZATIONS OF NEW WORLD QUAIL

In contrast to the grouse, in which sexual behavior patterns are closely related to and in general derived from aggressive behavior related to territorial establishment and maintenance, no such nucleus of basic social behavior patterns exists in the New World quail species. Thus, whereas in the grouse sexually active females can be selectively attracted to displaying males on the basis of male signals that perhaps originally served as male-to-male agonistic signals, the high degree of gregariousness and absence of well-defined social hierarchies in quail coveys have not promoted the evolution of elaborate male-to-male aggressive signals. Instead, a considerable number of social signals are typically present that have such functions as maintaining contact among members of a social unit, warning others of danger, and reassembling the group after forced separation.

Perhaps partly because of their vulnerability to various predators but also because of the ecological advantages of using acoustic rather than visual signals for communication, the quails in general have tended to specialize in vocalizations that serve to integrate their social requirements instead of evolving elaborate long-distance visual communication systems. There is certainly no question that species-specific body movements and postures do occur in many species (see table 24), but these are in general performed between specific individuals at close range, instead of being generally broadcast and widely visible, as is the case, for example, with the territorial display flights of ptarmigans, the flutter-jumps or drumming flights of various grouse, or the "air sac" flashing of the lekking grouse. One must therefore look for possible evolutionary origins of quail social signals among such sources as the basic sounds used by parents to coordinate brood activities and those uttered by young birds to maintain or regain contact with their parents. Stokes (1967) has shown that the "lost" call of bobwhite chicks develops with increasing age directly into the separation, or "scatter," call that serves to reassemble broken coveys and during the breeding season serves to reunite separated mates. With some modifications, this same separation call also serves in males of *Colinus* and *Callipepla* as the basis for the unmated male advertisement call. With slightly different modifications, the call is also used by paired birds during encounters with others and serves to repel them. Thus a single type of chick vocalization, through ontogeny and sexual or intensity modifications, acquires at least four fairly distinctive communication functions among bobwhite adults.

Regrettably little is known so far of the acoustical communication systems of the morphologically primitive species of New World quails that are found primarily in tropical forests, other than the fact that well-developed vocal communication systems (often involving duetting) do occur. Indeed, future studies may prove that these species are actually quite highly specialized in this regard, judging from the apparent complexity and diversity of the calls that have so far been described among them. Instead of trying to generalize from this group, it is more practical to examine the social behavior patterns and vocalizations of the more northerly and arid-habitat genera, such as *Colinus*, *Callipepla*, and related forms. Several species from this group have been well studied behaviorally, and some evolutionary trends in behavior and vocalization may readily be detected.

Judging from observations of all four species of *Callipepla* and the bobwhite and limited information on the mountain quail, a major part of the vocabularies of these species is concerned with the coordination of pair and flock activities (table 25), with the same calls serving to keep the pair intact during the breeding season as those used by the covey for that purpose during the rest of the year. This requirement for individual recognition of the mate's separation call can readily be demonstrated under controlled laboratory conditions. The separation call, or a modification thereof, also typically serves as the advertisement call of unpaired males. In this situation the call is usually uttered from a conspicuous and often regularly used location, but in spite of these characteristics it should not be regarded as typical territorial behavior (see chapter 5). In addition, calls that are uttered by members of the flock during foraging are the same as those used by males of those species that perform the "tidbitting" display (Domm, 1927), which evidently plays an important role in establishing and maintaining the pair bond.

All the American quails and Old World partridges studied so far have several well-developed alarm signals, which usually include distinctively different notes for ground and for aerial predators, as well as more general alarm and escape notes (table 25). Although a few species may assume silent "freezing" positions (e.g., bobwhites, harlequin quail), species of the genus *Callipepla* more typically respond to threats by fleeing on foot while uttering rapidly repeated alarm notes.

The sexual and agonistic vocalizations of quail are not especially numerous, which is not surprising in view of their poorly developed social hierarchies and lack of aggressive territoriality. In *Callipepla* species, males utter two different calls in agonistic (mostly male-to-male) situations. These include a series of rather soft and frequently repeated threat notes, as well as a single louder call, sometimes repeated, that is usually associated with neck-stretching and tossing the head backward varying amounts, thus exposing the distinctive throat markings (figure 17). This latter display

TABLE 25

SUMMARY OF ADULT VOCALIZATIONS IN THREE QUAIL AND PARTRIDGE SPECIES

	Bobwhite[1]	*California Quail*[2]	*Chukar Partridge*[3]
I. Flock and pair activities (both sexes)			
A. Covey or pair separation	2*	1	1
B. Covey or pair contact	2	1	1 (male only)
C. Feeding (and male tidbitting display)	1	1	1 or 2
Subtotals	5	3	3 or 4
II. Avoidance of enemies (both sexes)			
A. Flying predator alarm	1	1	1
B. Ground predator alarm	1	1	1
C. General alarm & escape	2*	2	1
D. All's well (male only)	—	—	1
E. Hand-held distress	1*	1	(same as B)
Subtotals	5	5	4
III. Sexual and agonistic			
A. Unmated male advertisement	1	1	1
B. Waltzing display (male)	—	—	1
C. Aggressive (mostly male)	3	2	2
D. Submissive (male & female)	1	1	1
E. Nesting (male & female)	1	1	1
F. Copulation	1 (female)	1 (both sexes)	1 (male)
Subtotals	7	5	7
IV. Parental (both sexes)	2†	1†	—?†
TOTALS	19*	14	14–15

1. Based in part on Stokes, 1967 (who considers four variants as separate calls)
2. Based in part on Williams, 1969
3. Based in part on Stokes, 1961

* Plus additional variants

† Excluding calls from other categories above

is one of the few in which sounds and body movements are closely integrated into a complex display in the New World quails. In the bobwhite the corresponding call (called the caterwaul) is not associated with head movements, but is more complex acoustically. This species also has a conspicuous frontal threat posture involving wing-spreading that is less highly developed in *Callipepla.*

In the American quails and Old World partridges, unlike most grouse, vocalizations are typically associated with copulation. In the quail species studied so far, these calls are uttered by the female and sometimes also by the male during treading. In the Old World genus *Alectoris* the male utters a copulation-intention call. Choate (1960) has reported the only copulation calls by grouse known to me, and states that calling by both sexes occurs during treading in the white-tailed ptarmigan. Watson and Jenkins (1964) state that the male red grouse does not call until copulation is completed and that the female remains silent.

As indicated in table 25, some fourteen or more calls (Stokes, 1967, reports twenty-four for the bobwhite) can be detected in the quail and partridge species so far studied, more or less equally divided among the categories of general social activities, avoidance of enemies, and sexual and agonistic signals. Sexual dimorphism in quail vocalizations is restricted, being generally limited to calls that serve to advertise the presence of unmated males or which are given only by males in agonistic situations.

It is of interest to compare these quail vocabularies with some reported for grouse species. One of the most complete surveys of grouse vocalizations is that of Watson and Jenkins (1964) for the red grouse, which is summarized in table 26. For a contrast with the monogamous grouse, in which all of the calls are common to both sexes, two lek-forming species of prairie grouse are also included in the table. Data on the sharp-tailed grouse are based on the observations of Lumsden (1965), whose study did not include possible female parental calls but is otherwise apparently comprehensive. Vocalizations of the pinnated grouse are generally so similar to those of the sharp-tailed grouse that they can be comparably organized, but no single paper adequately summarizes the call repertoire of this species. Some parental calls are mentioned by Gross (in Bent, 1932), while Lehmann (1941) and various other authors have discussed the sexual and agonistic calls of pinnated grouse. Evidently no special calls in this species serve to announce the presence of enemies; the birds typically freeze or squat silently, not giving their alarm notes until taking flight (Hamerstrom, Berger, and Hamerstrom, 1965; Berger, Hamerstrom, and Hamerstrom, 1963). Lumsden (1965) reported a possible preflight alarm note in the sharp-tailed grouse, but indicated that three silent alarm postures are usually assumed by birds when they are disturbed.

In contrast to quail, it may be seen not only that the lek-forming grouse have virtually no flock or pair integration vocalizations and very few calls that serve to provide a general alarm but also that there are a large number of male agonistic and sexually related calls. These calls are generally uttered less frequently or not at all by females. Apparent intensity differences make it difficult to judge how many male calls should be recognized, but this is to be expected considering the close relationship between male social structure and reproductive efficiency in these species.

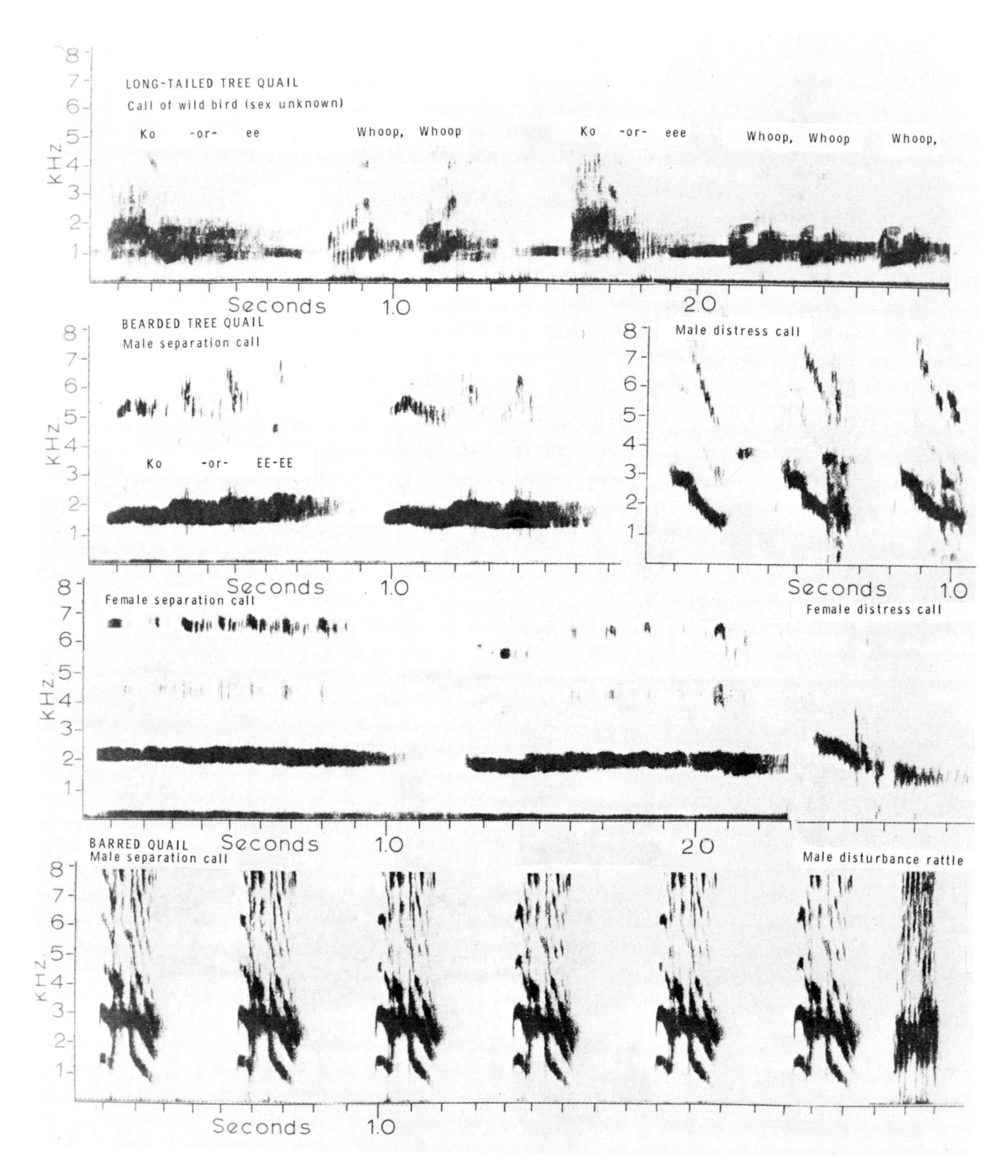

FIGURE 18. Representative sonagrams of calls typical of New World quails.

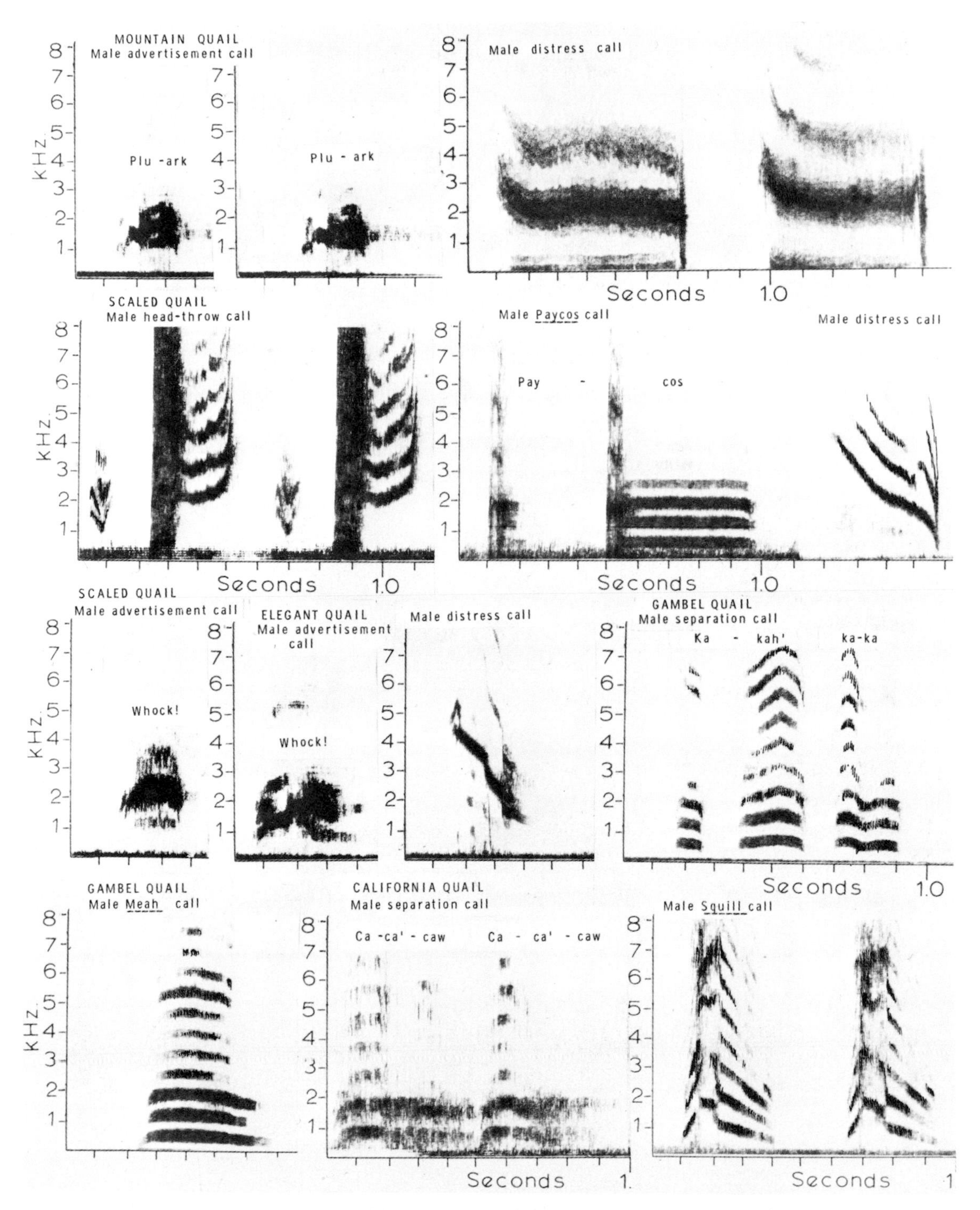

FIGURE 19. Representative sonagrams of calls typical of New World quails.

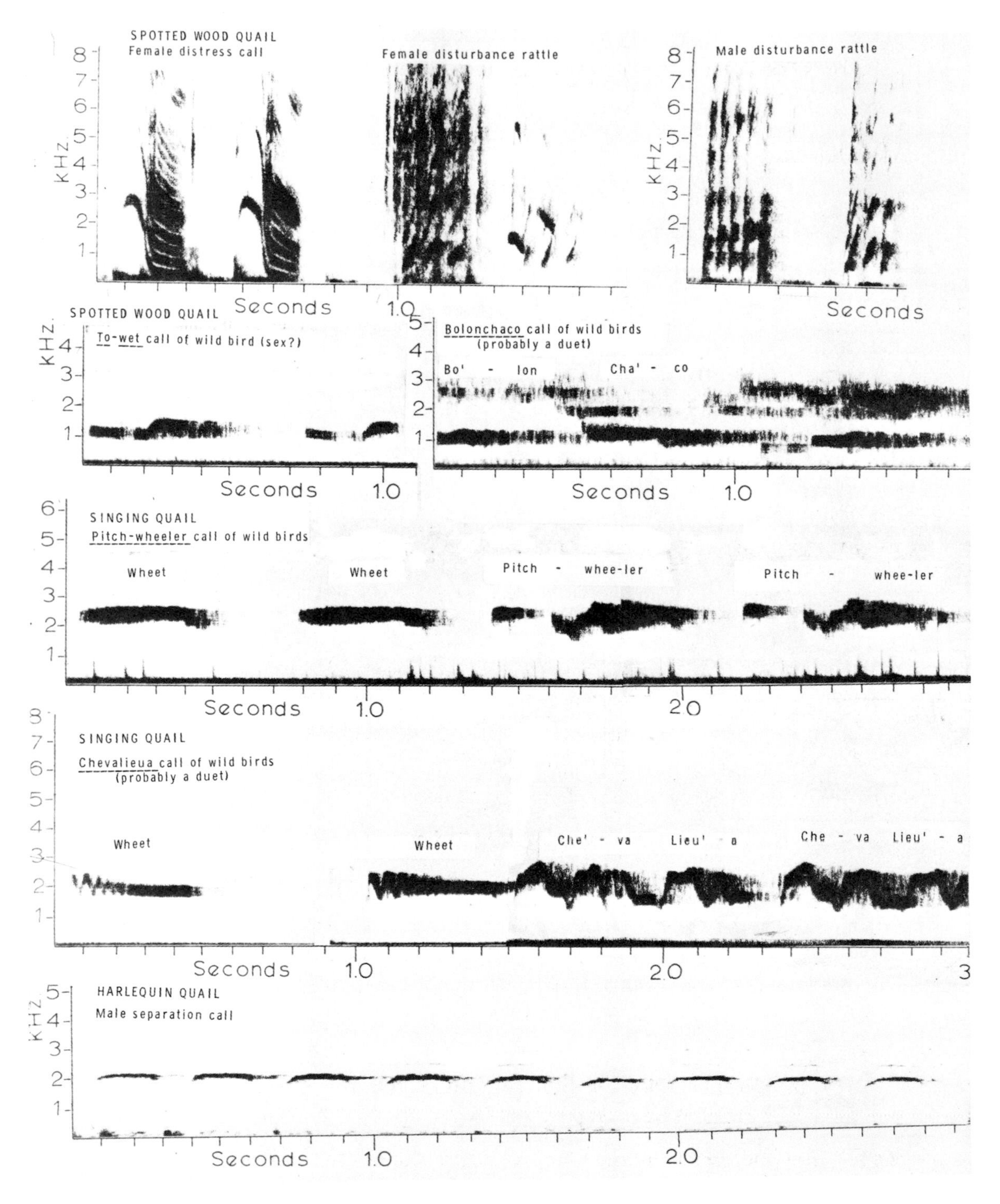

FIGURE 20. Representative sonagrams of calls typical of New World quails.

TABLE 26

Summary of Adult Vocalizations in Three Grouse Species

	Willow Ptarmigan (Red Grouse)[1]		*Sharp-tailed Grouse*[2]		*Pinnated Grouse*[3]
I. Flock or pair activities (both sexes)					
A. Flight intention	1				
B. Social contact	1				
Subtotal	2				
II. Avoidance of enemies (both sexes)					
A. Flying predator alarm	1		1?(preflight)		
B. In-flight alarm	—		1		1
C. Fleeing (& chase)	1				
D. Flying predator defense	1				
E. Hissing defense	1				
Subtotal	4		2		1
III. Sexual and agonistic					
A. Song (in flight/on ground) (both sexes)	2	A. Aggressive cackle (male & female)	1	A. Long cackle (mainly males)	1
B. Attack; attack-intention; threat (both sexes)	1- 3	B. Aggressive *Lock-a-lock* (both sexes)	1	B. Aggressive *Ca-ca'-caa* (males)	1
C. Sexual (both sexes)	1	C. Cooing (males)	1	C. Booming (males)	1
		D. Aggressive whine (males)	1	D. *Kwier* whine (males)	1
		E. Squeal & cork calls (males)	2	E. *Kliee/Kwaa/Kwah* calls (males)	1*
		F. *Chilk* & *Cha* (males)	2	F. *Pwiek/Pwark/Pwk* calls (males)	1*
		G. *Pow* (male courtship call)	1	G. *Pwoik* (male courtship call)	1
Subtotal	4- 6		9		5- 7
IV. Parental	2†		?		2- 3†
TOTALS	12-14		11+		8-11

1. Based on Watson & Jenkins (1964), all calls uttered by both sexes
2. Based on Lumsden (1965), female parental calls not included in study
3. Based on Gross (1928), Lehmann (1941) and personal observations

* Probably variants of whining and *pwoik* calls

† Excluding calls from other categories above

THE EVOLUTION OF SOCIAL SIGNALS IN PHEASANTS AND PARTRIDGES

Among the true pheasants and the Old World partridges the fundamental nucleus of galliform display patterns should be present, to which the kinds of social behavior found in the grouse, New World quails, and turkeys must somehow be related. Within this vast phasianine array of some 150 species, about two-thirds of the species are regarded as quails, partridges, or francolins (Perdicini), while the remainder comprise the true pheasants and peafowl (Phasianini). In addition to being generally larger and having more prevalent sexual differences in plumage and morphology, male pheasants are also usually crested and iridescent and have ornamental tails of various shapes, lengths, and patterns; and their feet are usually spurred. However, no single character unequivocally separates the pheasants from the partridge-like species, and indeed the pheasant group may actually be of polyphyletic origin, simply including those phasianine species that have for the most part abandoned monogamous mating characteristics for polygamous or promiscuous ones. To mention only one example of doubtful tribal relationships, there is a remarkable similarity between the downy young of the blood pheasants (*Ithaginis*) and those of the snow partridge (*Lerwa*) that is certainly suggestive of close affinities. It is also possible that similar male plumage characteristics have evolved independently in distantly related pheasant lines and have obscured phyletic relationships. It would thus seem that downy, juvenile, and female plumages might provide the best morphological indices of relationships and bases for generic recognition, with information on hybrid viability, fertility, and chromosomal or biochemical evidence useful supporting data. Male displays are so subject to selective pressures for species isolation that they are useless for such classification purposes, although they are nonetheless of interest in their diversity and their relationships to male plumage development and signal functions.

In spite of the remarkable species diversity to be found in male plumage patterns of the pheasants and their relatives, a surprising degree of similarity in the display motor patterns can be detected (Schenkel, 1958). Functions and motivations of these motor elements have no doubt been greatly modified to fit ecological needs or other adaptations, but nonetheless the display patterns to be found among pheasants, partridges, quails, and grouse are basically so similar as to suggest that fairly close evolutionary relationships may exist among the entire group. It is, for example, most difficult to find specific display features that can be used to separate these into tribes, subfamilies, or families according to their taxonomic treatment.

Starting with the uncertain but reasonable assumption that the partridges and true quails are more generalized in behavior and morphology than are the pheasants, the behavior of such well-studied Old World genera as *Coturnix*, *Perdix*, and *Alectoris* may perhaps serve as representative of this large group. Sexual plumage dimorphism is fairly slight in these forms, and species-specific display features would appear to be centered in the face, throat, breast, and flank regions. The tail and wings are for the most part specialized neither in pattern nor in shape, and in general do not contribute significantly to display. In at least two genera (*Alectoris* and *Excalfactoria*) a lateral display is present in which one wing is drooped, but in both these genera the wing involved is the one away from the object of the display, and thus the flank feathers are rendered more conspicuous (Harrison, 1965, Goodwin, 1953). Indeed, in such species the function of lowering the farther wing may simply be maintaining balance (Goodwin, 1953). Apparently, strong wing-lowering during lateral display is absent both in *Coturnix* and *Perdix*, which interestingly both lack specialized flank coloration. Throat patterning is well developed in *Excalfactoria*, *Alectoris*, and *Coturnix* and is probably displayed during calling or frontal displays in all these forms. The taxonomic distribution of the tidbitting display among the partridge-like forms is uncertain and seems to be unrelated to plumage morphology, but it occurs at least in *Coturnix*, (Schenkel, 1956), *Alectoris* (Goodwin, 1953; Stokes, 1961), and *Excalfactoria* (Harrison, 1965).

Judging from the observations of Stokes (1961) and Goodwin (1958), the genus *Alectoris* possesses several basic phasianid display elements, including lateral display and tidbitting. Tidbitting serves in this genus both as a low-level aggressive signal between males and as an important sexual signal of males toward females. The associated tidbitting call is also used by both sexes in directing their young to food. Other pheasant-like display postures include wing-flapping, a high-stepping posture, and a "rear approach" of the male to the female for copulation. Representative displays of *Alectoris* and *Perdix* are illustrated in figure 21.

The early studies on the behavior of the domestic form of red jungle fowl (*Gallus gallus*) have provided much of the basic terminology used to describe pheasant display patterns, and thus the domestic fowl might be considered a "type" example of phasianine display patterns. Some of the most complete studies on the behavior of the domestic fowl are those of Wood-Gush (1954, 1956). He reported that nearly all the male postures are used both in agonistic and courtship situations. As might be expected in a polygamous or promiscuous species, the female exhibits very few of these same displays and instead performs submissive or appeasement gestures. Apart from overt fighting and retreating, males perform a number of other gestures that probably reflect varying degrees of conflicting tendencies to attack, escape, or react sexually, according to ethological theory. One of these displays is "high-stepping," which is performed by the male in an erect stance as he advances on his opponent. During strutting the male droops both wings and raises his tail and ruff slightly. Stationary

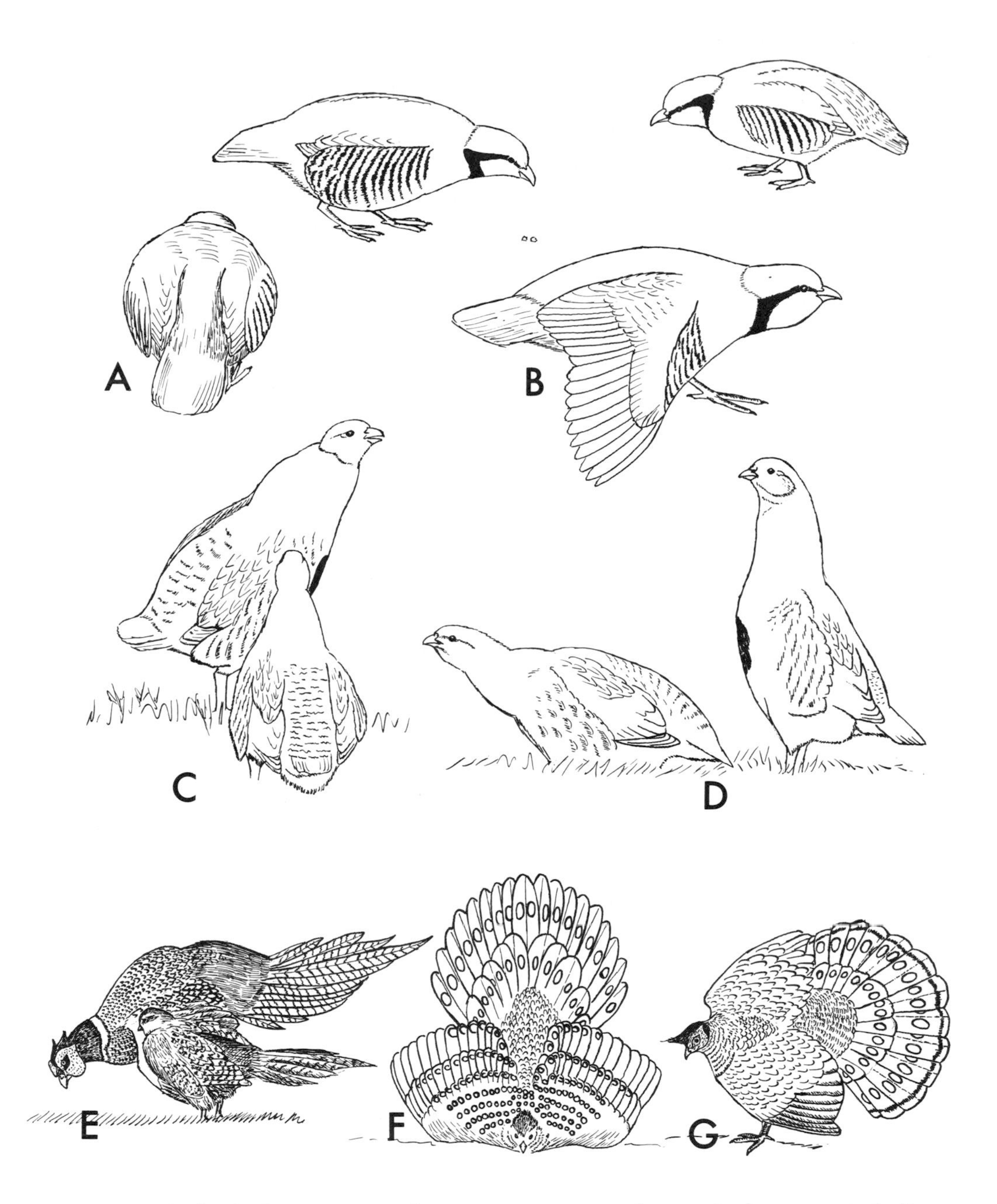

FIGURE 21. Male display postures of representative partridges and pheasants, including (A) chukar partridge tidbitting, (B) chukar partridge waltzing, (C) gray partridge in lateral courtship display, (D) gray partridge in precopulatory display, (E) ring-necked pheasant performing lateral wing display, (F) gray peacock pheasant in frontal display, and (G) Palawan peacock pheasant in lateral display. (After various sources.)

wing-flapping is performed with or without calling but with loud wing-clapping. A major display of domestic fowl is "waltzing" (Davis and Domm, 1942), which is composed of several components. These include circling the other bird, lateral display of the flanks and often the tail, and a wing display achieved by lowering the off-side wing towards the ground. Kruijt (1962) suggested that the evolutionary origin of wing display during waltzing resulted from a compromise of motor patterns reflecting tendencies to flap the wings aggressively and to fold the wings in association with escaping. Unlike the situation in partridges, wing display of pheasants seems to be limited largely to males, but it is present in females of the genus *Pucrasia* (Wayre, 1964).

Another basic phasianid display performed by domestic fowl is tidbitting (Domm, 1927), consisting of ground-pecking intention movements, which in some species are supplemented by calls. This may have had its evolutionary origin as aggressive pecking movements that are redirected toward the substrate, but in many species this activity has evolved into an important method of pointing out food sources to the young or the mate. Schenkel (1956) has described how the basic movements and calls as found in *Gallus* and *Phasianus* are increasingly modified through ritualization in *Polyplectron*, *Lophophorus*, and perhaps *Tragopan*, and are finally represented by the elaborate frontal display of *Pavo*, which typically occurs in the absence of actual food.

Other components of male agonistic and sexual display of the domestic fowl reported by Wood-Gush (1954, 1956) include ruffling the neck feathers, circular head-shaking, tail-wagging, preening, and whining. Two strictly sexual elements include "cornering," in which the male moves away from the female, partially crouches, and scratches or stamps with his feet (Kruijt, 1962). Stokes (1961) believes that cornering should be considered ceremonial nest-building, and Kruijt has made a similar suggestion. Finally, males perform the rear approach, in which the bird attempts to mount the female from behind. The domestic fowl lacks a well-developed frontal display, but during high-intensity threat the male exhibits "two-sided wing-lowering," while raising its ruff and directly facing its opponent (Kruijt, 1962).

Male displays of the pheasant species have been separated into two major classes, lateral and frontal (Beebe, 1926; Pocock, 1911). The lateral, or one-sided, display has also been called waltzing and wing display, and consists of several interrelated components. These include both a lateral orientation to the object of the display, and a variable lowering of one wing which except in the genera *Gallus* and *Pucrasia* (Wayre, 1964) is apparently always the nearer wing among the species of true pheasants. The tail is also usually raised, spread, or tilted, or combinations of these may occur, and the body may be tilted toward the object of the display, making the upper body surface and tail a major focus of attention for specific display features. Finally, there is a circling around the other bird, which may take the form of a rapid forward running or hopping (*Polyplectron*, *Rheinardia*, and *Chrysolophus*), sideways hopping movements (*Syrmaticus reevesii*), the appearance of a somewhat drunken waltz (*Gallus*), or a slow and stately walk (*Phasianus*, *Lophura*, and *Tragopan*). Published descriptions of these movements are not always clear, and intermediate or compound situations no doubt occur; thus the great argus (*Argusianus argus*) is said to begin displaying with a circular walk and strong foot-stamping around the female, then it suddenly rushes past her while performing lateral wing display (Seth-Smith, 1925). The final stage consists of stopping, opening and erecting one wing, then opening both wings and facing the female in the climactic frontal display (Lint, 1965). Pocock (1911) astutely discerned the significance of the asymmetry of the lateral display as an evolutionary precursor to the elaborate frontal display of several pheasant species. He points out that in *Polyplectron bicalcaratum* (and, as later reported, in *Polyplectron emphanum*) not only is the tail spread and tilted but also the more distant wing is raised and tilted in a manner that exposes the ocellated dorsal patterning. This essentially dorsal-lateral display may thus readily be modified into the typical frontal display, by the bird's stopping, calling the female with the tidbitting call, lowering its head to the ground, and orienting both wings and the vertically spread tail directly toward the hen. This tremendously impressive display places the burden of signal features on the upper wing surface, especially the secondaries, and on the tail and helps to account for the fantastic development of these feather areas in the great argus. In contrast, the crested argus (*Rheinardia ocellata*) lacks a clear-cut frontal display, and its long tail feathers are simply raised and spread vertically during the lateral display while both wings are lowered (Huxley and Bond, 1942). This species lacks specialized wing and mantle patterning, such as iridescence or ocelli, and the tail, although extremely long, is not modified in shape or coloration for frontal display.

It may thus be seen that the lateral display provides the probable evolutionary basis for the frontal display, which gains equal or greater importance in *Polyplectron* and *Argus* and finally altogether replaces it in peafowl (*Pavo*). It should be noted here that at least one other genus has a very similar frontal display, namely the monals (*Lophophorus*). Literature descriptions have not permitted me to judge whether the motor origin of the frontal display of these species corresponds to that of the genera just mentioned, but it is known that a similar lateral display with associated drooping of the near wing precedes the frontal display (Delacour, 1951). Lastly, in the peafowl and apparently also in the African peacock (*Afropavo*), there is no lateral display at all. Kruijt (1962) suggests that wing movements during frontal display of the Indian peafowl (*P. cristatus*) may represent a derivation of wing-shaking or wing-flapping, but there is no trace of asymmetry in the display and the focus of display features has centered on the back and tail coverts instead of on the wings or tail. In addition, since the head is not hidden behind the wings during frontal display as in the argus but is visible and held upright, the entire head and neck region have also become highly iridescent and specialized. The plumage and morphology of the African peacock likewise are correlated with display, during which the male and female sit on branches facing each other and bow their heads deeply, with their spread-out tails raised to an angle of forty-five degrees (Verheyen, 1962).

In addition to lateral and frontal displays, male pheasants exhibit a variety of other male display postures and movements (figure 21). Wing-flapping, such as might occur possibly as a displacement activity in many species, is highly ritualized in *Lophura* and *Syrmaticus* (Delacour, 1951), in both of which whirring sounds are generated. An actual display flight by males is evidently present in *Lophophorus* (Ali and Ripley, 1969). Shaking of the vaulted and often distinctively patterned tail occurs in *Lophura* and *Crysolophus*, and this exists in a modified version as vibration of the peafowl's erect train. Engorgement of the brightly colored bare facial skins occurs in several genera such as *Phasianus* and *Gallus* and reaches a maximum in the display of the Bulwer pheasant (*Lophura bulweri*). Male tragopans exhibit a rather different form of facial and throat engorgement, and in contrast to the forms just mentioned the males display them in a distinctly frontal orientation. Tidbitting not only occurs in *Gallus*, in which it was first described as such, but also in *Polyplectron* (Seth-Smith, 1914) and probably other genera. Schenkel (1956, 1958) has extensively summarized other evolutionary aspects of pheasant displays, particularly male calls, which have not been considered here. It would seem that in general the pheasants exhibit a much greater degree of conservatism in motor patterns than in the morphological features exhibited by these motor patterns, thus the same display performed identically by two species may be rendered species-specific by differences in male plumage characteristics.

INTERGROUP DISPLAY HOMOLOGIES

Although it is fairly safe to assume that lateral display with wing-lowering and the other similar postures of the Old World partridges are homologous to those of pheasants, it is more difficult and dangerous to make such comparisons between the pheasants and the New World quails. Although in the American species lateral display is certainly a fundamental aspect of both agonistic and sexual behavior, this may or may not be associated with a circling of the other bird, and in no species has an asymmetric wing-lowering been described. Rather, as in partridges, the flank feathers seem to be the center of signal concentration for lateral displays, and these are often asymmetrically fluffed on the side toward the opponent male or the female. Wing-flapping is common during agonistic situations among New World quails, and tidbitting also plays a major role in the pair-forming processes of several species. Also in common with the Old World partridges, tidbitting calls are used by both sexes to attract the young to food.

Male display postures of the grouse also show a considerable number of similarities to those of typical pheasants, some of which are undoubtedly only superficial. The stationary wing-flapping of ruffed grouse, for example, should probably be regarded as a modification of aerial display rather than homologous with the wing-flapping associated with crowing in various pheasant species. Tail-cocking and tail-spreading displays occur in virtually all species, but it is questionable whether these postures are equivalent in a homologous sense to similar displays of male pheasants. Symmetrical wing-drooping with tail-fanning and an associated strutting is particularly well developed in ptarmigans and *Dendragapus* species (Brooks, 1926; MacDonald, 1968), and in these species the postures closely resemble those of various pheasants. This similarity is increased in ptarmigans, which perform a waltz-like circling display, during which the wing nearer the other bird is strongly drooped, the spread tail is tilted, and the displaying bird circles the other closely while performing high-stepping movements (Watson and Jenkins, 1964). Somewhat similar asymmetrical lateral display with slightly drooped wings and a widely spread, tilted tail may be seen in the ruffed grouse (see figure 16), but in this species there are no associated circling movements. A circling of the female without asymmetrical wing-lowering occurs in both sharp-tailed and pinnated grouse as well as in black grouse (Hamerstrom and Hamerstrom, 1960). Tidbitting has apparently not yet been reported for any grouse species, although C. Braun (cited by Schmidt, 1969) observed probable tidbitting as a precopulatory display in white-tailed ptarmigan.

In conclusion, it might be mentioned that a number of male displays of the common turkey (*Meleagris gallopavo*) are strikingly similar to the strutting postures of various grouse as well as to the displays of peafowl. Turkey displays include tail-cocking and tail-spreading, symmetrical wing-drooping, and short forward runs associated with breathing sounds somewhat like those of *Dendragapus.* Although it is obviously not valid to use male displays as a basis for major evolutionary conclusions, the turkey's grouse-like sexual behavior provides no contradictory evidence to the view that the New World turkeys and the grouse might have evolved from common cracid-like ancestors inasmuch as there are no known fossil remains of typical pheasants from North America.

8

Aviculture and Propagation

THE rearing of grouse and quail for enjoyment, profit, or stocking in the wild has been an important aspect of grouse and quail biology. The very presence of chukar and gray partridges in North America, the occurrence of ruffed grouse in Newfoundland and Nevada, the presence of bobwhites, scaled quail, and California quail in Washington, and many other examples are ample testimony to the potential value of careful propagation and release programs. Between 1938 and 1968 a total of 110,663 bobwhites, 18,136 other native quails, 7,977 grouse, and 50,568 chukar partridges were released under Pittman-Robertson programs in the United States (based on a recent summary provided by the Bureau of Sports Fisheries and Wildlife). An additional but unspecified number of gray partridges was also part of the release program. Yeatter (1935) estimated that more than 260,000 of these birds had been released in North America by the 1930s.

The problems of keeping and breeding grouse in captivity are distinctly different from and much greater than those of propagating quails and partridges, and as a result relatively few persons have succeeded in keeping and breeding grouse in large numbers or with consistent success. This is largely a reflection of the greater sensitivity of grouse to various poultry diseases and parasites that are transmitted by ground contact, forcing the game breeder to keep the birds on wire-bottom cages where they can have no direct contact with the ground or their own droppings. A summary of the diseases and parasites of grouse and quails has been provided by Bump et al. (1947) and Stoddard (1931), respectively, although the treatments recommended have been greatly modified in more recent years.

Flieg (unpublished ms.) has summarized the difficulties of keeping grouse (and, to a lesser extent, quail) on the ground and the treatment or preventative measures for the most commonly encountered diseases and parasites. These include coccidiosis, enteritis, caecal worms, blackhead, and capillaria worms. Coccidiosis is caused by a protozoan parasite (*Eimeria*) that is a serious problem with both quails and grouse, but it can be prevented by adding Amprolium to the diet at the rate of three-fourths cup to twenty-five pounds of feed and can be treated with Sulmet. Intestinal inflammation, or enteritis, can be avoided by adding NF-180 to the food in the amount of one ounce per twenty-five pounds of food, although this reduces male fertility and therefore must be discontinued during the breeding season. Caecal worms (*Heterakis*) are probably more serious in grouse than in quail because of the more highly developed caeca of grouse, and a serious infection can be lethal. The use of Hygromix at the rate of one ounce to twenty-five pounds of food serves as an effective treatment for these worms as well as most other worm parasites. A related infection is enterohepatitis or blackhead (caused by *Histomonas*), which is often carried by *Heterakis* and affects both the liver and digestive tract. A preventive measure is Emtryl at the rate of three teaspoons per twenty-five pounds of feed, and higher doses can be used for treatment.

Probably the worst enemy of grouse in captivity is the cropworm (*Capillaria*), which, although not usually a serious threat to wild grouse, may cause severe losses in captive birds. It has been reported in the ruffed grouse, rock ptarmigan, sharp-tailed grouse, and pinnated grouse (Braun and Willers, 1967). It is apparently less serious in quail but has been reported to occur (Hobmaier, 1932). Flieg reported that one ounce of vitamin A premix to twenty-five pounds of food may be used to prevent and partially control cropworm, while a much more dangerous drug, Task, will serve as a more thoroughly effective treatment if used with extreme care.

Pullorum disease, a bacterial infection caused by *Salmonella*, and aspergillosis, a fungus disease of the respiratory tract, are other serious problems for the person who keeps grouse and quail. Both of these present difficult treatment problems, but Flieg reported some success in treating *Salmonella* infections with antibiotics such as Neomycin and Cosa Terramycin. Aspergillosis and similar fungal diseases may be avoided by adding copper sulfate to the drinking water or by treatment with a product of Vineland Poultry Laboratories called Copper-K, a combination of acidified copper sulfate and synthetic vitamin K (Allen, 1968). Staphylococcus infections can sometimes be treated effectively with Tylocine and B-complex vitamin preparations (McEwen, Knapp, and Hilliard, 1969).

Many of these problems can be avoided or minimized by keeping the birds on wire, but this poses new problems of providing grit and dusting places for feather maintenance and if the floor is unsteady may also reduce the probability of effective fertilization during copulation. The absence of natural vegetation for hiding and nest-building may further inhibit reproductive success in birds maintained on wire-bottom cages.

General principles of breeding game birds, especially quail and partridges, have been summarized well by Greenberg (1949). It is impossible to summarize all of the points made by him in the space available here, and only a few highlights might be mentioned.

EGG CARE AND INCUBATION

Eggs should not be held longer than a week before being placed in the incubator, and during storage they should be kept at a temperature of between 50 and 60 degrees Fahrenheit and a relative humidity of about 80–90 percent. Placing the eggs in plastic bags during storage improves their hatchability (Howes, 1968; Kealy, 1970), and they should be stored with the pointed end down. Tilting them or turning them daily during the preincubation storage period is also desirable. Incubation may be done in either a still air or forced air incubator, with the latter being generally preferred although considerably more expensive. In either case, the eggs should be rotated ninety degrees every three to six hours, or at a similar regimen, until the last few days of incubation when they are moved to hatching trays. Ideal incubation temperatures differ with the incubator type. Romanoff, Bump, and Holm (1938) stated that the ideal temperature for incubating bobwhite eggs is 103 degrees in still air incubators (60–65 percent relative humidity) during the first two weeks, and 99.5 degrees in forced air incubators (similar relative humidity) during that period. During the last two or three days of incubation the temperature should be slightly higher (0.25 to 0.5 degree) for best results, and there should be an increase in the availability of fresh air. Depending on the species, the final humidity should either be somewhat lower or higher than earlier in incubation, with higher humidity generally recommended for quail eggs. Chukar and gray partridge eggs are usually put in a still air incubator for the last few days (103 degrees) at a slightly higher humidity.

In their studies of prairie grouse, McEwen, Knapp, and Hilliard (1969) found that hand-turned incubators were unsatisfactory for grouse eggs and recommended using an incubator with automatic turning and a temperature for the first three weeks of 99.75 degrees F., with a wet-bulb reading of 82–86 degrees F. After the eggs are placed in the hatching incubator, they are held at a temperature of 99.5 degrees F. and a wet-bulb reading of 90–94 degrees F. Moss (1969) reported that ptarmigan eggs could successfully be hatched in a still air incubator provided that the humidity was held as high as possible. Bump et al. (1947) reported that still air incubators were preferred over forced air models for incubating ruffed grouse eggs. They recommended an incubation temperature of 103 degrees F. for still air models and 99.5 degrees for forced air machines, and a 60–65 percent relative humidity, with eggs being turned three to four times a day during the first twenty days. During the last few days of incubation the humidity and temperatures should be maintained at these same levels.

CHICK CARE

Following hatching, chicks must be provided with supplemental heat, either in the form of broody hens as foster mothers or artificial brooders. For artificial brooders, newly hatched chicks should initially be exposed to a brooder temperature of 95 degrees F., which is gradually reduced so that by the time the birds are about two weeks old the brooder temperature is around 70 degrees F. Newly hatched chicks should be provided with a high-protein food such as chick starter and in addition may benefit from finely cut fresh green leaves such as lettuce, endives, or dandelion. For many delicate species, the availability of live insect food such as meal worms (*Tenebrio*) may be crucial in inducing the young to begin eating. Shoemaker (1961) found that coating the worms with a vitamin-mineral concentrate avoided weakness in the legs (perosis), generally thought to be related to manganese deficiency. Dellinger (1967) indicated that he was able to stimulate feeding in harlequin quail chicks by sprinkling Purina Startina with hard-boiled eggs and finely chopped greens on a paper towel, to which he added small live and chopped up meal worms. For water, he recommended jar lids filled with water and marbles, with one-half teaspoon of Furacin or Terramycin added per quart of water as a disease preventative. Coats (1955) dipped meal worms into egg yolk or corn syrup, then dusted them with high protein starter mash, to initiate chick feeding.

Problems that might be encountered in the raising of grouse chicks have been discussed by a number of writers, including McEwen, Knapp, and Hilliard (1969), Fay (1963), and Bump et al. (1947). Fay recommended an initial brooder temperature of from 100 to 105 degrees at chick-level. He used various game bird starter feeds, as well as limited amounts of fresh green material. He also added soluble Terramycin to the drinking water at the rate of one teaspoon of powder to two gallons of water. McEwen, Knapp, and Hilliard found that water could effectively be provided to young chicks without the danger of drowning by using dripping siphon tubes at the eye-level of the chicks. The rate of dripping can be controlled by clamps, and the water falls through the mesh floor to be caught below the cage.

Howes (1968) recommended vaccination of young chicks for bronchitis and Newcastle disease if the birds were kept near other poultry by adding the vaccine to the drinking water. He also advised vaccination against pox.

Cannibalism, the pecking of chicks by one another, is frequently a serious problem, especially where crowding is necessary. Such pecking may be reduced by providing sufficient grit, a source of greens or other roughage at which the birds can peck, and a balanced diet. Trimming of the beak may also be necessary to prevent serious damage or even death when pecking becomes a major problem.

CARE AND HOUSING OF ADULTS

In the case of quail, considerable numbers of adults can usually be maintained in fairly small pens, although breeding is no doubt more successful when paired birds can be individually housed. The well-known McCarty pens, described by Greenberg (1949), provide a proved method for housing and breeding quail, chukar partridge, and gray partridge and are probably also suitable for certain grouse. Bobwhite quail can be effectively housed in such breeding compartments with two females per male; in spite of their monogamous pair bonds under natural conditions two females will readily tolerate each other in captivity. Recently a technique for artificial insemination of bobwhites has been developed (Kulenkamp, Coleman, and Ernst, 1967) which has produced fertility and hatchability rates as high as those achieved with natural mating.

Minimum space requirements for grouse are considerably greater than those for quail, because of both the generally larger sizes of the birds and their reduced social tolerance. McEwen, Knapp, and Hilliard (1969) recommended that at least thirty square feet of floor space per bird was required for minimizing conflicts among prairie grouse. Thus, a five-by-eighteen-foot pen would accommodate a maximum of one male and two female grouse, and a ten-by-eighteen-foot pen could serve for up to four or five birds. It is important when keeping grouse to provide enough natural cover or artificial hiding places for the female to retreat to when the male begins to become highly aggressive during the breeding season (Moss, 1969). McEwen,

Knapp, and Hilliard (1969) recommended the use of dusting boxes (with 5 percent Rotenone powder added) to control external parasites.

Probably the most complete summary of the problems of maintaining grouse in captivity is that provided by Bump et al. (1947) for the ruffed grouse. No doubt many of the techniques described for the ruffed grouse are equally applicable to other species. They found that breeding pens measuring 6 by 8 feet wide and 3 feet high were adequate for a single pair of grouse, with one end of the pen enclosed and the other end open wire mesh. They also noted that up to twenty birds could be maintained in pens measuring 8 by 32 feet, especially if ten-inch-high cross-boards were placed at 4-foot intervals to help establish territorial boundaries. A wintering flight pen measuring 25 by 110 feet was judged able to hold up to three hundred full-winged grouse and was constructed around a service room that facilitated feeding and watering the birds.

Greenberg (1949) has summarized the techniques generally used for the propagation of chukar and gray partridges. Elevated wire mesh breeding pens that have a three-by-eight-foot bottom area and are attached to a coop measuring forty by thirty-six inches are recommended for gray partridges. One of these breeding pens is presumably designed for a single pair of partridges, but in all likelihood an extra female could be added without seriously affecting fertility. Studies on chukar partridges summarized by Christensen (1970) indicate that a breeding ratio of three females per male was as effective as using only one or two females per male. Ground pens were generally more satisfactory than elevated ones with wire floors. Fertility and hatchability of eggs from birds that were two years old were higher than in those of younger birds.

Artificial lighting will stimulate earlier and increased egg production in chukar partridge as well as in most American quail species. Studies on the bobwhite (Kirkpatrick, 1955; Kirkpatrick and Leopold, 1952) indicate that artificial illumination of seventeen-hour photoperiods with as little as 0.1 foot-candles will produce egg-laying in bobwhites in from fourteen to forty-four days after the initiation of such lighting. Observations in my laboratory indicate that the scaled, Gambel, and California quails are also stimulated into reproductive activity by increased photoperiods.

RECORDS OF INITIAL PROPAGATION OF GROUSE AND QUAILS

Some species of American quails have been kept in captivity for so long that the earliest date of their propagation under such conditions is unknown. This would certainly apply to the bobwhite, Gambel quail, and California quail. Audubon quoted an account by John Bachman of an early success in propagation of the bobwhite in captivity, presumably in the early 1800s. Seth-Smith (1929a) reported that the scaled quail was first bred in London in 1913 and that the black-breasted bobwhite (*Colinus virginianus pectoralis*) was bred in 1912. This race of bobwhite and the elegant quail were first imported for the London Zoo in 1911 (*Proceedings Zoological Society of London*, 1912, p. 3). He also noted that the California quail had often been bred in English aviaries, and the elegant quail had also been bred in the London Zoological Gardens. The elegant quail was first bred there in 1912 (*Proceedings Zoological Society of London*, 1912, p. 911), probably its first propagation in captivity. The harlequin quail was perhaps first bred in France, in 1911 (Seth-Smith, 1929a). In the United States it was probably first bred by J. S. Ligon.*

The barred quail was apparently first imported into England in March of 1927 (*Proceedings Zoological Society of London*, 1927, p. 490) and was first imported into the United States in 1933 by K. C. Beck.† There is no definite record that either importation resulted in the birds' breeding. F. E. Strange obtained a pair of wild-caught birds in 1967, and these laid eggs during the next four years, with chicks first being hatched and reared in 1967.

Avicultural data on the remaining species of native quails are limited. The mountain quail has doubtless been kept in captivity for many decades, but I can find no definite record of the earliest breeding success in captivity. Grinnell et al. (1918) reported unsuccessful breeding attempts in the early 1900s, and F. E. Strange‡ first bred mountain quail in the 1930s, as did Ezra (1938). Recent summaries of mountain quail breeding techniques were provided by Schlotthauer (1967) and Bateman (1968).

For the endemic species of Mexican quails there is little information as to the extent of importation and propagation. The black-throated bobwhite was not listed by Seth-Smith (1929a) as having by then been imported into England, and the first record of importation into the United States that I know of was by C. H. Epps, Silverhill, Alabama, in the late 1960s. In 1970 I exported ten of these birds from Mexico, and during the same year F. E. Strange obtained three additional birds from a Mexican source. I have since hatched and reared young of this species.

I can find no record of the singing quail's having been exported from Mexico, and I know of only one instance of its ever having been kept in a zoo. A young bird was brought to the zoo at Tuxtla Gutierrez, where

*F. E. Strange, 1971: personal communication.
†F. E. Strange, 1971: personal communication.
‡F. E. Strange, 1971: personal communication.

it lived about three or four months.*

The spotted wood quail has probably only rarely been imported into the United States and has apparently never been bred in captivity. There was a single bird in the National Zoological Park in the late 1960s,† and F. E. Strange obtained a pair from Mexico in 1970. They are fairly commonly kept as cage birds by natives in some parts of Mexico, and one individual lived for twelve years in the Tuxtla Gutierrez zoo.‡

There are even fewer records of tree quails' being successfully exported from Mexico. John O'Neill§ informed me that he once saw and photographed a bearded tree quail in the Houston, Texas, zoo, and a buffy-crowned tree quail is presently in the San Diego zoo. I successfully exported five bearded tree quails from Mexico in 1970, and they are now in the care of F. E. Strange, Torrance, California. In 1971 one of these pairs constructed a nest and laid a total of sixteen eggs, from which six young hatched and four were raised, the first known breeding of any tree quail in captivity.

As an index to the relative frequency of quail breeding in captivity, the number of successful propagations listed in the first ten volumes of the *International Zoo Yearbook* have been totaled. These include fifty-two breedings recorded for the California quail, forty-one for the bobwhite, nineteen for the Gambel quail, seven for the scaled quail, and two each for the mountain and harlequin quails. During the same period only two North American species of grouse were reported bred, with two breedings each recorded for the willow and rock ptarmigans. Obviously, these records are highly incomplete and exclude all the private aviculturalists, who are responsible for most of the successful breedings of these birds, but they do provide at least a rough measure of the relative ease of propagation for these species.

Records of successful propagation of North American grouse are far fewer. Perhaps the earliest record of any grouse's being successfully maintained in captivity is that of W. L. Bishop (quoted by Bendire, 1892), who kept spruce grouse in captivity for some time. At present, very few spruce grouse are in captivity, and the only recent rearing success was reported by Pendergast and Boag (1971).

Blue grouse are seen in captivity almost as infrequently as spruce grouse, and the earliest report of successful rearing of this species I am aware of is that of Simpson (1935). Smith and Buss (1963) hatched and reared four blue grouse through their juvenal stages, while Zwickel and Lance (1966)

*M. Alvarez del Toro, 1970: personal communication.
†Kerry Muller, 1970: personal communication.
‡M. Alvarez del Toro, 1970: personal communication.
§1970: personal communication.

hatched and reared twenty-seven chicks from eggs taken in the wild.

A few records of sage grouse propagation exist, including those of Batterson and Morse (1948) and Pyrah (1963, 1964), who hatched and raised birds from eggs taken in the wild.

Ruffed grouse have probably been raised in captivity more frequently than any other grouse species. Edminster (1947) reviewed the history of this species' propagation in captivity and noted that the first instance of rearing birds from eggs taken in the wild came in 1903 but that A. A. Allen developed the basic techniques needed for successful propagation during the 1920s. Later work by the state game biologists of New York resulted in the rearing of nearly two thousand grouse, including birds of the tenth generation.

Success in rearing and propagating ptarmigans has been quite limited. Seth-Smith (1929b) indicated that willow ptarmigan and the related red grouse were successfully reared in England during the early 1900s, but that the rock ptarmigan had only rarely been kept in captivity. By that time, the pinnated grouse, sharp-tailed grouse, and ruffed grouse had also been maintained in captivity in recent years (Carr, 1969), with rock ptarmigan having been reared from eggs to maturity, and the willow and rock ptarmigans surviving well in captivity after having been caught as adults in the wild. Moss (1969) has described techniques used for hatching and rearing ptarmigan from eggs taken in the wild. He reported success in breeding captive stock over a several-year period, so that breeders four or more generations removed from wild birds have been obtained.

One of the earliest persons to propagate pinnated grouse in captivity was J. J. Audubon, who obtained 60 wild-caught birds in Kentucky. He indicated that many of these birds laid eggs, and a number of young were produced. The history of recent attempts to propagate prairie grouse has been summarized by McEwen, Knapp, and Hilliard (1969), who noted that it is only recently that any real success has been attained with pinnated grouse and sharp-tailed grouse. They have maintained individual greater prairie chickens and sharp-tailed grouse in captivity many years, with one male sharp-tail at least seven years old still vigorous and breeding, and one male pinnated grouse attaining six years of age. From more than forty-four hundred eggs laid by captive birds, 375 pinnated and sharp-tailed grouse were reared by them. Some of the greatest success in rearing prairie grouse in captivity has been by Lemburg (1962). He has been rearing sharp-tailed grouse since 1960 and greater prairie chickens since 1965, and he began raising lesser prairie chickens in 1966. During the last few years he has raised an average of 60 to 70 prairie grouse per year, and in some years has raised as many as 100 birds.

9

Hunting, Recreation, and Conservation

THERE can be little doubt that the grouse and quail provide the most important and most popular targets for more than ten million small-game hunters every year in North America (National survey, 1965). In much of the southeast, to go "bird" hunting simply means a day in pursuit of bobwhites, and likewise in New England "pa'tridge" hunting is regarded as the premier sport of all upland game hunting. These two species, the bobwhite and ruffed grouse, in 1970 were hunted in forty-seven states and eight provinces and are without question the most important of all North American upland game species (table 27). Although neither species was hunted during 1970 in Arizona or South Dakota, both have been legal game in South Dakota in recent years, and masked bobwhites originally occurred in southern Arizona, where they are now being restocked. In addition, the bobwhite occurs over much of Mexico and is an important game species in that country.

In table 27 is presented a list of the grouse and quail occurring north of Mexico, as well as the states and provinces in which they could legally be hunted during the 1970–71 hunting season, based on information available to the author. Of course, the length of the season and the daily limits varied greatly in different areas and in a few instances the total season lasted only a day or two. However, the list does provide a method of estimating the relative importance of the species as game. On this basis alone, the ruffed grouse might be judged most important, while the bobwhite is almost as widely hunted. Other species that are currently hunted in ten or more states and provinces are the sage grouse, blue grouse, spruce grouse, sharp-tailed grouse, chukar partridge, and gray partridge.

TABLE 27

STATES AND PROVINCES WHERE GROUSE AND QUAIL WERE LEGAL GAME IN 1970

Species	States and provinces
Sage grouse:	California, Colorado, Idaho, Montana, Nevada, North Dakota, Oregon, South Dakota, Utah, Washington, Wyoming, Alberta
Blue grouse:	Alaska, California, Colorado, Idaho, Montana, Nevada, New Mexico, Oregon, Utah, Washington, Wyoming, Alberta, British Columbia
Spruce grouse:	Alaska, Idaho, Minnesota, Montana, Washington, Alberta, British Columbia, Manitoba, New Brunswick, Ontario, Quebec, Saskatchewan
Willow ptarmigan:	Alaska, Alberta, British Columbia, Manitoba, Newfoundland, Ontario, Quebec, Saskatchewan
Rock ptarmigan:	Alaska, British Columbia, Newfoundland, Quebec, Alberta, Manitoba, Ontario, Saskatchewan (rare to infrequent in last four provinces listed)
White-tailed ptarmigan:	Alaska, Colorado, Alberta, British Columbia
Ruffed grouse:	Alaska, California, Delaware, Georgia, Idaho, Illinois, Indiana, Iowa, Kentucky, Maine, Maryland, Massachusetts, Michigan, Minnesota, Montana, Nevada, New Hampshire, New Jersey, New York, North Carolina, North Dakota, Ohio, Oregon, Pennsylvania, Rhode Island, Tennessee, Utah, Vermont, Virginia, Washington, West Virginia, Wisconsin, Wyoming, Alberta, British Columbia, Manitoba, New Brunswick, Newfoundland, Nova Scotia, Ontario, Prince Edward Island, Quebec, Saskatchewan
Prairie chicken:	Kansas (greater), Nebraska (greater), New Mexico (lesser), Oklahoma (both), South Dakota (greater), Texas (lesser)

TABLE 27 — (continued)

Sharp-tailed grouse:	Alaska, Colorado, Idaho, Michigan, Minnesota, Montana, Nebraska, North Dakota, South Dakota, Washington, Wisconsin, Wyoming, Alberta, British Columbia, Manitoba, Ontario, Quebec, Saskatchewan
Mountain quail:	California, Idaho, Nevada, Oregon, Washington, British Columbia
Scaled quail:	Arizona, Colorado, Kansas, Nevada, New Mexico, Oklahoma, Texas, Washington
Gambel quail:	Arizona, California, Colorado, Idaho, Nevada, New Mexico, Texas, Utah
California quail:	California, Idaho, Nevada, Oregon, Utah, Washington
Bobwhite:	Alabama, Colorado, Connecticut, Delaware, Florida, Georgia, Idaho, Illinois, Indiana, Iowa, Kansas, Kentucky, Louisiana, Maryland, Massachusetts, Michigan, Mississippi, Missouri, Nebraska, New Hampshire, New Jersey, New Mexico, New York, North Carolina, Ohio, Oklahoma, Pennsylvania, Rhode Island, South Carolina, Tennessee, Texas, Vermont, Virginia, Washington, West Virginia, Wyoming, British Columbia, Ontario
Harlequin quail:	Arizona only. A few may be taken in New Mexico during the general quail season.
Gray partridge:	Idaho, Indiana, Iowa, Minnesota, Montana, Nevada, New York, North Dakota, Oregon, South Dakota, Utah, Washington, Wisconsin, Wyoming, Alberta, British Columbia, Manitoba, Nova Scotia, Ontario, Prince Edward Island, Quebec, Saskatchewan
Chukar partridge:	California, Colorado, Idaho, Montana, Nevada, Oregon, Utah, Washington, Wyoming, British Columbia

A more meaningful but much more difficult method of evaluating the sporting value of each species is to try to estimate the annual hunter kill for all the states and provinces in which it is legal game. Such estimates are regularly made by most but not all state and provincial game agencies, but since the techniques used for these estimates vary greatly, the accuracy of the estimates varies as well. Nevertheless, in the belief that an inexact estimate is better than none at all, I have attempted to gather annual hunter-kill estimates for all of the species concerned (table 28). In some cases these were derived from annual reports of the game agencies or from technical or semitechnical periodic publications of these agencies, while in others they represent unpublished estimates that are normally used for management purposes or other functions. Because of the diversity of origins of the data, these sources are not indicated in the table, and clearly the estimates should be regarded only as general ones, in spite of the fact that they are not usually rounded off to the nearest thousand. Wherever possible, I have used and averaged figures from a several-year period rather than listed the most recently available single-year's data, since, for grouse in particular, there tend to be major yearly variations in hunter success.

TABLE 28

SOME ESTIMATED RECENT STATE AND PROVINCE HARVESTS, UNITED STATES AND CANADA

Alabama:	2,160,603 bobwhites in 1967.
Alaska:	Average harvests from 1952 to 1957 plus 1961, 93,971 ptarmigan, 59,306 total grouse (blue, spruce, ruffed, and sharp-tailed).
Arizona:	6,000 harlequin quail in 1969, average of 40 chukar partridges from 1962 to 1967, and 1,541,978 total other quail (scaled and Gambel) in 1968.
Arkansas:	400,000 bobwhites in 1967.
California:	3,200 sage grouse in 1969, average of 3,471 blue and ruffed grouse, 73,471 chukar partridges, and 2,432,557 quail (mountain, Gambel, and California) from 1963 to 1969.
Colorado:	1968 estimated kill of 13,107 sage grouse, 27,251 blue grouse, 3,382 white-tailed ptarmigan, 2,612 sharp-tailed grouse, 28,127 scaled quail, 4,469 chukar partridges, and 25,249 other quail (Gambel and bobwhite).

TABLE 28—(continued)

Connecticut:	No data on ruffed grouse; a few bobwhites (and released chukars) are killed annually.
Delaware:	No data (bobwhite only).
Florida:	2,500,000 bobwhites in 1968.
Georgia:	2,498,587 bobwhites in 1968. The annual ruffed grouse kill is about 2,500.
Idaho:	81,700 sage grouse and 105,600 forest grouse (spruce and ruffed) in 1969. In 1968, 110,000 total quail (mountain, Gambel, California, and bobwhite), and in 1969, 171,200 chukar partridges and 64,700 gray partridges.
Illinois:	Average of 2,020,840 bobwhites between 1958 and 1967; average of 9,716 gray partridges from 1961 to 1967.
Indiana:	911 ruffed grouse in 1966; 550,000 bobwhites in 1967; average of 6,960 gray partridges from 1963 to 1964.
Iowa:	720 ruffed grouse in 1968; 750,000 bobwhites in 1967. The annual gray partridge kill averages about 12,000.
Kansas:	46,000 greater prairie chickens in 1967; 3,000,000 scaled quail and bobwhites in 1968. No data on lesser prairie chicken (season closed between 1936 and 1969, 3-day season held in 1970).
Kentucky:	Average of 996,000 bobwhites from 1964 to 1967. The annual ruffed grouse kill is usually 30,000–35,000.
Louisiana:	700,000 bobwhites in 1968.
Maine:	273,033 total grouse (ruffed and spruce) in 1968. Ruffed grouse kill from 1955 to 1960 averaged 185,000.
Maryland:	No data (bobwhite and ruffed grouse).
Massachusetts:	12,936 bobwhites in 1962. Average yearly kill of ruffed grouse estimated at from 65,000 to 75,000.
Michigan:	Average kill of 356,000 ruffed grouse from 1955 to 1960. Sharp-tailed grouse harvest of less than 500 in recent years. No data on bobwhite, which is hunted in only a few counties.
Minnesota:	560,000 ruffed grouse in 1969; 8,833 gray partridges in 1966. No data on spruce grouse. Average sharp-tailed grouse harvest between 1965 and 1969 was 11,000 birds.
Mississippi:	1,250,000 bobwhites in 1967.

TABLE 28—(continued)

Missouri:	2,810,000 bobwhites in 1967.
Montana:	Average harvests between 1964 and 1968 were: sage grouse, 48,964; blue grouse, 53,441; spruce grouse, 33,227; ruffed grouse, 56,408; sharp-tailed grouse, 88,067; chukar partridge, 3,235; gray partridge, 93,717.
Nebraska:	49,000 prairie grouse (pinnated and sharp-tail) in 1969. An estimated total of 15,000 pinnated grouse were taken in 1967.
Nevada:	In 1967 the estimated harvest was 7,300 sage grouse, 408 blue grouse, 49,000 chukar partridges (including some gray partridges), and 72,898 total quail (mountain, Gambel, and California).
New Hampshire:	No data (ruffed grouse only).
New Jersey:	110,000 ruffed grouse and 111,000 bobwhites in 1969.
New Mexico:	Between 1958 and 1968 the average harvest was 1,700 blue grouse, 1,100 pinnated grouse, and 202,000 total quail, including an estimated 162,000 scaled, 36,000 Gambel, and 4,000 bobwhites.
New York:	Average harvest of 409,450 ruffed grouse between 1966 and 1969. No data on bobwhites or gray partridges.
North Carolina:	63,043 ruffed grouse in 1964; 2,500,000 bobwhites in 1968.
North Dakota:	Sage grouse harvest in 1964 was 100–200 birds. In 1969 the harvest was 5,014 ruffed grouse, 109,255 sharp-tailed grouse, and 69,142 gray partridges.
Ohio:	16,600 bobwhites in 1969; annual ruffed grouse kill estimated to be about 5,000. No recent data on gray partridges, but harvest probably less than in 1959, when 5,400 were taken.
Oklahoma:	Average pinnated grouse harvest from 1959 through 1968 was 7,700. In 1968, 3,326,000 scaled quail and bobwhites were harvested, of which an estimated 3,000,000 were bobwhites.
Oregon:	1968 sage grouse harvest was 51,700 and forest grouse (blue, spruce, and ruffed) harvest was 143,300. Blue grouse harvest estimated at 24,476 in 1960. The 1968 total quail

TABLE 28—(continued)

	harvest (mountain, California, and bobwhite) was 216,638, plus 123,000 chukar partridges and 72,500 gray partridges.
Pennsylvania:	1969 harvest was 25,000 bobwhites and 280,000 ruffed grouse.
Rhode Island:	Average harvest from 1958 through 1959 was 290 bobwhites and 530 ruffed grouse.
South Carolina:	1968 harvest was 2,500,000 bobwhites. The annual ruffed grouse kill is only 100 to 250 birds.
South Dakota:	1969 harvest was 95,000 prairie grouse and 7,500 gray partridges. In 1967 the pinnated grouse kill was estimated to be 10,000. Sage grouse harvest in 1966 and 1967 about 2,000 birds. Bobwhite harvest 500 in 1959.
Tennessee:	1968 harvest of 1,700,000 bobwhites. The annual ruffed grouse kill is about 15,000 birds.
Texas:	1968 harvest of 8,000,000 bobwhites and 2,000,000 scaled quail. No data on Gambel quail. Average annual lesser prairie chicken harvest from 1965 through 1969 was 275 birds.
Utah:	1967 harvest included 5,089 sage grouse, 17,527 forest grouse (blue and ruffed), 26,187 quail (Gambel and California), 48,906 chukar partridges, and 16,049 gray partridges.
Vermont:	No data (ruffed grouse only).
Virginia:	1,380,405 bobwhites in 1968. The annual ruffed grouse kill is about 85,000 birds in good years.
Washington:	Average harvests from 1964 through 1969 include 2,483 sage grouse, 162,400 blue grouse, 16,744 spruce grouse, 162,400 ruffed grouse, 113,551 chukar partridges, 25,100 gray partridges, 220,000 California quail, and a few hundred mountain quail, scaled quail, and bobwhites.
West Virginia:	1969 harvest was 66,000 bobwhites and 115,000 ruffed grouse.
Wisconsin:	289,960 ruffed grouse in 1969. Average gray partridge harvest between 1964 and 1968 was 31,835. No data on sharp-tail kill (season closed from 1965 through 1967).
Wyoming:	Sage grouse harvest from 1960 through 1969 averaged

TABLE 28—(continued)

	53,387 and forest grouse (blue, spruce, and ruffed) kill averaged 4,193 from 1964 through 1969. Sharp-tail kill from 1967 through 1969 averaged 739. Average 1960–69 harvest of 15,036 chukar partridges and 2,616 gray partridges. Bobwhite kill unknown but very limited.
Alberta:	Sage grouse harvest 272 in 1967; blue grouse harvest about 100 in 1960; ruffed grouse harvest averaged 52,795 between 1950 and 1956; sharp-tail harvest averaged 122,000 in 1966 and 1967. Gray partridge harvest averaged 104,985 between 1950 and 1956.
British Columbia:	Average harvests between 1964 and 1968 were 132,030 blue grouse, 133,362 spruce grouse, 361,293 ruffed grouse, 21,365 sharp-tailed grouse, 7,641 chukar partridges, and 13,352 quail (mountain and California).
Manitoba:	Average harvests between 1964 and 1969 were 14,922 spruce grouse, 9,709 ptarmigan, 56,973 ruffed grouse, 55,484 sharp-tailed grouse, and 6,265 gray partridges.
New Brunswick:	No data available.
Newfoundland:	From 25,000 to 50,000 ptarmigan are harvested annually, and since 1968 a small number of ruffed grouse have also been harvested. No data available from Labrador.
Nova Scotia:	The annual kill of ruffed grouse ranges from 50,000 to 65,000, and from 1,500 to 2,500 gray partridges are also harvested.
Ontario:	No data available.
Prince Edward Island:	About 500 ruffed grouse are taken annually.
Quebec:	No detailed estimates for any species (ptarmigan, ruffed grouse, spruce grouse, gray partridge), but annual grouse kill may approach 100,000 birds, since over 35,000 small game licenses were sold in 1969.
Saskatchewan:	Average harvests between 1962 and 1969 were 8,579 spruce grouse, 30,400 ruffed grouse, 129,000 sharp-tailed grouse, and 132,475 gray partridges.

By taking these individual state and provincial harvest figures and summing them by species (prorating totals in cases where several species were grouped together), it is possible to make a very tentative total annual harvest estimate for each species (table 29). These totals suffer from the fact that harvest data were not available to me from four of the smaller eastern states, two provinces, and the two Canadian territories. Nevertheless, with these numerous limitations in mind, a relative measurement of each species' probable hunting importance is possible. If these figures can be accepted, it would appear that nearly fifty million grouse, quail, and partridges are harvested every year in the United States and Canada, of which about 70 percent are bobwhites. Other quail which are clearly harvested in large numbers are the scaled, Gambel, and California quails. Not surprisingly, the ruffed grouse is the species with the largest estimated total hunter harvest, comprising nearly 70 percent of the total estimated grouse harvest of over five million birds.

It is of some interest that the chukar partridge and gray partridge now provide sport for hunters in sixteen states and eight provinces, and probably more than a million birds are now harvested annually. Indeed, in terms

TABLE 29

RELATIVE HUNTING IMPORTANCE OF GROUSE AND QUAIL SPECIES, UNITED STATES AND CANADA

	Open Season in 1970			*Estimated Annual Kill*		
	States	*Provinces*	*Total*	*States*	*Provinces*	*Total*
Sage grouse	10	1	11	250,000	few	250,000
Blue grouse	11	2	13	240,000	130,000	370,000
Spruce grouse	5	7	12	140,000	300,000	440,000
Ptarmigans	2	7	9	100,000	200,000	300,000
Ruffed grouse	33	10	43	2,700,000	1,000,000	3,700,000
Prairie chicken	6	0	6	85,000	0	85,000
Sharp-tailed grouse	12	6	18	255,000	200,000	455,000
Mountain quail	5	1	6	375,000	few	375,000
Scaled quail	8	0	8	3,600,000	0	3,600,000
Gambel quail	8	0	8	1,300,000	0	1,300,000
California quail	6	1	7	2,200,000	few	2,200,000
Bobwhite	37	2	39	35,000,000	few	35,000,000
Harlequin quail	1	0	1	6,000	0	6,000
Chukar partridge	9	1	10	650,000	8,000	658,000
Gray partridge	14	8	22	400,000	250,000	650,000
Totals				47,301,000	2,088,000	49,389,000

of numbers of states and provinces where it can be legally hunted, the gray partridge now ranks third (behind the ruffed grouse and bobwhite) among the most important sporting birds of this group.

Assuming that most of the ten million or more small-game hunters in the United States spend part of their time hunting grouse, quail, or partridge species and that nearly fifty million of the birds are harvested here yearly, then the average season kill per hunter is approximately five birds. For most species, this is no more than a single day's limit of birds. This fairly reasonable estimate would suggest that the estimated total nationwide kill may not be very far from the actual number and may indeed be conservative. The economic value of this harvest, in terms of dollars spent in pursuit of the sport, is even more difficult to judge, but on the basis of average expenditure figures provided by the National Survey of Fishing and Hunting it must probably amount to more than six hundred million dollars per year.

It is, of course, impossible to place a dollar value on any living creature; and the grouse and quail present a special esthetic quality for lovers of nature. Leopold (1949) beautifully stated this view as follows: "Everybody knows that the autumn landscape in the north woods is the land, plus a red maple, plus a ruffed grouse. In terms of conventional physics the grouse represents only a millionth of either the mass or the energy of an acre. Yet, subtract the grouse and the whole thing is dead."

Thus, the value of the grouse and quail to bird watchers is real, and, indeed, to this group perhaps the birds are at least as valuable as they might seem to hunters. To many people, the first bobwhite whistle is not only the harbinger of spring, it *is* the spring. To others, the muffled drum roll of ruffed grouse in a distant glade is anticipated as eagerly as the earliest hepatica blossom, and on the midwestern prairies the vernal predawn booming of prairie chickens at their ancestral leks is as rich a heritage as the big bluestem and Indian grass that then lie golden in the swales.

To be individually appreciated by humans, grouse and quail must first be seen. This is not to say that a white-tailed ptarmigan on an inaccessible mountain peak that has yet to be climbed is any the less valuable than the California quail that make their daily jaunts to a back-yard feeding station and can be observed from a living-room easy chair. To many, in fact, a ptarmigan on a mountain meadow, surrounded by dwarf alpine flowers and framed by a glacial cirque, is the very essence of the American wilderness and represents an esthetic value beyond measure. But for the average American, tied to a city job during the week and enclosed by a concrete jungle of maddening noise and confusion, there is a special attraction in being able to drive a few miles into the country in the hope of catching a glimpse of the local wildlife. To obtain some measurement of this relative accessibility of the grouse and quail to American bird watchers, I have extracted data from the annual Audubon Society Christmas counts, the distribution of which reflects in some measure the distribution of people in the country and their relative bird watching opportunities. For the twelve years from 1957 through 1968 I have tabulated (table 30) the number of years the various species of grouse, quail, and partridges have been reported by at least one party; the average number of birds seen on the highest yearly counts (excluding years when the species was not seen at all); and the highest individual count during the entire twelve-year period. For the years 1957 through 1962 (later summaries of this nature were not compiled), the average count of each species seen in all stations where the bird was reported at all has been calculated, providing a rough index to the population density and perhaps also to the relative sociality of each

TABLE 30

GROUSE AND QUAILS REPORTED ON AUDUBON CHRISTMAS COUNTS, 1957–1968

	1957–1968			*1957–1962*	
	Years Reported	*Average High Count*	*Highest Count*	*Average Count per Station**	*Average No. Stations*
Sage grouse	9/12	28.8	97	10.3	1.3
Blue grouse	10/12	3.4	14	3.4	2.3
Spruce grouse	7/12	3.6	8	3.0	7.3
Ruffed grouse	12/12	56.0	91	5.4	145.0
Willow ptarmigan	6/12	8.5	29	3.0	0.2
White-tailed ptarmigan	8/12	6.9	28	2.0	0.8
Rock ptarmigan	1/12	11.0	11	. . .	. . .
Sharp-tailed grouse	12/12	94.8	158	20.8	11.6
Greater prairie chicken	11/12	42.4	95	17.2	4.8
Lesser prairie chicken	5/12	144.0	443	48.0	0.2
Mountain quail	12/12	27.0	62	8.7	3.8
Elegant quail	1/12	5.0	5	. . .	. . .
Scaled quail	12/12	341.1	769	66.8	16.5
Gambel quail	12/12	370.3	725	57.1	12.6
California quail	12/12	1346.0	6854	179.3	41.0
Bobwhite	12/12	421.6	655	39.9	227.0
Harlequin quail	9/12	22.1	55	10.0	0.2
Chukar partridge	12/12	25.9	123	14.1	1.3
Gray partridge	12/12	266.4	552	28.6	25.5

*Excluding stations not reporting species

species. From these figures it may be seen that the bobwhite is the species most often encountered by American bird watchers in wintertime, with the ruffed grouse in second place and the California quail third. On the other hand, the great sociality of the California quail during winter and its consequently large covey sizes cause it to attain first place in average yearly high count among all stations recording it, the highest average count per station, and the highest individual count of any single station. The scaled, Gambel, and bobwhite quails also exhibit relatively high numbers of birds counted per station, which likewise reflects their covey-forming tendencies.

The other side of the coin is provided by the remaining forest grouse, sage grouse, prairie grouse, and ptarmigans, all of which were recorded by relatively few Christmas count groups, and generally were found only in small numbers. These species have a kind of "rarity appeal" that adds to their attractiveness for winter bird watchers, and their appearance on a daily check-list provides ample testimony to the effort expended in locating the birds. I personally can vividly recall snowshoeing across seemingly endless snow-covered fields of eastern North Dakota on one December day with a temperature of seven degrees below zero, in hopes of flushing a covey of gray partridge to add to the Christmas count.

Both hunters and nonhunting nature lovers can wholeheartedly agree to the need for conserving our irreplaceable grouse and quail. Perhaps too often the nonhunter might accuse the upland game sportsmen of "killing off our quail," or whatever the species concerned might be. With the present controls on hunting this is, of course, utter nonsense; every species included in this book has a relatively high reproductive rate associated with a comparable mortality rate, and under most circumstances hunting cannot measurably alter the mortality rate of the species. Far more important than the number of birds shot during the fall hunting season is the amount of winter food and cover available to support the survivors until the following breeding season. Except in rare circumstances, it is the simple presence or absence of adequate cover to provide the species' daily and annual needs that will determine whether or not a wild species can survive and prosper in an area. Unlike the situation with our migratory waterfowl, we cannot blame the people living somewhere else when our upland game populations diminish; the local environment is the critical factor in the success of upland game populations. In most cases this does not necessarily mean the retention of large wilderness areas. Bobwhites thrive on the "edge effect" produced by an interspersion of cultivated and uncultivated lands; and ruffed grouse benefit from local burning or cutting of too dense forests. What is serious, however, is widespread habitat disturbance or destruction during the nesting or brooding season or the reduction of adequate winter food and cover so that the birds are forced into marginal habitats and increasingly exposed to the elements and to predators.

We have only recently become fully aware of another threat to our wild populations that is unrelated to cover, hunting pressure, or any other of the classic concerns of game biologists. This is the threat of pesticides and their insidious ability to permeate the natural environment before we are really aware of the enormous damage that they might do. These particularly include the "hard" or persistent insecticides such as the chlorinated hydrocarbons which can remain in soils and living tissues for great lengths of time, becoming increasingly concentrated as they are passed progressively up the food chain. Since the grouse and quail feed primarily on plant material as well as some insects, they do not suffer from this "biological magnification" to the degree that is true of various predators, fish-eating species and the like. However, they do store materials such as DDT in their body fats, and not only might these materials cause physiological damage during times of fat utilization but also they may be passed on to human consumers or predators. So far, DDT levels high enough to affect eggshell thickness and, as a result, reduce hatchability have not been detected in either the grouse or the quail of North America. We need not compliment ourselves on this circumstance, however; sufficient damage has been done by DDT to our fish-eating birds and other avian predators such as the falcons to justifiably indict the pesticide industry and its apologists for a disaster of unprecedented magnitude.

Environmental pollutants of greater immediate threat to upland game birds are the organic mercury fungicides used for treatment of seed wheat and other grains. Since small grains are a major food source for prairie grouse, quail, partridges, and pheasants in the Great Plains states, the birds are likely to ingest considerable amounts of the fungicide when they consume treated grain. In the fall of 1969 the Alberta Department of Lands and Forests found it necessary to close the hunting season on pheasants and gray partridges because of the concentrations of mercury found in these birds, and the Montana Fish and Game Department similarly found sufficient concentrations of mercury to cause them to caution hunters against eating the birds. Unlike DDT, mercury poisoning is produced by far smaller concentrations and it operates directly on the central nervous system. The physiological effects of DDT on vertebrates are far less localized and a wide variety of organ systems and physiological processes are disrupted. The first case of closing a game bird season because of dangerous DDT levels in a game bird species occurred in 1970, when New Brunswick closed its season on woodcock.

This is a sad period in the history of North America for lovers of wildlife and the outdoors. We are witnessing the progressive extirpation of the greater prairie chicken from one state after another, and we must soon face the possibility that both the Attwater prairie chicken and the lesser prairie chicken will join the heath hen in the shadows of extinction. It also seems unlikely that the magnificent sage grouse will be able to withstand indefinitely the combined onslaughts of sage clearing and sage destruction through herbicide spraying, and it will be fortunate to survive the rest of this century. In Mexico, the fate of the tree quails and the spotted wood quail will become questionable as the cloud forests are progressively ravaged and the previously impregnable and mahogany-rich rain forests of eastern Chiapas are ripped apart by bulldozers, trucks and chain saws. A few short-term advances have been made and are properly rejoiced in, such as the establishment of several grassland refuges for prairie chickens, while at the same time the tide of increasing population and its associated degradation of our natural environment silently inches ever higher and begins to threaten our own survival.

We are not separate from our environment; each species we destroy and each habitat we ravage, whether by bulldozer or pesticides, represents one more bridge that we have burned in our own ultimate battle for survival. It is a melancholy thought that, after its compatriots had disappeared, the last surviving male heath hen in North America faithfully returned each spring to its traditional mating ground on Martha's Vineyard, Massachusetts, where it displayed alone to an unhearing and unseeing world. Finally, in the fall of 1931 it too disappeared. With it died the unique genes that reflected the sum total of the species' history, from Pleistocene times or earlier through uncounted generations of successful survival to the very last, when inbreeding, habitat disruption, fire, and disease inexorably tipped the balance of survival a final time. No one knows exactly how or when that last survivor died, and no bells tolled to mourn its passing. Indeed, only by the absence of its dirge-like booming the following spring was the heath hen's extinction finally established, and the bird that had been as much a part of our New England history as the Pilgrims was irrevocably lost.

Part II

Accounts of Individual Species

10

Sage Grouse

Centrocercus urophasianus (Bonaparte) 1827

OTHER VERNACULAR NAMES

SAGE hen, spiny-tailed pheasant, sage cock, sage chicken.

RANGE

From central Washington, southern Idaho, Montana, southeastern Alberta, southern Saskatchewan, and western North Dakota south to eastern California, Nevada, Utah, western Colorado, and southeastern Wyoming (modified from *A.O.U. Check-list*).

SUBSPECIES

C. u. urophasianus (Bonaparte): Eastern sage grouse. Resident from southern Idaho, eastern Montana, southeastern Alberta, southern Saskatchewan, and western North Dakota south to eastern California, south central Nevada, Utah, western Colorado, and southeastern Wyoming.

C. u. phaios Aldrich: Western sage grouse. Resident from central and eastern Washington south to southeastern Oregon.

MEASUREMENTS

Folded wing: Males, 282–323 mm; females, 248–79 mm. Using flattened wings, females range from 240 to 285 mm and males from 288 to 334 mm, with 290 mm a calculated best division point (Crunden, 1963).

Tail: Males, 297–332 mm; females, 188–213 mm.

IDENTIFICATION

Adults, 19–23 inches long (females), 26–30 inches long (males). The large size and sagebrush habitat of this species make it unique among grouse. Both sexes have narrow, pointed tails, feathering to the base of the toes, and a variegated pattern of grayish brown, buffy, and black on the upper parts of the body, with paler flanks but a diffuse black abdominal pattern. In addition, males have blackish brown throats, narrowly separated by white from a dark V-shaped pattern on the neck, and white breast feathers concealing the two large, frontally directed gular sacs of olive green skin. Behind the margins of the gular sacs are a group of short white feathers with stiffened shafts, which grade into longer and softer white feathers and finally into a number of long, black hair-like feathers that are erected during display. Males also have rather inconspicuous yellow eye-combs that are enlarged during display. Females lack all these specialized structures but otherwise generally resemble males. Their throats are buffy with blackish markings, and their lower throats and breasts are barred with blackish brown.

FIELD MARKS

The combination of sage habitat, large body size, pointed tail, and black abdomen is adequate for certain identification. Males take flight with some difficulty and fly with their bodies held horizontally; females take off more readily and while in flight their bodies dip alternately from side to side. When the bird is in flight the white underwing coverts contrast strongly with the blackish abdomen.

AGE AND SEX CRITERIA

Females may readily be separated from adult males by their weights and measurements (see above), by the absence of black on the upper throat, and by the fact that the white tips of the under tail coverts extend part way down the feather rachis (Pyrah, 1963). Crunden (1963) provides a sex and age key based on primary measurements.

Immatures (under one year old) resemble females but are paler, the outer primaries are more pointed and mottled than the others, the outer wing coverts are narrowly pointed instead of being unmottled dark gray and are marked with brown and white and have white tips (Petrides, 1942). Immatures also have light yellowish green toes, unlike the dark green toes of adults. Males do not usually achieve their full breeding condition their first year; subadult males have narrower white breast bands than do adults. The tail feathers of immature males are also blunter and are tipped with white. During their first fall immature birds have bursa depths in excess of 10 mm (averaging 18.9 mm in October), whereas adults have maximum bursa depths of 7 mm and average depths of 1.6 mm (Eng, 1955).

Juveniles have conspicuous shaft-streaks on their upper body feathers and tail feathers with white central shafts that spread out into narrow terminal white fringes (Ridgway and Friedmann, 1946).

Downy young (illustrated in color plate 61) have a distinctive "salt and pepper" appearance dorsally that is devoid of striping and consists of a mottled combination of black, brown, buff, and white. The head is whitish, spotted with brown and black in a fashion similar to blue grouse downies, and the underparts vary from grayish white to buff and brownish on the chest region, where a brown-bordered buff band is usually evident. The malar and nostril spots of this species are unique (Short, 1967), and a definite loral spot is also present. The broken pattern of dark markings on the forehead and crown found in this species probably corresponds to the black border that occurs around the brown crown patch in most other grouse (Short, 1967).

DISTRIBUTION AND HABITAT

At one time this species was found virtually wherever sagebrush (*Artemisia*, especially *A. tridentata*) occurred, throughout many of the western and intermountain states. In early times it occurred in fourteen or fifteen states and was the principal upland game species in nine (Rasmussen and Griner, 1938). However, overgrazing and drought contributed greatly to the species' near demise. By the early 1930s it was a major upland game species in only four states (Montana, Wyoming, Idaho, and Nevada), and by 1937 only Montana retained a regular open season. Restricted hunting was by then still permitted in Nevada and Idaho, but all other states had established closed seasons (Rasmussen and Griner, 1938). After 1943,

Montana also established a closed season which lasted nine years. The species became completely extirpated from British Columbia and New Mexico, although New Mexico has recently successfully reestablished the bird, and British Columbia has been attempting to do the same (Hamerstrom and Hamerstrom, 1961). There are no recent specimen records from Nebraska, although a few birds may occasionally stray across the Wyoming state line. There are no Oklahoma records since 1920 (Sutton, 1967).

A low ebb in sage grouse populations in the western states occurred in the middle to late 1940s. Idaho reported an upturn in populations after 1947, and, after four years of protection, reopened hunting in 1948. Nevada reestablished limited hunting in 1949, followed by Washington in 1950. Permit-only seasons were established by Wyoming in 1948 after eleven years of protection and by Utah in 1950. California opened one county (Mono) to hunting in 1950, after five years of protection. Judging from figures presented by Patterson (1952), the total United States kill in 1951 was less than 75,000 sage grouse.

Except for two years (1944 and 1945), Colorado maintained a closed season from 1937 until 1953, and in 1952 Montana held its first season since 1943. South Dakota began hunting sage grouse again in 1955 after nineteen years of protection, and in 1964 North Dakota held its first season since 1922. Alberta initiated a highly restricted season in 1967.

In the past decade the sage grouse has recovered sufficiently to be a major game species again in about five states. Hamerstrom and Hamerstrom (1961) reported estimated hunter-kill figures for 1959 of about 44,000 birds in Wyoming, 23,000 each in Idaho and Montana, 15,000 in Colorado, and 12,000 in Nevada, plus approximately 2,000 each in California, Washington, and Utah, totaling more than 100,000 for the country as a whole.

The most recently available hunter-kill estimates would indicate that the sage grouse is at least maintaining its population sufficiently to be a major game species in five states and of secondary importance in six more states and one province. The estimated 1969 kill in Idaho was 81,729 birds, and in 1968 total state-wide harvests were 55,361 in Wyoming and 53,462 in Montana. Colorado biologists estimated that 21,922 sage grouse were shot there in 1969. Kill estimates for Nevada and Utah were 11,765 (1968) and 11,109 (1969), respectively. Considerably smaller numbers are harvested annually in California (3,200 in 1969), Washington (2,300 in 1969), Oregon (4,760 in 1969), and South Dakota (about 2,000 in 1967). Currently Alberta and North Dakota each have kills of only a few hundred per year. The overall yearly harvest is thus currently about 250,000 birds. This harvest does not reflect so much a recent increase in grouse populations as increased hunting pressure and a recognition that limited harvests are not a controlling factor in protecting the security of the sage grouse.

Of the several published range maps for sage grouse (e.g., Aldrich, 1963; Edminster, 1954), that prepared by Patterson (1952) appears to be most representative of current distributional patterns. The range map that I have prepared is based largely on Patterson's map, but has been modified to take into account information such as that appearing in recent sage grouse status questionnaires. Patterson estimated that some 90 million acres of preferred sagebrush-grassland habitat existed in the early 1950s and that an additional 40 million acres of desert scrub habitat was also available to sage grouse. If the 90-million-acre figure is assumed to be currently representative, this would total about 140,000 square miles of preferred habitat. If an average population density of 10 birds per square mile might be assumed, the total sage grouse population might be roughly estimated at 1,500,000 birds. The present yearly harvest of 250,000 would then represent 17 percent of the total, which would not seem exorbitant.

In spite of this seemingly comfortable number of birds, it is difficult to be optimistic about the long-term future of the sage grouse in North America. The continued clearing of extensive areas of sage for irrigated farming, as has occurred widely in central Washington, and the expanded use of herbicides to improve grazing conditions are likely to further reduce sage grouse habitat and populations in future years. Schneegas (1967) estimated that five to six million acres of sagebrush have been removed in the last thirty years, a portent of things to come.

POPULATION DENSITY

Patterson (1952) estimated sage grouse densities by determining strutting ground sizes and numbers in two study areas that totaled 250 square miles. He reported an average of one strutting ground per 5.7 square miles, and a density of 12.5 males per square mile. This, of course, excluded all females and probably some immature males from consideration. Edminster (1954) thus calculated that the total spring population of sage grouse might be from 30 to 50 birds per square mile, or thirteen to twenty-one acres per individual. Rogers (1964) likewise reported that certain counties of Colorado support 10 to 30 birds per square mile in some sections, while the remaining habitat supports 1 to 10 birds per square mile.

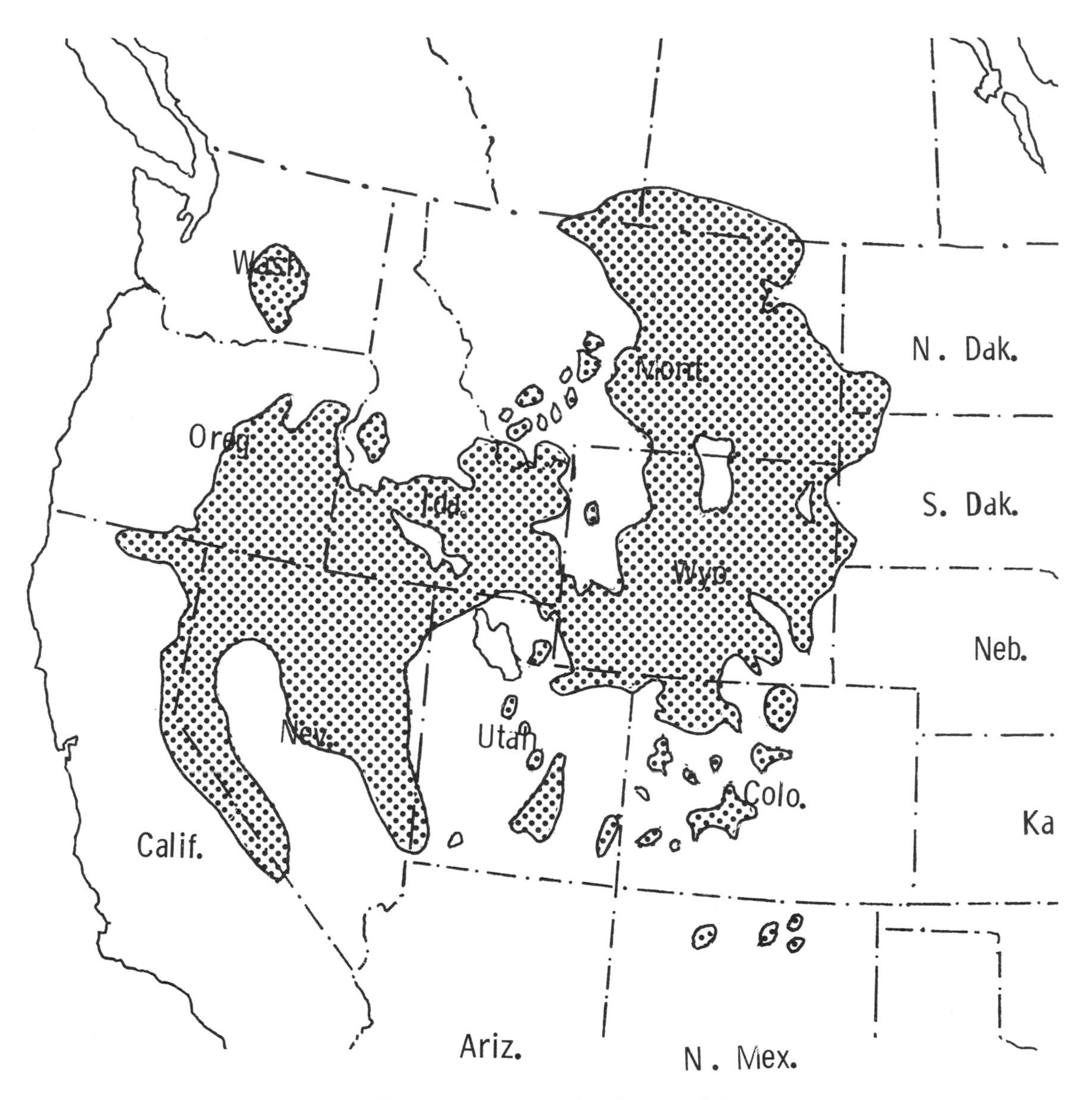

FIGURE 22. Current distribution of the sage grouse.

HABITAT REQUIREMENTS

Wintering Requirements

During winter, sagebrush provides not only nearly 100 percent of the food which is utilized by sage grouse but also important escape cover. Edminster (1954) pointed out that during winter sagebrush has the important attributes of being evergreen, tall enough to stand above snow, and highly nutritious. Rogers (1964) indicated that the best wintering areas in Colorado were those at the lowest elevations, where sagebrush was available all winter. Local topography may influence availability of sagebrush, because of snow cover, but sage grouse may be expected to occur wherever exposed sagebrush may be found through the winter period. Dalke et al. (1963) reported that wintering concentrations of sage grouse in Idaho usually occurred where snow accumulations were less than six inches deep, which occurred in areas some thirty to fifty miles from the habitats used during fall and spring. Black sage (*Artemisia nova*) is the preferred winter food in eastern Idaho but is often covered by snow.

Spring Habitat Requirements

In late winter, male sage grouse begin to leave their wintering areas and return to their traditional strutting grounds. Based on a total of forty-five strutting grounds classified by type of land area, Patterson (1952) found that eleven were on wind-swept ridges and exposed knolls, ten were on flat sagebrush areas with no openings, seven were on bare openings on relatively level lands, and the remaining seventeen occurred in seven other habitat types. Relatively open, rather than dense, sage cover is clearly the preferred habitat for strutting grounds, as indicated by a number of writers such as Scott (1942) and Dalke et al. (1963). The latter study reported that new strutting grounds could be readily established by clearing areas of one-fourth to one-half acre in dense stands of sage.

Nesting and Brooding Requirements

Patterson (1952) reported that 92 percent of the nests he found were under sagebrush plants, usually in cover from ten to twenty inches tall, and in drier sites where the shrub cover was less than 50 percent. In Utah, Rasmussen and Griner (1938) found that a related species, silver sage (*A. cana*), provided preferred nest cover, with plants of this species from fourteen to twenty-five inches tall providing cover for 33 percent of 161 nests, while the more common big sage (*A. tridentata*) of the same height category accounted for 24 percent of the nests. The highest nesting densities (up to 23 nests on 160 acres, or 1 nest per 6.95 acres) occurred in dense second-growth sagebrush. Klebenow (1969) found that 91 percent of 87 nests or nest remains were associated with three-tip sage (*A. tripartita*). In nesting habitats he noted that the sagebrush averaged only eight inches tall but that taller plants were preferred for nest sites. No nests occurred where the shrub cover exceeded 35 percent. In the best nesting areas, nest densities of up to 1 nest per 10 acres were found.

Brooding habitat requirements are evidently slightly different from sage grouse nesting requirements. Klebenow (1969) reported that 83 percent of the broods he observed were in big sagebrush but not in dense stands. All but three of ninety-eight broods recorded were seen in areas of less than 31 percent shrub cover. As the summer progressed, broods moved into moister areas that still contained green plant material, until by late August they had gathered near permanent water sites. However, available water in the form of green vegetation, rain, or dew evidently provides adequate moisture for sage grouse.

Observations of Martin (1970) in Montana indicated that in 158 locations, young broods used areas having less plant density and lower crown cover (nine to fifteen inches high) than did older broods or adults (seven to twenty-five inches high). Rogers (1964) also reports that low sage (seven to fifteen inches high) is preferred feeding, nesting, and roosting cover, while taller plants serve for nesting, shade, and escape cover. Spraying with the herbicide 2, 4-D in Montana greatly reduced summer usage by sage grouse, apparently by altering vegetational composition, particularly of favored food plants (Martin, 1970). Similarly, Peterson (1970) concluded that components of brood habitat for sage grouse include a diversity of forms and a density of sage ranging from 1 to 20 percent.

FOOD AND FORAGING BEHAVIOR

The importance of sagebrush as a food item for adult sage grouse is impossible to overestimate. Martin, Zim, and Nelson (1951) reported that sage made up 71 percent of the diet in 203 samples and that usage of animal material ranged from 9 percent in summer to 2 percent in spring and fall. Apart from sagebrush, vegetable food consists largely of the leaves of herbaceous legumes and weeds (collectively called *forbs*) and grasses, which are utilized primarily in late spring and summer (Edminster, 1954). Patterson (1952) reported that sage comprised 77 percent (of a total of 95.7 percent plant material) of foods found in 49 samples from adult sage grouse in Wyoming and 47 percent (of a total of 89 percent plant material) from 45 juvenile sage grouse analyzed. Evidently sage is taken in limited amounts even during the first month of life (Griner, 1939), although like all grouse, newly hatched chicks feed principally on insect life.

During early life, young sage grouse feed heavily on ants, beetles, and weevils and later add grasshoppers to their food intake (Patterson, 1952), although the total animal content of the diet drops from as much as 75 percent to less than 10 percent. The study of Klebenow and Gray (1968) indicates that insects predominate in the diet only during the first week of life, and thereafter forbs become the predominant food, with shrubs only gradually assuming a place of primary importance. The importance of forbs is also indicated by a study by Trueblood (1954), who found that this food category comprised from 54 to 60 percent of the major food items consumed by juvenile sage grouse in Utah and from 39 to 47 percent in adults. On lands partially reseeded to grass, he found that adults persisted in their preference for shrubs, while juveniles exhibited a preference for forbs and a strong aversion to grasses.

Martin's study (1970) has provided additional evidence of the value of a variety of forbs as a source of summer food for sage grouse. He found that, in a sample of 35 sage grouse collected from July to September, sagebrush totaled 34 percent of the food, while dandelion (*Taraxacum*) comprised 45 percent. Collectively, these plants plus two additional forb genera (*Trifolium* and *Astragalus*) contributed over 90 percent of the food material. Two California studies (Leach and Hensley, 1954; Leach and Browning, 1958) also indicate that weedy forbs such as prickly lettuce (*Lactuca*) and cultivated herbaceous broad-leaved plants such as clover and alfalfa play important roles as early fall food sources for sage grouse.

One of the most complete studies available on juvenile food requirements is the recent study of Peterson who analyzed the food of 127 birds up to twelve weeks of age. During that period, forbs comprised a total of 75 percent of the diet, and two genera (*Taraxacum* and *Tragopogon*) together made up 40 percent of the food consumed. Insect use declined from a high of 60 percent in the first week to only 5 percent by the twelfth week, and sagebrush was used very little by chicks before the age of eleven weeks.

MOBILITY AND MOVEMENTS

Seasonal Movements

The most complete study on seasonal movements of sage grouse so far available is that of Dalke et al. (1963). Patterson (1952) had previously summarized the literature on possible migratory movements of these birds, noting that in Oregon a winter migration to lower elevations was followed after nesting by a migration to summer ranges at eight-thousand-foot elevations. Possible winter movements of Wyoming and Montana birds into South Dakota were discussed by Patterson, and he mentioned a male that was banded in Wyoming and recovered the following fall still in Wyoming but some seventy-five air miles from its point of banding.

In mountainous country, wintering grounds of sage grouse are often some distance from spring and summer habitats, at considerably lower elevations. With the gradual regression of snow, male grouse on their wintering grounds begin working toward the strutting areas. Dalke et al. (1963) reported that these birds move in small flocks, flying short distances, during this migration. Many such birds in Idaho may move from fifty to one hundred miles along established routes before reaching their strutting grounds. Adult females evidently reach the strutting grounds about the same time as adult males or somewhat later. Patterson (1952) noted that males began to arrive on Wyoming strutting grounds as early as February and were followed in one or two weeks by females. Dalke et al. (1963) found that males and even females occupied grounds in late March or early April that were not yet free of snow. A rapid build-up of adult males occurred in early April, while subadult females arrived about a week after adult females, and subadult males did not appear in numbers until most of the females had already left the grounds in late April.

Movements of birds between strutting grounds is evidently fairly rare, both within one season and from year to year. Dalke et al. (1963) noted that of 78 adult males banded in 1959 and 1960, a total of 14 (18 percent) were observed later on grounds other than those where they had been banded. During the same two years 107 females were banded, and 6 of these were subsequently observed visiting other strutting grounds. Movements by males between strutting grounds covered distances of from 550 yards to 4.3 miles. Dominant males were only rarely involved in these movements, suggesting that the movements are the result of attempts by subordinant males to establish territories in various locations. Earlier, Dalke et al. (1960) had reported that 70 percent of banded sage grouse that were again observed on strutting grounds in the first three years were seen on their original strutting grounds and no others. Some master cocks occuppied nearly identical territories in successive years, while others lost their territorial positions.

It is not well known how far the females move from strutting grounds to build their nests, but current evidence would suggest that it is usually not very far. Klebenow (1969) noted that on one area of three-tip sage (a favored nesting cover) located more than a mile from the nearest strutting ground no nests were found and only one very young brood was seen. In each of two areas of big sage, nests were found within a half mile and at only slightly lower elevations. However, unpublished Colorado studies indicate that females regularly move three or four miles from a display ground to a nest site and may travel as far as seven miles.*

Following nesting, females gradually move their broods to places where food supplies are plentiful, usually in relatively moist areas such as hay meadows, river bottom lands, irrigated areas, and the like. Patterson (1952) estimated that family units break up and juveniles become relatively independent at about ten or twelve weeks, when they have completed their molt into juvenal plumage.

Spring dissolution of the strutting grounds by males is a gradual process, and some subadult males may remain after most adult birds have left for summer ranges (Dalke et al., 1960). However, Eng (1963) found that adult males were the last to leave the strutting area. These are usually at higher elevations, but the birds may move down into alfalfa fields near irrigated valleys. Schlatterer (1960) reported that the sequence of arrival of birds on the summering areas in Idaho was males, unproductive females, and productive females. In southern Idaho the summer brood range may be from thirteen to twenty-seven miles from the nesting grounds, a considerable movement for these recently fledged birds.

Fall movements toward wintering areas is likewise a gradual process, and the rates probably vary according to weather conditions. Pyrah (1954) reported that immature females were the first to leave for wintering areas, followed by mature females, then adult males. Immature males associated with immature and mature females. Dalke et al. (1963) reported that birds collected in flocks near water holes as freezing temperatures began and that movements were quite noticeable by the time the daily minimum dropped to twenty degrees Fahrenheit. Birds usually remained in a single place for several days then moved out in groups. By the time the first snows fell, flocks were usually composed of between fifty and three hundred birds in loose associations. During severe weather, flocks of up to one thousand birds could be seen, but in midwinter they normally consisted of less than fifty individuals, with old males often in groups of less than twelve.

Daily Movements

Daily movements and activity patterns of sage grouse have yet to be carefully documented, but some work with banded birds is of interest.

*Terry May, 1970: personal communication.

Lumsden (1968) noted the daily locations of several individually marked males on a strutting ground and confirmed that individual males returned daily to their specific territories. However, their territorial boundaries were rather ill defined and exhibited considerable overlap. On one occasion, when a cluster of hens formed about fifty-five meters from Lumsden's blind, six males left their usual territories and moved toward the hens, apparently maintaining their positions relative to one another. Of twenty-seven individually marked hens, 16 were observed later on the same display ground. Four were seen to visit the ground on three mornings, one was seen twice, and eight only once. Seven were observed mating, in each case only once, and none of these birds was seen again.

Males arrive on the strutting grounds long before dawn and early in the season may actually remain all night. Hens arrive before dawn and usually leave shortly after sunrise. After daybreak, immature males are the first to leave the grounds, followed by successively more dominant males and finally the master cock. The birds normally walk to feeding areas which may be within a half mile of the strutting grounds (Pyrah, 1954). Hens rarely return to the strutting grounds in the afternoon.

Observations on nesting hens by Girard (1937) and Nelson (1955) indicate that they normally leave their nests twice a day during incubation. Girard reported that these foraging periods occurred between 9:30 and 11:30 A.M. and between 2:00 and 3:00 P.M., whereas Nelson reported earlier morning and later afternoon periods. The feeding periods usually lasted between fifteen and twenty-five minutes, according to Nelson.

In late summer, sage grouse roost until about 6:00 A.M., forage until about 10:00 A.M., rest until about 3:00 P.M., forage again until 8:00 P.M., and finally go to roost again about 9:00 P.M. (Girard, 1937). Unlike the prairie grouse, sage grouse exhibit no fall display activities. During winter, daily movements of sage grouse have no definite pattern, and apart from foraging, much time is spent resting and preening. Roosting occurs on rocky outcrops (Crawford, 1960; Dalke et al., 1963).

REPRODUCTIVE BEHAVIOR

Prenesting Behavior

In a sense, the sage grouse may be regarded as the classic lek-forming species of North American grouse. Not only are the lek sizes the largest in terms of average numbers of males participating, but also the degree of segregation according to dominance classes is the most evident. Further, although Scott (1942) was by no means the first to describe the social strutting behavior of sage grouse, his study first recognized the complex social hierarchy of males and designated the most dominant males as master cocks. This term has since been applied to most other lek-forming grouse, such as pinnated grouse and sharp-tailed grouse.

As soon as traditional display grounds are relatively free of snow, male sage grouse begin to occupy them. In different years conditions may vary, but in the northern United States the birds are usually on their strutting grounds by late February or March. Most studies indicate that the first birds to occupy the grounds are the adult males, which may return to virtually the same territorial site that they occupied in previous years.

It might be assumed that the male behavior patterns exhibited on the strutting grounds perform two separate functions: proclamation and defense of territory on the one hand and attraction and fertilization of females on the other. Although natural selection thus operates through the differential successes of individual males in attracting females, it is of interest that apparently in all grouse the behavior patterns serving to attract females are derived directly from hostile behavior patterns associated with the establishment and defense of territory. As a result, relatively few of the displays performed by male grouse in lek situations serve strictly as male-to-female displays, but rather those postures and calls that function in territorial establishment are for the most part utilized in sexual situations as well. It is therefore generally impractical to separate fully signals associated with attack and escape (agonistic displays) from those which function sexually to attract females (epigamic displays). The resulting close relationship between relative individual success in performing territorial behavior (achieving male-male social dominance) and relative individual reproductive success (fertilization of females) provides a basic key to the understanding of social behavior in lek-forming grouse. This contrasts with the situation in socially displaying duck species, in which agonistic and sexually oriented displays are much more separable, probably because of the absence or insignificance of territoriality during pair-forming processes of waterfowl.

The fact that most male displays performed by lek-forming grouse are derived from hostile responses further complicates their dual role as sexual attractants. Female grouse must not only be attracted to these signals, but must in turn identify themselves as females in order to avoid attack by territorial males. This is usually achieved by submissive postures which in general are associated with inconspicuousness through slimmed plumage, silent movements, and general lack of male-like signals. Thus a kind of paradox may be seen in lek-forming grouse. Whereas in non-lek-forming species of grouse (e.g., ptarmigan) the females may perform fairly elaborate and often male-like displays, in the social species the degree of development of female display is perhaps inversely proportional to the relative development of male displays and other male signals. The role of the female in lek-forming grouse is therefore reduced to simply appearing on the lek, being attracted to particular males, and allowing copulation to occur. This last point is achieved by a precopulatory squatting display with wings partially spread, which is virtually identical in all grouse so far studied. In the sage grouse, where hens often cluster in groups around specific males (master cocks), fighting between hens may sometimes occur, but it is not likely that this occurs in other species.

Male Territorial Advertisement Behavior

Although strutting has been described by many writers, the accounts by Lumsden (1968) and Hjorth (1970) are by far the most complete and accurate. The following summary is therefore in large measure based on their descriptions. Lumsden and Hjorth have confirmed the basic findings of Scott (1942), who discovered the relationship of social dominance to sexual success, with master cocks representing the individuals maintaining a central territory that is selectively sought out by females for copulation. It is important to note, however, that the strutting behavior of master cocks differs in no obvious way from that of birds occupying lower social ranks, such as the secondary status "subcocks" and "guard cocks" or the peripheral attendant males. Strutting by nonterritorial yearling males is, however, poorly developed and may readily be recognized from that of older birds. Such immature birds probably represent the so-called "heteroclite" males described by Scott.

Overt fighting between males is largely but not entirely limited to the edges of territories. Fighting males typically stand ten to twenty inches apart, head to tail and nearly parallel to one another, with heads upright and feathers usually lowered. The tail may be raised or lowered and is sometimes shaken rapidly, producing a rattling sound that perhaps corresponds to the tail-rattling display of sharp-tailed grouse. Periodically the males attempt to strike each other with their nearer wing, but unlike the prairie grouse, males do not fly into the air and strike with their feet. The associated calls are *kerr* sounds, often in a series of eight to twelve repeated notes.

Overt fighting is less common in sage grouse than is ritual fighting, in which the same parallel posture is assumed but the birds remain virtually motionless. At times the birds may actually close their eyes as if sleeping in this posture, which Lumsden interprets as "displacement sleeping." When threatening, male grouse draw up the skin on the sides of the neck, thus erecting the filoplumes and increasing the exposed areas of white feathers. The tail may also be cocked and spread and the body held more upright when in such a threat posture. In general, the amount of white feathers exhibited by a male is a relative index of its aggressive tendencies. It is thus of interest that female grouse lack white areas and that the white neck area of yearling males is smaller than that of adults. When charging, the posture assumed by the adult male is strongly similar to that held during the strutting display. This would suggest that strutting represents a ritualized form of charging, in which the forward body movement component has been almost entirely lost.

When on territory and between strutting sequences, the male is usually in an "upright" posture (Hjorth, 1970), with tail cocked and spread, wings slightly drooped, neck feathers ruffled, and the esophageal pouch partly inflated and hanging in a pendulous fashion. In this posture he may jerk his head upwards and utter a soft snoring note that is apparently associated with the inhaling of air (Hjorth, 1970).

The strutting display ("ventro-forward" of Hjorth, 1970) is a complex sequence of stereotyped movements (figure 23) and sounds, which lasts about three seconds and which Lumsden has divided into ten stages. In the first stage the male assumes an erect posture with the tail fanned and held slightly behind the vertical, lowers his folded wings, and takes a step forward. The back is gradually raised, so that by stage two it is held at a forty-five-degree angle from the ground. The anterior neck feathers then suddenly part, exposing two olive green skin patches. The third stage begins as the bird opens his beak and apparently takes a breath. The pendent esophageal bag is then lifted and the skin patches disappear, another step forward is taken, and the folded wings are quickly drawn across the stiffened feathers at the sides of the neck as it is jerked upwards ("first vertical jerk" of Hjorth), producing a brushing sound. In the fourth stage the beak is shut, the wings are moved forward again, and the esophageal bag is lowered. In stage five the neck again swells, the oval skin patches are exposed a second time but again are not greatly inflated, and a second although silent backward stroke of the wings is performed. In stage six a third step forward is taken, the wings are again moved forward, the skin patches are somewhat more fully expanded, and the esophageal bag begins to move upward again. In stage seven the neck is diagonally extended ("second vertical jerk" of Hjorth), as the esophageal bag is strongly raised, nearly hiding the head, and the wings are again rubbed against the breast feathers as they make their third backward stroke. In stage eight the head is withdrawn into the erected neck feathers, the esophageal bag bounces downward, and the inflated bare skin patches form large oval bulges ("first forward thrust" of Hjorth), while the wings move forward and back a fourth time. In stage nine the

FIGURE 23. Sequence of the ventro-forward display of the sage grouse, from Hjorth, 1970. (Numbering represents cine frames exposed at the rate of twenty-four per second. Enlarged views illustrate the three forward thrusts from two angles.)

head is quickly withdrawn into the neck feathers so that it becomes completely concealed, compressing the esophageal bag so greatly that the skin patches bulge strongly outward in the shape of hemispheres ("second forward thrust" of Hjorth), and the wings complete a fifth backward stroke. Pressure on the trapped air in the esophagus is now suddenly released, causing the skin to collapse with two plopping sounds, and the head is moved upwards toward a normal position. In the tenth and final stage the head returns to the original starting position, the white neck feathers close over the bare skin areas, and the body returns to the stance assumed at the beginning of the display.

The major motor elements of the entire display sequence thus consists of several forward steps (Hjorth reported four to seven), five rotary wing movements, two brushing sounds of the wings against the sides of the breast and neck, and four increasingly greater inflations of the esophagus, with associated expansions of the colored skin patches. The predominant nonvocal sound is a "resonant squeaking, swishing" noise (Lumsden, 1968) that is followed by two plopping sounds. However, a call is also uttered, which Lumsden described as sounding like *wa-um-poo*, only the last part of which can be heard at any distance. Hjorth (1970) determined that there are actually four vocal notes produced, of which the second is the loudest.

The sage grouse lacks much of the pivoting action of the pinnated grouse's booming, but as Lumsden pointed out strutting is not a specifically frontal display. Although visually impressive when seen from the front, the long and colorful under tail coverts are also conspicuous signals when seen from behind. Lumsden found no strong tendency of males to face hens when performing their strutting displays, and often they faced directly away from them.

Apart from the fighting call and that which is uttered during strutting, only one other male call has been reported for sage grouse. Lumsden noted a deep grunting sound, which occurred both in threat situations and when near hens and often as a prelude to fighting. The same call was occasionally heard from hens. Hjorth (1970) called this vocalization a "grunting chatter."

The strutting behavior of males when hens are present is not noticeably different from when they are absent, except perhaps for the greater frequency of displays. Hen sage grouse typically gather together in tight groups near master cocks; from fifty to seventy hens have been seen in single clusters in large leks. Lumsden noted that, although hens clustered at twenty different locations during his observations, the groups nearly always formed near the most dominant male. Thus, hens are clearly attracted to specific males rather than to specific mating spots on a lek. Clusters of hens evidently serve as a sexual stimulus for females, and precopulatory squatting by one often provides an apparent stimulus for others to behave similarly. Males normally quickly mount any soliciting female, and copulation lasts only a few moments. Unlike other grouse, the male does not normally grasp the female's nape in his beak while mounted, perhaps because of the considerable disparity in size between the sexes.

Most studies indicate that the majority of copulations are achieved by only one, or at most two, males in any center of mating activity. Scott (1942) found that master cocks performed 74 percent of 174 observed copulations, Patterson (1952) found mating success similarly restricted to a few males, and Lumsden (1968) found that two males accounted for more than half of the 51 copulations he observed. However, Hjorth (1970) found that four males took part in the matings he observed on one lek.

Following copulation, the female usually runs a short distance forward, shaking her wings and tail for several seconds before starting to preen. Usually females leave the strutting grounds within a few minutes after copulation. Males usually remain in a motionless squatting position for several seconds after copulation, which Lumsden regards as a ritualized display posture that he believes may function to reduce disruption of the hen cluster.

In contrast to nearly every other North American grouse (the ruffed grouse is the only other case), the sage grouse lacks a flight display. Lumsden is probably correct in explaining this on the basis of the male's large size and poor agility, plus the fact that needs for territorial advertisement are reduced in sage grouse because of the large number of males usually present and the conspicuous nature of individual birds. Lumsden also believes that "call flights" by hens serve to advertise the location of the strutting ground. Such "quacking" calls are uttered by hens when flying toward the ground or when flying from one part of the ground to another. Occasionally the calls are also uttered when the hen flies away from the strutting ground. Lumsden also described a "wing-bar signal" display, which he states may be performed by females in flight prior to landing, perhaps functioning as a landing-intention signal. This display is sometimes, but not always, associated with a call flight, and is produced by drawing the white underwing coverts up over the leading edge of the wing so they are visible from above and behind the bird. A somewhat similar "shoulder-spot display" occurs in both sexes of sage grouse while on the ground. Lumsden regards this display as an expression of conflict, with fear as one of the components.

Calls of male sage grouse include the strutting call, grunt, and fighting call already mentioned, as well as a high-pitched and repeated *wut* note used as an alarm call (Lumsden, 1968). Males, especially yearlings, may

↑ Sage Grouse Lek

↓ Sage Grouse, Territorial Males

↑ Sage Grouse, Male

↓ Sage Grouse, Male and Female

C. G. Hampson

↑ Sooty Blue Grouse, Male

↓ Sooty Blue Grouse, Male

C. G. Hampson

↑ Dusky Blue Grouse, Male

↓ Dusky Blue Grouse, Male

↑ Dusky Blue Grouse, Male

↓ Dusky Blue Grouse, Male

↑ Dusky Blue Grouse, Female

New Mexico Dept. of Game & Fish

↓ Blue Grouse, Female

Robert Starr

↑ Canada Spruce Grouse, Female

↓ Canada Spruce Grouse, Female

E. M. Brigham, Jr.

↑ Canada Spruce Grouse, Male

↓ Franklin Spruce Grouse, Male

C. G. Hampson

S. D. MacDonald

Franklin Spruce Grouse, Male

J. Jehl

↑ Willow Ptarmigan, Male

↓ Willow Ptarmigan, Male

↑ Willow Ptarmigan, Male Song-flight

↓ Willow Ptarmigan, Male

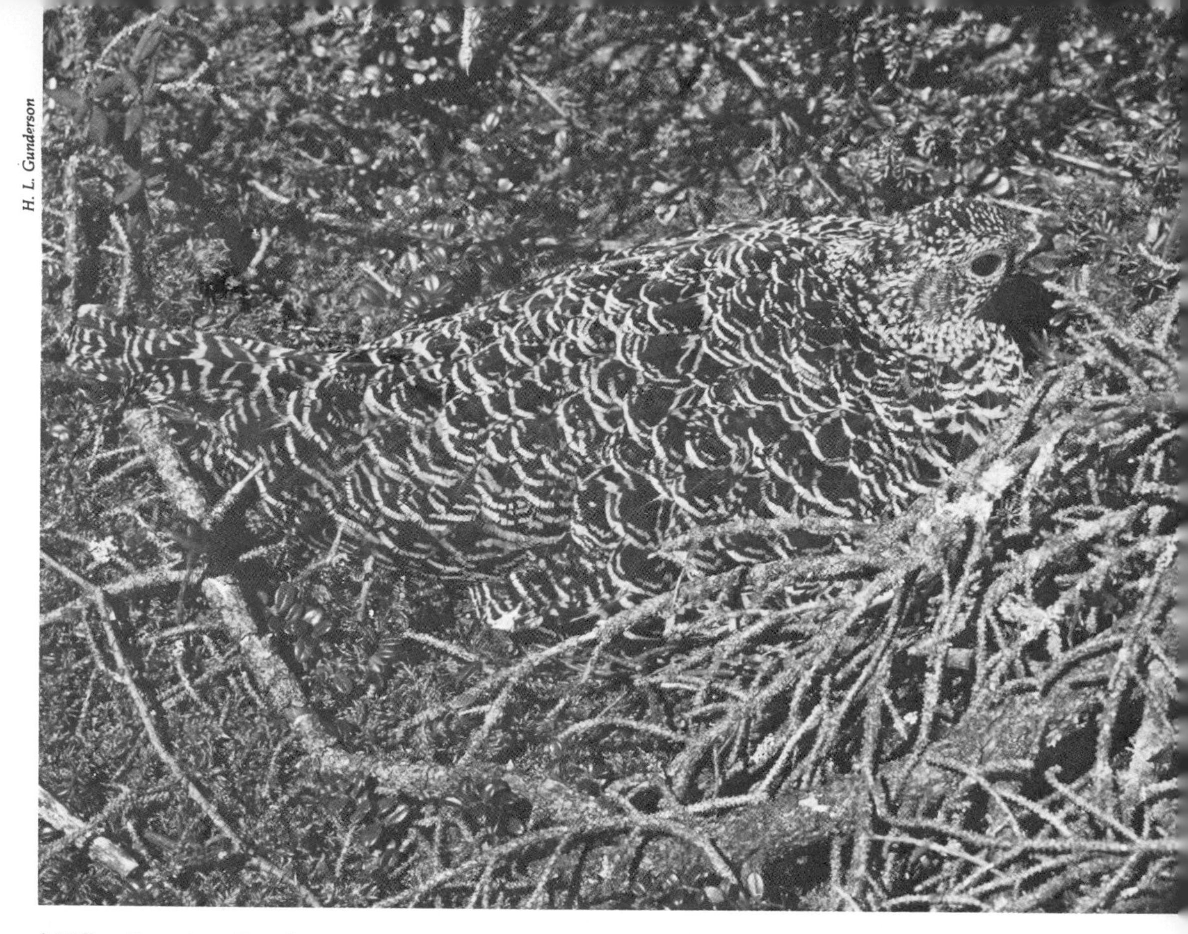

H. L. Gunderson

↑ Willow Ptarmigan, Female

↓ Rock Ptarmigan, Male

S. D. MacDonald

↑ Rock Ptarmigan, Female

↓ Rock Ptarmigan, Males Fighting

↑ Rock Ptarmigan, Male Calling

↓ White-tailed Ptarmigan, Male

Clait Braun

↑ White-tailed Ptarmigan, Female

↓ White-tailed Ptarmigan, Male in Spring

M. Martinelli

↑ White-tailed Ptarmigan, Fall Flock

↓ White-tailed Ptarmigan, Winter

Colo. Game, Fish & Parks Dept.

also utter a squawking note, perhaps as a flight-intention signal. Hens also have well-developed fighting notes, as well as whining notes in agonistic situations. Both sexes may also hiss when being handled, according to Lumsden.

Nesting Behavior

Once fertilization has been accomplished, the hen apparently leaves the strutting ground for nesting. There is no present evidence that a hen requires more than one successful copulation to complete her clutch. Patterson (1952) believed that females begin laying within a few days after mating, although Girard (1937) indicated that from 7 to 12 days may be taken up in locating a nest and in nest construction. This kind of delay would not seem to be normal, and Dalke et al. (1963) found a good correlation between actual and calculated hatching period by assuming that 10½ days would be required to lay an average clutch of eight eggs, and that 26 ½ days more would be required for incubation, for a total elapsed time of 37 days between mating and hatching.

Estimates of average clutch size usually range from 7 to 8 eggs. Patterson (1952) reported an average clutch size of 7.26 eggs in eighty nests during one year, and 7.53 eggs in seventy-four nests the following year. Griner (1939) reported an average clutch size of 6.8 eggs in Utah, Nelson (1955) reported 7.13 in Oregon, and Keller, Shepherd, and Randall (1941) reported 7.5 in Colorado. Patterson (1952) believed that a very limited amount of renesting might occur, judging from smaller late clutches and the presence of new nests near destroyed or deserted nests. Although Eng (1963) found a second peak of females on strutting grounds in late May, this was not reflected in a second late hatching peak, and he concluded that reduced male fertility late in the season prevents effective renesting.

Patterson's estimate (1952) of a twenty-five-to-twenty-seven-day incubation period for sage grouse has generally been supported by later workers such as Pyrah (1963), who utilized data from captive grouse. This contrasts with various earlier estimates of a twenty-to-twenty-four-day incubation period. Sage grouse appear to have a high rate of both nest destruction and nest desertion. Gill (1966) summarized data on fates of nests from eight different studies, which ranged in hatching success from 23.7 to 60.3 percent. Predator activity was responsible for a large part of the nesting losses, predators accounting for 26 to 76 percent of the lost nests of six studies summarized by Gill. Of a total of 503 nests represented, 47.7 percent were destroyed by predators. Coyotes, ground squirrels, and badgers are evidently among the more important mammalian predators, while magpies and ravens may be significant avian predators of nests.

EVOLUTIONARY RELATIONSHIPS

For reasons that have never been evident, taxonomists have traditionally regarded the sage grouse as closely related to the true "prairie grouse," namely the pinnated grouse and the sharp-tailed grouse. Not until the analysis by Hudson and Lanzillotti (1964) was it proposed that the sage grouse may have its nearest affinities with the "forest grouse" instead. Short (1967), using various lines of evidence, supported the view that *Centrocercus* probably evolved from an ancestral type similar to *Dendragapus* and that *D. obscurus* represents the nearest living relative of the sage grouse. Lumsden's analysis of behavior (1968) also presented this view, and he pointed out that the male sage grouse shares with the blue and spruce grouse the characteristic of having a white "V" marking on the throat that apparently has signal value at least in the sage grouse. Lumsden suggested that the sage grouse and blue grouse diverged from a common ancestral type that was a forest-dwelling bird, to which the spruce grouse and Siberian spruce (or sharp-winged) grouse (*Dendragapus falcipennis*) are the nearest modern equivalents. In contrast, Short suggested that the ancestral grouse was a woodland edge species, of which the earliest offshoot was a grassland-woodland form ancestral to *Tympanuchus*, followed later by separation of pre-*Dendragapus* and pre-*Centrocercus* types.

I believe that both adult and downy plumage characteristics stongly favor the view that *Dendragapus* and *Centrocercus* are closely related, and that the male sexual displays of sage grouse and blue grouse have many features in common. The evolution of lek behavior by the sage grouse produced some convergent similarities to the social displays of prairie grouse, but these should not be regarded as evidence for close common ancestry.

11

Blue Grouse

Dendragapus obscurus (Say) 1823

OTHER VERNACULAR NAMES

DUSKY grouse, fool hen, gray grouse, hooter, mountain grouse, pine grouse, pine hen, Richardson grouse, sooty grouse.

RANGE

From southeastern Alaska, southern Yukon, southwestern Mackenzie, and western Alberta southward along the offshore islands to Vancouver and along the coast to northern California, and in the mountains to southern California, northern and eastern Arizona, and west central New Mexico (*A.O.U. Check-list*).

SUBSPECIES (*ex A.O.U. Check-list*)

D. o. obscurus (Say): Dusky blue grouse. Resident in the mountains from central Wyoming and western South Dakota south through eastern Utah and Colorado to northern and eastern Arizona and New Mexico.

D. o. sitkensis Swarth: Sitkan blue grouse. Resident in southeastern Alaska south through the coastal islands to Calvert Island and the Queen Charlotte Islands, British Columbia.

D. o. fuliginosus (Ridgway): Sooty blue grouse. Resident from the boundary between Yukon and Alaska south through the mainland of southeastern Alaska, coastal British Columbia including Vancouver Island, western Washington, and western Oregon to northwestern California.

D. o. sierrae Chapman: Sierra blue grouse. Resident on the eastern slope of the Cascade Mountains of central Washington south into California and from southern Oregon south along the Sierra Nevada into California and Nevada.

D. o. oreinus Behle and Selander: Great Basin blue grouse. Resident in mountain ranges of Nevada and Utah.

D. o. howardi Dickey and van Rossem: Mount Pinos blue grouse. Resident on the southern Sierra Nevada from about latitude 37° N. to the Tehachapi range and west to Mount Pinos, where extremely rare.

D. o. richardsonii (Douglas): Richardson blue grouse. Resident from the southern Yukon and Alaska south through interior British Columbia to the Okanagan Valley and western Alberta to Idaho, western Montana, and northwestern Wyoming.

D. o. pallidus Swarth: Oregon blue grouse. Resident from south central British Columbia south through eastern Washington to northeastern Oregon.

N.b.: Three of the forms listed (*sitkensis, fuliginosus,* and *sierrae*) are sometimes specifically separated from the remaining ones and tend to have eighteen rather than twenty rectrices, yellowish rather than grayish downy young, and certain other minor structural differences.

MEASUREMENTS

Folded wing (unflattened): Adult males, 196–248 mm; adult females, 178–235 mm (adult males of all races average over 217 mm; females, under 216 mm).

Tail (to insertion): Adult males, 131–201 mm; adult females, 111–59 mm (adult males average over 150 mm; females, under 150 mm).

IDENTIFICATION

Adults, 17.2–18.8 inches long (females), 18.5–22.5 inches long (males). This is the largest of the coniferous-forest grouse of the western states and provinces. Sexes differ somewhat in coloration, but both have long, squared, and relatively unbarred tails (pale grayish tips usually occur in both sexes of all races except *richardsonii* and *pallidus,* which sometimes have suggestions of a pale tip). Upperparts of males are mostly grayish or slate-colored, extensively vermiculated and mottled with brown and black markings, and the upper wing surfaces are more distinctly brown. White markings are present on the flanks and under tail coverts, and feathering extends to the base of the middle toe. The bare skin over the eyes of males is yellow to yellow orange, and the bare neck skin exposed during sexual display varies from a deep yellow and deeply caruncled condition (in the *fuliginosus* group) to purplish and somewhat smoother (in the *obscurus* group). Females have smaller areas of bare skin and are generally browner overall, with barring of mottling on the head, scapulars, chest, and flanks.

FIELD MARKS

Blue grouse are likely to be confused only with the similar but smaller spruce grouse, the ranges of which overlap in the Pacific northwest. Male blue grouse lack the definite black breast patch of male spruce grouse. Female blue grouse have relatively unbarred, grayish underparts, as compared with the spruce grouse's white underparts with conspicuous blackish barring. A series of five to seven low, hooting notes is frequently uttered by territorial males in spring.

AGE AND SEX CRITERIA

Females may be recognized by barring on the top of the head, nape, and interscapulars which is lacking in adult males (Ridgway and Friedmann, 1946), and by the bases of the neck feathers around the air sacs, which are grayish brown rather than white. The sex of adults may be determined from the wings alone; females have a more extensively mottled brownish pattern on their marginal upper wing coverts; in males these feathers are gray, with little or no mottling (Mussehl and Leik, 1963).

Immatures (in first-winter plumage) may be recognized by one or more of the following criteria: the outer two primaries (retained from the juvenal plumage) are relatively frayed and more pointed (van Rossem, 1925) as well as being lighter and more spotted than the inner ones, the outer tail feathers are narrow and more rounded (up to 7/8th inch wide at ½ inch below tip, as opposed to being at least 1¼ inch wide in adults), and the tail is shorter than in adults (the maximum length of plucked feathers of juvenile males is 152 mm, of juvenile females, 134 mm, compared with 162 and 138 mm in adult male and female *fuliginosus* (according to Bendell, 1955b). Immatures of both sexes generally resemble adult females but may usually be recoginized by their pale buffy or white breasts, the absence of a gray area on the belly, and (except in *richardsonii* and *pallidus*) the absence of a gray bar at the end of the tail (Taber, in Mosby, 1963).

Juveniles (in juvenal plumage) may be distinguished by the conspicuous white (tinged with tawny) shaft streaks of the upperparts, wings, and tail, and the brown rectrices which may be mottled or barred and lack a gray tip (Ridgway and Friedmann, 1946). The juvenal plumage is carried only a very short time in this species as in other grouse, and the juvenal tail feathers are molted almost as soon as they are fully grown.

Downy young are illustrated in color plate 61. Considerable variation in downy coloration exists among the numerous races (Moffitt, 1938). Downy blue grouse lack the chestnut crown patch of spruce grouse, exhibiting instead irregular black spotting over the crown and sides of the head and a conspicuous black ear patch. The black head marking in young blue grouse also includes a central crown mark that connects with frontal spotting, two indefinite lateral stripes, and a faint brownish area posteriorly that is bordered by slightly darker markings (Short, 1967). This species is thus intermediate between the extreme type of head markings found in the sage grouse and the more *Lagopus*-like markings typical of the spruce grouse.

RANGE AND HABITAT

The over-all North American range of the blue grouse is closely associated with the distribution patterns of true fir (*Abies*) and Douglas fir (*Pseudotsuga*) in the western states (Beer, 1943). Its range more closely conforms with that of the Douglas fir than any other conifer tree species, but this is probably a reflection of both species' being closely adapted to a common climatic and community type rather than any likelihood of the blue grouse's being closely dependent on Douglas fir. The species actually occupies a fairly broad vertical range in the western mountains, breeding at lower elevations, sometimes as low as the foothills, and spending the fall and winter near timberline or even above it. Rogers (1968) reports that in Colorado the birds are usually to be found at between 7,000 and 10,000 feet but have been seen at elevations as low as 6,100 feet and as high as 12,400 feet, averaging about 9,000 feet. At least in the moist Pacific northwest, lumbering and fire produces a more open forest that improves the breeding habitat of blue grouse by opening the forest cover, but heavy grazing on lower slopes can be deleterious (Hamerstrom and Hamerstrom, 1961).

In contrast to several grouse species, no major range changes of importance have occurred in the blue grouse in historical times (Aldrich, 1963). In none of the states and provinces where the species occurs is it in danger of extirpation, although the southern populations in New Mexico and Arizona are relatively sparse and scattered.

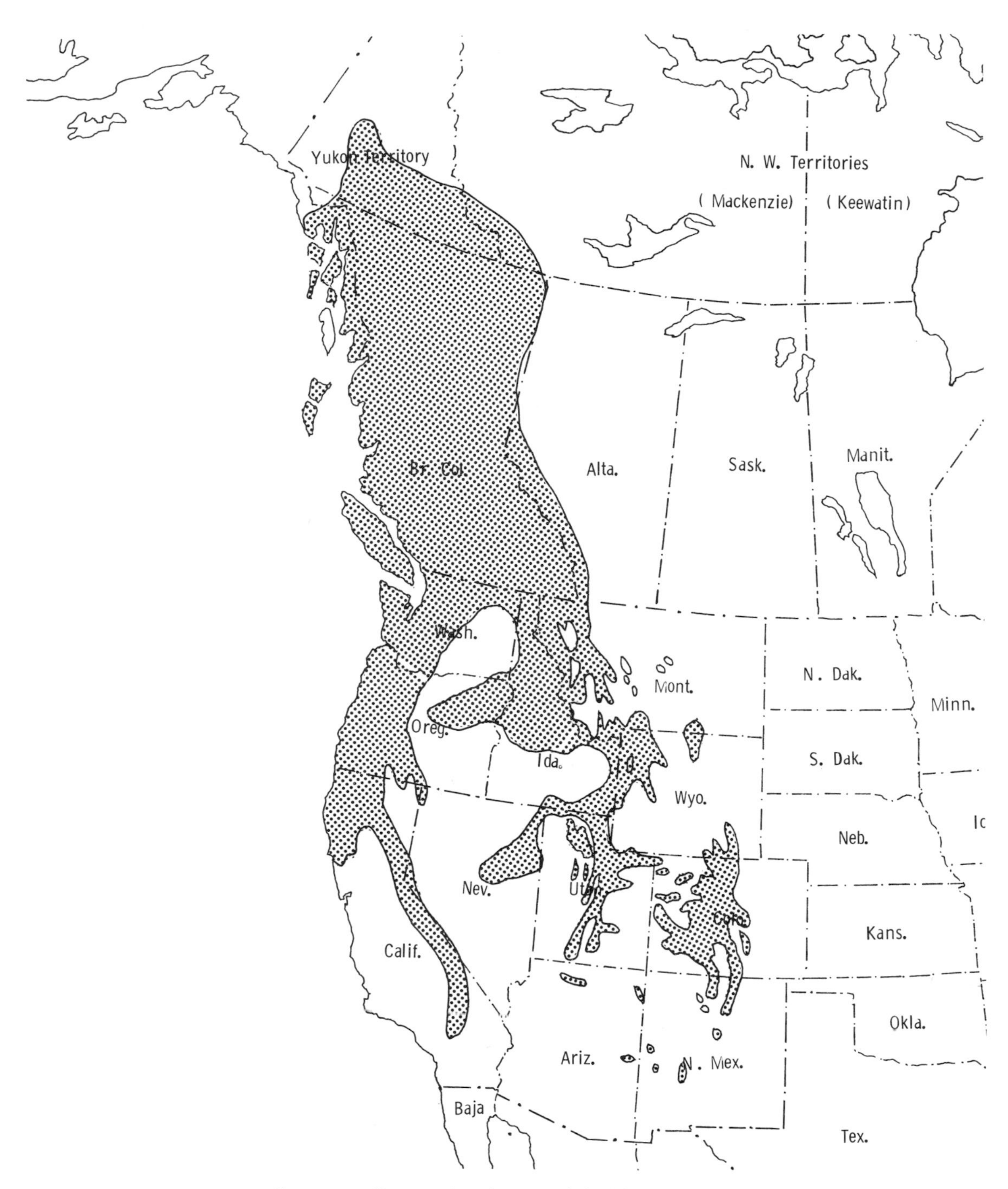

FIGURE 24. Current distribution of the blue grouse.

Although the blue grouse depends heavily on coniferous cover for wintering, its preferred habitat also includes a number of deciduous tree species, shrubs, and forbs. Foremost among broadleaf trees are aspens (*Populus*), and a variety of shrubs provide food and escape cover. Rogers (1968) summarized records of dominant trees, shrubs, forbs, and grasses associated with blue grouse observations in Colorado over a several year period. In all years, aspen was the dominant tree, snowberry (*Symphoricarpos*) was the dominant shrub, bromegrass (*Bromus*) was the dominant grass, and groundsel (*Senecio*) or vetch (*Astragalus*) were the dominant forbs. Trees recorded less frequently were juniper (*Juniperus*), spruce (*Picea*), Douglas fir, and ponderosa pine (*Pinus ponderosa*). Although hens and broods were sometimes seen in piñon pine (*Pinus edulus*) and juniper cover, summer concentrations of males were usually in open coniferous stands of spruce and fir. Rarely were blue grouse seen more than a mile from trees or shrubs, and females with broods were usually not far from water.

Similar observations on blue grouse habitat characteristics have been made in southern Idaho by Marshall (1946). There the vertical range used by the species extends from less than five thousand feet in ponderosa pine–Douglas fir forest, which is infrequently used by blue grouse, to subalpine forests reaching over eight thousand feet, which provide wintering areas for both sexes and summering habitats for males. In these higher ridges they use the conifers, especially Douglas fir, for both food and cover. In all but eight of twenty-five cases, the grouse were observed to land in conifers on being flushed, while the remainder landed on the ground. Of 159 observations of birds as to cover type, 87 were in Douglas fir, 41 were in subalpine cover, 25 were on banks of streams, and the remaining 6 were in grass or brush cover.

A study by Fowle (1960) on Vancouver Island provides comparable data for the coastal population of blue grouse. Summer habitat there consists of second-growth cover produced by fire and logging of Douglas fir forests. About 45 percent of the sample areas had no vegetation at all, while in the rest mosses, lichens, ferns, and grasses, as well as a variety of shrubs and forbs, made up most of the cover. Except near water, where alders (*Alnus*), willows (*Salix*), and dogwood (*Cornus*) occurred, trees were only in scattered groups. About 20 percent of the area was covered with important grouse foods, including bracken fern (*Pteridium*), willow, Oregon grape (*Mahonia*), blackberry (*Rubus*), huckleberry (*Vaccinium*), salal (*Gaultheria*), and cat's-ear (*Hypochaeris*). These plants made up a total of more than 90 percent of adult food samples, and over 80 percent of juvenile food samples.

By the end of September, the birds move up to higher slopes, and they winter in the coniferous zone (Bendell, 1955c), where they are found primarily in subalpine forests. Zwickel, Buss, and Brigham (1968) point out that winter habitat is probably determined more by cover type than by altitude per se and may occur in Washington at as low as four thousand feet, between the ponderosa pine and Douglas fir zones, with the critical factor apparently being the presence of interspersed Douglas and true firs.

POPULATION DENSITIES

Estimates of blue grouse population densities are difficult because of the cover inhabited by the species, and the generally solitary nature of the species. Rogers (1968) summarized results of grouse surveys from vehicles; over a three-year period in two study areas they averaged one grouse per 26.07 miles, ranging from 10.3 to 38.72 miles in various years.

Using a strip-count census method, Fowle (1960) counted adult grouse on Vancouver Island during two summers. In four areas totaling 272 acres he determined a density in 1943 of 2.6 acres per bird. Later work in the same area by Bendell and Elliott (1967) indicated that the density of territorial males in dense and sparse populations respectively was approximately 0.44 and 0.13 or less males per acre, or from about 2.3 to 7.7 acres per territorial male. Similar counts of territorial male blue grouse were made by Mussehl and Schladweiler (1969) in Montana on six study areas that were in part exposed to insecticide spraying. Numbers of territorial males on sprayed and unsprayed areas did not appear to differ and averaged about 1 male per 18 acres, ranging from 12 to 24 acres per male.

Whether the blue grouse is subject to population "cycles" is perhaps questionable, but at least major population fluctuations and corresponding changes in density evidently do occur. Fowle (1960) and Hoffmann (1956) summarized historical data on grouse populations during the 1900s but neither attempted to explain these fluctuations. Zwickel and Bendell (1967) hypothesized that population fluctuations in the species are related to the nutritional condition of females, as determined by the summer range conditions, which might affect chick survival and in turn determine subsequent autumn population densities. However, no relationship was found between the number of young in autumn and the breeding density in the following year. They suggest that the death rate or dispersal of juveniles between autumn and early spring is the single most important factor regulating breeding densities.

HABITAT REQUIREMENTS

Wintering Requirements

Primary wintering needs for the blue grouse appear to be sufficient trees to provide roosting and escape cover and a supply of needles from trees of the genera *Abies, Tsuga*, or *Pseudotsuga* as a source of food. Beer (1943) reports that adult blue grouse subsist almost entirely on needles from November through March. Needles, buds, twigs, and seeds of Douglas fir may all be eaten in winter, and needles, buds, and pollen cones of true fir are also used. Where both *Abies* and *Pseudotsuga* are present, the former appears to be preferred. Larches (*Larix*) may be used until its needles are shed, and various species of pines are used for their buds, pollen cones, and seeds. Marshall (1946) noted that 99 percent of the contents of nine birds killed during winter in Idaho consisted of needles and buds of Douglas fir. Interestingly, grit is evidently retained in the gizzard through the winter, in spite of the deep snow cover. Hoffmann (1956) reported that white fir (*Abies concolor*) provided favored winter roosts in California.

Spring Habitat Requirements

As the winter ends, both sexes begin to move downward from the coniferous zones, and males seek out areas suitable for territories. Bendell and Elliott (1966) analyzed the habitats used by both sexes of blue grouse on Vancouver Island from spring through August, classifying cover as "very open" (40 percent tree, log, stump, and salal cover) or "very dense" (100 percent woody cover). The relative grouse use in two types was 115 in very open cover compared to 18 in very dense cover. The use of the very dense cover was limited to some territorial males that apparently established territories there before it became so heavily vegetated and overgrown. The authors concluded that the blue grouse is better adapted to a dry habitat than is the ruffed grouse and may indeed have evolved from a grassland species. Supporting this view was their finding that young captive blue grouse required only about half as much water as captive ruffed grouse. They concluded that the breeding habitat of blue grouse might be defined as open and dry, with shrubs and herbs interspersed with bare ground.

In California, Hoffmann (1956) found that the persistence of snow cover determined the onset of hooting in spring and the transition to spring behavior in a study area where virtually no seasonal migration occurred. Blackford's studies (1958, 1963) on Montana provide additional information on territorial requirements for an interior population (*obscurus*) of this species. In this area, hooting occurred either at ground level or in trees during strutting. Strutting areas were in forest-edge habitats with combined grassy, open forest border and a dense coniferous stand. Occasionally, rocky outcrops occurred, and old logs were present on the forest floor. Blackford's observations established that earlier, widely reported differences in territorial defense and strutting behavior between coastal and inland populations of blue grouse are not in general absolute.

Yearling males may migrate downward to the breeding areas or may remain on the wintering areas through the summer. Bendell and Elliott (1967) estimated that about half of the yearling males moved to the summer range their first year. There they are silent, move about widely, and may be attracted to hooting territorial males. These authors observed two cases of territorial yearling males. Females may return to the same general area of the summer range in subsequent years, but are not nearly so localized in this respect as are males (Bendell, 1955c). Unlike males, females are not particularly aggressive to one another, and their home ranges may overlap. However, Stirling (1968) suggested that during the squatting and egg-laying periods females do become somewhat aggressive, and this behavior tends to scatter females and perhaps allows for a spacing of nests.

Nesting and Brooding Requirements

Surprisingly little has been written on specific nesting needs for blue grouse, perhaps because their nests are rather difficult to locate. Usually the nest is located near logs or under low tree branches and is fairly well concealed. Bendire (1892) stated that most nests are under old logs or among roots of fallen trees and are generally to be found in more open timber along the outskirts of the forest. He found one nest beside a creek in rye grass some two miles away from timber and another in an alpine meadow under a small fir tree, with no other trees within thirty yards. Bowles (in Bent, 1932) noted that nests are usually in very dry, well wooded sites, and they are often at the bases of trees or under fallen branches or some other shelter. However, they may be up to one hundred yards from trees, with little or no concealment. Lance (1970) found that nests were usually fairly near territorial males but well separated from the nests of other females.

Brooding habitat for blue grouse appears to be that which provides ample opportunities for the young to feed on insects and other invertebrates. Beer (1943) suggested that blue grouse usually nest in open situations where there will be an abundance of insect life for the newly hatched birds. For the first ten days, the young feed almost exclusively on animal material, especially ants, beetles, and orthopterans, according to Beer. As the young grow older, berries, such as currants (*Ribes*) and Juneberries (*Amelanchier*) are sought out, and the young birds and adults gradually move upwards as they follow the ripening berry crop.

Wing, Beer, and Tidyman (1944) reported that broods occupy home ranges that are characterized by semiopen vegetation and available water. Relatively open areas are used by newly hatched chicks, while older broods move into more densely vegetated areas. Mussehl (1963) found that brood cover in Montana is consistently low (averaging seven to eight inches), has little bare ground (8 to 20 percent), and is predominantly herbaceous in nature, with grasses next in importance, followed by low shrubs and forbs. Woody cover increases in importance for food and escape cover as the birds mature.

FOOD AND FORAGING BEHAVIOR

In spite of the rather broad geographic range of the blue grouse, its food requirements appear to be fairly consistent. Martin, Zim, and Nelson (1951) report that Douglas fir was the most important food item in 158 samples from the northern Rocky Mountains, and in 154 samples from the Pacific northwest Douglas fir and true firs provided the major food items. They also list a variety of herbaceous plants and sources of berries that are used in summer and fall. Judd (1905b) indicated that winter blue grouse foods include ponderosa pine, Douglas fir, true firs (*Abies concolor* and *A. magnifica*), and hemlocks (*Tsuga heterophylla* and *T. mertensiana*).

Beer (1943) analyzed over one hundred crops and gizzards of blue grouse mostly from Washington and Oregon, and noted that adult foods were 98 percent plant materials, with conifer needles comprising 63.8 percent, berries 17 percent, miscellaneous plant materials 17.2 percent, and animal material 1.7 percent of the specimens examined. Beer noted that the grouse reach the peak of their morning feeding by 7 A.M. and stop by 9 A.M. Later feeding periods are just before noon, during late afternoon, and particularly toward evening, when the most intensive foraging of the day occurs. Growing young feed more continuously than adults, but those of all ages forage most heavily during the last three hours of daylight. Similar observations were made by Fowle (1960), who noted that although feeding occurred through the day, the greatest amount of food was consumed in the evening after 6 P.M. Males often alternate feeding with hooting, but females with young evidently restrict their foraging to the evening. Fowle never saw wild grouse drink water and believed it might not be important when berries or other succulent foods are available.

Hoffmann (1961) noted that blue grouse in California rely during the winter almost entirely on needles of white fir (*Abies concolor*), which he analyzed for protein content. He found that needles from high in the tree had a higher protein content than those from lower branches but that no apparent yearly differences occurred over a three-year period during which the grouse population suffered a major decline.

MOBILITY AND MOVEMENTS

Seasonal Movements.

The altitudinal movements of blue grouse to coniferous wintering areas has been reported for most areas; the exception being Hoffmann's study in California (1956). Doubtless the horizontal distances involved in movements between summering and wintering areas differ greatly in various regions, but relatively little detailed information is available. One banding study by Zwickel, Buss, and Brigham (1968) in north central Washington indicates that autumn migrations of blue grouse may be fairly long. The longest movement recorded by a banded bird was 31 miles, which occurred in less than two months. Of 30 birds recovered, 50 percent had moved over 5 miles, and 30 percent were recovered over 10 miles from where they had been banded. In contrast, Mussehl (1960) reported a maximum fall movement of 3.4 miles in Montana, while Bendell and Elliott (1967) found a maximum fall movement of 10 miles on Vancouver Island. Zwickel, Buss, and Brigham speculated that at least some breeding females leave their broods behind and return to their previous wintering areas, which stimulates wandering by young birds and the possible colonization of new wintering areas.

Daily Movements.

Evidently relatively little daily movement is performed by adult male blue grouse from the time they arrive on the summer range and establish territories until they begin their fall movement back to the wintering areas. Males probably establish territories as soon as weather conditions permit, and maintenance activities such as foraging, dusting, and sleeping are all carried out within the territorial boundaries (Bendell and Elliott, 1967). Territorial size presumably varies inversely with population density. In dense populations of about 0.44 males to the acre, Bendell and Elliott estimated that territorial sizes averaged about 1.5 acre. In sparse populations of about 0.13 males to the acre, territories were at least 5 acres in size.

Similarly, female grouse probably exhibit little daily movement, at least after fertilization has occurred. Until then they presumably move about through the territories of males until sufficiently stimulated to permit mating. Various studies of marked broods (Mussehl, 1960; Mussehl and Schladweiler, 1969) indicate that prior to dispersal the broods move about relatively little, and individual brood ranges may overlap considerably.

REPRODUCTIVE BEHAVIOR

Territorial Establishment

Male blue grouse evidently become territorial immediately after their arrival on the breeding range (Blackford, 1963) or as soon as snow cover conditions permit (Hoffmann, 1956). Territorial site requirements are somewhat ill-defined and may vary locally or with subspecies. In Colorado, Rogers (1968) states that display sites may be in aspen-ponderosa pine, mixed fir and aspen, open and dense aspen, mixed shrubs, sagebrush, wheat fields, and on roadbeds, but preference is shown for fairly open stands of trees or shrubs. Physical features include earth mounds, rocks, logs, cut banks, and occasionally tree limbs. Preference is generally given to flat, open ground, although steep slopes are also used. Display sites may be near heavy cover, but this is normally used for escape rather than for display. Two observations were made of birds displaying at more than twenty feet, but ground display is typical of interior populations of blue grouse.

In contrast, Hoffmann (1956) found that in a California population (*fuliginosus*) the males normally hooted from the tops of white fir or sometimes from Jeffrey pine (*Pinus jeffreyi*) or lodgepole pine (*P. contorta*). Bendell and Elliott (1966, 1967), studying the same subspecies on Vancouver Island, found that many hooting sites were elevated areas on the ground and that territories included diverse cover types, with males hooting from virtually all types of cover within their territory. In dense cover with small openings, territories are related to the location of openings. Thickets within territories are used for resting and concealment. This combination of open areas for display and shelter in the form of fir clumps, logs, or stumps used for hiding and as observation posts provides the basic territorial requirements. Several display sites may be used within a single territory; Rogers (1968) noted that from two to eleven hooting sites for one bird have been recorded.

Territorial Advertisement

Territorial proclamation by male blue grouse is achieved by a combination of postures, vocalizations, and movements that are collectively called hooting. In spite of reported differences in hooting behavior among different populations, current evidence indicates that actual differences are few and tend to be quantitative rather than qualitative. Thus, the interior populations (dusky grouse) have much weaker hooting calls that are barely audible more than fifty yards away, whereas the coastal populations (sooty grouse) have strong hooting notes that carry several hundred yards. The former typically call from the ground but may use trees, while the latter more often call from tree limbs. The gular sac of dusky grouse males is generally purplish, while that of sooty grouse is more heavily wrinkled and yellowish. The eye-combs of dusky grouse are large and vary from yellow to a bright red under maximum stimulation; those of sooty grouse are smaller and usually are lemon yellow, but sometimes also become livid red (Bendell and Elliott, 1967).

During hooting the male partially raises and spreads his tail and opens the feathers of his neck to expose an oval gular sac that is surrounded by white-based neck feathers, forming a "rosette" pattern. Both wings are slightly drooped toward the ground. In this posture (called the "oblique" by Hjorth, 1970) the gular sac is partially inflated in a pulsing manner as up to seven but usually five (in the dusky grouse) or six (in the sooty grouse) *hoot* sounds are uttered in fairly rapid succession. These are repeated at frequent intervals. Bent (1932) reported intervals of 12 to 36 seconds between call sequences of *fuliginosus*, Steward (1967) determined a mean interval of 24.2 seconds in *sitkensis*, and Rogers (1968) noted intervals of from 6 to 23 seconds for *obscurus*. Such hooting is uttered at various times during the day, but is most prevalent in early morning and again in late evening, primarily between 3 and 5 A.M. and again between 7 and 10 P.M. (Bendell, 1955c). Hjorth (1970) noted that although in both subspecies groups the call sequence lasts about three seconds, the fundamental frequencies of dusky grouse calls (95 to 100 Hz.) are lower than those of sooty grouse (100 to 150 Hz.) and have much less amplitude. Males may periodically move about between hooting sites, and while walking they keep the head low and the tail cocked and spread, exposing the spotted under tail coverts ("display walking" of Hjorth, 1970).

Strutting Displays

When in the presence of another grouse, the male stands in an erect posture with his tail tilted toward the other bird ("upright cum tail-tilting" of Hjorth, 1970), the eye-combs enlarged, and the wing away from the intruder drooped in proportion to the amount of tail-tilting. In this posture the male may perform vertical head-jerking movements, with the gular sac nearer the intruder expanding in synchrony with these head movements (Hjorth, 1970). Hjorth also reported that these downward head movements ("bowing cum asymmetric apteria display") may be greater in the dusky grouse group than in sooty grouse.

In this erect and tilted-tail posture, the male typically advances toward the intruder. Bendell and Elliott (1967) stated that in the sooty grouse the head and neck are held broadside to the other bird in such a way as to be framed against the background of the dark tail. Rogers (1968) has a photograph of the comparable posture of a Colorado dusky grouse. The approach display is climaxed by a quick, arcing dash toward the other bird ("rush cum single hoot" of Hjorth, 1970), which is associated with maximal tail-cocking and spreading, extreme engorgement of the eye-combs, and a drooping of the wings so that they drag on the ground. In this posture the male jerks his head several times, then lowers it and runs forward with short, fast steps, terminating the run with a deep *oop* or *whoot* note. Rogers (1968) noted that this sound could be heard as far as 510 feet away, in contrast to the hooting series in Colorado grouse, which could not be heard beyond 105 feet. Bendell and Elliott (1967), as well as Hjorth (1970), observed that it is actually a double note, with a short squeal or whistle following the deeper sound. Hjorth (1967, 1970) noted that during the forward dash the male deflates his neck, turns his tail toward the other bird, and holds his neck in such a way that the cervical rosette is maximally exposed. The head is held low, the tail is twisted to provide maximum surface exposure, and the wing on the far side is increasingly drooped as the tail is twisted. After the call is uttered the bird gradually assumes a normal posture again.

If the other bird is a receptive female she may remain in place, and the male then displays about her, raising and lowering his body and jerking his head, always keeping the neck rosette and nearer eye-comb in full view of the female. After two or three minutes of such display, the male moves behind the female and attempts to mount her. During treading the male grasps the nape of the hen in his beak, and holds her body against his lowered wings as she squats. Following treading the male again assumes his upright display posture (Hjorth, 1970).

Flight Displays

The other primary aspects of display by male blue grouse involve fluttering or flying movements which have been variably ritualized to produce sound and advertise the presence of the male. They are difficult to classify, since they have been described differently by various observers. Blackford (1958, 1963) attempted to classify these aerial displays based on his observations in Montana, which may be summarized as follows:

"Wing-fluttering" is a brief flapping of wings as the bird rises about eight or ten inches in the air, producing relatively little noise. It may be performed by either sex, both on the ground and in trees.

"Wing-drumming" is the typical male display flight, or flutter jump. It is a short, vertical leap into the air as the bird beats his wings strongly a few times before descending. Often one wing is beat much more strongly than the other, producing a rotary movement ("rotational drumming") and causing the bird to make an incomplete turn before landing.

"Wing-clapping," so far noted only by Blackford, is an upward leap associated with a single, very loud wing note.

"Drumming flight" was distinguished by Blackford from normal wing-drumming by the fact that a circular flight some ten to twelve feet in diameter is made before landing again near the takeoff point.

Several other possible wing signals were noted by Blackford (1963), including a "double wing flutter," a "perching signal," an "explosive flush," and an "aerial signal." Since they have not been well studied or described by others, they need not be given further consideration here.

Vocal Signals

Male vocalizations other than the *hoot* and *oop* calls are relatively few, judging from most accounts. Rogers (1968) reported a "gobbling" sound uttered by a male after making a clapping wing-beating flight to a branch. This was followed by regular hooting sounds until a single two-note *ca-caw* was uttered about eighteen minutes later.

Female vocalizations reported by Blackford (1958) include an in-flight alarm call, *kut-kut-kut,* a low warning note uttered before flight, *kr-r-r,* and an "excitement" call, *kutter-r-r-r,* which fluctuates greatly in pitch. Rogers (1968) noted that the in-flight alarm call of females was the note most commonly heard. Female blue grouse also produce a "whinny" call that is highly effective in stimulating males to begin hooting and to move toward the source of the sound. Use of tape-recordings of such calls is an effective method of censusing blue grouse (Stirling and Bendell, 1966). Likewise, recorded chick distress calls evoke clucking responses from broody hens.

Stirling and Bendell (1970) have recently reviewed the behavior and vocalizations of adult blue grouse. They described and presented sonagrams of three male calls, including the hooting call, the *whoot* call associated with the rush display, and a growling *gugugugug* associated with attack. Females were believed to have two calls related to reproduction: the "whinny," related to copulation readiness and the "quaver call" or *qua-qua* that consists of a pulsed series of notes produced by breeding females just prior to the time that males reach maximal reproductive development, thus possibly synchronizing breeding cycles. Females also utter a "hard cluck" or *bruck-duck* call, which apparently serves as a threat signal.

Collective Display

Although blue grouse are regarded as a species which normally defends fairly large territories and displays in a solitary fashion, several observations of collective display have been made. Bendell and Elliott (1967) noted that of 420 territorial males studied, the average distance between nearest territorial neighbors in open cover was approximately six hundred feet. In 5 percent of the one thousand-foot circular areas they studied there were seven or eight hooting males, which were usually two hundred to five hundred feet apart and formed a "hooting group" that usually called in chorus. They regard such hooting groups as indicating a habitat favorable for territories rather than as a variant of lek behavior, since, they point out, blue grouse remain on their territories through the breeding season, in contrast to typical lekking grouse. However, Blackford's observations (1958, 1963) of collective display indicated that males will leave their territorial sites and cross over adjacent territories to perform in a "communal court." In one case he noted that at least two males, two females, and one bird of unknown sex converged on the territory of another male, where collective display occurred. This kind of temporary establishment of collective display areas by males which perhaps follow females into the territory of an unusually effective resident male might provide the evolutionary basis for typical lek behavior, provided that such "hooting groups" are more efficient in attracting females than are individual males displaying in a solitary fashion.

Nesting and Brooding Behavior

Since the male plays no role in nest defense, incubation, or brooding, the female undertakes these duties alone. Evidently nearly all females, including yearlings, attempt to nest (Zwickel and Bendell, 1967). Further, most hens that fail to produce a brood of young do so because of nest destruction rather than nest desertion. Zwickel and Bendell (1967) found that of thirty nests found, twelve hatched successfully, eight had been deserted, and ten had been destroyed. The deserted nests were attributed to human disturbance. In that area, foxes and weasels were suggested as principal nest predators. How much renesting might occur after nest destruction or desertion in blue grouse is still uncertain, but Zwickel and Lance (1965) reported two definite instances indicating that renesting might occur even when the first nest is destroyed late in incubation and that a second clutch can be started within about fourteen days after such destruction.

Zwickel and Bendell (1967) found that fifty-one nests contained 323 eggs, or an average clutch size of 6.3 eggs. Gabrielson and Jewett (1940) reported that nine nests contained 74 eggs, averaging 8.2 eggs per clutch. Zwickel and Lance (1965) indicate that the laying rate for blue grouse is 1.5 days per egg and that the incubation period is 26 days.

Upon hatching, blue grouse chicks become fairly independent of the female relatively soon. Zwickel (1967) found that chicks began to eat plant materials at one day of age, can fly at six to seven days of age, and by two weeks of age can fly up to sixty meters. No chicks older than eleven days were observed being brooded by the hen, and few over seven days old were seen being brooded. Contrary to other writers, Zwickel (1967) doubts that chilling by rain or cold days normally plays an important role in chick survival. Zwickel noted several calls of brooding females. When the chicks wailed loudly with their distress note, the females uttered a low brood call, *cu-cu-cu*. While foraging, hens produced a similar but less audible series of notes that Zwickel terms a contact call. When calling the brood together, the female sometimes produced a high-pitched *kwa-kwa-kwa* call which the chicks responded to by wailing. When the hen returned to her brood after a considerable absence she would cluck loudly or produce a high-pitched *kweer-kweer-kweer* which was audible for up to ¼ mile under favorable conditions. Zwickel concluded that vocal signals were highly important in maintaining brood organization and exhibited considerable plasticity to meet varying needs.

Evidently most chick losses occur during the first two weeks of age, according to Zwickel and Bendell (1967). These authors present data indicating that brood sizes for chicks up to fourteen days old average from 3.3 to 4.4 young, while brood sizes for chicks estimated to be older than forty-two days average 2.9 to 3.7 young. Mussehl's study in Montana (1960) indicated that the movements of eight marked broods for periods of nineteen to forty-seven days were restricted to areas having maximum diameters of 440 to 1,320 yards. During early July these broods primarily used a mixed grass-forb cover, but with gradual drying of the prairie forbs they moved into deciduous thickets for the remainder of their brooding period. Little use of montane coniferous forest was noted. By the end of

August most of the brooding range had been abandoned, and broods began to disperse. Juveniles then moved singly or in small groups, with individual birds making lateral movements of up to 2.1 miles, as they worked their way up toward the wintering ranges.

EVOLUTIONARY RELATIONSHIPS

The blue grouse presumably had its evolutionary origin in western North America, either in a coniferous forest situation or in a forest-grassland edge habitat. Jehl (1969) concluded that two species of *Dendragapus* occurred in the western United States in the late Pleistocene, one of which presumably directly gave rise to the modern blue grouse. I believe that the ancestral blue grouse probably originated in North America, whereas the ancestral spruce grouse may have had its origins in eastern Asia, only later coming into contact with the blue grouse.

It seems probable that the sage grouse also had its origin in the western part of North America and may be much more closely related to the blue grouse than the adult plumage patterns would suggest. The surprising similarities of the downy young would support this view, and the strutting behavior patterns of the two species are not greatly different. To a much greater extent than is usually appreciated, the breeding habitat of the blue grouse is relatively arid and open, and the bird is in no sense a climax coniferous forest species.

I would suggest that North America was invaded relatively early from Asia by a *Tetrao*-like ancestral type, which as it moved southward produced the more montane-dwelling blue grouse ancestor, and also the intermontane or valley-dwelling sage grouse ancestor. A second invasion brought the spruce grouse into North America, possibly as recently as late Pleistocene times.

Johnsgard 1973

12

Spruce Grouse

Dendragapus canadensis (Linnaeus) 1758
(*Canachites canadensis* in *A.O.U. Check-list*)

OTHER VERNACULAR NAMES

BLACK partridge, Canada grouse, cedar partridge, fool-hen, Franklin grouse, heath hen, mountain grouse, spotted grouse, spruce partridge, swamp partridge, Tyee grouse, wood grouse.

RANGE

From central Alaska, Yukon, Mackenzie, northern Alberta, Saskatchewan, Manitoba, Ontario, Quebec, Labrador, and Cape Breton Island south to northeastern Oregon, central Idaho, western Montana, northwestern Wyoming, Manitoba, northern Minnesota, northern Wisconsin, Michigan, southern Ontario, northern New York, northern Vermont, northern New Hampshire, Maine, New Brunswick, and Nova Scotia (*A.O.U. Check-list*).

SUBSPECIES (*ex A.O.U. Check-list*)

D. c. canadensis (Linnaeus): Hudsonian spruce grouse. Resident in east central British Columbia, central Alberta, central Saskatchewan, southwestern Keewatin, northern Manitoba, northern Ontario, northern Quebec, and Labrador south to central Manitoba, central Ontario, and central Quebec. Introduced into Newfoundland in 1964 (Tuck, 1968).

D. c. franklinii (Douglas): Franklin spruce grouse. Resident from southeastern Alaska, central British Columbia, and west central Alberta south through the interior of Washington to northeastern Oregon, central Idaho, western Montana, and northwestern Wyoming.

D. c. canace (Linnaeus): Canada spruce grouse. Resident from southern Ontario, southern Quebec, New Brunswick, and Cape Breton Island south to northern Minnesota, northern Wisconsin, Michigan, northern New York, northern New Hampshire, northern Vermont, northern and eastern Maine, New Brunswick, and Nova Scotia.

D. c. atratus (Grinnell): Valdez spruce grouse. Resident in the coast region of southern Alaska from Bristol Bay to Cook Inlet, Prince William Sound, and perhaps Kodiak Island (no recent records).

MEASUREMENTS

Folded wing: Males, 161–92 mm; females, 159–91 mm (males average 2 mm longer).

Tail: Males, 107–44 mm; females, 94–119 mm. (Adult males of all races average over 120 mm; females, under 110 mm.)

IDENTIFICATION

Adults, 15–17 inches long. A species that is associated with coniferous forest throughout its range. The sexes are quite different in coloration, but both have brown or blackish tail feathers that are unbarred and are narrowly tipped with white (*franklinii*) or have a broad pale brownish terminal band. The upper tail coverts are relatively long (extending to about half the length of the exposed tail) and are either broadly tipped with white (in *franklinii*) or tipped more narrowly with grayish white. The under tail coverts of both sexes are likewise black with white tips (males) or barred (females). Feathering extends to the base of the toes. Males are generally marked with gray and black above, with a black throat and a well-defined black breast patch that is bordered with white-tipped feathers. The abdomen is mostly blackish, tipped with tawny (laterally) to white markings that become more conspicuous toward the tail. The bare skin above the eyes of males is scarlet red; no bare skin is present on the neck. The females are extensively barred on the head and underparts with black, gray, and ochraceous buff in varying proportions; the sides are predominantly ochraceous and the underparts are mostly white.

FIELD MARKS

In the eastern states and provinces spruce grouse are likely to be confused only with the ruffed grouse, from which the spruce grouse can be readily separated by the unbarred tail and the presence of a lighter tip rather than a darker band toward the tip of the tail. The conspicuous black and white markings of the underparts of males will distinguish spruce grouse from blue grouse, and the predominantly white underparts of females will help to distinguish them from the generally similar female blue grouse.

AGE AND SEX CRITERIA

Females may be distinguished from adult males by their tawny to whitish throats and breasts, barred with dark brown (these areas are black or black tipped with white in males). Accurate determination of sex in most races is possible by using either the breast feathers (males' breast feathers are black tipped with white, those of females are barred with brown) or by the tail feathers (males have black rectrices, tipped and lightly flecked with brown; females' are black or fuscous, heavily barred with brown). In *franklinii* the breast condition is the same, but the tails of females are barred or flecked with buffy or cinnamon brown, while the males have uniformly black tails or black tails flecked with gray (Zwickel and Martinsen, 1967).

Immatures resemble adults of their sex but the two outer juvenal primaries are more pointed than the others and (at least in *franklinii*) are narrowly marked with buff rather than whitish on the outer webs (Ridgway and Friedmann, 1946). Ellison (1968a) also reported that the tip of the ninth primary in immature Alaskan spruce grouse is mottled and edged with brown, while in adults it is only narrowly edged with brown.

Juveniles resemble adult females but have white or buffy markings at the tips of the upper wing coverts, as well as on their primaries and secondaries. Their tail feathers are dark brown, barred, speckled, and vermiculated with lighter markings (Ridgway and Friedmann, 1946).

Downy young are illustrated in color plate 61. The downy plumage of this species more closely resembles *Lagopus* than does that of the blue grouse and has a discrete chestnut brown crown patch that is margined with black. Downy spruce grouse lack the feathered toes of ptarmigan; however, they are also more generally rufous dorsally and have less definite patterning on the back.

DISTRIBUTION AND HABITAT

The over-all geographic distribution of the spruce grouse is a transcontinental band largely conforming to that of the boreal coniferous forest (Aldrich, 1963). East of the Rocky Mountains, the species' range generally conforms with that of the balsam fir (*Abies balsamea*) and also the black and white spruces (*Picea mariana* and *P. glauca*). In the Rocky and Cascade ranges the bird's southern limit occurs well north of the limits of montane and subalpine coniferous forest, suggesting that other limiting factors are influential in that area. What role competition with blue grouse might play in limiting the western range of the spruce grouse is unknown.

Probably only in the southeastern limits of its range have the populations of spruce grouse undergone serious reduction. In Michigan, where the species was once common to abundant, it had become noticeably reduced as early as 1912 (Ammann, 1963a). They are now uncommon on the Upper Peninsula and rare in six counties of the Lower Peninsula, and hunting was last permitted in 1914. In Michigan they are more often found associated with jack pines (*Pinus banksiana*) than with spruces.

In Minnesota, the spruce grouse was fairly abundant in coniferous forests as late as 1880 but almost completely disappeared with the cutting of this forest (Stenlund and Magnus, 1951). Roberts (1932) believed that the species was doomed to be extirpated from the state "before many years have passed." However, by 1940 the second-growth forest that had grown following lumbering began to develop an understory of conifers (especially black and white spruce) and jack pine, and the spruce grouse again became common in several northern areas (Stenlund and Magnus, 1951). In recent observations reported by these authors, associated cover type was most commonly jack pine, followed in order by black spruce, balsam fir (*Abies balsamea*), and tamarack (*Larix laricina*). Of seventy-nine observations, 44 percent were made in cover that was completely evergreen, and 72 percent were in upland cover rather than in lowland or swamp cover. Shrader (1944) has also noted recent population gains in the spruce grouse in Minnesota following its near extinction.

The situation in Wisconsin for spruce grouse is apparently still extremely unfavorable. Scott (1943, 1947) has documented the historical changes in spruce grouse populations of that state. His map indicated that the species probably originally extended across northern Wisconsin from Polk to Marinette county, but as of 1942 was limited to about ten counties, with an estimated population of five hundred to eight hundred birds.

Finally, in southern Ontario, spruce grouse have nearly disappeared from the area south of Lake Nipissing (Hamerstrom and Hamerstrom, 1961).

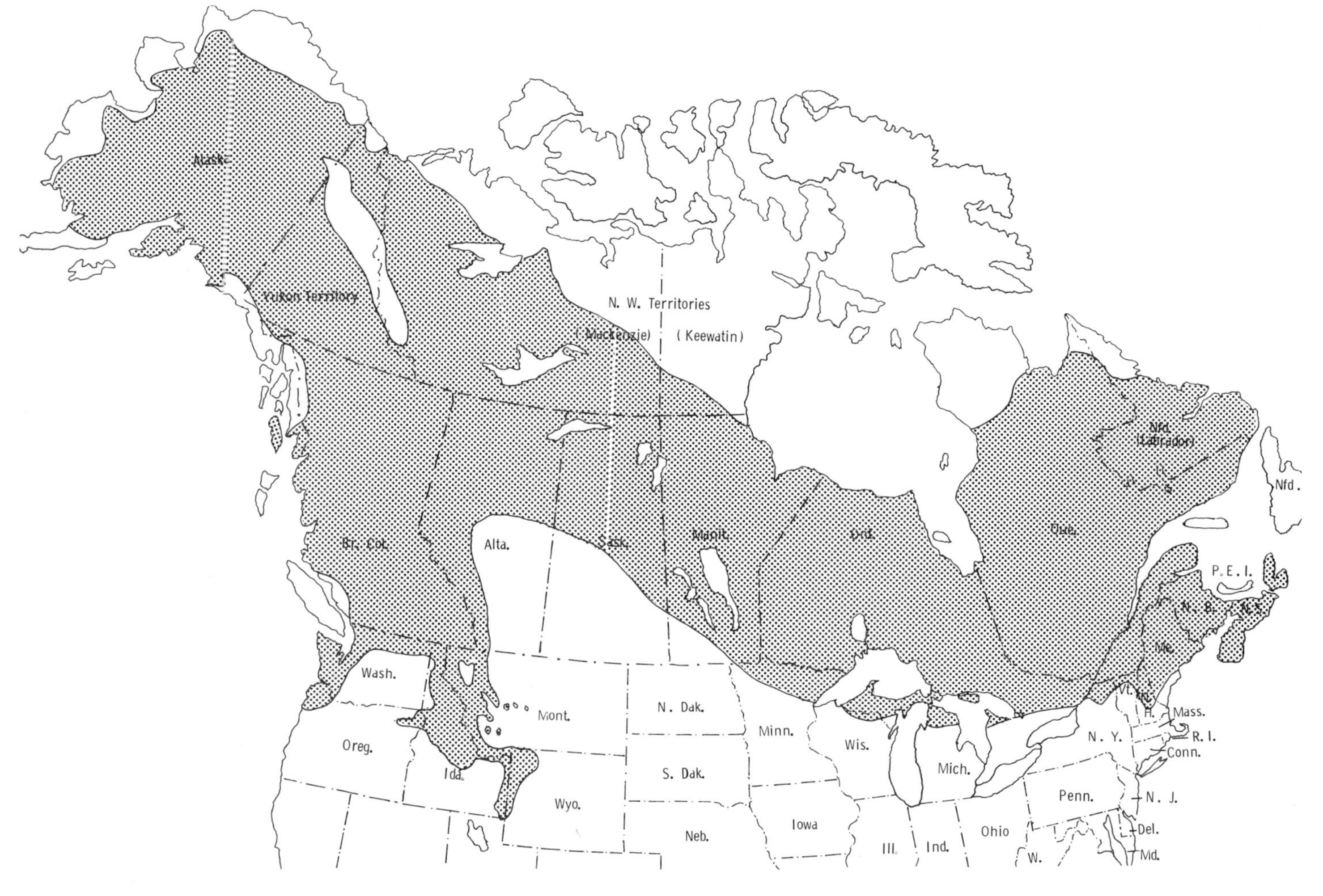

Figure 25. Current distribution of the spruce grouse.

Lumsden and Weeden (1963) pointed out that in the early 1960s spruce grouse had sufficiently high populations to be hunted in Maine, Montana, Washington, Idaho, Alaska, and all the Canadian provinces and territories except Nova Scotia (where it is protected) and Prince Albert Island (where it has been extirpated). In 1970, Minnesota allowed the hunting of spruce grouse as well, but it was still protected in Wisconsin, Michigan, New York, Vermont, and New Hampshire.

POPULATION DENSITY

Few estimates of population densities in spruce grouse are available. Robinson (1969) estimated a breeding density in northern Michigan of twenty to twenty-five birds (ten to twelve pairs) per square mile. Ellison (1968b) reported that a spring census of males in south central Alaska indicated a density of about ten males per square mile during two years and seven per square mile in a third year. He noted that this agrees with an estimate of seven males per square mile made by Stoneberg in Montana.

HABITAT REQUIREMENTS

A careful analysis of all the habitat needs of the spruce grouse remains to be done, but a recent study by Robinson (1969) provides a valuable analysis of summer habitat needs. By analyzing tree composition, as well as that of shrubs and low herbs, and comparing locations of spruce grouse sightings, a useful indication of habitat selection was obtained. Of 430 trees where spruce grouse were seen, 32 percent were spruces, although spruces (*Picea mariana* and *P. glauca*) made up only 3 percent of the tree cover. On the other hand, jack pines made up 91 percent of the tree composition but accounted for only 51 percent of the sightings. Pure stands of either jack pine or spruce were not used as much as mixed stands. In the shrub layer, young black spruces accounted for a larger proportion of spruce grouse sightings than would be expected from their relative abundance, while jack pines again provided a smaller proportion of sightings. Balsam firs at sighting points were more than seven times as abundant as at random sites. As to low vegetation, blueberry (*Vaccinium*), trailing arbutus (*Epigaea*), black spruce, and logs and stumps all were associated with higher than expected sightings of spruce grouse. In general, mature stands of either jack pine or spruce were not favored, apparently because of the lack of concealing cover at ground level. Robinson found that molting males used the same habitat in late summer as did females with broods and indeed were often seen accompanying broods. Robinson concluded that populations of spruce grouse in Michigan were highest in areas of boreal forest and jack pine forest. In one such area, the grouse selected habitats that had a mixture of spruces and jack pine, had a prevalence of young spruces in the shrub layer, and had a varied ground cover that included blueberries, trailing arbutus, and scattered stumps and logs.

In a comparable study of Alaskan spruce grouse, Ellison (1968b) noted that hilltops covered with white spruce, birch (*Betula*), and species of *Populus* were not a preferred habitat, although where an understory of alder was present some brood use and use by molting adults occurred in late summer. Two upland cover types provided preferred habitat. These were a white spruce and birch community with understories of grasses, spiraea, blueberry, and cranberry, and a black spruce community with a blueberry, cranberry, and lichen understory. Grouse sometimes also used dense lowland stands of black spruce, and broods were often found in stunted black spruce borders at the edges of bogs.

MacDonald (1968) noted that the habitat of the Franklin race of spruce grouse in Alberta consisted of lodgepole pine forests, with some clumps of aspen and poplar. Somewhat open stands of pines, some twenty to thirty feet tall, were evidently preferred areas for display by territorial males.

Winter habitat needs of the spruce grouse, to judge from their known food habits, consist simply of coniferous trees of various species that provide both food and cover requirements.

FOOD AND FORAGING BEHAVIOR

The survey by Martin, Zim, and Nelson (1951) indicated that spruce grouse in Canada and the northwest feed extensively on the needles of jack pine, white spruce, and larch and on the leaves and fruit of blueberries. A small fall and winter sample from British Columbia included a diverse array of berry species as well as lodgepole pine and spruce needles.

Jonkel and Greer (1963) analyzed crop contents during September and October in Montana and noted that western larch (*Larix occidentalis*) was an important early fall food but that it declined in use during October. Other important foods were needles of pine, spruce, and juniper, clover leaves, the fruits of huckleberry (*Vaccinium*), snowberry (*Symphoricarpos*), and white mandarin (*Streptopus*), and grasshoppers. A study by Crichton (1963) indicated that prior to snowfall in central Ontario, spruce grouse fed mostly on needles of jack pine and tamarack (*Larix laricina*) and the leaves of blueberries. After the shedding of the tamarack needles and the fall of snow, jack pine needles became almost the sole source of food in spite of a high availability of black spruce.

A seasonal analysis of spruce grouse foods in Alberta by Pendergast and Boag (1970) indicated that during winter lodgepole pine needles (*Pinus contorta*) made up nearly 100 percent of the food. In spring, the proportion of spruce needles to pine needles increased. The summer diet of adults was mostly ground vegetation, such as *Vaccinium* berries. In the fall the adults returned to feeding on conifers, but berries remained important. In contrast, chicks under a week old apparently subsisted entirely on arthropods. Later, they began to eat *Vaccinium* berries, but arthropods remained an important source of food through August. By October, the juveniles were starting to eat needles, and by November both the adults and young were using needles as a major food item.

A study in Alaska by Ellison (1966) yielded generally similar conclusions, except that the winter diet consisted primarily of needles of both black and white spruce. With spring, spruce was taken in decreasing amounts, and blueberry leaves, buds, and old cranberries were taken, as well as unripe crowberries (*Empetrum*). Summer food consisted largely of berries (crowberry, blueberry, and cranberry), and berry consumption continued into fall, as spruce needles again began to appear in the diet. Ellison reported that the protein content of spruce needles ranged from 5.7 to 6.3 percent, or about the same protein content as has been reported for Douglas fir and white fir.

MOBILITY AND MOVEMENTS

Spring Movements of Males.

Virtually the only detailed information on spruce grouse movements so far available is that provided by Ellison (1968b), who used radio transmitters to obtain movement data. He found that all adult males but only some yearling males established territories and became relatively sedentary. Those birds that were considered territorial remained localized on from 3 to 21 acres of forest during late April and most of May. Immature males considered nonterritorial occupied "activity centers" of from 6 to 16 acres during this time but also made fairly long trips of up to 1.25 miles from these centers, frequently entering the territories of other males in the process, evidently being attracted to them by displaying males. Interestingly, Ellison noted that in each year of the study, juvenile males tended to establish territories on the periphery of territories held by especially active territorial males, a tendency reminiscent of "hooting groups" of blue grouse, which has also been noted in ruffed grouse (Gullion, 1967). The actual estimated territorial size of four adult males ranged from 4.6 to 8.9 acres and averaged

6.9 acres. After May 21, these same males occupied larger home ranges of from 4.5 to 29.6 acres, averaging 20.1 acres. Considering four immature and territorial males as well, the maximum sizes of the home ranges of all eight males was 61 acres, while three of five nonterritorial males moved about over areas of 270 to 556 acres.

REPRODUCTIVE BEHAVIOR

Territorial Establishment

Ellison (1968b) reported that spruce grouse males established their territories and activity centers in stands of fairly dense spruce or stands of spruce and birch with trees some forty to sixty feet tall. Stands of trees up to eighty feet tall, with dense undercover, were sometimes used by non-territorial males but apparently were not suitable for territorial purposes. MacDonald (1968) indicated that pines from twenty to thirty feet tall which were not too closely spaced were preferred display sites. Stoneberg (1967) stated that of four males he studied, three displayed in small openings in dense forest, while one was in less dense forest. He estimated that the four marked males he studied had home ranges of ten to fifteen acres. Two remained in very localized sites during the display period, while one of the other two used several display sites within a twenty-five-yard radius, and the last moved about extensively and used no specific sites. However, this last bird was the only one that had no female on his territory at the time. MacDonald thought that males have favored display sites within their home ranges but that the latter are too large to have definite boundaries except in areas of contact with adjacent males.

Both Stoneberg and Ellison reported that display flights (drumming flights or wing-clapping flights) were performed in openings rather than in dense forest. Ellison described the ground vegetation of such openings as low, rarely more than 1.5 feet in height, and usually consisting of mosses, lichens, and *Vaccinium* species.

Territorial Advertisement

Several detailed accounts of strutting behavior are now available. Displays of the Franklin race of spruce grouse have been described by Stoneberg (1967) and MacDonald (1968), and those of the nominate race by a number of writers, including Bishop (in Bendire, 1892), Breckenridge (in Roberts, 1932), Harper (1958), and Lumsden (1961a). Only a few differences appear to be present in the two forms, as will be noted below.

The basic male advertisment or "strutting" display consists of a standing posture ("upright" of Hjorth, 1970). In this posture the tail is cocked at an angle of from about 70 to 90 degrees, exposing the white-tipped under tail coverts that are held out at varying angles, the neck is fairly erect, the wings are slightly drooped, and the crimson eye-combs are engorged. The throat feathers are lowered to form a slight "beard," and the lateral black neck feathers are lifted as are the lower white-tipped feathers at the sides of the neck and the upper breast. No bare skin is exposed, but the pattern of feather erection is much like that of the male blue grouse. Lumsden has noted that the esophagus is evidently slightly inflated as well, but no hooting sound is normally heard. However, an extremely low-pitched sound (ca. 85–90 Hz.) may be produced by male spruce grouse (Stoneberg, 1967; Greenewalt, 1968). Stoneberg heard series of such notes ranging from one to four, and I have heard similar sounds coming from boxes containing several recently trapped males and females. MacDonald likewise heard hooting sounds apparently produced by a male when it rushed toward a female. However, Hjorth (1970) questioned on anatomical grounds whether male spruce grouse can produce such low-pitched sounds, believing that reports of such calling were the result of confusion with blue grouse hooting.

When in the strutting posture, the male usually walks forward with deliberate paces, typically spreading the rectrices on the opposite side as it raises each foot, making the spread tail asymmetrical ("display walking cum tail-swaying" of Hjorth, 1970). This lateral tail movement, which produces a soft rustling sound, may also occur when the bird is not walking, as has been noted by Stoneberg as well as by me. A similar display is tail-fanning, in which the rectrices of both sides are quickly fanned and shut again. This also produces a rustling sound and may occur during walking or when the bird is standing still, often alternating with tail-flicking. On one occasion I saw a male performing tail-fanning before a female as it uttered a series of low hissing notes that started slowly and gradually speeded up, with a fan of the tail accompanying each note. Lumsden (1961a) described this as occurring when a male observed his reflection in a mirror. Michael Flieg* informed me that a similar tail-fanning during calling is typical of the capercaillie.

When approaching a female in the strutting posture, the male may perform several displays that have been given different names by various writers. One is a vertical head-bobbing, which may grade into or alternate with ground-pecking (Harper, 1958; Lumsden, 1961a; Stoneberg, 1967; MacDonald, 1968). During the pecking movements the male faces the

*Michael Flieg, 1970: personal communication.

female and often tilts its head to the side, thus exposing both combs to her view. Wing-flicking may likewise occur at this time (Stoneberg); Harper also noticed what appeared to be wing-beating movements suggestive of the ruffed grouse's drumming.

Two other major male displays occur in the situation of close approach to a female by the male. These are the "neck-jerk" display described by Lumsden, which MacDonald preferred to call the "squatting" display; and the "tail-flick" described by Lumsden, but which Stoneberg calls the "head-on rush."

The tail-flicking, or head-on rush, display (called the "rush cum momentary tail-fanning" by Hjorth, 1970) is apparently homologous to the short forward rush of the male blue grouse. It begins with the male's making several short and rapid steps toward the female, stopping a few inches away, partially lowering its head, and suddenly snapping its tail open with a swishing sound. The wings are simultaneously lowered to the ground, and a hissing vocalization is uttered, followed by a high-pitched squeak. The wings are then withdrawn leaving the alulae exposed, the tail is closed, and the head is tipped downward with the neck still extended diagonally. In this rigid posture the tail is fanned a second time and is held open longer. During this display the male is usually oriented so that his head faces the female, exposing to her view the visual effect of the eye-combs, fanned tail, and contrasting breast coloration. In the Franklin race the white-tipped upper tail coverts are made conspicuous by the tail movements, but they are not evident in the nominate race. MacDonald noted that during this display (which he described under the general tail-swishing display) a single, soft hooting noise could be heard at very close range.

The squatting display is performed by the male as a possible precopulatory signal according to Lumsden, and MacDonald agreed with this interpretation but notes that it is sometimes omitted from the sequence. As the male approaches the female, the head-on rushes (or arcing rushes, since MacDonald indicates that the male may move in arcs in front of the hen) increase in frequency until he is quite close to her. After watching her intently for several seconds, the male sinks to the ground in a squatting position, with neck stretched, head nearly parallel to the ground, and tail held in a vertical and partially spread position, while the wings are slightly spread and lowered. This display has been observed only once by the writer, to whom it closely resembled the "nuptial bow" of pinnated grouse, which serves as a precopulatory display in that species. Hjorth (1967) illustrates the posture and agrees that it is homologous to the nuptial bow of prairie grouse. He believes that it is stimulated when the male's displays elicit neither attack nor pairing behavior.

Squatting as described by MacDonald probably does not correspond to the typical head-jerk as described by Lumsden and Stoneberg, since MacDonald mentions no actual head-jerking movements and I likewise noted none during one observation of the squatting display. Lumsden mentions seeing repeated, sudden upward movements of the head, first to one side, then to the other, as well as occasional circular head movements. With each upward movement the tail was fanned open and again shut, producing the usual rustling sound. Stoneberg noted two types of head-jerking movements, one of which was a rapid tossing of the head from one side to the other for up to three seconds, pausing and repeating it, with the tail kept vertical and the head near the ground. A slower type of head-jerking was associated with strutting, when the bird would stop, facing the female, and jerk the head from one side to the other while fanning or flicking his tail.

Aggressive male displays of the spruce grouse consist of at least two postures. MacDonald reports that then two males meet at a distance the resident territorial male sleeks his plumage, raises his tail, and flashes the lateral rectrices and upper tail coverts, uttering a series of gutteral notes. These notes no doubt correspond to the calls I heard from a male when I interrupted his strutting, which Lumsden describes as harsh hissing sounds. Stoneberg describes the rapid notes as "throaty kuks." The male then runs toward the opponent with the head low, neck extended, and the tail down (Lumsden's "head and tail down" display posture), with the wings held slightly away from the flanks. MacDonald found that such behavior was enough to cause a trespasser to fly away or at least to fly into a tree. When a mounted male is used or a mirror is set up, actual attack behavior may be elicited. Stoneberg found that by placing bright red pieces of felt on a male skin, he was able to elicit strong attack behavior. The male approached the skin with plumage sleeked except for the chin feathers, paused, then leaped at the skin, beating his wings and pecking at the head and breast. After a second attack, the male had succeeded in removing the combs as well as the feathers and skin from the neck and upper breast.

Aerial Displays

In contrast to the terrestrial displays of spruce grouse, some population variation may occur in the aerial displays of males. Lumsden has summarized the observations of aerial display by the nominate subspecies, which apparently consists of several variations. One of these is a short, vertical flight from a few to about fourteen feet in the air, drumming on suspended wings, and fluttering back to the ground. This behavior is closest to the typical flutter-jump of prairie grouse. More commonly, however, the male flies either vertically upward or horizontally toward a tree perch, checks its flight, and either lands on the perch or drops back to earth. If it lands on the elevated perch it may stay there varying lengths of time; Lumsden reports periods as short as ten seconds and as long as four minutes. The flight back down is always performed in the same manner, by dropping steeply downward until the bird is about four to six feet from the ground, then swinging the body into a nearly vertical position, and descending on strongly beating wings toward the ground. Although the drumming sound produced by the wing-beats can be heard as far as two hundred yards away, neither Lumsden nor Ellison (1968b) reported any wing-clapping sounds by males of this race, nor have other prior observers. Apparently no vocal calls are uttered during the flight.

Descriptions of the aerial display flights of the Franklin race are somewhat at variance with this general situation. Stoneberg (1967) states that the downward phase of the flight is as Lumsden described except that during the final drop to the ground two loud sounds are produced, apparently by clapping the wings together. Once Stoneberg heard wing-clapping before the bird landed in a tree, and in two of forty-five cases only one rather than two clapping sounds were produced. The wing-clapping display was most commonly heard near sunrise and sunset but often could be heard during the middle of the day as well. Stoneberg believed that cool temperatures favored the display.

MacDonald's observations of wing-clapping are unusually complete, and he regarded the display as being an advertisement of the location of territorial males. He noted that the wing-clap flight was never started from the ground but always from some elevated site. Flying out from a branch some ten to twenty feet high, the male moves on shallow wing-beats through the trees, with tail spread and tail coverts conspicuous. On reaching the edge of a clearing, he rises slightly, makes a deep wing-stroke, and brings the wings together above the back, producing a loud cracking sound. A second clap follows as the bird drops vertically toward the ground. The male soon selects another branch overhead and begins the sequence again. MacDonald noted that a resident male wing-clapped in the presence of an intruder, and after it had driven it away, began a sequence of vigorous displays and wing-clapping.

According to MacDonald, the vertical flight to a perch may be followed by display on the perch prior to launching into the wing-clapping display. He reported that after alighting on a branch and prior to the wing-clapping flight, the male may perform either or both of two different displays. These include a short rush along the branch followed by a spreading of the wings and tail, closing them, and again spreading the tail, apparently a variant of the tail-flicking display. A second display consists of three or four shallow wingstrokes, like the drumming of a ruffed grouse, producing a similar thumping sound.

Vocal Signals.

Two distinct vocal signals of males have been mentioned; one of these is the low-pitched "hoot" of a male in a sexual situation. These calls may be uttered as single notes or may occur in a series of notes roughly half a second apart (Greenewalt, 1968). They are notable for their extremely low-frequency characteristics of less than 100 Hz.

Males also utter a series of rather gutteral notes in aggressive situations. When I placed an adult and immature male in a box together, both birds produced such calls. These usually consisted of two preliminary low, growling *kwerr* notes, followed by from two to eight more rapidly repeated *kut* notes. Ocasionally the two types of calls were uttered independently of one another. In the younger male the calls were given at a noticeably higher pitch than in the adult male.

Female spruce grouse produced at least three different types of notes under caged conditions. The loudest and highest pitched was a repeated squealing or whining *keee'rrr* call that resembled the distress call of various quail species. Females also uttered a softer series of *pit, pit, pit* notes when disturbed and a fairly low-pitched gutteral *kwerrr,* which presumably correspond to the two types of agonistic male notes mentioned above. When in a tree looking down on a human or other potential enemy, females utter a series of clucking sounds that quickly reveal their presence. Bent (1932) described these as *kruk, kruk, kruk* sounds, and a *krrrruk* that no doubt corresponds to the *kwerrr* note mentioned above. In-flight alarm notes have not been reported.

Nesting and Brooding Behavior

There is no evidence that the male spruce grouse participates in nest or brood defense, although males may often be seen with females and well-grown broods in early fall. I observed this in southern Ontario during September of 1970, when at least four males were seen associated with females and broods. However, no attempt was made by the male to defend the brood; instead he simply appeared intent on displaying to the adult female.

Nests of the spruce grouse are usually situated in a well-concealed location, often under low branches, in brush, or in deep moss in or near spruce thickets. Ellison (Alaska Dept. of Fish and Game, *Game Bird Reports*, vols. 7–9, 1966–68) reported on nineteen nest locations, fourteen of which were in open, mature white spruce, birch, or spruce-birch-alder acotones, while two were in open black spruce, two were in moderately dense black spruce, and one was in a mixture of alder and grass. Of twenty-one nests he found, the clutches ranged from 4 to 9 eggs, and averaged 7.4. Tufts (1961) reported clutch sizes for thirty-nine nests, which ranged from 4 to 10 eggs and averaged 5.8. Robinson and Maxwell (1968) could find no authenticated record of a clutch of more than 10 eggs, and concluded that earlier estimates of larger clutches were in error. One instance of definite renesting has been found by Ellison (*Game Bird Reports*, vol. 9, 1968). Pendergast and Boag (1971) have reported the incubation period to be twenty-one days.

Robinson and Maxwell (1968) noted that when hens had chicks younger than ten days old (when fledging occurs) the female is highly aggressive and may make threatening movements that resemble male strutting behavior. If the attack fails to deter the intruder, a "sneak" distraction display resembling a "broken-wing act" may occur but without actual injury-feigning. In the case of hens with older broods, females may utter warning calls, but by that time they are much less aggressive toward intruders.

EVOLUTIONARY RELATIONSHIPS

Short's recommendation (1967) that *Canachites* be merged with *Dendragapus* appears to me to be fully warranted, for reasons which he outlined. It would seem that the nearest living relative to the spruce grouse is *Dendragapus ("Falcipennis") falcipennis*, the Siberian spruce or sharp-winged grouse, since it not only occupies a very similar habitat but evidently has nearly identical courtship displays (Short, 1967; Hjorth, 1970). Some similarities in courtship characteristics between the spruce grouse and the blue grouse are also evident, including the short run toward the female followed by a single-note call, the production of very low-pitched hooting sounds, the tail-fanning displays, and the drumming flight behavior. Some interesting features of the male spruce grouse display also suggest affinities with the capercaillie. These include the general posture, the erection of the chin feathers to form a "beard," and calling with simultaneous tail-fanning. The general plumage appearance of both sexes is also very similar in these two species and the Siberian spruce grouse. Similarities between the display of the capercaillie and the Siberian spruce grouse have also been noted (Kaplanov, in Dement'ev and Gladkov, 1967).

It seems probable that the evolutionary origin of the spruce grouse was in eastern Asia, where separation into two populations gave rise to the Siberian spruce grouse and the North American spruce grouse, the latter of which gradually moved southward and eastward through boreal forest and western coniferous forests. Contacts in the west with early blue grouse stock may have provided the selective pressure favoring the evolution of conspicuous upper tail covert patterning and wing-clapping during aerial display as sources of reinforcement of isolating mechanism differences between these two related types. There is apparently no fossil record of either *"Canachites"* or *"Falcipennis"* except for a late Pleistocene specimen from Virginia, whereas typical *Dendragapus* fossil remains are known from several localities in the western states (Jehl, 1969).

13

Willow Ptarmigan

Lagopus lagopus (Linnaeus) 1758

OTHER VERNACULAR NAMES

ALASKA ptarmigan, Alexander ptarmigan, Allen ptarmigan, Arctic grouse, red grouse (Scotland form), Scottish grouse, white grouse, white-shafted ptarmigan, willow grouse, willow partridge.

RANGE

Circumpolar. In North America from northern Alaska, Banks Island, Melville Island, Victoria Island, Boothia Peninsula, Southampton Island, Baffin Island, and central Greenland south to the Alaska Peninsula, southeastern Alaska, central British Columbia, Alberta, Saskatchewan, Manitoba, central Ontario, central Quebec, and Newfoundland (*ex A.O.U. Check-list*).

NORTH AMERICAN SUBSPECIES (*ex A.O.U. Check-list*)

L. l. albus (Gmelin): Keewatin willow ptarmigan. Breeds from northern Yukon, northwestern and central Mackenzie, northeastern Manitoba, northern Ontario, and south central Quebec south to central British Columbia, northern Alberta and northern Saskatchewan, and the Gulf of St. Lawrence in Quebec. Wanders farther south in winter.

L. l. alascensis Swarth: Alaska willow ptarmigan. Breeds from northern Alaska south through most of Alaska. Winters in southern part of breeding range.

L. l. alexandrae Grinnell: Alexander willow ptarmigan. Resident on the Alaska Peninsula south to northwestern British Columbia.

L. l. ungavus Riley: Ungava willow ptarmigan. Resident in northern Quebec and northern Labrador south to central Ungava.

L. l. leucopterus Taverner: Baffin Island willow ptarmigan. Resident from southern Banks Island and adjacent mainland to Southampton and southern Baffin islands; wanders farther south in winter.

L. l. alleni Stejneger: Newfoundland willow ptarmigan. Resident in Newfoundland.

L. l. muriei Gabrielson and Lincoln: Aleutian willow ptarmigan. Resident in the Aleutian Islands from Atka to Unimak, the Shumagin Islands, and Kodiak.

MEASUREMENTS

Folded wing: Adult males, 182–216 mm; adult females, 168–214 mm (males average 190 mm or more; females [except Baffin Island race] average less than 190 mm).

Tail: Adult males, 108–35 mm; adult females, 94–139 mm (males average 118 mm or more, females 116 mm or less).

IDENTIFICATION

Adults, 14–17 inches long. All ptarmigan differ from other grouse in that (except during molt) their feet are feathered to the tips of their toes (winter) or base of their toes (midsummer) and their upper tail coverts extend to the tips of their tails. The primaries and secondaries of all the North American populations of this species are white in adults throughout the year, while in winter all the feathers are white except for the dark tail feathers, which may be concealed by the long coverts. Males have a scarlet "comb" above the eyes (most conspicuous in spring) and during spring and summer are extensively rusty hazel to chestnut with darker barring above except for the wings and tail. The tail feathers are dark brown tipped with white except for the central pair, which resemble the upper tail coverts. In summer females lack this chestnut color and are heavily barred with dark brown and ochre. In autumn the male is considerably lighter, and the upperparts are heavily barred with dark brown and ochraceous markings, lacking the fine vermiculated pattern found in males of the other ptarmigans at this season. The female in autumn is similar to the male but is more grayish above and more extensively white below. In winter both sexes are entirely white except for the tail feathers, of which all but the central pair are dark brownish black. In addition, the shafts of the primaries are typically dusky and the crown feathers of males are blackish at their bases. In first-winter males and females the bases of these feathers are grayish.

FIELD MARKS

The dark tail of both sexes at all seasons separates the willow ptarmigan from the white-tailed ptarmigan but not from the rock ptarmigan. In spring and summer the male willow ptarmigan is much more reddish than the rock ptarmigan, and although the females are very similar, the willow ptarmigan's bill is distinctly larger and higher and is grayish at the base. In fall males are more heavily barred than are male rock ptarmigan, and females likewise have stronger markings than do female rock ptarmigan. In winter males lack the black eye markings that occur in male rock ptarmigan, but since this mark may be lacking in females, the heavier bill should be relied upon to distinguish willow ptarmigan.

AGE AND SEX CRITERIA

Females lack the conspicuous bright reddish "eyebrows" of adult males, are more grayish brown and more heavily barred on the breast and flanks than are males, and lack the distinctive rusty brown color of males in summer. In fall, females are somewhat grayer above and more heavily barred on the breast and flanks than are males. In winter they are like males but the concealed bases of the crown feathers are more grayish (Ridgway and Friedmann, 1946). They can be fairly accurately identified at this time by their brown rather than black tail feathers and central upper tail coverts and by certain wing and tail measurements (Bergerud, Peters, and McGrath, 1963).

Immatures in first-winter plumage tend to have the tip of the tenth primary more pointed than the inner ones, but this is not so reliable as the fact that (1) there is little or no difference in the amount of gloss on the three outer primaries of adults, whereas immatures have less gloss on the outer two primaries than on the eighth, and (2) there is about the same amount of black pigment on primaries eight and nine (sometimes more on eight than on nine) of adults whereas juveniles have more on the ninth than on the eighth (Bergerud, Peters, and McGrath, 1963).

Juveniles may be identified by the fact that their secondaries and inner eight primaries are grayish brown with pale pinkish buff margins or barring. However, the late-growing outer two primaries are white, often speckled with black, like the first-winter flight feathers that soon replace the secondaries and inner primaries.

Downy young are illustrated in color plate 61. Willow ptarmigan downies are reported (in the Scottish population) to be darker on both the dark and lighter areas, and have less clear-cut margins between these areas than downy rock ptarmigan (Watson, Parr, and Lumsden, 1969). These authors mention other differences that may also serve to separate downy young of these two species, although these may not apply equally well to North American populations. For example, in the Labrador populations, birds under three weeks are almost impossible to identify as to species, although young willow ptarmigan are slightly darker and somewhat greenish instead of yellowish on the underparts (Bendire, 1892). After three weeks they may be distinguished by differences in the bill.

DISTRIBUTION AND HABITAT

The North American breeding range of the willow ptarmigan is primarily arctic tundra, although it extends southward somewhat in alpine mountain ranges and in tundra-like openings of boreal forest (Aldrich, 1963). The basic habitat consists of low shrub, particularly willow or birch, in lower or moister portions of tundra. Weeden (1965b) has characterized the general breeding habitat of willow ptarmigan as follows: Typical terrain is generally level or varies to gentle or moderate slopes but frequently is at the bottom of valleys. Vegetation is relatively luxuriant, with shrubs usually three to eight feet high, and scattered through areas dominated by grasses, hedges, mosses, dwarf shrubs, and low herbs. The birds usually occur at the upper edge of timberline, among widely scattered trees, or may occur somewhat below timberline where local treeless areas occur.

Because of the relatively minor effect man has had on tundra to date, there have been few if any major evident changes in the total range of the species.

POPULATION DENSITY

Ptarmigans are among the arctic-dwelling species that exhibit major fluctuations in yearly abundance and are believed by some to exhibit cyclic

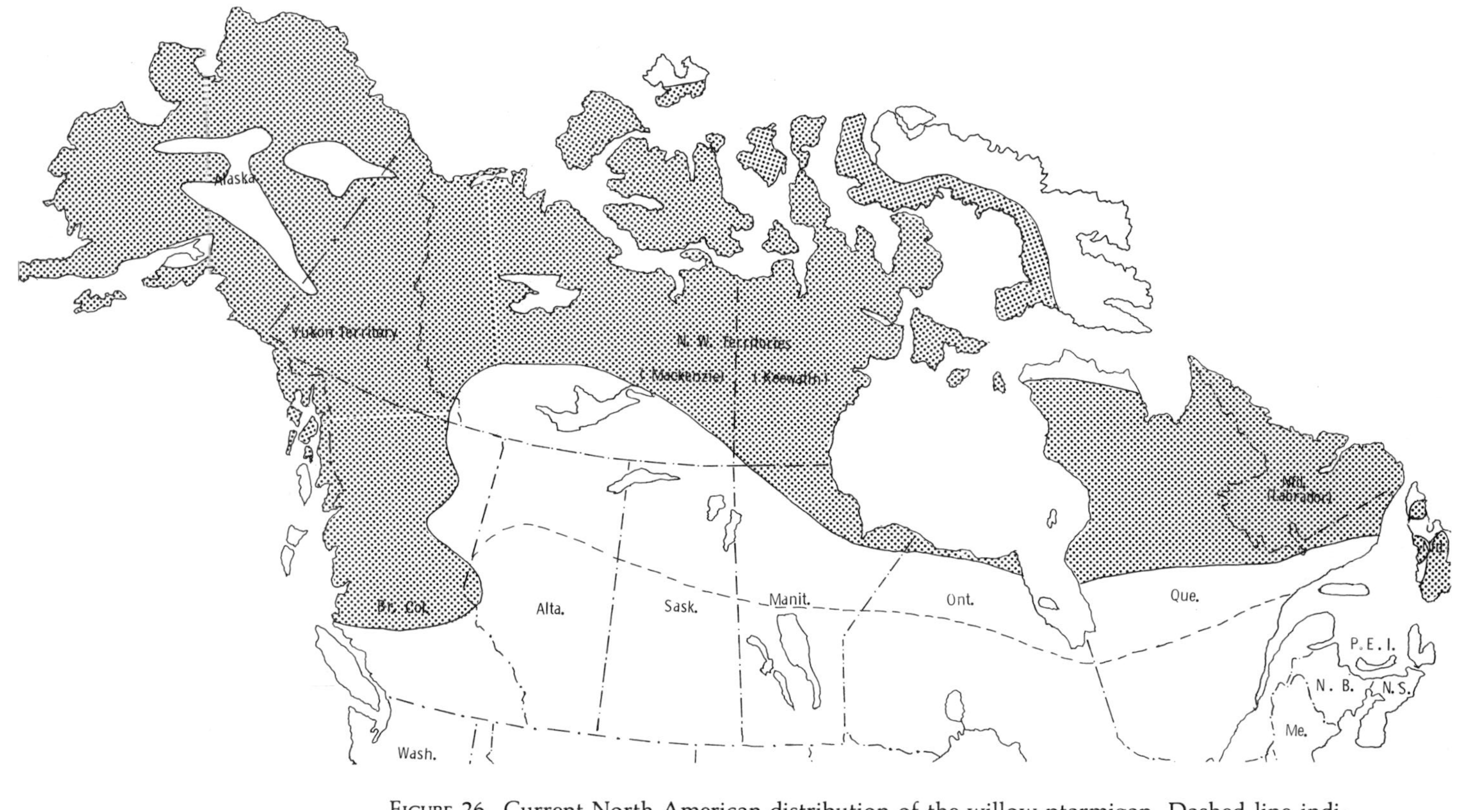

FIGURE 26. Current North American distribution of the willow ptarmigan. Dashed line indicates normal southern wintering limits.

population changes (Buckley, 1954). In any case, major changes in population density do occur, thus estimates of density may vary greatly by year as well as by locality. Weeden (1963) summarized estimates of population density for various areas in Canada. These estimates ranged from less than 1 adult per square mile (2.5 square miles per adult) to as many as 8 adults per square mile, with the sparser densities generally based on large areas that include much unfavorable habitat. He also reported (1965b) that a study area of 0.75 square miles had spring populations of males ranging from 38 to 150 males during seven years of study, which represents from 3.2 to 12.3 acres per male. Somewhat comparable density figures have been reported from Newfoundland (Mercer and McGrath, 1963), who estimated spring 1962 populations on Brunette Island of from 147 to 207 birds per square mile, depending on technique used. Considerable population work has been done on the Scottish red grouse (now generally considered conspecific with the willow ptarmigan) by Jenkins, Watson, and Miller (1963), who estimated spring densities of from 4.5 to 9 acres per pair.

HABITAT REQUIREMENTS

Wintering Requirements

Weeden (1965b) reported that winter habitat of willow ptarmigan consists of willow thickets along streams, areas of tall shrubs, and scattered trees around timberline and burns, muskegs, and river banks below timberline. Bent (1932) noted that in winter willow ptarmigan move to interior valleys, river bottoms, and creek beds, where there is available food in the form of tree buds and twigs of willows (primarily), alders and spruces, and such berries and fruits as can be found above the snow. Godfrey (1966) indicates that during winter the birds may be found well south of treeline, in muskegs, lake and river margins, and forest openings.

Spring Habitat Requirements

Weeden (1965b) stated that male habitat preferences for territorial establishment include shrubby and "open" vegetation, with the plants lower than eye-level for ptarmigan. Elevated sites such as rocks, trees, or hummocks are used by males during display. Resting areas are provided by small clumps of shrubs at the edges of open areas.

At least in Scotland, territorial establishment occurs during fall, although territories may be abandoned temporarily during winter if snow conditions require it. In Alaska some fall display and calling also occurs (Weeden,

1965b). Continued residence, however, is not typical in Alaska or probably in any part of the North American breeding range, since considerable seasonal movement is typical. Thus, local topography, as it affects snow deposit and rate of snow thaw exposing territorial sites, may have considerable effect on territorial distribution of birds in arctic North America.

Nesting and Brooding Requirements

Requirements for willow ptarmigan nest sites are apparently fairly generalized. Brandt (in Bent, 1932) reported that nesting may occur anywhere from coastal beaches to mountainous areas, and nests may be placed beside drift logs, in grass clumps, under bushes, in mossy hummocks, or similar sources of screening protection. Weeden (1965b) indicates that the nest is usually protected from above and the side by shrubby vegetation, while one side borders an open area. The nest is located within the periphery of the male's territory.

Brooding habitat is similar to nesting habitat, according to Weeden (1965b), with chicks using areas of very low vegetation, while older broods use shrub thickets for escape cover. Maher (1959) noted that broods used a variety of habitats with good cover and were common on upland dwarf-shrub and hedge tundra, as well as being sometimes found in riparian shrub and willow shrub at the bases of hills.

FOOD AND FORAGING BEHAVIOR

At least in Alaska, the most important single food source for willow ptarmigan is provided by willow buds and twigs. Weeden (1965b) noted that this source provided almost 80 percent of winter foods found in 160 crops from interior Alaska, and Irving et al. (1967) also indicated that winter foods consisted almost entirely of the buds and twig tips of willow. Weeden noted that dwarf birch buds and catkins were second in importance, and Irving et al. similarly found that in wooded areas some birch catkins and poplar buds are taken. West and Meng (1966) found that 94 percent of the winter diet of willow ptarmigan from northern Alaska consisted of various willow species, and 80 percent was from a single species (*Salix alaxensis*). They also noted that some birch may be used, but although alder is often available and has a higher caloric content than willow, it is seldom used.

One exception to the general winter diet of willow for North American willow ptarmigan has been noted, in Peters's study (1958) of the Newfoundland population. He found that the winter diet consists almost entirely of the buds and twigs of *Vaccinium* species, the buds and catkins of birch and alder, and the buds of sweet gale (*Myrica*).

With spring, the willow ptarmigan's dependence on willow declines in Alaska, and in addition to the leaves of willow, the birds begin to eat a larger variety of leafy materials (Weeden, 1965b). Peters (1958) also noted a spring return to fruit and leafy materials and the berry seeds of crowberry (*Empetrum*) and *Vaccinium*.

Summer foods in Alaska consist of various berries, especially blueberries, willow and blueberry leaves, and the tips of horsetail (*Equisetum*), which grows in willow thickets near streams (Weeden, 1965b). Peters noted that crowberries, blueberries, and the leaves of *Vaccinium* species, especially *V. angustifolium*, provided major sources of summer foods in Newfoundland.

In the fall, as the berry supplies are exhausted and leaves fall from *Vaccinium* bushes, the ptarmigan in Newfoundland return to a diet of buds and twigs again (Peters, 1958). The same situation applies to Alaska, although it is willow rather than *Vaccinium* buds and twigs that are resorted to (Weeden, 1965b). Irving et al. (1967) found a gradual increase in total crop contents of Alaskan ptarmigan from October to January, followed by reduced contents until April. This population migrates southward in October and November and northward from January until May. Evidently feeding is related to changing patterns of daylight, rather than to temperature cycles or to the cycle of migratory activities.

MOBILITY AND MOVEMENTS

The willow ptarmigan and its relative the rock ptarmigan are perhaps the most migratory of all North American upland game. Snyder (1957) reports that the willow ptarmigan is migratory to a considerable extent, occasionally wandering as far as the southern parts of the prairie provinces, northern Minnesota, the north shore of Lake Superior, southern Ontario, and southern Quebec. To some degree these southern movements may be related to unusually dense populations in the northern areas (Buckley, 1954). Evidently considerable differential movement according to sex occurs in Alaska (Weeden, 1964). At Anaktuvuk Pass, for example, most wintering willow ptarmigan are males, while many of the wintering birds in timbered areas to the south are females. Likewise, alpine-fringe areas of the Alaska Range and the Tanana Hills are utilized mostly by males during winter, while females are to be found abundantly in the Tanana Valley (Weeden, 1965b). Weeden (1964) suggested that this differential movement may represent a dispersal mechanism or serve as a means of reducing food competition or perhaps indicate that females may survive better in forested areas under winter conditions.

Irving et al. (1967) have documented the migration of willow ptarmigan through Anaktuvuk Pass in the Brooks Range. Although few ptarmigan nest there, some 50,000 birds pass through this point each year. The fall migration reaches a peak in October and is over by December, while the spring migration starts in January and early February, subsides in March, and is renewed in April. The early fall migrants are mostly juvenile males and females, whereas the number of adult males gradually increases to a maximum in March, or two months later than the maximum movement of juvenile males. The authors reported no clear indication of cyclic changes in population numbers annually. A spruce forest area occurring thirty-five miles south of the pass is one of the areas used for wintering, and breeding occurs on the north slope of the Brooks Range and beyond to the Arctic coast. Some of these breeding areas may not be occupied until late in May.

So far, virtually nothing is known of daily movements of willow ptarmigan, and such information will require detailed studies of individually marked birds. Jenkins, Watson, and Miller (1963), studying red grouse, found the birds to be remarkably sedentary in this nonmigratory population. Of 739 birds banded as chicks, only 5 were recovered more than 5 kilometers away that season, and some of this movement may have been caused by the birds' being driven for hunting purposes. Of 290 birds banded as chicks but recovered as adults, 230 were recovered within 1.5 kilometers of the point of banding. It would thus appear that willow ptarmigan move only as far as is necessary to maintain a source of food and cover during the coldest parts of the year. Weeden (1965b) reported that a male and its mate were both found a year after they were banded as adults, defending a brood about 100 yards away from the original point of banding a year previously, which would attest to considerable site fidelity in this species. Bergerud (1970b) reported that females are more mobile than males, with one banded female moving 61 kilometers in about three months.

REPRODUCTIVE BEHAVIOR

Territorial Establishment

Most observations of territorial behavior in this species derive from studies of the red grouse in Scotland by A. Watson, D. Jenkins, and their associates. Likewise, display descriptions are also based on this population, unless otherwise indicated.

Territorial behavior and the success of territorial establishment appears to be a crucial factor in the biology of red grouse populations, judging from the work of Jenkins and Watson. Territories in red grouse are established in early fall, and the numbers of such territorial males that can be accomodated on a habitat apparently limits the density of the breeding population. Nonterritorial males are forced out of the preferred areas into marginal habitats, where they are more heavily exposed to predation, starvation, and disease. However, such losses play little if any role in the success of the population. Since juvenile birds are rarely able to attain territorial status their first fall, early territorial establishment would favor reproduction by mature males.

Territorial establishment in the North American willow ptarmigan is presumably in spring, although some fall display and calling by males may occur (Weeden, 1965b). However, it is not until late April or May that the willow ptarmigan have acquired their striking nuptial plumage, which presumably provides important visual signals for territorial proclamation and attraction of females. Weeden (1965b) has made the interesting point that whereas the male willow ptarmigan undergoes courtship in this bright brown and white plumage, the rock ptarmigan is still in completely white plumage during courtship, which perhaps provides important visual distinction for species recognition between the two species.

Territorial size has been studied intensively by Jenkins, Watson, and Miller (1963) for red grouse. They found that in each year, some individual territories were larger than others, but in years of high grouse populations the territories in general averaged smaller than in years when grouse were few. Territories selected by previous residents were usually larger than could later be defended against newly colonizing juvenile birds. Sketched maps presented by these authors indicate that territorial size rarely exceeded a maximum diameter of three hundred yards, and most were much smaller. One study area of 56 hectares (138 acres) supported twenty-four territorial males (two of which were unmated) in 1961, thus territorial sizes averaged 5.7 acres in the area during that year. In 1960, sixteen males (two unmated) occupied the same area, and in 1958 there were over forty territorial males (ten unmated) on it. For the study areas as a whole, the breeding density over the years varied from one pair to about 5 acres, in 1957 and 1958, to about one pair per 15 acres in 1960.

Agonistic and Sexual Behavior

In contrast to the species considered previously, it is almost impossible to differentiate completely between male and female behavior patterns in the ptarmigan. This is primarily a reflection of their monogamous or nearly monogamous pair bond and a subsequent reduction of sexual selection pressures for dimorphic behavior patterns. Watson and Jenkins (1964) have provided a detailed account of behavior patterns in the red grouse which will be summarized here in the belief that their findings should apply to the North American willow ptarmigan with little or no modification. Although they also discuss displacement activities, distraction behavior, comfort and maintenance activities, and other aspects of behavior, only those patterns directly concerned with reproduction will be mentioned here.

Agonistic behavior patterns of males associated with establishment and defense of territories include sitting on an exposed lookout, such as a hillock or stone, where most of the territory can be seen. Intruders are approached in an attack-intention posture characterized by erect combs, the head and neck stretched forwards, the body near the ground, the wings held in the flanks, and the bill open. Prior to such an approach the bird may fan its tail and droop its wings in a manner resembling the waltzing display. A lesser type of threat consists of standing in one place and uttering *kohway* and *kohwayo* calls. Still weaker threat consists of standing and uttering a *krrow* call, which in turn grades into watchful behavior, flight intention, and finally fleeing by running or flying away.

Several kinds of aggressive encounters may occur. Brief encounters may last only a few seconds and involve birds of either sex, which may or may not occur on a territory. "Jumping" is a communal encounter that also is not limited as to sex and not related to territory. In this, two or more aggressive birds will begin to jump about with wing-flapping, causing them to become more fully separated. Prolonged chases may occur when a dominant male follows a subordinate bird for great distances, often beyond its territory, and may in fact kill or wound it. "Facing" occurs when two equally dominant birds face each other with combs erect, heads forward, and wings flicking, with neither one showing signs of retreat. When actual fighting occurs the birds usually do not face each other, but rather face in the same direction and strike each other from the side with their bills, wings, and feet. "Walking-in-line" consists of two birds' walking side by side some twenty inches apart. While so walking they utter *kohway* and *ko-ko-ko* calls that indicate attack intention, and they may also utter the *krrow* threat call. Such a display by two males often occurs at the edges of territories, while hens may perform the same display anywhere in the territory. Occasionally the display occurs outside breeding territories, where up to five or six birds may participate.

Sexual patterns involve pair formation behavior and copulatory behavior. Pair formation is achieved by the males' advertising their territories, and the females' being attracted to the more vigorous males. On arriving on a territory, the female may utter a *krrow* call and look for a displaying male to approach. If there is none, she may fly to another territory, until a resident male makes a song-flight landing near her and begins to strut towards her. The female then flees but may be driven back to the territory by the male. Sexual activity occurs in Scotland every month but is most common from February to April, and many pair bonds that had been established earlier are only temporary and may be easily disrupted. When in breeding condition, the male has a highly conspicuous red eye-comb that can be erected to about one centimeter. Although the hen's combs are much smaller and paler, they can also be erected.

The male's approach to another bird of either sex is essentially a threatening one, and in the case of a receptive hen the response is one of submissive gestures. Thus the sexual differences in display are not so much qualitative as they are quantitative, in terms of relative dominance and submission. Sex recognition is probably also achieved by the different voice, plumage, and comb development of the hen.

The postures performed by a male in the presence of a female but not elicited in the presence of other males may be considered "courtship" displays. Watson and Jenkins (1964) list five such displays: tail-fanning, waltzing, rapid-stamping, bowing, and head-wagging.

Tail-fanning is performed by a male when approaching a hen. While cocking his tail, he may fan it with a rapid flick, at the same time lowering his wings and scraping the primaries on the ground as he moves forward. In this stage, the wings are drooped equally and the tail is not tilted. Often the male moves in a slight curve in front of the female, or he may pass in front of her alternately from both sides. Sometimes the undertail coverts are exposed by his turning away from the hen. Such movements grade into "waltzing," during which the male circles the female closely, pivoting around her with short, high steps and drooping the wing nearer her, at the same time tilting the tail to expose its upper surface more fully to her view. The body may be tilted toward the hen as well. During "rapid-stamping" the male runs toward the hen with his tail slightly fanned, his neck thickened and arched, and his head held low with the bill wide open. In this posture he might pass close beside the other bird and appear to be attacking her, but the differences in wing and neck positions make it possible to distinguish these two types of behavior easily. If the hen does not flee and mounting does not occur at that time, the male will often raise and lower his head, with his body still held low, the tail partly fanned and the nape feathers raised, in a display called "bowing."

The last of the courtship displays is head-wagging, which both sexes perform. The bird crouches near its mate, extends its neck forward, and quickly wags its head in lateral fashion, exposing its eye combs and twisting its head slightly with each wag. When a hen approaches a cock, the male may also crouch low, erect his combs, and lower his head, producing a posture strongly suggestive of the precopulatory "nuptial bow" of prairie grouse. Although both sexes perform head-wagging, it is not a mutual display, and instead the birds often perform it alternately. When the female performs it, the male may attempt to mount her. However, during actual solicitation, the female crouches without head-wagging, opens her wings, and holds her head up. The male immediately mounts, drooping his wings around the hen during copulation. Afterwards, the male utters several threat calls, displays strongly for a few minutes, and often moves to a lookout post.

Vocal Signals

Watson and Jenkins (1964) describe fifteen different vocal signals of adults which are uttered by both sexes, although the hen's calls may be recognized by their higher pitch. Song flight, or "becking" is uttered as the bird takes off, flies steeply upwards for thirty feet or more, sails, and then descends gradually while fanning its tail and beating its wings rapidly. On landing the bird may stand erect, droop its wings, fan its tail, and bob its head. During the ascent phase the call is a loud, barking *aa*, while a *ka-ka-ka-ka* is uttered some eight to twelve times with gradually slower cadence. After landing a gruffer and slower call *kohwa-kohwa-kohwa* (also interpreted as *go-back*, *kowhayo*, and *tobacco*), is uttered for a varying length of time. Hens and nonterritorial males do not fly as high or call as loudly as territorial males, and no doubt this call is important in territorial proclamation.

In calling on the ground, a similar signal is uttered, often from a song post such as a stone. The bird stretches his neck diagonally upwards and utters a vibrating *ko-ko-ko-ko-krrrrr*, up to about twenty syllables, increasingly faster toward the end. Such calls may be used to threaten approaching animals or birds flying overhead and are largely but not entirely territorial advertisement.

During attack, the birds utter a *kowha* sound, like the last part of the flight song but without preliminary notes. It may be given during attack, when trying to mount hens, or immediately after copulation. A similar call, *koway*, is an attack-intention, or threat, call and is rapidly repeated as a series of hurried notes. A variant is *kohwayo*, also repeated, but indicating less aggressiveness than the last call. Still less aggressive notes are *krrow* and *ko-ko-ko*, the latter representing a flight-intention call. This call is given by a bird about to fly or one being handled by a human and may stimulate other birds to take flight.

When a grouse is charging another bird, a single note, *kok*, may be uttered, especially by the chased bird. The same call may be used as an in-flight alarm note. A similar *kok* note serves as a mammalian predator alarm note, while a *chorrow* note serves for an aerial raptor warning signal. A sexual note, *koah*, the emphasis on the first syllable, is used between members of a pair when crouching and head-wagging, when examining nest sites, or when bathing. Hens may also utter it when a nest is approached, but hissing is more often elicited under these conditions. Hissing may also occur when a bird is being handled. A *krow* note is used during distraction display by parents, causing the young to crouch, while a *korrr* or *koo-ee-oo* serves as a call to chicks, especially those uttering distress calls. Finally, a harsh, chattering *krrr* note is used as a defense against avian predators that are attacking the bird or its family.

Watson and Jenkins report that the distress *cheep* of chicks is uttered until the young are nearly full-grown, but that it gradually changes to a *kyow* note and finally to the adult *krrow* and probably then serves as a contact call. Even newly hatched chicks will utter a chattering call which evidently is aggressive in nature and apparently develops into the adult "ground song." By the age of ten to twelve weeks, the male begins to acquire a voice that differs from that of females, resembling more the voice of an adult cock.

Nesting and Brooding Behavior

The only available analysis of nest-site selection behavior is that of Jenkins, Watson, and Miller (1963) for Scottish red grouse. They studied 163 nests, nearly all of which were in heather cover (*Calluna*). The average height of the heather cover was twenty-seven centimeters, compared with a mean cover height of seventeen centimeters. Most nests were partly overhung with vegetation, but 17 percent were completely uncovered and 12 percent were completely covered. Most were on hard, well-drained ground, and 67 percent were on flat ground. Most were shallow scrapes, sparsely lined with various plants, including grasses and heather. Usually the nests were within five hundred feet of grit sources, water, and mossy or grassy areas where the chicks could feed. The clutch size of this population varied in different years and in different study areas from 6.1 to 8.1 eggs (average of 395 nests was 7.1 eggs) and the estimated incubation period was twenty-two days.

Some comparable information is available for North American willow ptarmigan. Kessel and Schaller (1960) reported that 5 nests in Alaska had 6 to 7 eggs, averaging 6.8. Eight clutches from northern Alaska in the Denver Museum average 7.8 eggs. Bergerud (1970b) reported an average clutch of 10.2 eggs in 106 Newfoundland nests. Nests containing up to 17 eggs would appear to be the work of at least two females. The incubation period of the North American birds is likewise twenty-one to twenty-two days, and the egg-laying interval is somewhat greater than twenty-four hours (Westerskov, 1956). Bergerud (1970b) judged that in Newfoundland renesting probably accounted for between 12 and 18 percent of the young produced.

Unlike the other species of ptarmigan, the male typically remains with the female through the incubation period and assists in brood defense. Jenkins, Watson, and Miller (1963) reported that the percentage of broods observed with both parents in attendance ranged from 61 percent to 90 percent in various years and areas. In good breeding years, most broods were attended by both parents until they were at least two months old, while in poor breeding years 30 to 40 percent were not attended by parents at any stage. The percentage of parents observed performing distraction display ranged from 4 to 72 percent. Individual brood sizes ranged to as many as 12, and averages varied greatly in different years. Roberts (1963) reported an average brood size of 6.3 chicks for Alaskan willow ptarmigan. This figure is higher than any yearly average reported by Jenkins, Watson, and Miller, whose highest reported brood size was 5.2 for one study area in 1960.

EVOLUTIONARY RELATIONSHIPS

Evolutionary relationships of the genus *Lagopus* as a whole would seem to be very close to both *Dendragapus* and *Tetrao*, as Short (1967) has already suggested. It is perhaps impossible to judge which of these two genera *Lagopus* most closely approaches, and presumably all three genera differentiated from common stock at about the same time.

Relationships within the genus *Lagopus* represent another problem. The white-tailed ptarmigan differs from the rock and willow ptarmigans in several respects, which have been enumerated by Short (1967), and it is clearly the most isolated of the three species. Höhn (1969) suggested such an early offshoot of ancestral white-tailed ptarmigan stock in North America, with which I am in agreement. Höhn judged that the willow and rock ptarmigan ancestral stock also diverged in North America, with the rock ptarmigan moving east to Greenland and both species moving west across the Bering Strait into Eurasia. This kind of speciation model seems unlikely to me, as I can visualize no major barriers that might have allowed for separation of ancestral willow and rock ptarmigan stock in northern North America. It seems more likely to me that one of these types developed in Eurasia and the other in North America after a splitting of common gene pools and after secondary contact the rather marked ecological differences between them allowed the development of the extensive geographic contact between them that now exists. In contrast, Johansen (1956) suggested that the genus *Lagopus* originated in Asia and reached North America at an early date, during which the ancestral white-tailed ptarmigan separated from pre-*mutus* stock.

In a strictly behavioral sense, I would regard the willow ptarmigan as more primitive than the other two ptarmigan, in both of which a breakdown on strong pair bonds and a tendency toward polygamy may be seen. It seems probable to me that the evolution of mating patterns in the grouse was from an originally monogamous situation to a polygamous or promiscuous one, rather than to believe that the monogamous situation of the willow ptarmigan is derived from a non-monogamous mating type. The retention of monogamy or near monogamy in the ptarmigans seems to me to be an ecological artifact, resulting from the greater needs for intensive parental care in an arctic situation than in a subarctic or temperate one, in which the duties of incubation and brood-rearing can be more effectively undertaken by the female alone. This latter arrangement thus frees the male to fertilize a potentially larger number of females, and these resulting reproductive advantages have led to reduced pair bonds or to promiscuous matings. It is curious, however, that the willow ptarmigan, rather than the rock ptarmigan, has more strongly retained a monogamous and prolonged pair bond, since the rock ptarmigan has an even more northerly breeding distribution and must nest under equally severe breeding conditions. Arnthor Gardnarsson* has found that in Iceland the males suffer a much higher rate of predation by gyrfalcons than do females, apparently as a result of the male's more conspicuous plumage during the breeding season. The mating system there is an essentially promiscuous one, since the females do not closely associate with males or their territories. Such differential sexual predation pressures might account for the rock ptarmigan's less strongly monogamous mating system and the reduced period of contact between the sexes.

*Arnthor Gardnarsson, 1970: personal communication.

14

Rock Ptarmigan

Lagopus mutus (Montin) 1776

OTHER VERNACULAR NAMES

ARCTIC grouse, barren-ground bird, Chamberlain ptarmigan, Dixon ptarmigan, Nelson ptarmigan, Reinhardt ptarmigan, rocker (in Newfoundland), snow grouse, Townsend ptarmigan, white grouse.

RANGE

Circumpolar. In North America from northern Alaska, northwestern Mackenzie, Melville Island, northern Ellesmere Island, and northern Greenland south to the Aleutian Islands, Kodiak Island, southwestern and central British Columbia, southern Mackenzie, Keewatin, northern Quebec, southern Labrador, and Newfoundland (*A.O.U. Check-list*).

NORTH AMERICAN (excluding Greenland) SUBSPECIES
(*ex A.O.U. Check-list*)

L. m. evermanni Elliot: Attu rock ptarmigan. Resident on Attu Island, Aleutian Islands.

L. m. townsendi Elliot: Kiska rock ptarmigan. Resident on Kiska and Little Kiska islands, Aleutian Islands.

L. m. gabrielsoni Murie: Amchitka rock ptarmigan. Resident on Amchitka, Little Sitkin, and Rat islands, Aleutian Islands.

L. m. sanfordi Bent: Tanaga rock ptarmigan. Breeds on Tanaga and Kanaga islands, Aleutian Islands.

L. m. chamberlaini Clark: Adak rock ptarmigan. Resident on Adak Island, Aleutian Islands.

L. m. atkhensis Turner: Atka rock ptarmigan. Resident on Atka Island, Aleutian Islands.

L. m. yunaskensis Gabrielson and Lincoln: Yunaska rock ptarmigan. Resident on Yunaska Island, Aleutian Islands.

L. m. nelsoni Stejneger: Alaska rock ptarmigan. Resident in northern Alaska and northern Yukon south to the eastern Aleutians, the Alaska and Kenai peninsulas, and Kodiak Island and east to the western Yukon.

L. m. rupestris (Gmelin): Canada rock ptarmigan. Breeds from northern Mackenzie, Melville Island, northern Ellesmere Island, and southern Greenland south to central British Columbia, southern Mackenzie, southern Keewatin, Southampton Island, northern Quebec, and Labrador.

L. m. dixoni Grinnell: Coastal rock ptarmigan. Resident on the islands and coastal mainland of the Glacier Bay region of Alaska and on the mountains of extreme northwestern British Columbia south to Baranof and Admiralty islands.

L. m. welchi Brewster: Newfoundland rock ptarmigan. Resident in Newfoundland.

MEASUREMENTS

Folded wing: Adult males, 172–202 mm; adult females, 163–95 mm (males average 9 mm longer than females).

Tail: Adult males, 97–120 mm; adult females, 85–115 mm (males of all races average 104 mm or more, females usually average under 104 mm).

IDENTIFICATION

Adults, 12.8–15.5 inches long. Both sexes carry blackish tails throughout the year, and although the scarlet comb of males is most evident during the spring, it is also apparent to some extent through the summer. In the summer males are extensively but rather finely marked with brownish black and various shades of brown and lack the rich chestnut tone of male willow ptarmigan. In summer females are more coarsely barred and are generally lighter overall but have somewhat finer markings than do female willow ptarmigan. Females have definite barring extending to the throat and breast, rather than having these areas finely barred or vermiculated as in males. In autumn males are generally pale above, with tones of ashy gray predominating (tawny brown predominating in some Aleutian races), and females at this time have relatively more brown and fewer black markings, plus a sprinkling of white winter feathers. Both sexes in winter are mostly white with blackish tails, and males (but not all females) have a black streak connecting the bill with the eye and extending somewhat behind the eye.

FIELD MARKS

The smaller, relatively weaker, and entirely black bill of the rock ptarmigan is sometimes detectable in the field and serves to separate this species from the willow ptarmigan in all seasons. In the winter, the presence of a black line through the eyes is also diagnostic, but its absence does not exclude this species. For plumage distinctions useful in separating the willow and rock ptarmigans, see the account of the preceding species. During the breeding season the rock ptarmigan is found in higher, rockier, and drier country than the willow ptarmigan, but they may occur together during winter and intermediate periods. In all seasons the dark tail distinguishes the rock ptarmigan from the white-tailed ptarmigan.

AGE AND SEX CRITERIA

Females lack the reddish "eyebrows" of adult males and in summer are more heavily barred wtih dark markings both above and below. In autumn the barring is reduced in the female, which is still somewhat more heavily marked than the grayish and finely vermiculated male. In winter the sexes are nearly identical, but females usually lack the black stripe through the eye that is present in males (Godfrey, 1966).

Immature females are browner and more narrowly barred with blackish brown above and on the breast than are adult females in autumn (Ridgway and Friedmann, 1946). The pointed condition of the outer primaries has been reported to be an unreliable indicator (Weeden, 1961). Instead, young rock ptarmigan may be distinguished by the fact that in adults the ninth primary (second from outside) has the same amount of pigment as the eighth, or less, whereas immature birds have more pigment on the ninth (Weeden and Watson, 1967).

Juveniles may readily be recognized by the presence of at least one brown primary or secondary feather (the eighth primary is the last to be molted). These feathers are typically mottled with pale buff (Ridgway and Friedmann, 1946).

Downy young are illustrated in color plate 61. The downy young are usually paler throughout than those of willow ptarmigan, and the crown is lighter and more chestnut-colored than the blackish brown crown of the willow ptarmigan (Watson, Parr, and Lumsden, 1969). See willow ptarmigan account.

DISTRIBUTION AND HABITAT

The most arctic-adapted of all the grouse, the rock ptarmigan is more widely distributed in the high arctic than is the willow ptarmigan. It also extends south to Hudson Bay during the breeding season, and undertakes considerable southward movement during winter, sometimes occurring as far south as James Bay. Unlike the willow ptarmigan, the rock ptarmigan breeds as far north as Ellesmere Island and on adjacent Greenland to its northern limits at approximately 83 degrees north latitude. Also unlike the willow ptarmigan, this species can survive in the rocky desert-like habitat of the high arctic which may be a limiting factor in the northern distribution of the willow ptarmigan. Weeden (1965b) reports that typical breeding terrain of the rock ptarmigan consists of moderately sloping ground in hilly country, such as the middle slopes of mountains. Typically, the vegetation is fairly complete, but may be sparse on the highest and driest slopes. Shrubs are usually from one to four feet tall and are concentrated in ravines or other protected sites, while most plants are usually less than one foot tall. Many creeping or decumbent woody plants are typical, as well as rosette forms, while sedges and lichens are usually abundant. Breeding terrain rarely extends below the upper limits of timberline, and usually occurs from one hundred to one thousand feet above timberline in hilly country.

There have probably been few changes in the distribution of rock ptarmigan in historical times, since it is the species least likely to be affected by human activities. Considerable population fluctuations are known to occur, but those occurring in Greenland and Iceland have been interpreted as representing a ten-year cycle. Buckley (1954) concluded that ptarmigan populations in Alaska are also cyclic in nature, but adequate data to prove this view are not yet available (Weeden, 1963).

POPULATION DENSITY

Weeden (1963) has summarized population density figures for rock

FIGURE 27. Current North American distribution of the rock ptarmigan. Dashed line indicates normal southern wintering limits.

ptarmigan based on various studies in the Northwest Territories. These estimates range from as many as 8 adults per square mile to 4,000 adults on 12,500 square miles. Based on a five-year intensive study on a fifteen-square-mile study area in Alaska, Weeden (1965a, 1965b) reported yearly spring densities of males varying from 5.9 to 11.3 per square mile. Slightly lower estimates of female populations were obtained for the same period.

In a study of Scottish ptarmigan, Watson (1965) estimated spring populations to be as high as one pair per 2 to 3 hectares (approximately 5 to 7.5 acres) in peak years on the best habitats. However, unlike the fairly uniform heather (*Calluna*) habitats favored by red grouse, the arctic-alpine breeding vegetation is typically more varied, and an area of 100 or more acres rarely contains no unfavorable habitat. Thus, extrapolations of local density figures to large areas is unprofitable; this also helps explain the wide differences in densities reported on small, favorable areas and those estimates based on large regional surveys. Watson (1965) estimated that in peak years, spring numbers on his study area of 1,220 acres were as high as fifteen to eighteen birds per 100 hectares (247 acres), and as low as five in one year.

HABITAT REQUIREMENTS

Wintering Requirements

In Alaska, rock ptarmigan winter in such locations as shrubby slopes at timberline, in large forest openings where shrubs, especially birch, project above snow level, and, rarely, in riparian willow thickets (Weeden, 1965b). Watson (1965) noted that in Scotland the birds moved down from their arctic-alpine breeding grounds into a moorland zone of heather that was used by red grouse during the breeding season. Ptarmigan can scratch through a few inches of soft snow to reach plants, but Watson did not find them burrowing under the snow to forage. Local variations in topography caused areas to be blown fairly free of snow periodically, exposing food plants, and the birds will move from one such area to another in search of food. Little if any competition for food between ptarmigan and red grouse was noted by Watson, since the two species remained almost completely separated during winter. As mentioned in the willow ptarmigan account, considerable separation of the sexes occurs in North American willow and rock ptarmigans during winter, with males remaining in more alpine-like habitats, while the females tend to move into relatively protected situations.

Spring Habitat Requirements

Territorial requirements for the rock ptarmigan consist of a larger proportion of relatively open vegetation than is the case for willow ptarmigan (Weeden, 1965b). Some territories contain no shrubs at all, and males utilize rocks, knolls, or similar elevations for territorial display and for resting. Watson (1965) reported that ptarmigan were most common where large boulders or outcrops occurred on stunted heath or a mixture of stunted heath and grassy vegetation. The birds rarely took territories on pure grassland, tall heaths, bogs, or stone fields without healthy vegetation. Favorite areas for territorial establishment were usually on varied heaths or a mixture of varied heaths and grasses. The highest territorial densities occurred on areas of nearly continuous heath broken up by large boulders, slightly lower densities were found on scattered patches of heath, and much lower densities occurred on areas of continuous heath with only a few boulders present. Territorial densities were lowest on bare, gravelly places with only scattered vegetation and boulders.

Nesting and Brooding Requirements

Nest sites for the rock ptarmigan may have less overhead concealment than those of willow ptarmigan, but some overhead protection is usually present (Watson, 1965). Parmalee, Stephens, and Schmidt (1967) indicated that the nesting habitat is usually dry and rocky and sometimes is barren and high but may consist of wet tundra sites with heavy vegetation where willow ptarmigan also breed.

Brooding habitat is similar to nesting habitat, but broods tend to gather in swales on ridges and upper slopes (Weeden, 1965b). They avoid dense shrubs and after beginning to fly at ten or eleven days of age escape by flying out of sight over knoll ridges.

FOOD AND FORAGING BEHAVIOR

The best source of information on rock ptarmigan food habits in North America is that of Weeden (1965b), based on 482 crop samples from interior Alaska. Winter foods there consist primarily of dwarf birch buds (*Betula*) and catkins, followed by willow buds and twigs (*Salix*). Dried leaves of shrubs extending above the snow are also taken in limited quantities.

Spring foods, based on relatively few samples, appear to consist of a variety of plant materials, including the new growth of shrubs, horsetail tips (*Equisetum*), and a small amount of birch and willow materials. Summer foods include an even greater array of plant foods, which consist largely of leaves and flowers in early summer and berries and seeds later on. Blueberries (*Vaccinium*), crowberries (*Empetrum*), and mountain avens (*Geum*) provide important food sources during this time. During fall, blueberries and heads of sedges (*Carex*) are important, and dwarf birch begins to assume the great importance that will continue throughout winter.

Reporting on birds taken on Baffin Island, Sutton and Parmelee (1956) noted that in the crops of eight adults taken in May about 60 percent of the total food materials consisted of buds and twigs of willow, 32 percent was the leaves and twigs of dryas (*Dryas*), and the remainder consisted of *Saxifraga*, *Draba*, and the galls of willows. A newly hatched chick had eaten leaves of crowberry (*Empetrum*).

Moss (1968) has made an interesting nutritional comparison of rock ptarmigan foods taken by birds of the Icelandic and Scottish populations. In Iceland, the birds have a diet predominantly of twigs of willow, leaves of dryas, the leaves and bulbils of *Polygonum*, which are relatively high in nitrogen and phosphorus, and berries of *Empetrum*, which are high in soluble carbohydrates. By comparison, the Scottish ptarmigan subsist on a relatively nutrition-poor diet of heather (*Calluna*), *Vaccinium*, and *Empetrum*. Correlated with this is the fact that in Iceland the ptarmigan have an average clutch size of about 11 eggs, whereas in Scotland the clutch is usually 6 to 7 eggs, averaging 6.6. The average clutch size in Alaska, based on studies made by Weeden (1965a), is essentially the same as in Scotland. Significant annual differences in clutch sizes do occur in Alaska and apparently also in Scotland, but they have not yet been adequately correlated with population density or food quality. Lack (1966) has suggested such a possible correlation between clutch size and heather conditions. Watson (1965) believed that annual differences in clutch sizes were unimportant compared with variations in chick survival. At least in the red grouse, chick survival may be related to the physical condition of the hens as determined by food supplies.

A possibly significant point related to food supplies and reproductive success is the fact that although the rock ptarmigan is the most northerly breeding of the ptarmigans, it is considerably smaller than the willow ptarmigan. Likewise, the alpine-breeding white-tailed ptarmigan is much smaller than either the rock or the willow ptarmigan, in contrast to what might be expected with arctic-breeding birds (Bergmann's principle). The possibility exists, therefore, that smaller body size in the rock and white-tailed ptarmigans is an adaptation to reduced food supplies and has evolved relatively independently of selective pressures related to environmental temperatures. Yet Irving (1960) reported that willow ptarmigan collected in arctic localities of Alaska averaged ninety grams heavier than those from subarctic points some six hundred miles south. Further, winter birds tended to be heavier than summer birds, and males, which averaged ten to forty grams heavier than females, wintered in more hostile environments.

Whereas Irving (1960) found that the willow ptarmigan at Anaktuvuk Pass are migratory, the rock ptarmigan there are not, and in winter they feed on high, rounded slopes where low vegetation is exposed. Also, although willow ptarmigan often retreat with their crops filled with from fifty to one hundred grams of food to burrows some one and a half to two feet under the snow, this behavior is apparently not typical of rock ptarmigan. Manniche (cited in Bent, 1932) does indicate that in Greenland the birds may spend the night in holes about twenty centimeters deep on the lee side of rocks or in narrow snow-filled ravines in the rocks. MacDonald (1970) noted that the birds would dig roosting forms deep enough that only their heads remained above the snow, or would use the depressions caused by humans walking across the snow.

MOBILITY AND MOVEMENTS

The relatively large heart size (Johnson and Lockner, 1968) of the rock ptarmigan suggests that it may be capable of considerable movements, but there is little detailed information on actual daily or seasonal movements in the species. Snyder (1957) stated that the bird is migratory to an appreciable degree in arctic Canada, and Weeden (1964, 1965b) reported that some low altitude wintering grounds of the species are at a minimum of ten, and probably fifteen to twenty, miles from the nearest alpine breeding areas. Weeden believed that, at least in the lower parts of the wintering range, rock ptarmigan move in an unpredictable fashion. By March and April, however, movements are quite limited and consist of visits to various feeding areas separated by distances of up to half a mile or more, the stay at each area lasting varying lengths of time. Irving (1960) reported that at Old Crow, Alaska, wintering birds might convene from a nesting area some thirty miles in diameter, but no actual evidence for a regular migratory pattern was indicated. Bent (1932) indicated that although the majority of the rock ptarmigan withdraw from the northern limits of their summer range, they do not usually retreat beyond the southern limits of their breeding range. Nelson (cited in Bent, 1932) reports a regular fall evening migratory movement across Norton Sound, via Stuart Island, and a comparable spring flight in April.

Weeden (1965b) noted that in Alaska the rock ptarmigan disappear from their wintering areas at low altitudes in March and April and that in 1962 the first migrants arrived at their Eagle Creek breeding ground study areas on March 29. This movement continued through April, and during April males begin establishing territories in advance of the arrival of most hens. In the study area, located northeast of Fairbanks, egg laying begins in the second to the fourth week of May. Farther north at Old Crow and Anaktuvuk Pass the males become territorial in late April and May. By comparison, the first flocks of rock ptarmigan which Parmalee, Stephens, and Schmidt (1967) saw on Victoria Island arrived in mid-May and were all males. The first territorial flights were noted on May 19, and the first female was seen May 23. Fresh eggs were noted from June 3 until late June, or nearly a month later than in central Alaska. Interestingly, the weights of spring males collected on Victoria Island averaged about one hundred grams more than Irving reported for Anaktuvuk Pass and Old Crow, and females averaged about ninety grams heavier.

REPRODUCTIVE BEHAVIOR

Territorial Establishment.

The period of breakup of winter flocks and establishment of territories probably varies greatly by locality and year. In Scotland, Watson (1965) noted that this behavioral transition occurs with the coming of spring thaws and sunny weather, which may be as early as the first part of January or as late as the end of April. In North America, where the birds usually move out of their breeding areas during the winter period, there is probably a fairly short lag between the arrival of the males on the breeding ground and the establishment of territories. The observations of Parmalee, Stephens and Schmidt (1967) indicate that this lag may be as short as a few days. Both yearling and adult male ptarmigan participate in territorial establishment; Weeden (1965a) found that the percentage of first-year ptarmigan in male breeding populations varied from 41 to 67 percent. Yearling females comprised from 17 to 75 percent of the breeding populations, and there was no evidence of any nonbreeding by females.

Agonistic and Sexual Behavior.

MacDonald's recent observations (1970) on Bathurst Island indicated that there individual males may defend surprisingly large areas of about one square mile, which include several lookout prominences adjacent to moist hummocky tundra with heavy vegetation. From these points the male watches for other ptarmigan, attacking males and courting females. During the early stages of territoriality the male spends much of his time advertising his location with song flight displays. As his aggressiveness increases, the size and brilliance of his eye-combs also increase. Territorial males, on seeing a rival male, engage in aerial chases with tails spread, combs erected, and their bodies rocking from side to side while in flight. Aerial chases of females were not seen by MacDonald but have been reported by Weeden.

The basic territorial advertisement display of the rock ptarmigan is the song flight. MacDonald noted that the height of this display flight varies from as little as about 4 feet early in the season to an estimated 250 feet observed in a highly aggressive male. The display may be performed spontaneously or may be elicited by a disturbance of some kind within hearing or visual range of the male. The bird typically leaps into the air, uttering a loud, belching call, and swiftly flies forward and upward with alternate wing-flapping and sailing. At the end of the climbing flight, the male sets his wings, fans his tail, and begins an upward soaring glide until he finally reaches stalling speed. At this point he swells his neck and begins to utter a series of staccato, belching notes. As the bird begins his descent on bowed wings a second series of belching notes is uttered and he slowly parachutes downward toward the ground. Just before landing the male tilts his spread tail vertically downward, and as he alights he quickly cocks it back upward to a near-vertical position. The wings are held to the side of the body and are drooped toward the ground, as the male stands with an erect neck or runs forward a short distance while uttering a staccato call. Then the male's neck is deflated, the primaries are lowered so that they drag on the ground, and the tail is fully spread while being tilted at an angle of forty-five degrees. Next, the bird begins a short forward run, simultaneously extending his neck and making a single, slow bowing movement with his head. When a female is newly present on his territory, the male may run in an arc toward her, tilting his tail toward her and extending one wing away from her. The head is also tilted toward the female, exposing the enlarged eye-combs. After a female has become established on a male's territory, this ground display is omitted. Females evidently gradually associate themselves with a specific male and his territory, initially following the male in flight and later being followed by the male. MacDonald noted that at least one male mated with three females in one season, all of which nested in the male's territory.

When two territorial males meet, violent fights may ensue. Threats may be uttered as the birds sleek their plumage, inflate their necks, and close their tails so that they are nearly hidden. The crown may be raised or lowered, and the combs erect or concealed. During attacks the birds attempt to grasp each other with their bills, while striking with the wings. Often feathers from the neck may be pulled out, and sometimes the eye-combs are torn.

Pair formation in rock ptarmigan is apparently a gradual process, judging from MacDonald's observations. He noted that while the resident male drives other males off his territory, the female becomes more submissive and dependent on him, relying increasingly on the male to warn her of danger. When near the female he continuously utters a contact call consisting of ticking notes, which change to a ratchet-like alarm call when alert to possible danger. When a female is thus alerted, she flushes and is immediately followed by the male, which may perform a song flight before landing. As the male returns to the female following the song flight he may perform the head-bowing and tail-tilting display described earlier. He typically circles the female at a distance of up to two feet, with his head held low, his wings dragging, and his tail tilted toward her. Apparently he attempts in this manner to direct the female into a tundra depression, seemingly trying to induce the female to crouch in it. In four observed instances of copulation, the female crouched in such a depression, partially extending her wings and exposing her white wrists. The male then stepped on her back and pecked at her nape but did not grasp her neck feathers. Rather, he remained with his body in a rather upright posture during copulation, finally bending forward and walking off her back over her shoulder. Then, with his head lowered and held forward, his tail spread and held vertically toward the female, and his wings dragging, he walked in a circular path around the female, with his combs greatly enlarged and his bill open. The female remained crouched for a time, then stood up, shook her plumage, and preened. In two cases the female ran from the male before he completed his postcopulatory display, while in one case the male circled around her twice while the female remained crouched.

MacDonald obtained some data indicating that males were more highly attracted to mounted specimens of females that had piebald brown and while plumage than to whiter females, which is of special interest since females molt into their brown nuptial plumage much earlier than males, which remain white and highly conspicuous throughout the pair-forming period.

Vocal Signals

MacDonald (1970) reported that although the territorial male has at least six different vocalizations, the sounds nearly defy description. In

↑ Sage Grouse, Males

↓ Sage Grouse, Male and Females

Ken Fink

↑ Sage Grouse, Male and Female

↓ Sage Grouse, Male and Female

Ken Fink

↑ Sooty Blue Grouse, Male

↓ Sooty Blue Grouse, Female

Ken Fink

Clait Braun

↑ Dusky Blue Grouse, Male

↓ Richardson Blue Grouse, Male

Ken Fink

↑ Canada Spruce Grouse, Male

↓ Franklin Spruce Grouse, Male

Ken Fink

↑ Canada Spruce Grouse, Male and Female

↓ Canada Spruce Grouse, Male

↑ Willow Ptarmigan, Male in Spring

↓ Willow Ptarmigan, Female

Bruce Porter

Ken Fink

↑ Willow Ptarmigan, Male and Chicks

↓ Rock Ptarmigan, Male and Female

↑ Rock Ptarmigan, Male

↓ Rock Ptarmigan, Female

Ken Fink

↑ White-tailed Ptarmigan, Male

↓ White-tailed Ptarmigan, Male and Female

Bird Photographs Inc., Ithaca, N.Y.

↑ Ruffed Grouse, Male

↓ Ruffed Grouse, Female

Ken Fink

Alan Nelson

↑ Ruffed Grouse, Female

↓ Greater Prairie Chicken, Male and Female

Greater Prairie Chicken, Male

↑ Greater Prairie Chicken, Male

↓ Lesser Prairie Chicken, Male

Alan Nelson

↑ Sharp-tailed Grouse, Male

↓ Sharp-tailed Grouse, Male

Alan Nelson

C.G.Pritchard

all cases, they appear to be variations of pulsed clicking sounds that resemble the noise produced by drawing a stick over the slats of a picket fence. The predominant frequencies are low, which is of interest in view of the fact that MacDonald discovered a seemingly unique membranous, inflatable sac on the dorsal side of the trachea in males. During vocalizations, not only the esophagus but presumably also this tracheal air sac may be inflated, which would facilitate the amplification of low-frequency sounds. The value of low-frequency sounds to the rock ptarmigan would seem to be correlated with the apparently large territories that they hold and associated with their long-distance visual signals in the form of the black and white plumage pattern.

MacDonald also noted that female rock ptarmigan produce at least three different vocalizations, which he described as whining, clucking, and a high-pitched screech, the latter apparently being an alarm call. He also noted a hissing produced during nest defense.

Nesting and Brooding Behavior.

Female ptarmigan locate their nests within the territorial boundaries of the male. In Scotland at least, the numbers of females associated with territorial males is rarely more than 50 percent (Watson, 1965), thus few if any males are normally likely to acquire more than one female. Weeden (1965b) reports that in Alaska two females may sometimes mate with a single cock, and presumably both hens nest within the territorial area of the male. To what extent the male defends the female and her nest is still not very clear for the rock ptarmigan. Höhn (1957) described how, when two female rock ptarmigan were shot, the male quickly approached and displayed to the corpses, but this kind of behavior clearly does not belong in the category of female defense. Weeden (1965b) noted that about one brood in twenty will have a male in attendance, but he never observed any actual brood defense by males. However, MacDonald (1970) reported several cases of brood defense by males, including both attack and distraction behavior.

Rock ptarmigan females build simple, shallow nests, the depressions often being little more than might be caused by the weight and movements of the brooding hen (Weeden, 1965b). Clutch sizes vary considerably by locality and by year. Weeden (1965a and unpublished *Game Bird Reports* vols. 7 to 10) noted clutch sizes varying annually between 1960 and 1969 from 6.4 to 9 eggs, and the average size of 195 clutches was 7.2 eggs. In the more arctic-like environment of Victoria Island, Parmelee, Stephens, and Schmidt (1967) found three nests, two containing eleven and one containing thirteen eggs, suggestive of somewhat larger clutch sizes at

←
Row 1: Blue, sage, and spruce grouse. Row 2: White-tailed, rock, and willow ptarmigans. Row 3: Ruffed, pinnated, and sharp-tailed grouse. Row 4: Chukar and gray partridges.

higher latitudes. Judging from Weeden's data (1965a), about two-thirds of the nests hatch during an average year. Renesting is apparently not common enough to affect over-all productivity. Weeden (1965a) provided data indicating an average brood size in August of 5.3 for 208 broods, with yearly averages ranging from 4.8 to 6.1 between 1960 and 1964. By comparison, Watson found that the average size of full-grown broods between 1945 and 1963 was from 1.2 to 6.2 young. Watson found that, on the average, 38 percent of the females went broodless each year, but in different years it varied from none to over 80 percent. Weeden (unpublished Alaska Fish and Game Department *Game Bird Report*, vol. 8, 1967) reported that between 1963 and 1966 60 percent of 130 year-old females were seen with young, while 77 percent of 185 older females were observed with young; thus, incubating or brooding efficiency evidently increases with age of the female.

The female is highly attentive to her young and when disturbed by humans utters a throaty *krrr* during distraction behavior (Sutton and Parmelee, 1956). When calling chicks toward her, she utters a clucking *kit* or *krit* call. Weeden (1965b) indicates that by imitating the distress peeping of a chick, he could elicit a low, crooning note that carried up to one hundred yards and helped locate broody hens.

Weeden (1965b) noted that one brood seen in 1960 moved about forty two hundred feet in five days, while another was found only about fifty feet from the point where it had been seen ten days before. In the case of two broods that were seen again after twenty-eight days, one had moved about fifty feet and the other family seventy-eight hundred feet. In general, the broods stayed within an area of about one-half square mile but did not appear to be attracted by the male's former territory. By late July, most broods had moved to areas higher than the nesting sites, congregating on moist and gentle slopes where sedges, grasses, forbs, and low shrubs predominated in the vegetation. Weeden also found several indications of transfer of individual chicks between broods. Hens which have lost their clutches or broods join the flocks of males that gather on high, rocky ridges or in streamside willow thickets. As the broods mature, they tend to combine, and these flocks in turn attract groups of males and nonproductive hens. In time, flocks of fifty to three hundred individuals may build up. However, at the same time, there is some calling and displaying among the males and an apparent resurgence of territoriality. The possible significance of this fall behavior is still unknown.

EVOLUTIONARY RELATIONSHIPS

Some general statements as to the evolutionary history of the ptarmigans have been mentioned under the willow ptarmigan account. In addition it might be noted that the rock ptarmigan is not only the most northerly and most widely distributed of all the ptarmigans but also might perhaps be considered as most representative of an ancestral ptarmigan type adapted for high arctic breeding. From such a type the evolution of an alpine off-shoot, as represented by the white-tailed ptarmigan, and a subarctic type, represented by the willow ptarmigan, might easily be imagined.

15

White-tailed Ptarmigan

Lagopus leucurus (Richardson) 1831

OTHER VERNACULAR NAMES

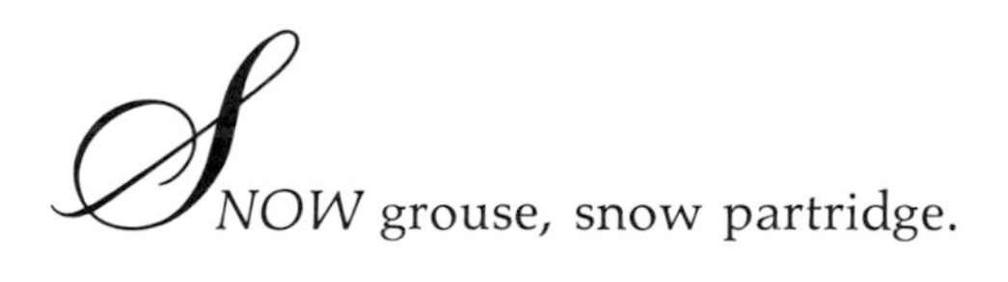

NOW grouse, snow partridge.

RANGE

From central Alaska, northern Yukon, and southwestern Mackenzie south to the Kenai Peninsula, Vancouver Island, the Cascade Mountains of Washington, and along the Rocky Mountains from British Columbia and Alberta south to northern New Mexico (*A.O.U. Check-list,* 1957).

SUBSPECIES (*ex A.O.U. Check-list*)

L. l. leucurus (Richardson): Northern white-tailed ptarmigan. Resident above timberline from northern Yukon, western Mackenzie, British Columbia, and west central Alberta south to the northern border of the United States.

L. l. peninsularis Chapman: Kenai white-tailed ptarmigan. Resident above timberline from south central Alaska to Cook Inlet and the Kenai Peninsula, extending east and southeast to Glacier Bay and White Pass.

L. l. saxatilis Cowan: Vancouver white-tailed ptarmigan. Resident above timberline on Vancouver Island, British Columbia.

L. l. rainierensis Taylor: Mount Rainier white-tailed ptarmigan. Resident above timberline in Washington from Mount Baker south to Mount Adams and Mount St. Helens.

L. l. altipetens Osgood: Southern white-tailed ptarmigan. Resident above timberline in the Rocky Mountains from Montana south through Wyoming and Colorado to northern New Mexico.

MEASUREMENTS

Folded wing: Adult males, 164–94 mm; adult females, 155–92 mm (males average 5 mm longer than females).

Tail: Adult males, 85–109 mm; adult females, 83–98 mm (males average 8 mm longer than females).

IDENTIFICATION

Adults, 12–13.5 inches long. In any nonjuvenal plumage the white tail will serve to separate this species from the other two ptarmigans. Adult males in summer plumage are vermiculated and barred or mottled with black, buffy, and white dorsally, with a buffy or pale fulvous tone predominating on the lower back, rump, and upper tail coverts, and the underparts are mostly white. Unlike the other ptarmigans, the wings as well as the tail (except for the central pair of feathers) are completely white at this season. Females are similar in plumage but have a heavily spotted and more yellowish color dorsally. In the fall both sexes are mostly pale cinnamon-rufous above, with fine spotting and vermiculations of brownish black and with a lighter head and neck. A few breast feathers are usually marked with white, and the abdomen, undertail coverts, tail, and wings are white. In the winter both sexes are pure white except for a black bill, eyes, and claws.

FIELD MARKS

A small alpine ptarmigan with white wings and tail in summer, or an entirely white plumage in winter, is of this species. It is usually extremely difficult to see against a lichen-covered rocky background and is therefore overlooked unless forced to fly.

AGE AND SEX CRITERIA

Females exhibit eye-combs (unlike the two other ptarmigan species) virtually identical to those of adult males, but in summer hens are more coarsely and regularly barred with black and rich ochraceous buff markings on their brownish back and side feathers, while feathers of males in these areas are finely vermiculated with brown and black. In addition, although males retain their white lower breast, abdomen, and undertail coverts through the summer, females have yellowish buffy brown feathers with some black barring present in these areas (Braun and Rogers, 1967a). In the autumn differences between the sexes diminish, but for a time females retain a few of their coarsely barred nuptial plumage feathers, especially on the nape, sides, inner wing, and upper tail coverts. In winter birds of both sexes are identical in plumage but may differ slightly in wing length, length of the outer five primaries, and outer rectrix length (Braun and Rogers, 1967a). In spring, males can be recognized by their distinctive black-tipped head and neck feathers, which provide a "hooded" effect that is lacking in females as they gradually acquire their brown, black, and yellow nuptial plumage (Braun, 1969).

Immatures may be recognized by the pigmentation of their two outer primaries (Taber, in Mosby, 1963). If black pigment occurs on either the ninth or tenth primary the bird may confidently be called an immature. Likewise pigmentation on the outer primary covert is an indication of an immature bird, whereas lack of pigmentation in these areas is typical of adults (Braun and Rogers, 1967a).

Juveniles have tail feathers that are yellowish brown centrally or white with mottled brown edges (Ridgway and Friedmann, 1946). Until they are all molted, the secondaries and inner eight primaries are also brownish in juveniles (see willow ptarmigan account).

Downy young are illustrated in color plate 61. Downy white-tailed ptarmigan are the least rufous dorsally of all the ptarmigans and have only a suggestion of the usual chestnut crown with its black margin. The two black dorsal stripes are also indistinct, and instead the back has an indefinite blending of buff, gray, sepia, and black shades. The feathered toes will separate downies of this species from any non-*Lagopus* forms.

DISTRIBUTION AND HABITAT

The current distribution of the white-tailed ptarmigan in North America closely conforms to that of alpine tundra, although it does not extend southward along the Cascade and Sierra ranges into Oregon or California,

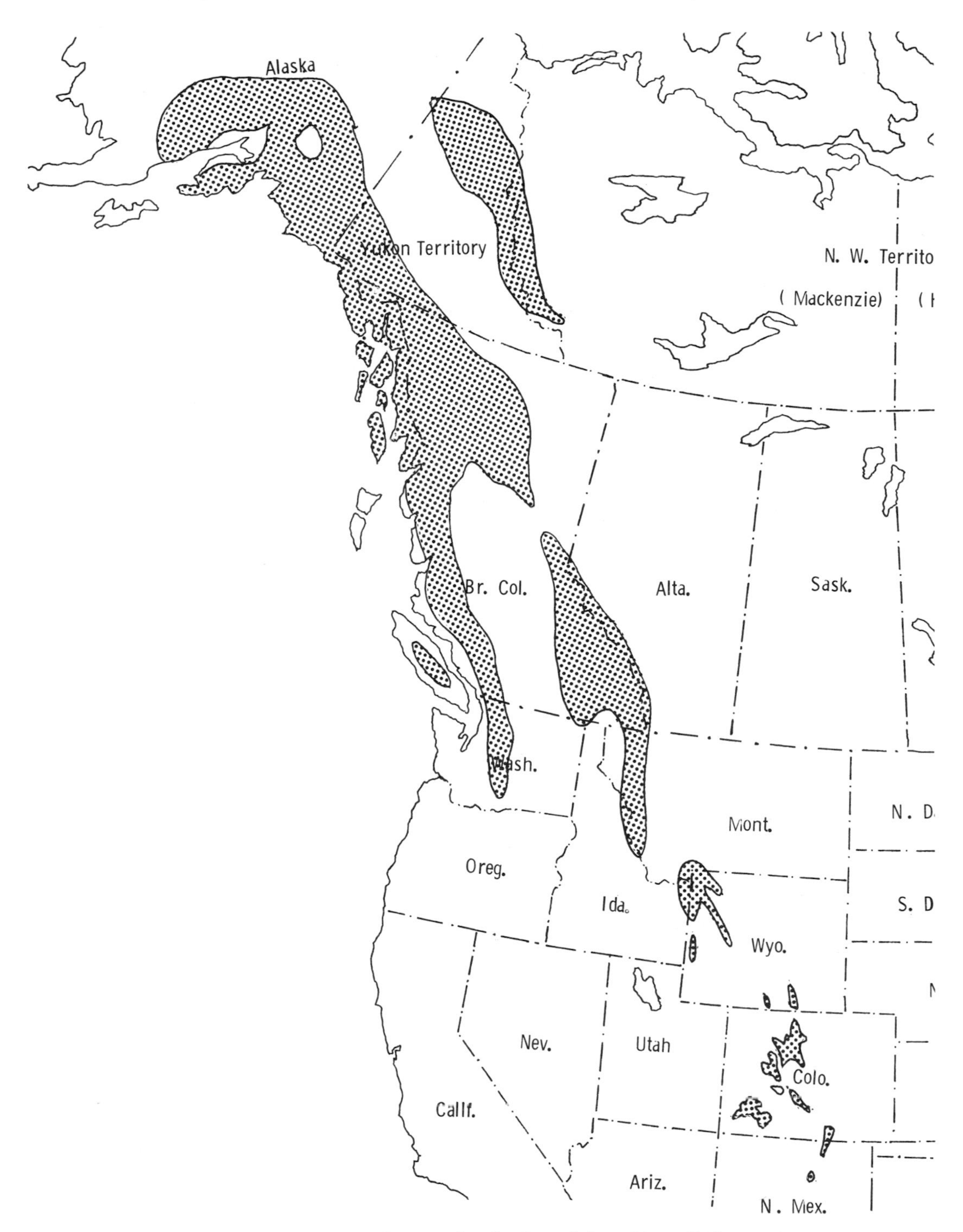

FIGURE 28. Current distribution of the white-tailed ptarmigan.

nor does it apparently include the Brooks Range of northern Alaska, both of which would seem to provide suitable habitat opportunities for the species. In the Rocky Mountains of the western states the range of the species is highly disjunctive because of the limited elevations above timberline, and it must be presumed that these southern populations became isolated during Pleistocene times. These southernmost populations are probably the ones most vulnerable to possible extirpation. Ligon (1961) noted that although the New Mexican range of this species once included all the alpine ridges of the Sangre de Cristo range from Lake Peak to the Colorado line, the birds are now found only on a few peaks near the Colorado line. Braun (1970) reported finding them on Costilla Peak in 1970, and has also verified their occurrence on Baldy Peak near Santa Fe. Braun (1969) concluded that although the birds may once have occurred in Oregon, Idaho and Utah, their recent natural occurrence in these states is unproved. Attempts have recently been made to introduce them in northeastern Oregon.

Except for Alaska, Colorado is the state with perhaps the greatest amount of white-tailed ptarmigan range in the United States. Rogers and Braun (1968) estimate that more than four thousand square miles in the state are occupied by this ptarmigan.

Weeden (1965b) reported that typical terrain of this species consists of steep slopes and ridges, often around cirques and stony benches, where ledges, cliffs, and outcrops commonly occur. The vegetation is generally sparse, with shrubs nearly absent and dwarfed when present. The birds in Alaska are usually from five hundred to two thousand feet above timberline. In Montana, Choate (1963) found that ptarmigan are not present in timber or in shrubby vegetation more than eighteen inches high. Rather, they prefer areas of rocks and moist ledges with alpine vegetation that is low-growing but well developed. Rocks from six to twenty-four inches in diameter provide optimum habitat, since they provide protection from bad weather and cover from visual predators. Ptarmigan are never found in boggy areas or areas where the vegetation is taller than the birds themselves. They usually frequent gently sloping areas where moisture is abundant and vegetation is present. Preferred cover plants, which also are among the most important food plants, including willow, heath (*Phyllodoce* and *Cassiope*), and mosses.

Braun (1969, 1970) concluded that in Colorado the distribution and abundance of alpine willow is the key factor determining ptarmigan distribution. Willow not only represented the majority of the ptarmigan's food from late September until May, but its occurrence in snow-free areas in late May is an essential component of breeding territories.

POPULATION DENSITY

Relatively little information is available on breeding densities. Choate (1963) reported the over-all density of breeding birds on a 2-square-mile plot at 17.5 birds per square mile, but if unsuitable habitats are excluded from consideration, the density could be calculated as 50 breeding birds per square mile. On study areas totaling 8.41 square miles, Rogers and Braun (1968) reported 52 and 56 breeding pairs plus 11 to 25 unmated birds in 1966 and 1967, or 15.2–15.5 birds per square mile. In 1968 there were 55 pairs and 21 unmated males on areas totaling 6.93 square miles, or 19.2 birds per square mile, and in 1969 there were 60 pairs and 28 unmated males on 8.41 square miles, or 17.8 birds per square mile (*Colorado Game Research Review*, 1968 and 1969).

HABITAT REQUIREMENTS

Wintering Requirements

Braun (1969, 1970) reported that wintering areas for ptarmigan in Colorado must contain alpine willows (*Salix nivalis* and *S. anglorum*), and alpine areas lacking this species cannot support ptarmigan for prolonged periods. Braun and Pattie (1969) reported that the Beartooth Plateau of Wyoming almost completely lacks willow in this timberline zone, and willow stands that do occur are snow-covered during winter. The birds evidently do not occur there or in certain northern New Mexico peaks where willow is also absent (Braun, 1970).

Spring Habitat Requirements

Braun (1969, 1970) reports that the presence of willow is essential to the habitat characteristics of successful male territories. In Colorado, breeding territories are adjacent to the spruce-willow alpine timberline (krummholz) zone, and also include small windblown areas. In the Beartooth area of Wyoming, this combination of habitat characteristics in the alpine zone is lacking, thus the area is apparently unsuitable as a breeding ground (Braun and Pattie, 1969). In Colorado, territories are established in suitable habitats where the snow is gone by early May (Braun, 1969).

Nesting and Brooding Requirements

Nest site characteristics for the white-tailed ptarmigan are evidently rather broad, judging from the diversity of nest sites that have been found (Schmidt, 1969). Probably more important than specific nest sites during the summer period is the accessibility of suitable brooding areas. Brooding areas for females and suitable summering areas for post-territorial males as well as unsuccessful hens occur where the vegetation is short and where rocks six inches or larger cover more than 50 percent of the ground surface (Braun, 1969). The vegetation of suitable meadow areas adjacent to rock fields consists principally of sedges (*Carex*) and forbs such as *Geum* and *Polygonum.* During late summer, adults and young move to snow accumulation areas between the summering and wintering habitats, which provide the last source of green plants in the alpine zone and also offer the best protection for intermediate-plumaged birds.

FOOD AND FORAGING BEHAVIOR

Weeden (1967) has reported on the analysis of 167 crops of this species collected from Colorado to Alaska. Winter foods of Alaskan populations differ from those in Colorado in that alder (*Alnus*) catkins are an important part of the winter diet, with willow (*Salix*) and birch (*Betula*) of secondary importance. In contrast, Colorado ptarmigan subsist largely on the buds and woody twigs of various alpine willows (Quick, 1947). Weeden attributed this difference to the increased availability of alder in northern areas, and to possible competition from other species of ptarmigan in Alaska.

May and Braun (1969) reported that among forty winter food samples from Colorado, willow occurred with a 100 percent frequency, but alder also occurred in samples from areas where that species was locally abundant. Coniferous food sources (*Picea, Pinus, Abies*), although readily available, are rarely taken in winter (May, 1970). With spring, a diversity of green leaves and flowers are consumed, although willow remains the most important food. The leaves and flowers of *Potentilla, Ranunculus, Saxifraga,* and *Dryas,* all of which are high in protein, were other important spring foods. During summer a diverse array of leaves and seeds are also consumed, and the bulbils of *Polygonum viviparum* are an important summer food for adults. During their first two weeks juveniles feed largely on invertebrate foods, then they too begin to feed extensively on these bulbils. Gradually willow gains importance over *Polygonum* for both juveniles and adults, and eventually the birds go back to a diet consisting almost entirely of *Salix* buds and twigs (May and Braun, 1970; May, 1970).

MOBILITY AND MOVEMENTS

Relatively little is known of white-tailed ptarmigan movements, but certainly little lateral movement is normally typical. During winter, the birds typically descend to the edge of treeline, where food is more readily available. In Colorado, ptarmigan gather in flocks of five to thirty birds in high alpine basins where willows are abundant (Quick, 1947). Single birds also sometimes occur in alpine fir (*Abies lasiocarpa*), limber pine (*Pinus flexilis*), or on steep rock slopes during winter but when flushed usually drop down into the snow basins below. Weeden (1965b) indicated that in Alaska most birds remain above the timberline, feeding in areas such as steep cliffs, ridge topes, and benches that are blown fairly free of snow. In parts of southwest Colorado the birds go to low valleys every winter regardless of snow cover (Braun and Rogers, 1967b). During early winter in Colorado, flocks of up to fifty ptarmigan can be found in areas containing available willow, but later the sexes tend to segregate, with males occurring nearer timberline and females remaining in the larger willow expanses at lower elevations (Braun and Rogers, 1968). Birds may move as much as a mile in a day during winter and up to fifteen miles on a longer basis (Braun and Rogers, 1967b).

In spring, Colorado ptarmigan move back up to the breeding areas, which in the case of males may be a distance of less than a mile. Movements of both sexes are very restricted during the breeding and nesting periods, with birds rarely moving more than five hundred yards (Braun and Rogers, 1967b). When broods appear, males and broodless females move uphill into higher rocky summering areas that may be up to 2 miles from nesting areas, where the birds once again become fairly sedentary. Hens may also move their broods as much as 1/3 mile to such summer brood-rearing areas (Braun and Rogers, 1967b). Subadult males and unsuccessful hens move considerably farther than adult males or brooding females, and fall movements of females may exceed 10 miles (Braun, 1969).

Daily movements probably differ considerably according to sex, age, and time of year and with varying weather conditions. Minimal daily movements may occur among brooding females caring for young chicks. Schmidt (1969) noted that one brood moved about eight hundred yards in ten hours, and another moved three hundred to four hundred yards in three hours. Similarly, males on breeding territories move very little. Schmidt found in 1967 that males had an average territory size of 19 acres, with maximum use occurring in 5.3 acres, and in 1968, with a better sample, territories averaged 36 acres, with maximum use in a 9.5 acre area. These territorial areas were used over a 2½ month period, during the entire pair bond period.

REPRODUCTIVE BEHAVIOR

Prenesting Behavior

Virtually all that is known of the reproductive behavior of the white-tailed ptarmigan consists of the work of Schmidt (1969), which as of this writing is still unpublished. The following summary is based on Schmidt's observations.

Territorial Establishment

With the return of the males from their timberline wintering areas to the alpine breeding grounds, territories were gradually established, which ranged in size from 16 to 47 acres. Within these fairly large defended areas, which overlapped slightly, males were usually to be found in areas of maximum use of from 3.2 to 15.7 acres. Males typically returned to their same territories of past years, and females usually returned to the same territory and the same male each spring. Territorial activity was not strong until the arrival of the females on the breeding areas, and males would often feed together until that time.

Males were typically monogamous, and Schmidt found that although males were sometimes found with two females, this was less common than seeing unpaired males. Territories were usually held by males at least twenty-two months old, with subadults successful in obtaining territories only if they were vacated by older birds. Territorial defense and proclamation became spirited in late April or early May when the females arrived, and the pair-forming period occurred at the same time. The most intensive territorial activity was typically in very early morning or after feeding in the evening, but during foggy periods or snow squalls activity was intense, apparently as a result of restricted visibility.

Male Territorial and Pair-forming Behavior

Male displays and calls may be discussed according to whether they serve the dual purpose of warding off other males from the territory and attacting females, or whether they are performed only in a sexual situation. The basically agonistic territorial signals may be considered first.

Schmidt classified the territorial behavior of males into three general types, the "scream flight," "ground challenging," and intimidation displays, noting, however, that they form a continuum of functions and have certain merging characteristics. The male scream flight, which corresponds to the song flight of willow ptarmigan, consists of the birds' taking off and uttering a raucous call containing four syllables, *ku-ku-KIIII-KIIERR*, lasting about one second and being repeated at intervals of about one to three seconds. Choate (1960) had noted that this flight was sometimes characterized by a steep rise followed by a shallow glide, which Schmidt did not see. This display clearly attracted females and warned rival males of the territorial location. However, the display was sometimes seen in midsummer after territories had been abandoned, and females sometimes uttered a homologous call while the male was calling or when defending chicks.

Ground challenging was uttered from convenient calling posts, and the associated call varied considerably in emphasis, such as *duk-duk-DAAK-duk-duk* or *DAAK-DAAK-duk-DAAK-duk-duk-duk*. Some "long ground screams" closely resemble the flight scream in their last four notes. Intimidation displays performed on the ground included two major postures. These were a flat posture assumed during running and an upright threat posture held during slow walking or while standing still. During these displays the eye-comb was exposed by raising crown feathers and low clucking sounds were typically uttered. During territorial border disputes males would usually face one another at distances of from five to thirty feet in the upright postures, sometimes making short flights while calling. Aerial chases occurred occasionally.

With the arrival of females on a territory, the responses of resident males changed. Males would chase the individual females that entered their territories and perform several specific postures and calls. The "courtship chase" and associated strutting was much like an aggressive attack toward another male, but the head was held more upright, the tail and undertail coverts were more strongly lifted, the breast feathers were fluffed, and the wings were slightly drooped. When the female attempted to escape from the approaching male, chases typically ensued.

Males sometimes varied their strutting approach to females with a "slow approach" and a rhythmic "head-bowing," that resembled the ground-pecking "displacement" display of male spruce grouse, but the bill was lowered only part way toward the ground. Frequently, a "waltzing" display was performed by the male as he approached the female and attempted to circle in front of her. While so doing, he tilted the tail toward the female and dragged both wings, with the wing nearer the female held lower than the more distant one. This waltzing display lasted from one to five seconds and was usually repeated several times in a twenty- to forty-second interval. No calling was heard during this display.

Evidently pair-formation was achieved by the repeated performance of these displays, after which the female followed the male closely, the two birds feeding and resting at the same times. While the female fed, Schmidt heard the male utter "assurance clucks" from fifty to eighty times a minute. When the female rested near the base of a rock, the male typically stood on the top of that rock or an adjacent one.

Copulation and the associated behavior patterns were observed only a few times, and occurred just prior to the period of egg laying and incubation. On one occasion Braun (cited in Schmidt) observed an apparent instance of precopulatory invitational "tidbitting," during which the male pecked the ground and uttered a series of low-pitched clucking sounds that stimulated the female to rush over and join in the pecking. As the pair began pecking head to head, the male raised his head, exposed his eye-combs, fluffed his feathers, and drooped his wings. He then began bowing his head over the female while uttering "churring sounds." Then he walked around the female and grabbed her nape, causing the hen to drop to the ground with her neck extended forward. When mounting and during copulation the male lowered his wings and crouched down on the female. When released, the female ran forward in several short dashes, stopping between dashes to shake. The postcopulatory display of the male resembled normal strutting, but the wings were more strongly drooped, and the bird walked in slow steps. In each of four cases, the male moved from ten to fifty feet before resuming normal feeding. In one case, several short dashes were made by the male as well.

One other display noted by Schmidt was "tail-wagging," which apparently occurred as a displacement activity in times of stress. Schmidt found that it occurred in adults of both sexes and in young only six weeks old. Females typically performed tail-wagging when approached by a courting male but only when approached from the side or behind. Displacement feeding movements were also noted in stress situations.

Vocal Signals

In addition to the several calls mentioned earlier, Schmidt noted several other vocal signals. Hissing sounds were emitted by females when defending the nest, and when performing distraction displays the female typically uttered a harsh *craaow* note that apparently served as an alarm call to the chicks. Females also uttered a loud *brrrt*, apparently of similar function. When the young were older, females uttered "alert calls," running to the cheeping distress calls of young and uttering high clucks in an upright alert posture. Females also uttered soft contact calls in the presence of their broods and while pecking made cackling noises that served to attract the young. Schmidt noted that such functional tidbitting behavior had earlier been reported for both willow ptarmigan and sage grouse. It is of interest that so far only in the white-tailed ptarmigan has tidbitting been reported as an adult display pattern, where it possibly serves as a precopulatory attraction signal.

Nesting and Brooding Behavior

Relatively few nesting studies have been made on this species. Choate (1963) reported on eleven nests in Montana that had from 3 to 9 eggs, averaging 5.2. Bradburry (1915) mentioned six Colorado nests containing from 5 to 7 eggs. Braun (1969) noted that nineteen nests in Colorado had from 4 to 7 eggs, averaging slightly under 6. Choate (1963) found one known instance of renesting in Montana, and Braun (1969) concluded that renesting was also probable in Colorado. He estimated an egg-laying interval of slightly under one and one-half days and an incubation period of twenty-two to twenty-three days.

Choate (1963) found an incubation success of 70 percent for nests studied in Montana, and a hatching success of 85.5 percent of eggs observed. Braun (1969) reported a nearly identical hatching success of 81.1 percent in Colorado.

The male apparently normally remains with the female until the time of hatching, judging from observations of Schmidt and Braun in Colorado, although Choate (1963) indicated that the pair bond may last only two or three weeks. Females regularly perform strong nest and brood defense displays, and Schmidt (1969) noted that males may also defend the nest site. Early in the incubation period, a female disturbed from the nest typically skitters over the ground for from ten to fifty feet, with her wings dragging and her head low in a distraction display. As hatching approaches the female is more likely to remain at the nest, hissing and spreading her wings. Schmidt never found a male defending a brood, but female brood defense may take several forms. She may attack the intruder, with expanded eye-combs and exposed white carpals, running with the wings extended and head raised and uttering hissing sounds. When the chicks were still very young the female often performed distraction behavior and lead the intruder from the brood. When the chicks were older, the female usually uttered "alert calling" or would place herself between the observer and the brood, running back and forth and hissing. When they were from ten to twenty-one days old the chicks could fly from 20 to 150 feet, after which they would run and utter cheeping calls. Loud calls were also uttered by lost chicks, which gradually changed to hoarse *cheer-up* sounds in older birds. When captured, birds up to twelve months old would sometimes utter similar sounds.

Concentration of females with broods occurred on certain favored areas that provided a combination of rocky habitat and an abundance of low, rapidly growing herbaceous vegetation. Brood mixing commonly occurred on such areas. Hens remained with well-grown young through the autumn period, as the birds gradually moved closer to wintering areas (Braun, 1969).

EVOLUTIONARY RELATIONSHIPS

General comments as to the ptarmigan relationships have already been made earlier (see willow ptarmigan account). Recent authorities (Höhn, 1969; Braun, 1969) appear to be agreed that the white-tailed ptarmigan must have been derived from a relatively early offshoot of ptarmigan stock that became isolated in western North America. Braun agreed with Johansen (1956), who thought that the white-tailed ptarmigan originated from ancestral stock of *Lagopus mutus* which arrived very early in North America. Judging from plumage characteristics of downy young as well as adults, I would favor the view that such a separation of pre-*leucurus* stock occurred before a subsequent splitting of gene pools that gave rise to the modern rock and willow ptarmigans; thus I believe that these two species are more closely related to one another than either is to the white-tailed ptarmigan. Differences in bill size among the three species where they occur together in Alaska and western Canada may be advantageous in reducing foraging competition; thus, indirectly, selection for differences in body size among the three species may have occurred. Weeden (1967) has already suggested that winter foods taken by white-tailed ptarmigan in Alaska may be influenced by competition from the two other species of Alaskan ptarmigans.

16

Ruffed Grouse

Bonasa umbellus (Linnaeus) 1776

OTHER VERNACULAR NAMES

BIRCH partridge, drummer, drumming grouse, long-tailed grouse, mountain pheasant, partridge, pine hen, pheasant, tippet, white-flesher, willow grouse, wood grouse, woods pheasant.

RANGE

Resident in the forested areas from central Alaska, central Yukon, southern Mackenzie, central Saskatchewan, central Manitoba, northern Ontario, southern Quebec, southern Labrador, New Brunswick, and Nova Scotia south to northern California, northeastern Oregon, central Idaho, central Utah, western Wyoming, western South Dakota, northern North Dakota, Minnesota, central Arkansas, Tennessee, northern Georgia, western South Carolina, western North Carolina, northeastern Virginia, and western Maryland. Recently introduced in Nevada and Newfoundland (modified from *A.O.U. Check-list*).

SUBSPECIES (*ex* Aldrich and Friedmann, 1943)

B. u. umbellus (Linnaeus): Eastern ruffed grouse. Resident in wooded areas of two regions, from east central Minnesota, southern Wisconsin, and southwestern Michigan south to central Arkansas, extreme western Tennessee, western Kentucky, and central Indiana (this population sometimes separated as *B. u. mediana* Todd 1940), and from central New York and central Massachusetts south to eastern Pennsylvania, eastern Maryland (formerly), and New Jersey.

B. u. monticola Todd: Appalachian ruffed grouse. Resident from southeastern Michigan, northeastern Ohio, and the western half of Pennsylvania south to northern Georgia, northwestern South Carolina, western North Carolina, western Virginia, and western Maryland.

B. u. sabini (Douglas): Pacific ruffed grouse. Resident of southwestern British Columbia (except Vancouver Island and the adjacent mainland) southwest of the Cascade Range, through west central Washington and Oregon to northwestern California.

B. u. castanea Aldrich and Friedmann: Olympic ruffed grouse. Resident of the Olympic Peninsula and the shores of Puget Sound south to western Oregon.

B. u. brunnescens Conover: Vancouver Island ruffed grouse. Resident of Vancouver Island and adjacent mainland south to Puget Sound and north at least to Lund.

B. u. togata (Linnaeus): Canadian ruffed grouse. Resident from northeastern Minnesota, southern Ontario, southern Quebec, New Brunswick, and Nova Scotia south to northern Wisconsin, central Michigan, southeastern Ontario, central New York, western and northern Massachusetts, and northwestern Connecticut.

B. u. affinis Aldrich and Friedmann: Columbian ruffed grouse. Resident from central Oregon northward, east of the Cascades through the interior of British Columbia to the vicinity of Juneau, Alaska (not recognized in *A.O.U. Check-list*).

B. u. phaia Aldrich and Friedmann: Idaho ruffed grouse. Resident from southeastern British Columbia, eastern Washington, and northern Idaho south to eastern Oregon and on the western slopes of the Rocky Mountains to south central Idaho.

B. u. incana Aldrich and Friedmann: Hoary ruffed grouse. Resident from extreme southeastern Idaho, west central Wyoming, and northeastern North Dakota south to central Utah, northwestern Colorado (rarely), and western South Dakota.

B. u. yukonensis Grinnell: Yukon ruffed grouse. Resident from western Alaska east, chiefly in the valleys of the Yukon and Kuskokwim rivers, across central Yukon to southern Mackenzie, northern Alberta, and northwestern Saskatchewan.

B. u. umbelloides (Douglas): Gray ruffed grouse. Resident from extreme southeastern Alaska, northern British Columbia, north central Alberta, central Saskatchewan, central Manitoba, northern Ontario, and central Quebec south, east of the range of *affinis* and *phaia*, to western Montana, southeastern Idaho, extreme northwestern Wyoming, southern Saskatchewan, southern Manitoba, southern Ontario, and across south central Quebec to the north shore of the Gulf of St. Lawrence, probably to southeastern Labrador.

MEASUREMENTS

Folded wing: Adult males, 171–93 mm; adult females, 165–90 mm (males of all races average 178 mm or more; females usually average under 178 mm).

Tail: Adult males, 130–81 mm; adult females, 119–59 mm (males average more than 147 mm; females average less than 142 mm).

IDENTIFICATION

Adults, 16–19 inches long. Both sexes have relatively long, slightly rounded tails that are extensively barred above and have a conspicuous subterminal dark band. The neck lacks large areas of bare skin, but both sexes have dark ruffs. Feathering of the legs does not reach the base of the toes; the lower half of the tarsus is essentially nude. Both sexes are definitely crested, but the feathers are not distinctively colored. In addition males have a small comb above the eyes that is orange red and most evident in spring. Most races (*castanea* is perhaps the only exception) exist in both gray and brown phases, which appear with the first-winter plumage. Otherwise, little seasonal, sexual, or age variation occurs. The birds are generally wood brown above, with blackish ruffs (less conspicuous in females and immatures) on the sides of the neck, and with small eye-spot markings on the lower back and rump (less conspicuous in females). The tails of both sexes have seven to nine alternating narrow bands of black, brown, and buff, followed by a wider subterminal blackish band that is bordered on both sides with gray and is less perfect centrally in females and some (presumably first-year) males. In winter, both sexes develop horny pectinations on the sides of their toes, which are more conspicuous than in most other species.

FIELD MARKS

The fan-shaped and distinctively banded tail and neck ruffs of both sexes make field identification easy. The birds usually take off with a conspicuous whirring of wings, and in spring males are much more often heard drumming than they are seen.

AGE AND SEX CRITERIA

Females have shorter tails than do males (see above) and their central tail feathers lack complete subterminal bands near the middle of the tail. A mottled pattern on the central tail feathers (which occurs in about 15 percent of the population) can indicate either sex, but a bird with this characteristic is twice as likely to be a male as a female (Hale, Wendt, and Halazon, 1954). Females also have little or no color on the bare skin over the eye, whereas in males this area is orange to reddish orange (Taber, in Mosby, 1963). Davis (1969a) reported that the length of the plucked and dried central rectrices provides a 99 percent effective means of determining sex of both adult and immature ruffed grouse, but specific separation points for these groups vary with populations.

Immatures can be identified by the pointed condition of their two outer primaries, especially the outermost one. Davis (1969a) stated that during the hunting season the condition of the tenth primary was useful for determining age of nearly 60 percent of the birds, with only a 2 percent error. However, the presence of sheathing at the base of the outer two primaries (adults) or on the eighth but not the ninth or tenth primaries (immatures) separated 79 percent of the birds examined with a 3 percent error. Immature males can be distinguished from adults by their shorter central tail feathers (length of plucked feather, 159 mm or less, compared to at least 170 mm in adults) as well as various other criteria (Dorney and Holzer, 1957). Ridgway and Friedmann (1946) report that the two outer primaries of immatures have outer webs that are pale fuscous and mottled or stippled with lighter buff, instead of being buff or whitish with darker brown markings.

Juveniles resemble the adult female but have barred tail feathers that lack the heavy subterminal band and have the gray tips poorly developed (Ridgway and Friedmann, 1946). Juveniles also have white rather than buff chins and primaries with more mottling on their outer webs (Dwight, 1900).

Downy young are illustrated in color plate 61. Downy ruffed grouse can readily be identified by the restriction of black on the head to an elongated ear-patch that is narrowly connected to the eyes and a few midcrown spots. The crown is otherwise a uniform ochraceous tawny, gradually blending with the buffy face color. The back lacks definite patternings and varies from russet or dark brown dorsally to pale buff or yellow ventrally.

DISTRIBUTION AND HABITAT

The distribution of the ruffed grouse in North America covers a surprising variety of climax forest community types, from temperate coniferous rain forest to relatively arid deciduous forest types. The unifying criterion, however, is that successional or climax stages include deciduous trees, especially of the genera *Betula* and *Populus*. For example, the range of the balsam poplar (*Populus balsamifera*) bears a surprising similarity to that of the ruffed grouse, as does that of the paper birch (*Betula papyrifera*). Aldrich (1963) correlated racial variation in the ruffed grouse with major plant formations. He indicated that *togata* occurs in northern hardwood-conifer ecotone area, *umbellus* and *monticola* in eastern deciduous forest, *mediana* in oak-savanna woodland, *umbelloides* in typical boreal forest, *yukonensis* in northern or "open" boreal areas, *incana* in drier montane woodlands and aspen parklands, *brunnescens, castanea,* and *sabini* in the Pacific coast rain forest, and *phaia* in the corresponding wet interior forest. The relatively drier montane woodlands of the Pacific northwest are occupied by *affinis.* Not only is there a correlation between the relative wetness or dryness of these general habitat types and associated darkness or paleness of the body plumage, but there are also some relationships between climate or vegetation and color phases. The gray phase of ruffed grouse is typically associated with northern areas or higher altitudes, while the reddish brown color phase is more characteristic of southern and lower altitude populations. Gullion and Marshall (1968) have discussed the ecological significance of color phases in ruffed grouse, and they suggest that gray-phase birds are perhaps physiologically better adapted to cold than are red-phase ones, and predominate in conifers and aspen-birch forest of these colder areas. They also suggest that gray-phase birds may be less conspicuous in boreal forests, while in the hardwood forests where raptors have poorer hunting conditions and mammalian predators are more important the color phase may not be significant. However, their data indicate that gray-phase birds survive relatively better in hardwood than do red-phase ones, and both phases survive better in hardwoods than in conifers.

Gullion (1969) has pointed out that on a continent-wide basis, the areas of highest population density of ruffed grouse correspond to the distributional patterns of aspens (*Populus* spp.), which he related to winter as well as summer food use by adults, as well as their value as brooding habitat. Weeden (1965b) reported that ruffed grouse habitat in Alaska typically contains large amounts of aspen and usually also contains white spruce (*Picea glauca*) and white birch (*Betula papyrifera*). Where ruffed and spruce grouse

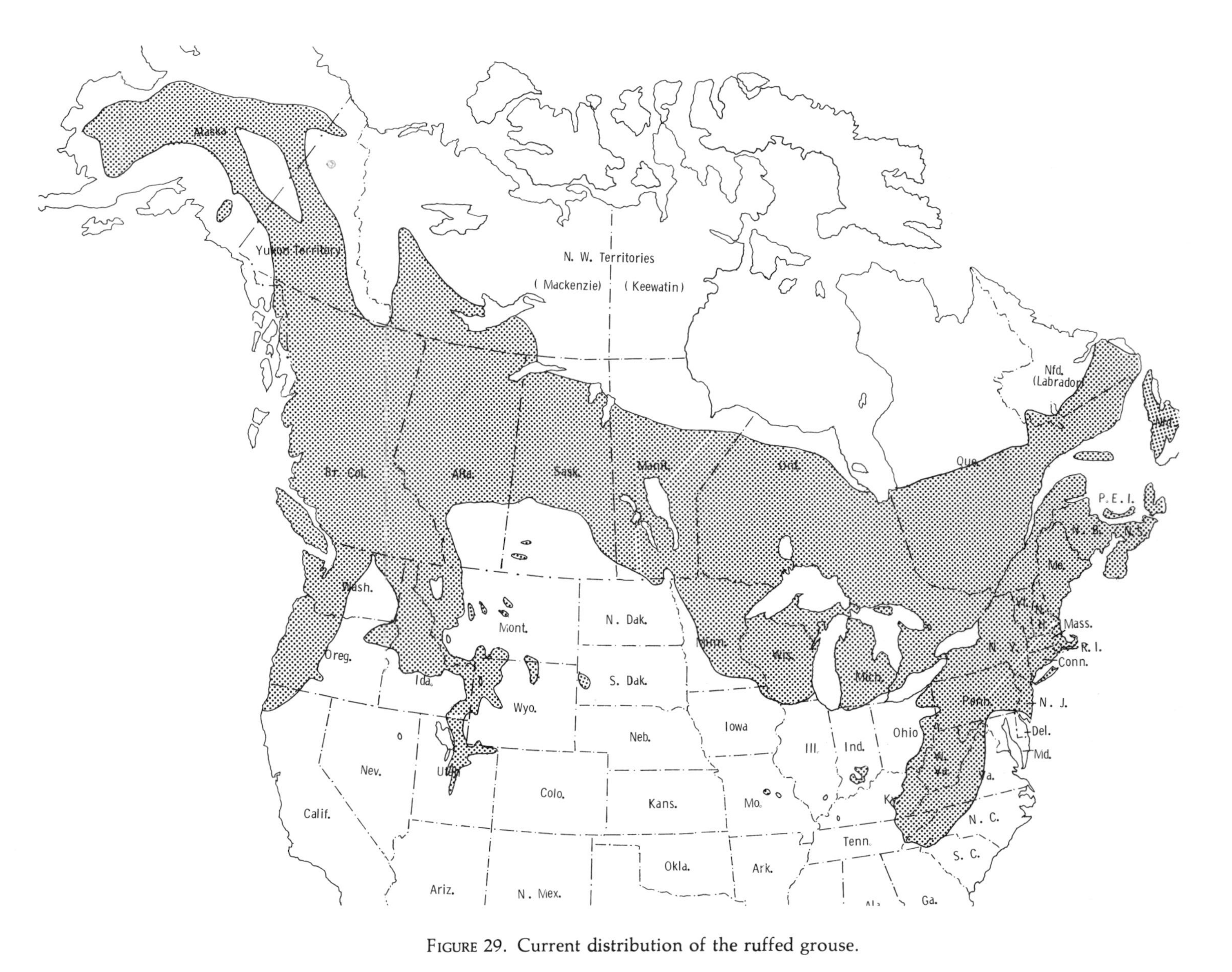

FIGURE 29. Current distribution of the ruffed grouse.

occur together in Alaska, the ruffed are found in earlier stages of succession, frequenting edges, shrubby ravines, and similar openings. Likewise in southern Ontario I have noticed that both species may be found within a hundred yards or less of one another, but ruffed grouse are always associated with birch or poplar, while spruce grouse are usually to be found under coniferous cover such as jack pine.

Edminster (1947) has analyzed the general shelter requirements of the ruffed grouse in the northeastern states according to vegetational succession stages. Open land types dominated by herbaceous plants provide some food sources for grouse but are of secondary importance. Overgrown fields with shrubs and saplings include single-species stands of high quality quaking aspen cover (*Populus tremuloides*), pin cherry (*Prunus*), scrub oak (*Quercus*), or alder (*Alnus*) cover of moderate quality, and low quality gray birch or hardhack cover. Other important cover types include mixed-species stands of hardwood shrubs and trees and mixtures of hardwood and coniferous species. Slashings following lumbering activities produce an early stage dominated by many shrubs and herbaceous species, especially blackberries and raspberries (*Rubus* spp.), of considerable value to grouse. A later, thicker stand of saplings and taller trees is of less value, especially for young birds.

Older forest stands in the northeast include hardwood types, mixed hardwoods and conifers, and predominantly coniferous forest types. Edminster reported that younger hardwood stands have better undercovers for grouse than older stands and that scattered openings improve the value of either age class. Pasturing also may affect the undercover development. Edminster believes that hardwoods with about 20 percent coniferous species provides better cover than pure hardwood stands and that those woodlands with from 20 to 70 percent conifers provide both food and cover at all seasons, although summer cover may be imperfect. Predominantly conferous stands of trees may be food-deficient in younger stages, but in mature stands with a hardwood understory this is not the case.

A study by Dorney (1959) in Wisconsin provides some additional information on grouse-forest relationships. Dorney also reported that mixtures of hardwoods and conifers have greater ruffed grouse use than do hardwoods alone, but Wisconsin grouse appear to be less dependent on conifers for cover than is the case in New York. A heavy shrub understory is needed by grouse for drumming sites, and an absence of shrubs in young hardwood stands causes rapid loss of drumming territories.

Gullion (1969) reported that in Minnesota young aspen stands first become habitable by adult ruffed grouse about four to twelve years after regeneration following logging or fire, when the trees are twenty-five to thirty feet tall and the stem densities are less than six thousand per acre.

Grouse continue to use the habitat throughout the year for the next ten to fifteen years, until stem densities drop below about two thousand per acre. Older stands of aspen provide important winter food in the form of male flower buds besides providing nesting habitats.

The importance of small clearings in deciduous forest, as found by Edminster, was proved by Sharp (1963), who established a number of small clearings ¼ to 1 acre in size in half of a 1,470-acre pole timber forest. These changes were initiated in 1950, and during the next five years from seven to twenty-one broods used the managed area, while two to three used the unmanaged portion of the forest. After ten years, the openings in the forest had filled in, and the value of the area for brood use had declined.

Probably the over-all range of the ruffed grouse has not changed greatly in historical times. Slight additions to the range have occurred with introductions. Wild-trapped grouse from Nova Scotia, Wisconsin, and Maine have apparently been successfully introduced into Newfoundland (Tuck, 1968), and they have also been successfully introduced in the Ruby mountain range of northeastern Nevada (McColm, 1970).

Restrictions in ranges have occurred in a number of states, as indicated by Aldrich (1963). Although it once occurred in northeastern Nebraska, the ruffed grouse is now completely extirpated from the state. It is also gone from northeastern Kansas and northeastern Alabama (*A.O.U. Check-list*, 1957). However, a specimen was recently collected in Jackson County, Alabama (*Audubon Field Notes*, 21:15, 1967). The population in Missouri was probably never high and may have declined to less than one hundred birds by the 1930s, although recent attempts at reintroduction have had some success (Lewis, McGowan, and Baskett, 1968). By 1930 the once extensive Iowa population was also nearly gone except for a remnant in northeastern Iowa. This population still persists in good numbers locally, and hunting for the first time in 45 years was allowed in 1968 (Klonglan and Hlavka, 1969). In Ohio, where grouse once ranged over the entire state, a low ebb was reached about 1900, and the species was protected for thirty-two of thirty-four years following 1902 (Davis, 1969b). Remnant populations occur in southern Illinois, where the species is protected. The species is also protected where it occurs in northwestern South Carolina, which is at the extreme southern limit of the species' range. Although limited to a small area of southern Indiana, the grouse population there has been fairly stable for the last two decades and is distributed through about eleven hundred square miles in five counties. In 1965 the first limited season was held since 1937.

POPULATION DENSITY

Grouse populations have been intensively studied in New York by Bump et al. (1947), who reported breeding densities of from eight to twenty-two acres per bird near Ithaca and from twenty-one to thirty-eight acres per bird in the Adirondacks. Maximum fall densities in the two areas ranged from five to twenty acres in various years. Gullion (1969) estimated that maximum breeding densities in Minnesota allowed by territorial behavior are one pair (i.e., 1 territorial male) per eight to ten acres, although normal area-wide densities are more commonly 4 to 6 birds per one hundred acres. Slightly lower breeding densities of 2 to 4 birds per one hundred acres occur in Ohio (Davis, 1968). Porath (cited in Klonglan and Hlavka, 1969) estimated a spring breeding density of 30 to 35 birds per square mile (4.5 to 5.3 birds per one hundred acres) in northeastern Iowa, while late summer populations were approximately 90 to 135 birds per square mile in the same area. In Indiana, Thurman (1966) reported a spring density of 18 males per square mile.

Consideration of ruffed grouse densities are not complete without mention of the well-known cycles of population abundance that have been reported for several grouse species but are especially often attributed to the ruffed grouse. Keith (1963) has made an intensive survey of population fluctuations in a variety of birds and mammals in northern North America, and his conclusions appear to be well founded. He believed that the ruffed grouse has undergone fairly synchronous ten-year population cycles at local, regional, and continental levels over most of its North American range with the exception of the eastern United States and New Brunswick. His book summarizes population density figures from a variety of studies in Minnesota, Michigan, and Wisconsin that indicate peak-year fall densities of from 123 to 180 birds per square mile in Michigan and up to 353 birds per square mile in Minnesota. The average ratios between densities of peak years and those of the subsequent low ones range from a ratio of 3:1 to as much as 15:1, with twelve such estimates averaging about 8:1.

In seven studies of local grouse populations, the ruffed grouse had peak populations or initial declines the same year as prairie grouse and spruce grouse, in two cases the ruffed grouse peaked or declined a year before the others, and in four cases the other grouse peaked or began declines one to three years before the ruffed. Likewise, at state or provincial population levels, the ruffed grouse peaked or began declines the same year as the prairie grouse in six of fourteen cases, while in six cases the other grouse peaked or declined one to three years before the ruffed grouse, and in the remaining two cases the ruffed grouse peaked or began its decline a year before the others (Keith, 1963).

Wintering Requirements

Although the ruffed grouse is one of the most temperate-adapted of all North American grouse, as indicated by its distribution in the southeastern states, it is well adapted to withstand cold weather. Edminster (1947) indicates that cold weather alone, if not accompanied by snow or sleet, does not materially affect grouse survival. However, during stormy weather, the grouse resort to coniferous trees or to roosting beneath the snow, where they may remain several days. Although the birds are rarely if ever frozen into such snow roosts, they become highly vulnerable to predation by mammals such as foxes, and Edminster reported mortality rates from 25 to 100 percent higher than normal during a year of unusually heavy snow-roosting activity.

Although conifers provide valuable winter roosting cover for ruffed grouse in New York, the birds continue to rely on hardwood trees for their food, particularly buds and twigs of such trees as poplars, apples (*Malus*), birches, oaks, and cherries (*Prunus*). When available, understory shrubs and vines such as grapes (*Vitis*), greenbrier (*Smilax*), laurel (*Kalmia*), blueberry (*Vaccinium*), and wintergreen (*Gaultheria*) also provide important sources of winter food and cover (Edminster, 1947).

Spring Habitat Requirements

The spring habitat needs of ruffed grouse appear to be closely tied to ecological situations associated with suitable drumming sites, or "activity centers" (Gullion and Marshall, 1968). Within a general activity center, a specific display site, or "drumming stage" must be present, and Gullion and Marshall believe that two factors govern the choice of such a site. These are the presence of a number of forty-to-fifty-year-old aspens near or within sight of a drumming log and also a tradition of occupancy of the site by male grouse. They concluded that the presence of aspens is the most important aspect of cover which regulates the choice of activity centers, and they found strong relationships between cover types and male survival. Males survived best in hardwoods completely lacking evergreen conifers (which is in contrast to conclusions mentioned earlier by Edminster), but the presence of spruce and balsam fir (*Abies balsamea*) did not reduce survival. However, survival did decrease as the density of mature pines increased, and male grouse did not survive as well in edge situations as in uniform forest types.

Boag and Sumanik (1969) gathered evidence supporting the view that ruffed grouse do not select drumming sites at random, but that the nature of the surrounding vegetation plays an important role. Comparing eighty drumming sites with ninety-eight similar sites that were not used, they found shrub sizes greater at used than unused sites, and canopy coverage as well as the frequency of young white spruce trees was higher at used sites. Only at used sites was aspen the predominant tree species in the tree layer. They believe that selective pressure for the male to choose open and visually effective sites for drumming is counterbalanced by selection favoring sites protected from predators. The result has been selection favoring sites which give the males sufficient height above the ground from which to observe other grouse or large ground predators, sufficient openings in the shrub layer to see at least twenty yards in most directions, and sufficient canopy and stem coverage to screen the birds from aerial predators. These conditions are met in Alberta by those areas where the density of young hardwood trees and the density and canopy coverage of young spruce are the highest.

The specific drumming stage is usually but not always a log, thus the presence of logs in suitable habitats is an important component of spring ruffed grouse habitat. Palmer (1963) analyzed forty drumming logs in Michigan that had been regularly used by male grouse. Of the total, thirty-four were old, decayed conifers, primarily pines. Males always drummed near the larger end of these, usually about 5 feet from the end. The logs ranged from seven to twenty-one inches in height at the drumming position, and none was shorter than 5.5 feet long. Vegetation over 8 feet high was significantly more dense near the logs than in the surrounding cover, and among the larger shrubs, speckled alder (*Alnus incana*) comprised about three-fourths of the sampled stems. In general, drumming sites were associated with ground vegetation less dense, and large shrub and tree cover more dense, than was typical of the surrounding general vegetation.

Several studies have indicated that a male grouse may utilize more than one log in his territory for drumming purposes, but one is typically favored. Gullion (1967a) called this log the "primary log," and designated additional drumming sites as "alternate logs." Disturbance may force the bird to use yet other "secondary logs." Logs and activity sites may also be classified as perennial if they are used through the lifetimes of a succession of grouse, or transient if they are used by one grouse and not used again for several years by other birds. Although perennial logs apparently supply the appropriate ecological conditions that attract male grouse, Gullion and Marshall (1968) have found that male grouse using such sites suffer higher mortality as an apparent result of predators' learning the locations of favored display areas.

Nesting and Brooding Requirements

Habitats selected by female grouse for nesting have been analyzed by Edminster (1947), based on the study of 1,270 nests in New York. Medium-aged stands of hardwoods, with a few conifers, was most commonly used for nesting habitat, followed by medium-aged stands of mixed hardwoods and conifers. When consideration is given to relative cover availability, slashings were also found to be of importance as grouse nesting habitat in New York. Middle-aged stands of hardwoods or mixed stands were found to be considerably more valuable as nesting habitat than were mature forest habitats. As to specific nest sites, the bases of trees appeared to be the most favorable site, being used about two-thirds of the time. Most of these trees were hardwoods, and nearly all were of considerable size. Most of the remaining nest sites were at the bases of tree stumps, under logs, bushes, or brush piles. Edminster concluded that nest sites are chosen to provide a combination of visibility, protection, an escape route, and proximity to edges and to satisfy an apparent desire for sunlight. The undergrowth nearby is usually open and the canopy density is also relatively open. More than half of the nests were within fifty feet of a forest opening, often the edge of a road. Slope considerations are evidently not important, except that steep slopes are avoided.

Gullion (1967b, 1969), summarizing research done at Cloquet, Minnesota, reported that female grouse probably begin a search for a clone of male aspen trees after mating, near which they locate their nests. These trees are then used by the incubating hens for foraging during incubation.

Brood habitat analyses have also been made by Edminster (1947). Based on studies of 1,515 broods in New York, it was clear that females with broods showed a preference for brushy habitats, especially overgrown land, followed by slashings. Hardwood stands that have been "spot-lumbered" exhibited a high brood usage, as has been later confirmed by studies in Pennsylvania by Sharp (1963). At the same time, hardwood forests continue to receive heavy use from adult grouse (males and broodless females) during the summer, while mixed woods and coniferous forest types serve for escape from extreme heat and summer storms.

FOOD AND FORAGING BEHAVIOR

Korschgen (1966) has analyzed the nutritional value of seasonal foods of ruffed grouse in Missouri and concluded that high-protein foods are taken in greatest amounts during summer, foods high in fat and carbohydrate were taken most during winter, and the largest amounts of mineral sources were taken during times of reproduction. Evidently grouse select food to fulfill seasonal nutritional needs. Korschgen summarized the principal ruffed grouse foods indicated by twenty-four published studies. Aspen and poplars are listed as principal foods in seventeen of these studies, birch in eleven, and all other food sources were mentioned less often, with apple, grape, sumac, beech, and alder all being listed in several studies. In analyses of foods from six areas in the eastern United States, Martin, Zim, and Nelson (1951) list aspen as being of first or second importance in five areas, and lacking only in samples from the Virginia Alleghenies. Other plants listed in several studies are clover, greenbrier, hazelnut, and grape.

Winter foods of the ruffed grouse consist largely of buds and twigs of trees. Edminster (1954) lists the following major winter sources of such foods: birches (several species), apple, hop hornbeam (*Ostrya*), poplar, cherry, and blueberry. In the Cloquet area of Minnesota, aspens (*Populus tremuloides* and *P. grandidentata*) are usually the most important source of winter foods, and with the appearance of the male catkins in late winter these trees provide the most nutritious food source available to ruffed grouse as long as snow is on the ground (Gullion, 1969).

A study in Utah by Phillips (1967) indicated that chokecherry (*Prunus virginiana*) was the most preferred winter food, followed closely by aspen and maple (*Acer*). Aspen was also the second most important fall food, but hips from roses (*Rosa*) had higher usage. In Ohio, Gilfillan and Bezdek (1944) found that the fruit and leaves of greenbrier (*Smilax*) had high winter use, as well as aspen buds, fruit of dogwood (*Cornus*), grape (*Vitis*), sumac (*Rhus*), beech (*Fagus*), and other plants. Winter food in Maine, as reported by Brown (1946), consisted primarily of buds of aspens, followed by buds and leaves of willows, catkins and buds of hazelnut (*Corylus*), and the buds of wild cherry and apple.

Following winter, as ground vegetation is exposed, food consumption of ruffed grouse becomes more diversified, but at least in New York the buds of poplar, birch, cherry, hop hornbeam, and blueberry are still consumed well into May (Edminster, 1947). Likewise in Maine the buds and catkins of poplar are a primary spring food, in addition to buds and catkins of birch, willow buds, and the leaves of strawberry (*Fragaria*) and wintergreen (*Gaultheria*). In Minnesota, male grouse sometimes continue to feed almost entirely on the male catkins of aspens long after snow melt allows succulent evergreen herbaceous plants to become available (Gullion, 1969). Quaking aspen in this region is preferred over big-toothed aspen by a ratio of more than 2 to 1.

The diet of adult grouse changes drastically in early summer as berries and fruits become available (Edminster, 1947). These fruits include strawberries, raspberries and related species of the genus *Rubus*, cherries, blueberries, and Juneberries (*Amelanchier*). Insects comprise a small percentage of adult foods at this time, rarely if ever exceeding 10 percent.

In contrast, the basic food of ruffed grouse chicks for at least the first week or ten days of life consists of insects. Bump et al. (1947) reported that 70 percent of the food taken in the first two weeks consists of insects, compared to 30 percent during the third and fourth week, and dropping to 5 percent by the end of July. Ants are among the most frequent food items, but a variety of other insect types, including sawflies, ichneumon flies, beetles, spiders, grasshoppers, and various caterpillar species make up the remainder of chick foods from animal sources. As dependence on insects declines with age, the amount of plant foods, particularly sedge achenes and the fruits of strawberries, raspberries, blackberries, and cherries increases correspondingly (Bump et al., 1947).

Fall foods for juvenile and adult birds include a variety of fruiting shrubs, such as viburnums, dogwoods, thorn apples, grapes, greenbriers, sumacs, and roses (Edminster, 1954). The availability of many of these persists into winter, when they supplement the standard diet of buds, twigs, and catkins.

Gullion (1966) has emphasized that the abundance of data on fall food intake by gamebirds, is often misleading in that the diversity of foraging indicated during that time of year is not representative of the critical dietary sources needed for the population's survival through the winter. Thus, the availability of a winter source of male catkins of birch, alder, hazel, and particularly aspen is probably the most important single factor influencing the wintering abilities of ruffed grouse. Gullion believed that quantitative or qualitative difference in these winter foods might account for major population fluctuations in Minnesota ruffed grouse. Lauckhart (1957) had earlier pointed out that periodic heavy seed crops in trees may sap the nutrients from buds and stems for a several-year period between such crops, causing a nutrient deficiency for animals highly dependent on these trees. The usual cycle of aspen seed crops is four to five years; thus an interaction of this cycle and some other factor or factors might account for the ten year grouse "cycle." Clearly this idea has great promise and should be investigated thoroughly before being discarded.

The importance of water, either in the form of standing water, dew, or succulent plants, also should not be overlooked for reffed grouse. Bump et al. indicate that captive grouse can easily survive for at least twelve days without food if they are provided with water but in the absence of both food and water will live only a few days. Since most grouse foods contain considerable water, it is probable that the birds can normally survive indefinitely in the absence of standing water.

MOBILITY AND MOVEMENTS

Ruffed grouse do not perform any movements that might be considered migratory, although there are some seasonal variations in mobility. Little movement is normally exhibited by ruffed grouse broods prior to the brood's breaking up and dispersing; Chambers and Sharp (1958) reported that the cruising radius of most marked broods was no more than a quarter mile. With the dispersal of the broods, more than half of the juveniles moved distances of more than a mile, in one case up to 7.5 miles. Similarly, Hale and Dorney (1963) reported that about one-fourth of the juveniles they banded had moved more than 1 mile from the banding site at the time of recovery. One grouse they banded as a three-month-old juvenile was shot thirty-one days later some 12 miles from the banding site. Apparently these fall movements were independent of population densities and were unrelated to so-called "crazy flight" behavior, during which young grouse may make long and erratic movements apparently related to inexperience and perhaps fright.

By winter, movements of both young and adult grouse decline, and the birds become virtually sedentary by spring. Hale and Dorney (1963) found that males banded on drumming sites were highly sedentary and normally returned to the same site each year. Chambers and Sharp (1958) likewise reported that grouse become sedentary as they mature, with males only rarely moving more than one-fourth mile, while females sometimes moved more than a mile. Hale and Dorney likewise reported that, except during winter, females were consistently more mobile than males. Gullion and Marshall (1968) noted a high degree of fidelity by adult male ruffed grouse not only to a particular territory but also to a specific display site. Only about 36 percent of 168 males that lived at least twelve months or longer moved to another log during their drumming lifetimes, and such movements averaged only about three hundred feet. At least 20 males, however, moved to new activity centers.

Movements by female ruffed grouse during the spring season are of equal interest and have been studied by Brander (1967). By studying the daily movements of three females in early May, Brander found that the females moved from their established winter home ranges of seven to twenty-six acres towards male drumming sites, apparently stimulated by the drumming behavior, particularly drumming sounds. One female was apparently attracted to three different males on different days before copulation occurred, and the pair remained together no more than a few hours. Since the male continued to drum after her departure, Brander concluded that the ruffed grouse mating pattern should be regarded as a promiscuous one. He estimated that the three females each remained in a state of receptivity for only four days, ending the day before the first egg was laid. The hen located her nest in each case within the area of her movements of the previous week to ten days. As mentioned previously, the female usually seeks out a clone of male aspen near which she establishes her nest (Gullion, 1969).

REPRODUCTIVE BEHAVIOR

Territorial Establishment and Advertisement

According to Bump et al. (1947), captive male grouse begin to exhibit aggressiveness as early as the first of March, although they have sometimes been seen strutting on warm days in winter. Edminster (1947) reported that drumming has been heard every month of the year and every hour of the day and night, but the most intensive drumming in New York occurs in early spring during late March and April, tapering off in May.

The two basic aspects of male reproductive display are drumming ("wing-beating" of Hjorth, 1970) and strutting ("upright," "bowing," and "rush" sequence of Hjorth, 1970). There is no doubt that drumming is primarily an acoustic display and serves to advertise the location of the male in fairly dense forest cover. Strutting, however, is a predominantly visual display, and is probably not normally released except in the visual presence of another grouse or similar stimulus. Undoubtedly both displays are essentially agonistic or aggressive in origin, serving for territorial proclamation and establishment of dominance. Since drumming is the basic means of territorial advertisement, it will be discussed first.

The motor patterns of the drumming display (Figure 30) are well described in Bent (1932) and many other references and need little amplification here. The male typically stands on a small log, facing the same direction and at virtually the same location on each occasion. With his tail braced against the log and his claws firmly in the wood, he begins a series of strong wing-strokes. These strokes, which start slowly at about one second intervals, rapidly speed up, with a complete series lasting about eight (Allen, in Bent, 1932) to eleven seconds (Hjorth, 1970). Hjorth found that in a sample of drumming displays from Alberta there were consistently forty-seven wing-strokes, while one from Ohio has 51. Aubin (1970) noted that among six ruffed grouse studied in southwestern Alberta the number of wing-strokes varied only from forty-four to forty-nine in his samples and was even more consistent for individual birds.

Allen hypothesized that the muffled drumming sound produced by the wings resulted from the forward and upward thrust rather than the return stroke. This strong forward thrust produces a counter pressure that forces

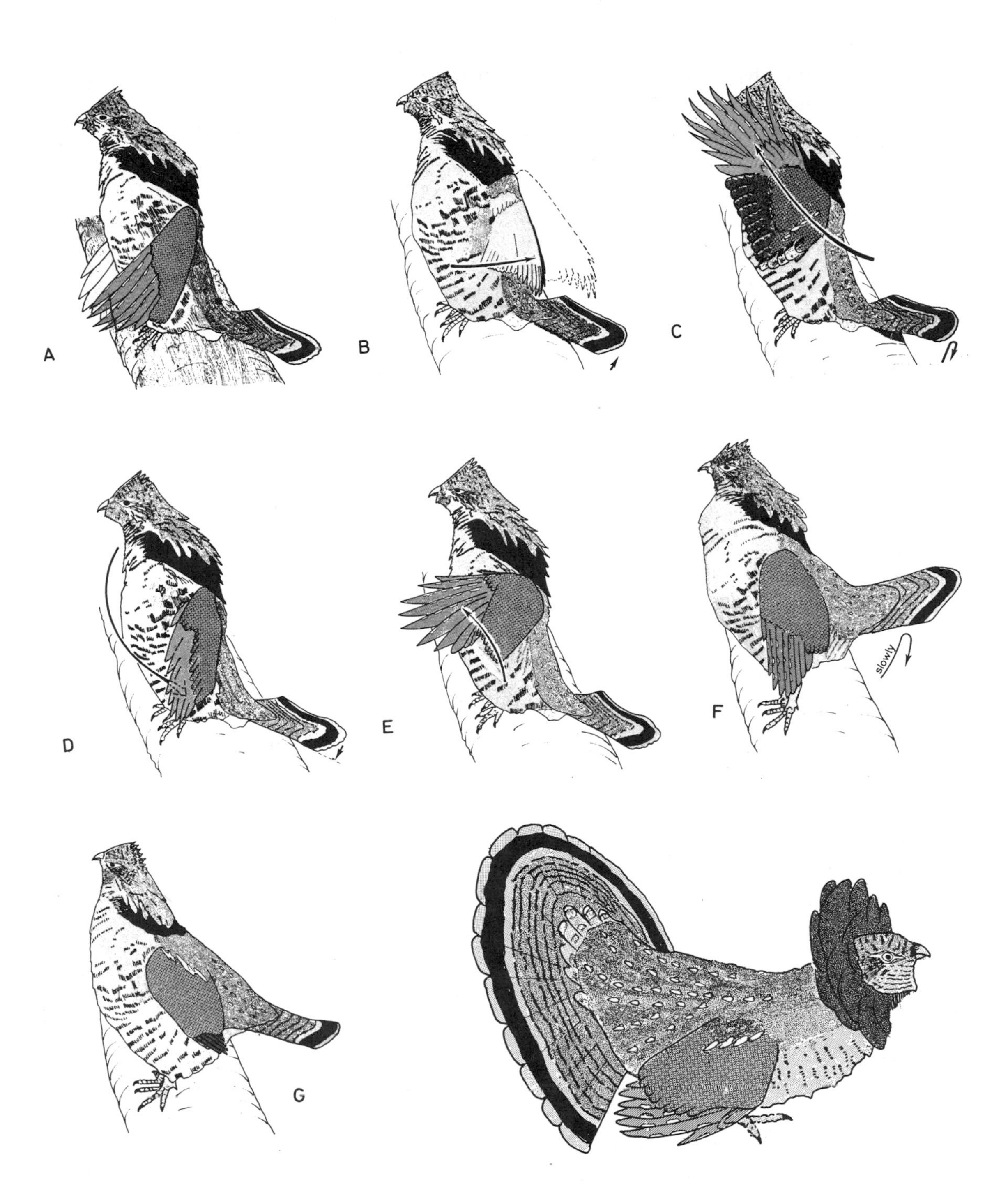

FIGURE 30. Sequence of the drumming display (A–G), and the final stage of the rush display of the ruffed grouse (from Hjorth, 1970).

the bird backward, thus explaining the need for the brace provided by the tail and the importance of clutching the log with the claws. At the end of the last stroke this pressure is released, and the bird tips forward on its perch. As Allen noted, the wings do not touch each other during the drumming, and the noise simply results from air compression, which accounts for the dull throbbing nature of the sound. Recently, Hjorth (1970) has advanced the idea that the downstroke rather than the upstroke may be responsible for this sound.

Drumming usually begins well before daylight and may continue until somewhat after sunrise. It usually begins again about an hour before twilight and may continue until dark (Bump et al., 1947). The usual interval between drumming displays is three to five minutes, but this interval varies from a few seconds to much longer periods.

As noted earlier, most males use a single log on which to drum, but some may use more than one. Bump et al. (1947) reported an average of 1.33 logs per male used by 1,173 grouse, Aubin (1970) found that from 1.5 to 1.7 logs per male were used in different years and independently of population densities, while as noted earlier Gullion and Marshall (1968) noted a certain amount of movement in display sites of male grouse.

Gullion (1967a) found that only a few male grouse establish drumming logs their first fall, and a few also fail to become established the following spring. Most birds occupying logs in his study area were full adults, at least twenty-two months old. He also found a hierarchy of dominance among males. An established male on a drumming log is a "dominant drummer," and within his activity center a second, or "alternate" drummer may occur and take over the site of the dominant drummer if it is killed. Nearby rivals on adjacent activity centers are called "satellite drummers," but these are fairly rare. However, other males are "nondrummers," and drum infrequently or not at all. These are presumably young grouse that have been unable to establish drumming sites.

Gullion (1967a) also found "activity clusters" of males, consisting of from about four to eight males occupying sites in fairly close proximity. These seem to represent an expanded collective display ground, similar to those that have been described for blue grouse.

Gullion reported that males remain closely associated with their display sites during the summer and that fall drumming may approach or even exceed spring drumming activity. At least a few young males, no older than seventeen to twenty weeks, may become established at this time.

Male Strutting Behavior

Presumably the normal releaser for strutting rather than drumming is the appearance of another grouse near the display log. Edminster (1947) indicates that the drumming male will then strut very slowly toward the intruder, with tail erect and spread. The ruffs on the side of the neck are raised ("upright cum ruff display" of Hjorth, 1970), and the male begins to emit hissing sounds that parallel the tempo of the drumming display. With each hiss the head is lowered and shaken in a rotary fashion ("bowing cum head-twisting and panted hissing" of Hjorth, 1970), giving the impression of a locomotive getting underway (Bump et al., 1947). The display ends with a blur of head-shaking and hissing, followed by a short, quick run toward the other bird as both wings are dragged along the ground ("rush cum prolonged hiss" of Hjorth, 1970). Photographs of this display suggest that in the early stages it is oriented laterally, with the tail and upper part of the body tilted toward the object of the display and the head turned in the same direction. However, the short rush is in a shallow arc toward the other bird (Hjorth, 1970). The similarities of this display to the short rushes of the blue grouse and the spruce grouse are clearly evident. Unlike the spruce grouse, however, the tail is neither shaken nor fanned to produce sound.

Bump et al. (1947) described a "gentle phase" following the strutting phase, which in turn was followed by a "fighting phase" of males. However, their data do not support such a strict interpretation of male behavior patterns nor would such a sequence seem biologically probable. The strutting behavior of males serves equally well as a preliminary threat display toward other males prior to fighting and as a preliminary to attempted copulation with females. The means by which males recognize the sex of intruders on their territories is still uncertain, but in all likelihood there is a differential sexual response of males and females to strutting in another bird. Hjorth (1970) gave the posture associated with this reputed "gentle phase" the name "slender upright cum head-shaking."

The period of receptivity of females is apparently only from three to seven days (Bump et al., 1947; Brander, 1967) and probably is terminated as soon as a successful copulation is achieved. Assumption of the typical receptive posture of grouse, with the wings drooped and slightly spread and the tail slightly raised, while the body feathers are depressed, will stimulate copulation attempts by the male.

Vocal Signals

Hissing is performed by both sexes. Males hiss during their head-shaking and short-rush displays, and females hiss when defending a brood (Bump et al., 1947). Females also utter a squeal during distraction display and quiet their hiding chicks with a downward-inflected scolding note. After any danger is past, they call the brood together with a low, humming call (Bump et al., 1947). Adult grouse of both sexes utter a startled *pete-pete-pete* note, and a chirping *perrck* note which Bump et al. attributed to "curiosity." A variety of "conversational" notes are also present.

Chicks have four principal call-notes, according to Bump et al. (1947). These include alarm calls, two different notes uttered by scattered chicks, and a warning signal of several descending notes that is uttered by older chicks.

Nesting and Brooding Behavior

Typical nest sites for the ruffed grouse have already been mentioned earlier in the discussion of nesting requirements. Bump et al. (1947) report that the female lays her eggs at an average rate of two eggs every three days, thus taking seventeen days to complete an average clutch of eleven eggs. The attachment of the female to the nest increases as the clutch size increases, but incubation does not begin until the last egg is laid. The period of incubation is from twenty-three to twenty-four days, but low environmental temperatures may delay hatching a few days beyond this time. Bump et al. report that during incubation the female will leave the nest for from twenty to forty minutes, or only rarely longer, to feed. Evidently feeding may occur twice each day under normal conditions, but during stormy weather the bird may remain on the nest continuously. Much enlarged "clocker" droppings are typical of incubating females; these are usually found in the vicinity of nests near the usual foraging areas.

Bump et al. (1947) report that although the average clutch size for 1,473 first nests was 11.5 eggs, 149 renesting attempts averaged only 7.5 eggs. Since no cases of second renesting attempts were found, they estimated that the maximum number of eggs that a female might lay in a single season is about 19. There is no evidence that second broods are ever raised by this or any other species of grouse in North America.

Female ruffed grouse exhibit strong nest and brood defense tendencies and will often resort to a disablement display, feigning a broken wing, especially prior to hatching time. Following hatching, the female more often stands her ground, spreads her tail, and assumes a posture similar to the male's strutting posture as she hisses or utters squealing sounds. When the chicks gain the power of flight after ten or twelve days, the usual response of both hen and chick is to fly when disturbed. By mid-September, when the chicks are twelve or more weeks old, the families begin to break up and dispersal of the juvenile birds begins.

EVOLUTIONARY RELATIONSHIPS

In his revision of grouse genera, Short (1967) merged the monotypic genus *Bonasa* with the Eurasian genus *Tetrastes*, which contained two species of "hazel grouse." The two Eurasian species lack neck ruffs but otherwise are very similar to the ruffed grouse, and Short considered that, of the two, the European hazel hen (*T. bonasia*) is nearest to the North American ruffed grouse. The habitat of this bird in Europe is one of mixed hill woodlands and thickets, and it is especially prevalent in aspen and birch, which strongly suggests a common ecological niche. The winter diet of the Siberian hazel hen (*T. b. sibiricus*) consists of from 70 to 80 percent buds and catkins of birches (Dement'ev and Gladkov, 1967), which further attests to the strong ecological similarities of these species and certainly suggests a common evolutionary descent.

In contrast to the ruffed grouse, the hazel hen is apparently monogamous and forms a pair bond that lasts at least until hatching and sometimes beyond. An additional behavioral difference is that the male display consists largely of whistling calls (Dement'ev and Gladkov, 1967). There is no drumming display, but apparently an aerial display involving the whirring of wings does occur (Hjorth, 1970). It would seem that the evolution of a promiscuous mating system, development of nonvocal acoustical signals rather than reliance on vocal whistles, and the correlated ritualization of aerial display flights into a sedentary drumming display all occurred after the separation of ancestral ruffed grouse stock.

Short (1967) concluded that the nearest relationships of the genus *Bonasa* (in the broad sense) are with *Dendragapus* and that the former genus probably arose from pre-*Dendragapus* stock. I agree that modern species of *Dendragapus* or *Tetrao* probably represent the nearest living relatives of *Bonasa*.

17

Pinnated Grouse

Tympanuchus cupido (Linnaeus) 1758

OTHER VERNACULAR NAMES

PRAIRIE chicken, prairie cock, prairie grouse, prairie hen.

RANGE

Current resident of remnant prairie areas of Michigan, Wisconsin, and Illinois and from southern Manitoba southward to western Missouri and Oklahoma and portions of the coastal plain of Texas. Also (*pallidicinctus*) from southeastern Colorado and adjacent Kansas south to eastern New Mexico and northwestern Texas.

SUBSPECIES

T. c. cupido (Linnaeus): Heath hen or eastern pinnated grouse. Extinct since 1932. Formerly along the East Coast from Massachusetts south to Maryland and north central Tennessee.

T. c. pinnatus (Brewster): Greater prairie chicken. Currently limited to several small isolated populations in Michigan, Wisconsin, and Illinois and to the grasslands of extreme southern Manitoba, northwestern Minnesota, North Dakota, South Dakota, Nebraska, Kansas, and western Missouri.

T. c. attwateri Bendire: Attwater prairie chicken. Currently limited to a few isolated populations along the coast of Texas from Arkansas and Refugio counties to Galveston County, and inland to Colorado and Austin counties.

T. c. pallidicinctus (Ridgway): Lesser prairie chicken. Currently limited to arid grasslands of southeastern Colorado and southwestern Kansas southward through Oklahoma to extreme eastern New Mexico and northwestern Texas. Recognized by the *A.O.U. Check-list* (1957) as a separate species.

MEASUREMENTS

Folded wing (greater prairie chicken): males, 217–41 mm (average 226 mm); females, 208–20 mm (average 219 mm).

Folded wing (lesser prairie chicken): males, 207–20 mm (average 212 mm); females, 195–201 mm (average 198 mm).

Tail (greater prairie chicken): males, 90–103 mm (average 96 mm); females, 87–93 mm (average 90 mm).

Tail (lesser prairie chicken): males, 88–95 mm (average 92 mm); females, 81–87 mm (average 84 mm).

IDENTIFICATION (Greater Prairie Chicken)

Adults, 16–18.8 inches long. Both sexes are nearly identical in plumage. The tail is short, somewhat rounded, and the longer under (but not upper) tail coverts extend to its tip. The neck of both sexes has elongated "pinnae" made up of about ten graduated feathers that may be relatively pointed (in *cupido*) or somewhat truncated (other races) in shape and are much longer in males than in females. Males have a conspicuous yellow comb above the eyes and bare areas of yellowish skin below the pinnae that are exposed and expanded during sexual display. The upperparts are extensively barred with brown, buffy, and blackish, while the underparts are more extensively buffy on the abdomen and whitish under the tail. Transverse barring of the feathers is much more regular in this species than in the sharp-tailed grouse, which has V-shaped darker markings and relatively more white exposed ventrally.

IDENTIFICATION (Lesser Prairie Chicken)

Adults, 15–16 inches long. In general like the greater prairie chicken, but the darker, blackish bars of the back and rump typical of greater prairie chickens are replaced by brown bars (the black forming narrow margins), the breast feathers are more extensively barred with brown and white, and the flank feathers are barred with brown and dusky instead of only brown. Males have reddish rather than yellowish skin in the area of the gular sacs and during display their yellow combs are more conspicuously enlarged than those of greater prairie chickens. As in that form, females have relatively shorter pinnae and are more extensively barred on the tail.

FIELD MARKS

The only species easily confused with either the greater or lesser prairie chicken is the sharp-tailed grouse, which often occurs in the same areas where greater prairie chickens are found. Sharp-tailed grouse can readily be recognized by their pointed tails, which except for the central pair of feathers are buffy white, and by their whiter underparts as well as a more "frosty" upper plumage pattern, which results from white spotting that is lacking in the pinnated grouse.

AGE AND SEX CRITERIA (Greater Prairie Chicken)

Females may readily be recognized by their shorter pinnae (females of *pinnatus* average 38 mm, maximum 44 mm, males average 70 mm, minimum 63 mm) and their extensively barred outer (rather than only central) tail feathers. The central crown feathers of females are marked with alternating buffy and darker cross-bars, whereas males have dark crown feathers with only a narrow buffy edging (Henderson et al., 1967). In the Attwater prairie chicken the pinnae of females are about 9/16 inch (14 mm) long, while those of males are over 2 inches (53 mm), according to Lehmann (1941).

Immatures may be recognized by the pointed, faded, and frayed condition of the outer two pairs of primaries (see sharp-tailed grouse account). The pinnae length of first-autumn males is not correlated with age (Petrides, 1942).

Juveniles may be recognized by the prominent white shaft-streaks, which widen toward the tip, present in such areas as the scapulars and interscapulars.

Downy young are illustrated in color plate 61. Downy greater prairie chickens are scarcely separable from those of lesser prairie chickens (see that account) and also resemble young sharp-tailed grouse. However, prairie chickens have a somewhat more rusty tone on the crown and the upper parts of the body and richer colors throughout. There are usually three (one small and two large) dark spots between the eye and the ear region and several small dark spots on the crown and forehead. Short (1967) mentions, however, that at least some downy specimens of *attwateri* have only one or two tiny postocular black markings, which thus would closely approach the markings of downy sharp-tailed grouse.

AGE AND SEX CRITERIA (Lesser Prairie Chicken)

Females may be identified by their lack of a comb over the eyes and their brown barred undertail coverts, which in males are black with a white "eye" near the tip (Davison, in Ammann, 1957). Males have blackish tails, with only the central feathers mottled or barred, while the tails of females are extensively barred (Copelin, 1963).

Immatures can usually be identified by the pointed condition of the two outer pairs of primaries. The outermost primary of young birds is spotted to its tip, while that of adults is spotted only to within an inch or so of the tip. In addition, the upper covert of the outer primary is white in the distal portion of the shaft, whereas in adults the shafts of these feathers are entirely dark (Copelin, 1963).

Juveniles are more rufescent than the corresponding stage of the greater prairie chicken or the adults. The tail feathers are bright tawny olive and have terminal tear-shaped pale shaft-streaks (Ridgway and Friedmann, 1946).

Downy young (not illustrated) are nearly identical to those of the greater prairie chicken (Short, 1967) but are slightly paler and less brownish on the underparts. On the upperparts the brown spotting is less rufescent and paler, lacking a definite middorsal streak (Sutton, 1968).

DISTRIBUTION AND HABITAT

The original distribution of the pinnated grouse differs markedly from recent distribution patterns; without doubt it is the grouse species most affected by human activities in North America. Aldrich (1963) identified the habitat of the now extinct eastern race of pinnated grouse, the heath hen, as fire-created "prairies" or blueberry barrens associated with sandy soils from Maryland to New Hampshire or Maine. The presence of oak "barrens" or parklands may have also been an integral part of the heath hen's habitat, particularly in providing acorns as a source of winter foods (Sharpe, 1968).

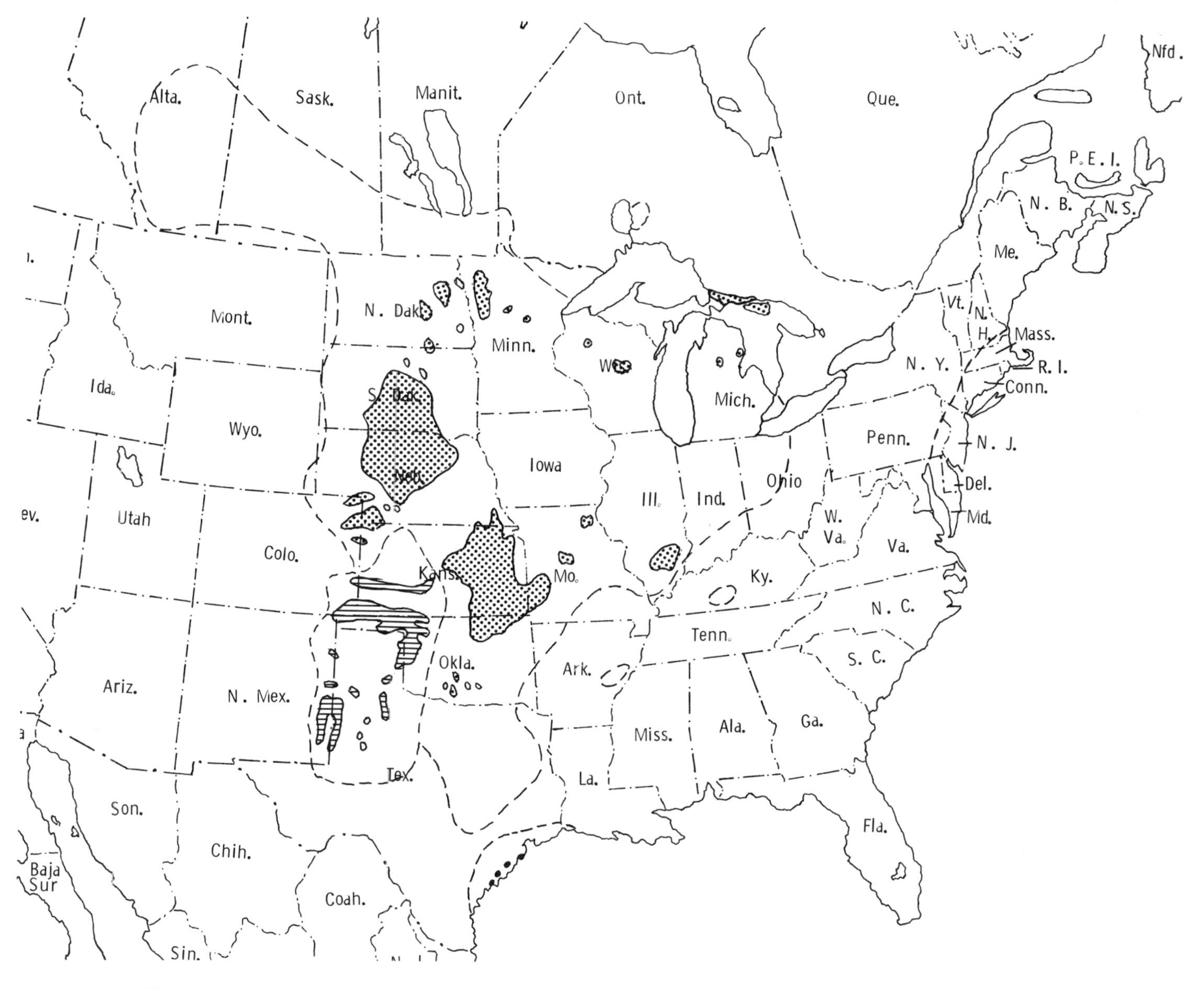

FIGURE 31. Current distribution of the lesser (hatched), greater (shaded), and Attwater (inked) prairie chickens. Original or recent distributions of these forms and the heath hen are indicated by dashed lines.

The range of the coastal Texas race, the Attwater prairie chicken, once extended over much of the Gulf coastal prairie from Rockport, Texas northward as far as Abbeville, Louisiana, an area of more than six million acres (Lehmann and Mauermann, 1963). The lesser prairie chicken once occupied a large area of arid grasslands, with interspersed dwarf oak and shrubs or half-shrub vegetation (Aldrich, 1963; Jones, 1963). The birds occurred over an extensive area from eastern New Mexico and the panhandle of Texas northward across western Oklahoma, southwestern Kansas, and southeastern Colorado. Over this area they were found on two major habitat and soil types, the sand sage—bluestem (*Artemisia filifolia–Andropogan*) shrub grasslands of sandy areas and the similarly sand-associated shin oak–bluestem (*Quercus havardi–Andropogon*) community (Jones, 1963; Sharpe, 1968). The greater prairie chicken originally occurred in the moister and taller climax grasslands of the eastern great plains from approximately the 100th meridian eastward to Kentucky, Ohio, and Tennessee, and northward to Michigan, Wisconsin, Minnesota, and South Dakota (Sharpe, 1968). Sharpe suggested that the presence of oak woodlands or gallery forests throughout much of this range, and the more extensive oak-hickory forests to the east of it may have been an important part of the greater prairie chicken's habitat. Their absence in the western and northwestern grasslands may have made those areas originally unsuitable for prairie chickens. Probably a winter movement of no more than 250 miles to woody cover was typical, according to Sharpe.

With the breaking of the virgin prairies in the central part of North America, and their conversion to small grain cultivation, the prairie chickens responded greatly and moved into regions previously inhabited only by the sharp-tailed grouse (Johnsgard and Wood, 1968). Thus they moved into northern Michigan and southern Ontario, into northern Wisconsin and much of Minnesota, into the three prairie provinces on Manitoba, Saskatchewan, and Alberta, and westward through all or nearly all of North Dakota, South Dakota, and Kansas to the eastern limits of Montana, Wyoming, and Colorado. At the same time the lesser prairie chicken may have undergone a temporary extension northward into western Kansas, northeastern Colorado, and extreme southwestern Nebraska, where it may have been geographically sympatric for a relatively few years with the greater prairie chicken (Sharpe, 1968). However, their habitat requirements are quite different (Jones, 1963), and no natural hybrids between these forms have ever been reported.

During several decades the greater prairie chicken survived extremely well in these interior grasslands, where remaining native vegetation provided the spring and summer habitat requirements and the availability of cultivated grains allowed for winter survival. Eventually, however, the percentage of land in native grassland cover was reduced to the point that these habitat needs could no longer be provided, and the species began to recede from much of its acquired range and to seriously decline or become eliminated from virtually all of its original range. The sad history of this range restriction and population diminution has been recounted in various places and by many writers (Johnsgard and Wood, 1968). Space does not allow a detailed review of these changes, and all that will be attempted here is a statement of the current range and status of the three extant subspecies.

Of the three races, the Attwater prairie chicken is clearly in the greatest danger of extinction. The race became extirpated from Louisiana in about 1919, and between 1937 and 1963 the Texas population declined from about 8,700 to 1,335 birds (Lehmann and Mauermann, 1963). The remaining population suffers from a badly distorted sex ratio, intensified farming practices, predators, fire exclusion, pesticides, bad drainage practices, and relatively little area set aside specifically for their protection. The purchase of 3,420 acres of land in Colorado County by the World Wildlife Fund in the mid-1960s may be the best hope for the retention of a remnant population. By 1965, when the total Texas population was estimated to be from 750 to 1,000 birds, the estimated refuge population was 100 birds. Lehmann (1968) provided the most recent summary of the status of this bird currently available. As of 1967 an estimated 1,070 birds occupied some 234,000 acres, which represents a habitat loss of 50 percent since 1937 and a population reduction of 85 percent during the same time. No hunting of Attwater prairie chickens is allowed in Texas.

The present range of the lesser prairie chicken centers in the panhandle of northern Texas, but also includes parts of New Mexico, Oklahoma, Kansas, and southeastern Colorado (Copelin, 1963). In Oklahoma the present occupied range consists of 2,391 square miles, and from 1933 to the early 1960s there have been only two years (1950, 1951) when the species could be legally shot (Copelin, 1963). Currently, however, the species is legal game in seven counties, with a nine-day 1970 season. Copelin estimated the 1960 population in Oklahoma to be 15,000 and 30,000 in spring and fall respectively.

In Texas lesser prairie chickens have been almost continuously protected since 1937, but in spite of this protection the populations have declined seriously in recent years as a result of overgrazing, aerial pesticide spraying, and altered farming practices (Jackson and DeArment, 1963). The estimated Texas population in 1963 was no more than 3,000 birds. In 1967, after thirty years of protection, limited hunting of lesser prairie chickens was again established, and seasons were also held in 1968 and 1969. The 1967 Texas population was approximately 10,000 birds, and the average annual kill through 1969 has been 275 birds. In contrast, the very small Colorado population of lesser prairie chickens may have increased in recent years; Hoffman (1963) reports an increase of from 6 to 104 males on censused display grounds between 1959 and 1962.

In Kansas the distribution and population of the lesser prairie chicken have not been as thoroughly analyzed as in the other states, but Baker (1952, 1953) reported that the drought of the 1930s nearly eliminated the bird from the state. He found that the birds were limited to sandy lands in fourteen counties south of the Arkansas and Cimmaron rivers but did not estimate total population size. The lesser prairie chicken population in these western counties was first protected by a closed season in 1903, which was followed by a period of closed or greatly restricted seasons until the early 1950s (Baker, 1953). In 1970 the lesser prairie chicken was legally hunted over most of its Kansas range on a three-day season. This was the first hunting season that Kansas had established on lesser prairie chickens since 1935. A 1963 population estimate for Kansas was 10,000–15,000 birds (Sands, 1968), and the population has apparently remained at a static level during the last ten years.

The range of the lesser prairie chicken in New Mexico is currently limited to about five counties and centers around Roosevelt County. Except for closed seasons in 1957 and 1959, the species has been legal game every year since that time. The total yearly kill has averaged 1,153 from 1958 through 1968, with a maximum of 2,918 and a minimum of 519 birds. The most recent year for which data are available is 1968, when 776 birds were taken. The New Mexico population is thought to be between 8,000 and 10,000 birds (Sands, 1968).

The total population of the lesser prairie chicken may thus be estimated as a few hundred in Colorado, possibly three thousand in Texas, perhaps fifteen thousand in Oklahoma, ten thousand to fifteen thousand in Kansas, and eight thousand to ten thousand in New Mexico. These estimates would suggest a total population of from thirty-six thousand to forty-three thousand for the bird's entire range.

The status of the greater prairie chicken is almost as alarming as that of the lesser. It now may be regarded as virtually extirpated from all of the Canadian provinces (Hamerstrom and Hamerstrom, 1961). Christisen (1969) has provided a useful summary of the bird's status in the United States. Considering the form's probable original range, it has been extirpated as a breeding species from Iowa, Ohio, Kentucky, Texas, and Arkansas. The birds were gone from Ohio before 1930, and from Kentucky, Texas, and Arkansas at even earlier dates. The last nesting prairie chickens in Iowa were seen as late as 1952 and stray birds as late as 1960 (Stempel and Rogers, 1961). The estimated population in Indiana diminished from more than four hundred males occupying thirty-three booming grounds in 1942 to four males on a single booming ground by 1966. Christisen (1969) indicates a current estimated total Indiana population of only ten birds.

In Illinois the situation is only slightly better. Although protected since 1932, the population trend has been downward, and an estimated 300 birds remain in the state (Christisen, 1969). The birds are gone from their original ranges in southern Wisconsin and Michigan and persist in small pockets farther to the north, where their total populations are estimated at 1,000 and 200 birds, respectively. In Minnesota the species is also gone from most of its acquired range, and it has been fully protected since 1942, when an estimated 58,300 birds were taken. During its population peak in 1925 an estimated 411,900 birds were killed; by comparison the recent statewide population is estimated at 5,000 (Christisen, 1969).

Virtually all of Missouri might be considered as original greater prairie chicken range (Johnsgard and Wood, 1968), but between the early 1940s and the mid-1960s the species' range diminished from twenty-five hundred square miles to nine hundred square miles, and from nearly fifteen thousand to about seven thousand birds (Christisen, 1967). The birds were last hunted in 1906, and in the last few years the population trend has been upwards, with an estimated ten thousand birds present in the late 1960s (Christisen, 1969).

Colorado, Wyoming, and North Dakota all represent areas of acquired range for the greater prairie chicken. Only eastern Colorado and eastern Wyoming were ever occupied by the birds; June (1967) reports that in Wyoming it is now limited to Goshen County but once occurred also in Laramie County. Its population probably numbers in the hundreds. In Colorado, where it is also protected, the best populations occur in Yuma and Washington counties (Evans and Gilbert, 1969). The most recent state-wide population estimate is 7,600 birds (Christisen, 1969). In North Dakota the birds have been protected since 1945, although prairie chickens are sometimes shot during the sharp-tail season. It arrived in the state in the 1880s, peaked in the early 1900s, and began to decline in the 1930s. Between 1938 and 1942 from 29,000 to 47,000 birds were harvested yearly, and the estimated total population ranged from 300,000 to 450,000 (Johnson, 1964). The present and declining state population is approximately 1,800 birds (Christisen, 1969).

South Dakota's prairie chicken distribution largely represents acquired range, since the species probably originally extended not much farther than the location of the present city of Yankton. No harvest figures are available for the early years of this century, but the populations were probably comparable to those of North Dakota during the same era. In both states the drought of the 1930s brought about a severe decline in the number of prairie chickens which probably lasted for much of that decade. Since 1942, prairie chickens and sharp-tails have been hunted every year, with an average combined harvest of about forty thousand birds, sometimes in excess of one hundred thousand. However, prairie chickens are not nearly so abundant as they once were, and they are now largely limited to relatively few counties (Janson, 1953; Henderson, 1964). The highest populations occur in Jones County, where the native grasslands still occupy about 68 percent of the land area and cultivated lands occupy 30 percent; woody cover in South Dakota's prairie chicken range covers less than 1 percent of the total area (Janson, 1953). The 1967 harvest of prairie chickens was about ten thousand birds, and the declining state population is approximately one hundred thousand birds (Christisen, 1969).

In Nebraska the species probably originally occurred in the eastern part of the state, but it is now largely limited to the central portion, where it occurs along the eastern and southern edges of the sandhills, where native grasses and grain crops are in close proximity and provide both summer and winter habitat needs (Johnsgard and Wood, 1968). The state's population is relatively static, and this species as well as the more common sharp-tailed grouse have been regularly hunted, except in the case of the small and isolated population in southeastern Nebraska, which is an extension of the large Flint Hills population of eastern Kansas. In 1967 the estimated Nebraska harvest was fifteen thousand birds, and the state's recent total population was estimated at one hundred thousand birds (Christisen, 1969).

The heart of the greater prairie chicken's present range is in eastern Kansas, amid the bluestem (*Andropogon*) prairies that extend from the Oklahoma border in Chautauqua and Cowley counties to near the Nebraska border in Marshall County (Baker, 1953). This zone includes an easternmost zone of interspersed natural grassland and croplands, a zone of sandy soils associated with natural grasslands and wooded hilltops, a zone of flinty, calcareous hills and associated native grasslands, and a transition zone between these hills and the cultivated lands to the west. In the best areas for prairie chickens, the ratio of natural grasslands to cultivated feed crops is roughly two to one (Baker, 1953). Prairie chickens have been given protection in Kansas periodically since 1903. The population apparently underwent a marked decline in the early 1940s, followed by an increase to the end of that decade, when fifty thousand birds were conservatively estimated to be present in the state (Baker, 1953). In 1967 some forty-six thousand birds were harvested, and an estimated seven hundred and fifty thousand were believed present in the late 1960s (Christisen, 1969), suggesting that the Kansas population is by far the most secure of any state's.

The only remaining state still supporting greater prairie chickens is Oklahoma. They probably once inhabited all of eastern Oklahoma, but they are now largely restricted to the northeastern corner of the state north of the Arkansas River. Besides occurring in eight of these northeastern counties, birds have apparently been successfully restocked in four more southerly and westerly counties (Sutton, 1967). In contrast to all other states, the population trend in Oklahoma for prairie chickens is upward (Christisen, 1969), and in both 1967 and 1968 between thirteen thousand and fourteen thousand were killed. In contrast, the 1959 to 1968 average yearly kill was under eight thousand birds. Although Oklahoma has not invested in prairie chicken refuges, its successful restocking program combined with a policy of converting marginal timberlands and agricultural lands to natural grasslands has evidently been the major reason for the recent improvement in greater prairie chicken populations.

In summary, it would seem that the total collective populations for the three extant prairie chicken forms might be one thousand for the Attwater, fifty thousand for the lesser, and perhaps up to a million greater prairie chickens, with three-fourths of the last-named confined to the state of Kansas. Only in Kansas, Oklahoma, Nebraska, and South Dakota can the greater prairie chicken populations be considered safe, and in South Dakota the population is declining. Paradoxically, in none of these states is land being set aside by public agencies for prairie chicken populations, although this has been done for marginal populations in Indiana, Michigan, Illinois, Wisconsin, Missouri, and North Dakota (Christisen, 1969).

POPULATION DENSITY

Population density estimates for prairie chickens vary greatly for different areas and in general probably reflect the deteriorating status of the species, with declining populations being studied more intensively than the relatively few healthy or increasing populations. Grange (1948) estimated a spring prairie chicken population in Wisconsin of 1 prairie chicken per 110 acres in 1941 and 1 per 138 acres in 1942, or between 4 and 6 birds per square mile. In 1943, the prairie chicken range in Missouri likewise averaged 4.8 birds per square mile. In South Dakota's best remaining prairie chicken habitat of six counties, spring population densities of from 2 to 4 birds per square mile occur (Janson, 1953).

In contrast, Baker (1953) studied several flocks of prairie chickens in high-quality Kansas range on a study area covering about 3½ square miles. Two flocks used this area exclusively, while two other flocks used it in part.

Spring numbers of one flock varied over a three-year period from 15 to 104 birds, while a second flock varied from 15 to 43 birds during these three springs. A third flock consisted of about 20 birds. Using conservative figures, an average spring population of at least 50 birds must have been dependent on the area, or at least 14 birds per square mile. During population "highs," the spring density may have reached about 50 birds per square mile for the study area as a whole, and even more if only the composite home range areas are considered.

Data on male spring densities for the lesser prairie chicken are available from Oklahoma (Copelin, 1963). Over a six-year period on four different study areas having display grounds, the densities of males per square mile varied from 1.5 to 18.31 and averaged 7.4 males. Earlier figures available from one of these study areas for the 1930s indicated densities of from about 15 to nearly 40 males per square mile. Hoffman (1963) reported that male densities on three areas in Colorado increased from 0.8 to 5.8 males per square mile over a four-year period in this marginal part of the species' range. In Texas, Jackson and DeArment (1963) noted that numbers of males on a 100,000 acre area reached as high as 600 birds in 1942 (about 4 birds per square mile) but more recently have averaged about 200 males. These data would collectively indicate that spring densities of males in favorable habitats may exceed 30 per square mile, but probably average less than 10. Similarly, Lehmann (1941) reported spring densities of about 10 birds per square mile for the Attwater prairie chicken in Texas for the late 1930s. A 1967 survey of this population indicated that 645 birds were present on about 136,000 acres, or a density of 210 acres per bird (3 birds per square mile).

HABITAT REQUIREMENTS

Wintering Requirements

The winter requirements for pinnated grouse seem to center on the availability of a staple source of winter food, rather than protective cover or shelter from the elements. Lehmann (1941) reports for the Attwater prairie chicken that the birds moved into lightly grazed natural grassland pastures by mid-November and remained there until spring. In Oklahoma, Copelin (1963) found that the lesser prairie chickens used cultivated grains, especially sorghum, extensively during two winters. In the following winter, when production in the shin oak grassland pastures was apparently high, the birds remained in this pastureland area. During the following two winters increased usage of cultivated grains occurred, particularly in late winter when snow was nearly a foot deep for a week or longer, and shocked grain sorghum was then extensively utilized.

Edminster (1954) concluded that grainfields represent an important part of present-day prairie chicken habitat, with corn providing the best winter habitat, provided that it is either shocked or left uncut. Sorghum, like corn, stands above snow during the winter and thus is almost as valuable. Robel et al. (1970) confirmed the importance of sorghum in winter for Kansas prairie chickens. Other small grains such as wheat and rye are utilized whenever they can be reached by the birds during winter.

In contrast to the sharp-tailed grouse and nonprairie grouse, there is little evidence that the pinnated grouse ever resorts to buds as a primary source of food during winter. Martin, Zim, and Nelson (1951) list the buds and flowers of birch as a minor source of winter food for pinnated grouse from the northern prairies but found them of far less importance than cultivated grains or wild rose (presumably rose hips). Edminster (1954) lists the buds of birches, aspens, elm, and hazelnut among items used in the northern range during winter, but so long as grain or other seed sources are available this would not appear to be critical to winter survival. Mohler (1963) reported that the best winter habitats for prairie chickens in the Nebraska sandhills were areas where cornfields were located near the extensive and lightly grazed grasslands of the larger cattle ranches, providing a combination of available food and grassy roosting cover.

Spring Habitat Requirements

The habitat requirements of the lesser prairie chicken for display ground locations have been summarized by Copelin (1963). He reported that the males always selected areas with fairly short grass for display grounds and that the grounds were usually located on ridges or other elevations. In sand sagebrush habitat, display grounds on the other hand were located in valleys on short-grass meadows if the sagebrush on adjacent ridges was tall and dense. Lehmann (1941) noted that of several hundred Attwater prairie chicken booming grounds studied, most were on level ground or slightly below the adjacent land surface, but they typically consisted of a short-grass flat, about an acre in extent, surrounded by heavier grassy cover.

Ammann (1957) has provided similar observations for the greater prairie chicken in Michigan. He noted that of sixty-five prairie chicken and ninety-five sharp-tail display grounds observed, 47 percent were located on elevated sites and only four were in depressions. Of ninety-seven Michigan prairie chicken grounds studied in 1941, twenty-seven contained some woody growth other than sweet fern or leather leaf, while of sixty-five grounds studied since 1950 only two contained a sparse stocking of woody cover. Prairie chickens evidently will not tolerate as much woody cover on their booming grounds as will sharp-tailed grouse.

Robel et al. (1970) found that booming grounds in Kansas were associated with clay pan soil types, and the birds remained on these sites for some time after display activities ceased, feeding on succulent green vegetation, especially forbs. With the coming of hot summer weather, the steep limestone hillsides received greater use, probably because of the availability of shade for loafing. Lehmann (1941) likewise reported that heavy shrub cover provides shade for hot summer days, protection against predators and severe weather, and a source of fall food.

In a comparison of habitat requirements of greater and lesser prairie chickens, Jones (1963) found that both forms preferred level or elevated sites associated with short grasses. Plant cover differences were not significant, but greater prairie chickens tolerated somewhat taller vegetation than did the lesser (a mean of 15.1 cm versus 10.4 cm). Anderson (1969) reported that greater prairie chickens preferred grass cover less than six inches tall for their booming grounds, the combination of short cover and wide horizons apparently being far more important than the specific cover type present on the land.

Nesting and Brooding Requirements

Ammann (1957) indicated that of thirteen prairie chicken nests found in Michigan, eight were in hayfields, one was in sweet clover, three were in wild land openings, and one was located on an airport. All of the nests were in fairly open situations. Hamerstrom (1939) has similarly reported on twenty-three prairie chicken nests in Wisconsin. Eleven of these were in grass meadows near drainage ditches, three were in dry marshes or marsh edges, three were in openings or edges of jack pine–scrub oak woods, three were in scattered mixtures of brush, small trees, and grass, two were in small openings in light stands of brushy aspen or willow, and one was in rather dense mixed hardwoods. Both of these studies indicate the importance of grassy, open habitats for prairie chicken nests. Hamerstrom, Mattson, and Hamerstrom (1957) and Yeatter (1963) have both emphasized the importance of mixed natural grasslands or substitutes in the form of redtop (*Agrostis alba*) plantings as nesting and rearing cover types for prairie chickens. Yeatter (1963) correlated a decline in redtop production and prairie chicken populations in Illinois and found that birds nesting in redtop had a nesting success as high as or higher than those using pastures, idle fields, or waste grasslands.

Schwartz (1945) also provided information on nest site preferences in greater prairie chickens, and noted that of fifty-seven nest locations, 56 percent were in ungrazed meadows. Half the remainder were in lightly grazed pastures, while the others were in sweet clover, fencerows, sumac, old cornfields, or barnyard grass. The usual proximity of nests to booming grounds has led Schwartz, Hamerstrom (1939), and Jones (1963) to comment on this relationship. However, Robel et al. (1970) found considerable movements between booming grounds by females and questioned that the location of booming ground has any major influence on female nesting behavior. He found that nineteen nest sites averaged 0.68 miles from display grounds, and ranged up to 1.13 miles away. Jones (1963) noted that all of the nine greater prairie chicken nests he found were located near pastures or old fields that had a large number of forbs into which the broods were taken following hatching.

Lehmann (1941) reported that of nineteen Attwater prairie chicken nests found, seventeen were in long-grass prairie, one was in a hay meadow, and one was in a fallow field. All of them were located in the previous year's grass growth, and fifteen were in well-drained situations, often on or near mounds or ridges. Twelve were located near well-marked trails, such as those made by cattle. All of the nests were roofed over with grassy vegetation, and most had good to excellent concealment characteristics. Copelin (1963) reported on nine lesser prairie chicken nests in Oklahoma and Kansas. None of these occurred among shrubs more than fifteen inches high, and seven were located between grass clumps, particularly little bluestem (*Andropogon scoparius*). Two were under bunches of sage, and one was under tumbleweed. Shin oak shrubs from twelve to fifteen inches tall were associated with five of the nests.

Following hatching, females with broods typically moved to somewhat heavier cover than was utilized for nesting. Copelin (1963) noted that only one brood of lesser prairie chickens was found in the low shinneries of oak, but twenty-seven were seen in oak motts, which are clumps of oak four to twenty feet tall in stands up to one hundred feet in diameter. Oak motts provide better shade than do oak shinneries. In the absence of oak, the birds moved into cover provided by sagebrush or other bushy plants. Lehmann (1941) likewise found a movement of both young and old Attwater prairie chickens toward cover that provided a combination of shade and water. The importance of free water for prairie grouse is questionable (Ammann, 1957), but certainly in moister habitats the availability of succulent plants, insects, and shade all contribute to the value of the area as rearing cover.

Yeatter's (1943, 1963) studies in Illinois indicated that females with newly hatched young feed mainly in redtop fields and to some extent in small grain or grassy fallow fields. They also move along ditch banks and field borders, where there is heavier cover. In Missouri, females take their young to swales that provide cover in the form of slough grass, where a combination of shade, protection, and easy movement is present. As the birds grow older, they gradually move to higher feeding grounds such as grainfields or stubble but still return in the heat of the day to rest in the shade provided by shrubs, large herbs, or trees.

FOOD AND FORAGING BEHAVIOR

Winter foods of the prairie chicken are virtually all from plant sources (Judd, 1905b; Schwartz, 1945). Judd indicated that the prairie chicken consumes only about half as much mast as does the ruffed grouse, consisting mostly of the buds of poplar, elm, pine, apple, and birches. It also consumes some hazelnuts (*Corylus*) and acorns, which it swallows whole. In most parts of the bird's present range, however, grain is much more important than buds as winter food. As noted earlier, corn and sorghum represent major winter foods for the species, with corn more important in northern areas and sorghum of increasing importance farther south.

Korschgen (1962) found that in Missouri corn kernels and sorghum seeds are the primary winter foods, with corn remaining important well into spring. In late spring, soybeans (*Glycine*) exceed corn in usage, with the leaves being consumed first and later the seeds and seed pods. Sedge (*Carex*) flower heads are also important in the spring diet, as are grass leaves. Two cultivated grasses, oats and wheat, are heavily depended on in summer, first for their leaves and later for their grains. Korean lespedeza (*Lespedeza*) foliage is used almost throughout the year, but especially from July through September. In September ragweed (*Ambrosia*) seeds begin to appear in the diet and are used to a limited extent until February.

On a year-round basis, Judd (1905b) reported that animal foods (mostly grasshoppers) constitute about 14 percent and plant foods 86 percent of the greater prairie chicken's diet. Martin, Zim, and Nelson (1951) stated that during summer the animal portion may reach 30 percent but in winter and spring is as little as 1 to 3 percent. Lehmann (1941) found that adults of the Attwater prairie chicken consume about 88 percent plant material and 12 percent insect food, with seeds and seed pods alone comprising more than 50 percent of the materials eaten. In contrast to the high percentage of cultivated grains found in most studies of the greater prairie chicken, native plants found in lightly grazed pastures provided the major food items listed by Lehmann. These included ruellia (*Ruellia*), stargrass (*Hypoxis*), bedstraw (*Galium*), doveweed (*Croton*), and perennial ragweed

(*Ambrosia*) as well as many other less important species.

Jones's study (1963) of the greater and lesser prairie chickens in Oklahoma brought out some striking differences in foods taken in study areas about two hundred fifty miles apart. The percentage of insects consumed was much higher in the case of the lesser prairie chicken (41.8 and 48.6 percent average yearly volume in two habitats) than was true of the greater prairie chicken (8.2 and 20.8 percent average volume in two habitats). The remainder of the food of both species consisted of seeds and green vegetation, with the latter usually comprising more volume than the former. Both species fed in grassy cover, but whereas lesser prairie chickens preferred mid-length grasses for foraging, the greater was found feeding more frequently in short grasses. Jones also reported (1964b) that during the six-month period when plants were important food items, the half-shrub cover type (associated with sandy soils) was used for foraging for five months, and the short-grass cover type (associated with clay soils and used for display purposes) was heavily used only during April. Copelin (1963) reported that the relative use of sorghum in winter was closely related to the amount of snow cover, with large flocks moving to grainfields when snow was about a foot deep for a week or more. When such snow is present, lesser prairie chickens regularly make snow roosts (Jones, 1963), suggesting a fairly recent climatic adaptation to the warmer climates typical of the bird's present range.

MOBILITY AND MOVEMENTS

An early analysis of greater prairie chicken seasonal movements was made by Hamerstrom and Hamerstrom (1949) for the Wisconsin population. They suspected that little movement occurred during summer, especially during the brood-rearing period. However, during autumn considerable movement does occur, and some slight migratory movements may exist. Autumn movements of up to twenty-nine miles were established using banded birds, which perhaps correspond to the "fall shuffle" of quail or the general fall dispersion of young birds known for other grouse. Most of the longer movements found were those of females; six of the eight females recovered had moved at least three miles, while eighteen of thirty males had moved less than three miles.

During winter, prairie chickens typically occur in large packs formed by mergers of the fall packs. In Wisconsin these consist of up to one hundred to two hundred birds, which become progressively less mobile in the most severe weather. During very bad weather the birds move very little and may scarcely leave their winter roosts. Roosting sites in the Hamerstroms's study area were often from a quarter to a half mile from feeding fields and were seldom more than a mile and a quarter away.

By February, the winter packs begin to break up and the males start returning to their booming grounds. The Hamerstroms found that fifty-six banded males usually moved less than two miles from their winter feeding grounds to their booming grounds (fifty of fifty-six birds), while the remaining six males moved from two to eight miles. Apparently many males winter at feeding sites which are the nearest available ones to their booming grounds, and in late winter some daily movements between these locations may occur. During spring there is little movement on the part of males; the birds may roost on their territories or within a few hundred yards of it. Sources of water, shade, dusting places, and loafing sites are often within a half mile. Following the termination of display activities, the males may remain close to their booming grounds for much of the summer.

More recent studies of movements of greater prairie chickens have been made by Robel et al. (1970) in Kansas, using radio telemetry. They established monthly ranges for thirty-nine adult males, thirty-seven adult females, and thirty-one juveniles. Movements of adult males were greatest in February, as the birds began to visit their booming grounds and also had to search somewhat harder for food. Flights of a mile or more between feeding areas and display grounds were sometimes seen, and there was also some movement between display grounds. Immature males, however, exhibited their greatest movements in late February and March, with the later flights undertaken largely between display grounds as the birds unsuccessfully attempted to establish territories at various grounds. During April and May both adults and immatures exhibited reduced movements, with the birds remaining closely associated with specific booming grounds. Maximum movements of females occurred in April, during the time of peak male display. Females often visited several different booming grounds, with movements of up to 4.8 miles being recorded. One female that attempted to nest three times was fertilized at a different booming ground prior to each nesting attempt. Summer movements by both sexes were minimal, as the birds molted and females were rearing broods. However, during fall, longer movements again became typical, especially among juveniles. Three juvenile males moved distances of from 2.7 to 6.7 miles during October and November, but comparable data for females are not available. However, daily movements of females during that time averaged farther than those of males (808 yards versus 660 yards).

Monthly movements of the prairie chickens studied by Robel et al. (1970) reflect this seasonal behavior pattern. Summer monthly ranges of adult males were greatest in June (262 acres), fairly small in July (132 acres), and smallest in August (79 acres). In fall and winter the monthly ranges increased from 700 to almost 900 acres from November to February and reached 1,267 acres in March then decreased sharply and were at a minimum of 91 acres in May. Data for juvenile males indicated a similar monthly mobility pattern for the year. On a daily basis, adult males were most highly mobile in February (with an average daily movement of 1,121 yards), and they decreased their daily mobility through August (320 yards per day). The movements increased again in fall and through the winter averaged from 600 to 700 yards per day until February. During the period of February through September, adult females had average daily movements of from 332 to 928 yards. Juveniles of both sexes had daily movements rather similar to those of adult males, being least extensive in August and increasing to a peak in March.

Comparable data for the lesser prairie chicken are not available, but Copelin (1963) does provide some observations on mobility. He also found that movements were most limited in summer and most extensive in winter. The summer range of a female and her brood was estimated to be from 160 to 256 acres, or somewhat less than the estimates of monthly summer mobility in greater prairie chicken females. On the basis of observations of 114 banded birds, 79 percent were found within 2 miles of their point of capture, and 97.4 percent were within 4 miles. The maximum known distance of movement was 10 miles. In common with the Hamerstroms's study, he found that juveniles often moved considerable distances between their brood ranges and display grounds the following spring, with all of fourteen birds moving at least 0.5 mile, and two moving nearly 3 miles. Considering birds captured in fall and winter and observed the following spring on display grounds, he found that juvenile birds tended to move farther than adults during this time and that juvenile hens moved farther than juvenile males. Forty juvenile males moved an average distance of 0.93 miles and twenty adult males moved an average of 0.46 miles. Six juvenile hens moved an average distance of 2.12 miles and one adult hen moved 3.75 miles.

Lehmann (1941) provided some observations on seasonal movements in the Attwater prairie chicken which in general support the studies already discussed. He noted a summer movement of adult and fairly well grown young from nesting areas into heavier summer cover that provided shade and water, followed by a sedentary state until fall. At this time, from September onward, the birds moved out of some pasturelands and into others that provided winter food and cover conditions. During this time, large concentrations of up to 250 to 300 individuals were sometimes seen, in addition to many smaller flocks of 8 birds or fewer. These winter packs break up late in January, when males begin to display.

REPRODUCTIVE BEHAVIOR

Territorial Establishment

As in the sharp-tailed grouse, fall establishment of territories and associated fall display occurs regularly in the pinnated grouse. Copelin (1963) noted that during the fall old male lesser prairie chickens reestablish territories that they held during the spring, and although young males visit the booming grounds, they are apparently not territorial. In the greater prairie chicken an active period of fall display is likewise a regular phenomenon, at least in Missouri (Schwartz, 1945), Michigan (Ammann, 1957), and various other states, although Hamerstrom and Hamerstrom (1949) did not regard it as typical in Wisconsin. Whether or not the females regularly visit the grounds during fall is not so important as the fact that territorial boundaries are reestablished by mature and experienced males, and young males learn the locations of these display grounds. During the following spring some shifting about may occur as deaths among the males during the winter remove some territory holders, but the basic structure of the booming ground is probably formed during fall display.

The average size of the lek, in terms of participating males, is similar to that of sharp-tailed grouse. Lehmann (1941) indicated that for five Attwater prairie chicken grounds studied over a three-year period, the average yearly numbers of participating males ranged from 7.2 to 8.4. Grange (1948) indicated that on seventeen display grounds in Wisconsin in 1942, an average of 6.9 males were present. In Nebraska, an average of about 9 male prairie chickens is typical of booming grounds (Johnsgard and Wood, 1968). Generally similar figures have been indicated for Missouri (Schwartz, 1945) and Illinois (Yeatter, 1943). The largest reported booming grounds were those noted by Baker (1953) for Kansas; he observed one ground containing approximately 100 males.

Copelin (1963) summarized numbers of male lesser prairie chickens on display grounds in Oklahoma from 1932 to 1951. For a total of 64 grounds studied over varying periods of years, the average number of males present was 13.7 and was as high as 43. These grounds occurred on a study area of sixteen square miles, and in different years from as few as 8 to as many as 40 display grounds were found on this study area. The average figure of 24 display grounds would indicate that good lesser prairie chicken habitat might support about 1.5 active display grounds per square mile. Baker

(1953) indicated that 6 greater prairie chicken booming grounds were present on a study area of about 3.5 square miles of excellent range in Kansas, or 1.7 grounds per square mile. Most other studies indicate a greater scattering of display grounds for the greater prairie chicken, which may be in part a reflection of the effective acoustical distances associated with the male vocal displays. The lower-pitched booming calls of the greater prairie chicken presumably are effective over greater distances than are the homologous "gobbling" calls of the lesser prairie chicken, and this might affect spacing characteristics of display grounds.

Male Display Behavior

Since the basic sexual and agonistic behavioral patterns of the greater, lesser, and Attwater prairie chickens are virtually alike, a single description of motor patterns will be given, with comments on any differences that might occur, based on Sharpe's comparative analysis of the three forms (1968).

Booming is the collective term given to the sequence of vocalizations and posturing of greater prairie chicken males that serve both to announce territorial residence to other males and to attract females. During booming, the tail is elevated, the pinnae are variably raised to a point that may be almost parallel with the ground, the wings are lowered while held close to the body, and the primaries are spread somewhat. The bird then begins a series of foot-stamping movements (about twenty per second according to Hjorth, 1970), during which he moves forward a relatively short distance, followed by a multiple snapping of the tail in three rapid fanning movements. At the same time as the tail is initially clicked open and shut, a three-syllable vocalization ("tooting" of Hjorth, 1970) begins, lasting almost two seconds and sounding like *whoom-ah-oom*, with the middle note of reduced amplitude. During the second note a rapid and partial tail-fanning also occurs and the "air sacs" are partially deflated. During the third note the esophageal tube is again inflated and the lateral apteria or "air sacs" are maximally exposed. Simultaneously, the tail is rather slowly fanned open and again closed. Sharpe (1968) indicated that in the lesser prairie chicken a single, exaggerated tail-spreading movement occurs during the first phase of booming and the latter tail-spreading elements are lacking. He estimated that the maximum amplitude of the fundamental harmonic during booming is about 300 cycles per second (Hz) in the greater and Attwater prairie chicken and about 750 Hz in the lesser prairie chicken. In addition, the vocalization phase of the lesser lasts about 0.6 seconds, as opposed to nearly 2 seconds in the greater. The associated call ("yodelling" of Hjorth, 1970) sounds more like a "gobble" and has two definite syllables plus a terminal humming sound. However, "low-intensity" booming may have up to four syllables. Hjorth (1970) has distinguished a variant of the lesser prairie chicken's gobbling call which he called "bubbling," but it appears to be an incomplete and less stereotyped version of the more typical call and posture and probably corresponds to Sharpe's "low intensity booming." In contrast to the greater prairie chicken, male lesser prairie chickens frequently utter their booming displays in an antiphonal fashion ("duetting" of Hjorth, 1970), with up to ten displays being performed in fairly rapid sequence. An additional visual difference between the displays of the two forms is that the exposed gular sac of the lesser is mostly red, whereas those of the greater and Attwater prairie chickens are yellow to orange (Jones, 1964a; Lehmann, 1941).

A second major display of prairie chickens is flutter-jumping. It is performed in the same fashion by this group as by sharp-tailed grouse and no doubt serves a similar advertisement function. Unlike that of the sharp-tail, however, most prairie chicken flutter jumps have associated cackling calls ("jump-cackle" of Hjorth, 1970). Sharpe (1968) found that calls occurred during twenty-seven of thirty flutter jumps in Attwater prairie chickens, sixteen of twenty in lesser prairie chickens, and seventeen of twenty in greater prairie chickens. He noted that flutter-jumping is especially typical of peripheral males when hens are present near the middle of the display ground.

When defending territories against other males, several display postures and calls are typically seen. Ritualized and actual fighting, such as Lumsden (1965) described for the sharp-tailed grouse, is commonly seen, often with short jumps into the air and striking with the feet, beak, and wings. Between active fights, the males will commonly "face off," lying prone a foot or two apart and calling aggressively. Associated calls during facing off include a whining call much like that of sharp-tails, and similar more nasal "quarreling" note (Sharpe, 1968) that sounds like *nyah-ah-ah-ah*. Grange (1948) describes the "fight call" as a very loud, raucous *hoo'-wuk*. Apparent displacement sleeping, displacement feeding, and "running parallel" displays have also been noted by Sharpe at territorial boundaries. A white shoulder spot is often evident in such situations and Hjorth (1970) noted that in both sexes of lesser prairie chickens this may frequently be observed.

When a female enters a male's territory, his behavior changes greatly. Booming is performed with high frequency and extreme posturing, particularly as to pinnae erection and eye-comb enlargement. The eye-combs of all three forms are a bright yellow, but those of the lesser prairie chicken are relatively larger than those of either the greater or Attwater prairie chicken.

Between booming displays, the male will sometimes stop and "pose" before the female while facing her, but most booming displays are not oriented specifically toward the hen. Rather, the male circles about her and all aspects of his plumage are visible to her.

In the presence of females, when they are either nearby or at some distance, a characteristic *pwoik* call ("whoop" of Hjorth, 1970) is frequently uttered (Lehmann, 1941). Sharpe reports that this call is very similar in both the greater and Attwater prairie chickens, but in the lesser it is higher pitched and sounds like *pike* ("squeak" of Hjorth, 1970). It lasts for a shorter duration (0.23 seconds compared to about 0.4 seconds in the larger forms) and the greatest sound amplitude occurs at about 1,000 Hz, rather than 550 to 600 Hz.

All three forms of prairie chickens perform the "nuptial bow" ("prostrate" of Hjorth, 1970), which Hamerstrom and Hamerstrom (1960) originally described for the greater prairie chicken. They regarded it as a sexual display that often precedes copulation and yet is not a prerequisite for it. Sharpe (1968) found that the same applies to the Attwater and lesser prairie chickens, and in all three the display has the same form. The male, while actively booming and circling about a nearby female, suddenly stops, spreads his wings, and lowers his bill almost to the ground while keeping his pinnae in an erect posture. He may remain in this posture for several seconds as he faces the female.

When females are ready for copulation they squat in the typical galliform manner, with wings slightly spread, head raised, and neck outstretched. When mounting, males grasp the female's nape, lower their wings on both sides of her, and quickly complete copulation. After copulation, females usually quickly run forward a few feet then stop to shake. Males lack any specific postcopulatory displays and often begin booming again within a few seconds.

Vocal Signals

In addition to the booming, whining, quarreling, and *pwoik* calls already mentioned, pinnated grouse have several other vocal signals. Many cackling sounds are also uttered. Sharpe (1968) recognized a "long cackle" that consists of several individual notes spaced about 0.2 seconds apart and sometimes lasting several seconds. The notes uttered during flutter-jumping are essentially the same as these individual long-cackle sounds. Lehmann (1941) has listed several variants of these cackling calls and combinations of *pwoik* and cackling notes, and he also mentions several other notes. These include calls sounding like *kwiee*, *kwerr*, *kliee*, *kwoo*, and *kwah*. In the absence of comparative study and sonagraphic analysis, their possible functions cannot be guessed. Hjorth (1970) has noted that between flutter-jumping or booming the male often utters an indefinite staccato cackle, and during territorial confrontations it may produce cackling sounds that range from whinnies to whining cackles and explosive cackling sounds.

Nesting and Brooding Behavior

Following mating, the female begins to lay a clutch almost immediately; indeed, it is probable that she has already established a nest scrape prior to successful copulation. She may move a considerable distance away from the ground to her nest site and may actually nest nearer to another booming ground than to that at which copulation occurred (Robel et al., 1970). Robel et al. found that females had to visit a ground for an average of three consecutive days before copulation occurred, but did not return thereafter except perhaps for renesting attempts. Lehmann (1941) and Robel both found that renesting birds laid progressively smaller clutches, and sometimes up to two such attempts were made. The average clutch size of first clutches is about twelve to fourteen eggs for the lesser (Copelin, 1963), Attwater (Lehmann, 1941), and greater prairie chickens (Hamerstrom, 1939; Robel et al., 1970). Later clutches, probably the result of renesting, often have only seven to ten eggs.

Eggs are laid at the approximate rate of one per day, with occasional lapses of a day, so that it may take about two weeks to complete a clutch of twelve eggs (Lehmann, 1941). Incubation may begin the day before the laying of the last eggs or several days after the last egg is laid, according to Lehmann. Apart from two feeding and resting periods in early morning and late afternoon, the female incubates constantly. The incubation period is probably 23 to 26 days in all three forms (Lehmann, 1941; Schwartz, 1945; Coats, 1955; W. W. Lemburg*).

The process of pipping may require up to forty-eight hours, during which the female appears highly nervous and the nest is apparently extremely vulnerable, because of the noises made by the chicks and the odors of the nest (Lehmann, 1941). Normally, the nest is deserted within twenty-four hours after the last chick is out of its shell. Females with young chicks typically perform decoying behavior with heads held low and wings drooping and nearly touching the ground, uttering a low *kwerr, kwerr, kwerr* (Lehmann, 1941). After the young are able to fly well, both the hen and brood typically flush when disturbed.

Chicks less than a week old may be brooded much of the time, possibly

*W. W. Lemburg, 1970: personal communication.

up to half the daylight hours (Lehmann, 1941). However, older chicks are brooded only at night, during early morning hours, and in inclement weather. Broods typically remain with females for six to eight weeks, when families gradually disintegrate. There is also considerable brood mixing, as when separated chicks join the broods of other females, even if the young are of different ages.

EVOLUTIONARY RELATIONSHIPS

The close and clearly congeneric relationships of the pinnated grouse to the sharp-tail have already been mentioned in the account of that species. Thus, comments here will be restricted to the relationships among the four forms of pinnated grouse. Short (1967) has already dealt extensively with the criteria advanced by Jones (1964a) for considering the lesser prairie chicken as specifically distinct from the greater prairie chicken. Since then, Sharpe (1968) has found some male behavioral differences between the lesser prairie chicken and the two surviving races of *cupido*. These differences consist of acoustic differences (higher frequencies in the lesser), time differences (more rapid and shorter displays in the lesser), and some motor differences (one versus two tail movements during booming in the lesser). A few other contextual and orientational differences were also found, but Sharpe admitted that these differences may be attributed largely to size differences in the birds and possible selection related to aggressive behavior patterns rather than being the result of reinforcement for species differences during some past period of sympatry. He concluded that the lesser should be considered an "allospecies" to emphasize its greater difference from *T. c. pinnatus* than that exhibited by *T. c. attwateri*. This may well be the most effective way of handling questionable allopatric populations, but it is not used elsewhere in this book and has not been generally adopted.

It would seem that the living forms of pinnated grouse and those which have recently become extinct were all derived from some ancestral grouse associated with deciduous forest or its edge, since the original ranges of the lesser and greater prairie chickens as well as the extinct heath hen all had affinities with oak woodlands or oak-grassland combinations. The Attwater prairie chicken, on the other hand, is apparently associated with pure grassland vegetation. The separation of the ancestral stock of the lesser prairie chicken probably occurred during an early glacial period, and subsequent adaptation during postglacial times to an unusually warm and dry grassland habitat in the southwestern states has accounted for its smaller size and generally lighter coloration. More recent separation of gene pools no doubt brought about the separation of the east coast (heath hen) and Gulf coast (Attwater) populations from the interior form, but the behavioral and morphological differences among these are minimal.

18

Sharp-tailed Grouse

Tympanuchus phasianellus (Linnaeus) 1858

(Pedioecetes phasianellus in *A.O.U. Check-list)*

OTHER VERNACULAR NAMES

*B*RUSH grouse, pintail grouse, prairie grouse, prairie pheasant, sharptail, speckle-belly, spike-tail, spring-tail, white-belly, white-breasted grouse.

RANGE

Currently from north central Alaska, Yukon, northern Mackenzie, northern Manitoba, northern Ontario, and central Quebec south to eastern Washington, extreme eastern Oregon, Idaho, northeastern Utah, Wyoming, and Colorado, and in the Great Plains from eastern Colorado and eastern Wyoming across Nebraska, the Dakotas, northern Minnesota, northern Wisconsin, and northern Michigan.

SUBSPECIES

T. p. phasianellus (Linnaeus): Northern sharp-tailed grouse. Breeds in northern Manitoba, northern Ontario, and central Quebec. Partially migratory.

Fred Jenson, courtesy G. A. Allen, Jr.

Ruffed Grouse

Ruffed Grouse

Fred Jenson, courtesy G. A. Allen, Jr.

↑ Ruffed Grouse, Female

↓ Ruffed Grouse, Male

Fred Jenson, courtesy G. A. Allen, Jr.

↑ Sharp-tailed Grouse, Male Dancing

↓ Sharp-tailed Grouse, Males Dancing

↑ Sharp-tailed Grouse, Males ↓ Sharp-tailed Grouse, Males Fighting

↑ Greater Prairie Chicken, Male

↓ Greater Prairie Chicken, Male

↑ Greater Prairie Chicken, Male and Female

↓ Greater Prairie Chicken, Male and Female

↑ Greater Prairie Chicken, Males Fighting

↓ Greater Prairie Chicken, Males Fighting

↑ Lesser Prairie Chicken, Male

↓ Lesser Prairie Chicken, Male

Roger Sharpe

Roger Sharpe

↑ Lesser Prairie Chicken, Male

↓ Lesser Prairie Chicken, Males

Roger Sharpe

Buffy-crowned Tree Quail Habitat, Chiapas

↑ Bearded Tree Quail

↓ Barred Quail, Male

↑ Barred Quail, Chick

↓ Barred Quail, Juvenile

Scaled Quail, Male *New Mexico Dept. of Game & Fish*

↑ Scaled Quail, Male Frontal Display

↓ Scaled Quail, Male Head-throw

Mountain Quail, Male

T. p. kennicotti (Suckley): Northwestern sharp-tailed grouse. Resident in Mackenzie from the Mackenzie River to Great Slave Lake.

T. p. caurus (Friedmann): Alaska sharp-tailed grouse. Resident in north central Alaska east to the southern Yukon, northern British Columbia, and northern Alberta.

T. p. columbianus (Ord): Columbian sharp-tailed grouse. Resident from north central British Columbia and western Montana south to eastern Washington, eastern Oregon (now nearly extirpated), northern Utah, and western Colorado. Formerly extended to Nevada and New Mexico.

T. p. campestris (Ridgway): Prairie sharp-tailed grouse. Resident from southeastern Manitoba, southwestern Ontario, and the Upper Peninsula of Michigan to northern Minnesota and northern Wisconsin. Formerly extended to northern Illinois.

T. p. jamesi (Lincoln): Plains sharp-tailed grouse. Resident from north central Alberta and central Saskatchewan south to Montana (except the extreme west), northeastern Wyoming, northeastern Colorado, and western portions of Nebraska, South Dakota, and North Dakota. Formerly extended to Kansas and Oklahoma.

MEASUREMENTS

Folded wing: Adult males, 194–223 mm; adult females, 186–221 mm (males of all races average 202 mm or more; females, 201 mm or less).

Tail: Adult males, 110–35 mm; adult females 92–126 mm (males average 4 mm longer than females).

IDENTIFICATION

Adults, 16.4–18.5 inches long. The sexes are nearly identical in plumage. The tail is strongly graduated in both sexes, with the central pair of feathers extending far beyond the others, but the tips are not pointed. Both sexes are feathered to the base of the toes, and males have an inconspicuous yellow comb (somewhat enlarged during display) and pinkish to pale violet areas of bare neck skin that are also expanded during display, though not to the degree found in prairie chickens. Both sexes have inconspicuous crests, and the head and upperparts are extensively patterned with barring and spotting of white, buffy, tawny brown, and blackish. White spotting is conspicuous on the wings, and the relative amount of white increases toward the breast and abdomen, which are immaculate. The middle pair of tail feathers is elaborately patterned with brown and black, but the others are mostly white. The breast and flanks are intricately marked with V-shaped brown markings on a white or buffy background.

FIELD MARKS

The grassland, edge, or scrub forest habitat of this species varies considerably throughout its range, but the bird is basically to be found in fairly open country, where its pale, mottled plumage blends well with the surroundings. In flight the white underparts are conspicuous, as is the whitish and elongated tail. On the ground, the birds have a much more "frosty" appearance than do prairie chickens, which are generally darker and lack definite white spotting.

AGE AND SEX CRITERIA

Females may be identified with about 90 percent reliability by a transverse barring pattern on the central tail feathers, compared with the more linear markings of males. Also, the crown feathers of males have alternating buff and dark brown cross-bars, whereas the male crown feathers are dark with buffy edging (Henderson et al., 1967).

Immatures are identified by the usual character of pointed outer primaries. Ammann (1944) suggested that a comparison of the eighth and ninth primaries as to relative amounts of wear (equal or little wear on either in adults, greater wear on the ninth in immatures) is the most suitable method of judging age in prairie grouse.

Juveniles have white rather than buffy throats and have shorter median tail feathers than do adults. The lateral tail feathers of juveniles are more buffy, mottled and speckled with brown, while the median two pairs have broad, buffy central stripes (Ridgway and Friedmann, 1946). White shaft-streaks are conspicuous on the upperparts as well.

Downy young are illustrated in color plate 61. Downy sharp-tailed grouse have a clearer and paler mustard yellow color overall than do prairie chickens of the same age and lack the rusty tints of that species. There is the trace of a median black crown line and a few small crown spots, but only one or two black spots between the eyes and the ear region are present in this species.

DISTRIBUTION AND HABITAT

This species together with the pinnated grouse comprise the "prairie grouse" of North America. Such a designation for the sharp-tailed grouse

FIGURE 32. Current (shaded) and recent (dashed) distributions of the sharp-tailed grouse.

is not wholly accurate, for the original distribution of the species included not only grassland habitats but also sagebrush semidesert (*T. p. columbianus*), brushy mountain subclimax communites (*T. p. jamesi*), oak savannas and successional stages of deciduous and mixed deciduous-coniferous forests of the eastern states (*T. p. campestris*), and brushy habitats of boreal forests from Canada through Alaska (*phasianellus, caurus,* and *kennicotti*), as summarized by Aldrich (1963). Two of the races have suffered considerably from habitat changes associated with man's activities. One of these is the Columbian sharp-tailed grouse, which has been reduced in a remnant distribution pattern to the point that it is wholly eliminated from California, virtually gone from New Mexico, rare in Utah, Nevada, and Oregon, uncommon to rare in Oregon and Washington, and generally uncommon in Colorado, Wyoming, and Montana (Hamerstrom and Hamerstrom, 1961).

The prairie race of sharp-tailed grouse has similarly been extirpated from Illinois, Iowa, and southern portions of Wisconsin and Minnesota and is in danger of extirpation in the northern parts of these states (Hamerstrom and Hamerstrom, 1961). In the Lower Peninsula of Michigan, introduced sharp-tails probably reached their greatest distribution by 1950 (Ammann, 1957), and by the early 1960s only a few hundred birds could be counted on display grounds (Ammann, 1963a). On the Upper Peninsula the sharp-tail population had decreased at least nine percent between 1956 and the early 1960s, primarily through habitat losses (Ammann, 1963a). In Minnesota the general population trend appears to be downward, as a result of improved farming practices as well as increased reforestation and tree-farming activities (Bremer, 1967). Hamerstrom and Hamerstrom (1961) report that the Wisconsin population is in greater danger than those in Minnesota and Michigan as a result of fire protection, forest succession, pine plantations, and modern farm practices. The Canadian populations of this race in Ontario, Manitoba, and eastern Saskatchewan appear to be in relatively good condition.

The plains sharp-tail, with its extensive range from northern Alberta to North Dakota and northern New Mexico, has apparently suffered the least of the United States races, and still supports legally hunted populations in three provinces and six states. However, it is gone from northwestern Oklahoma and western Kansas, and its range in eastern Colorado has shrunk appreciably (Johnsgard and Wood, 1968).

The remaining, predominantly Canadian, populations of sharp-tailed grouse are evidently in relatively satisfactory condition.

HABITAT REQUIREMENTS

General habitat characteristics of the prairie race of sharp-tailed grouse have been analyzed by Grange (1948) for Wisconsin and by Amman (1957) for Michigan. Grange concluded that sharp-tailed grouse are abundant in areas covered from 25 to 50 percent by wooded vegetation, and Ammann indicated that from 20 to 40 percent woody cover is ideal, preferably with the trees in scattered clumps rather than widely scattered. Sparse or bare patches in the ground cover should not exceed half of the total, and the area of suitable open habitat in wooded vegetation should not be less than a square mile, in the opinion of Ammann. According to him, ideal summer sharp-tail habitat on a square mile unit should include an open portion of about 6 percent of the total area that would be a display site, loafing and foraging habitat for adult males and broods, and roosting sites for displaying males. About half of the area should consist of scattered large shrubs and trees, especially aspens. Heavy ground cover is needed for roosting, nesting, and feeding, while lighter ground cover serves for resting, dusting, and feeding, especially by broods. The remaining 44 percent of the cover should consist of an alternating series of small (ten-acre) brushy clearings and heavier second-growth timber stands of mixed hardwoods and conifers, which serves as a source of winter browse and protection from severe weather as well as escape cover. The scattered small clearing provides additional nesting and brood-rearing habitat and winter roosting opportunities. Paper birch (*Betula papyrifera*) and aspen (*Populus tremuloides*), especially the former, represent major winter food sources when snow cover prevents foraging on grains or other similar foods.

Although these habitat needs may apply to the prairie sharp-tailed grouse, they are clearly not strongly applicable to the Columbian and plains races, which occur in semidesert scrub and relatively dry grasslands, respectively. For the Columbian race at least, shrubs and small trees are important habitat components only during the late fall and winter, while during the rest of the year weed-grass cover types as well as cultivated crops such as wheat and alfalfa provide important food and cover requirements (Marshall and Jensen, 1937). Likewise, Hart, Lee, and Low (1952) list a variety of grasses and herbs as important components of Columbian sharp-tail habitat in Utah. Similarly, the plains sharp-tailed grouse inhabiting the sandhills of central Nebraska and the comparable sand dune areas of north central North Dakota are relatively independent of extensive tree cover (Aldous, 1943; Kobriger, 1965). In the late fall and winter these birds resort to foraging on rose hips and willow buds in the North Dakota sandhills (Aldous, 1943), while in Utah the buds of maples and chokecherries are major sources of winter foods (Marshall and Jensen, 1937). According to Edminster (1954), a minimum of 5 percent brush cover to total land surface is tolerable to sharp-tails in North Dakota.

Wintering Requirements

Grange (1948) reported that sharp-tails do not roost in trees overnight during winter; instead they utilize snow burrows which they scratch out in fairly dense marsh or swamp vegetation or sometimes in open stands of tamarack or spruce in northern Wisconsin. During snowless periods roosting usually occurs in dense and fairly coarse marshy vegetation.

Ammann's observations (1957) for Michigan sharp-tails are similar. During fall, the birds concentrate in "packs" on grain plantings near their summer habitat and may continue to use grain as long as it remains available. When the snow is deep and grain becomes unavailable, the catkins, twigs, and buds of trees such as paper birch, aspen, Juneberry, hazel, and bog birch are preferred, as well as the fruit of mountain ash, sumac, common juniper, rose, and black chokecherry. Of all these, the buds and catkins of birch and aspen are especially important, particularly birch. A wide variety of grains is taken if they are available, including wheat, buckwheat, field peas, corn, barley, soy beans, millet, and rye. Thus the availability of grain or native food sources in the form of fruiting shrubs or deciduous trees is an important component of winter habitat.

The presence of adequate snow during unusually severe weather conditions may be important to sharp-tails. Marshall and Jensen (1937) found that movement to maple-chokecherry cover in Utah was related to snow depth; there the birds could feed on buds and roost under the snow unless it crusted heavily, when they preferred to roost above the snow in brushy cover. Some deaths by freezing have been reported when strong winds were associated with low winter temperatures and no snow was available for roosting (Edminster, 1954).

Spring Habitat Requirements

Ammann (1957) reported on the general cover characteristics of ninety-five sharp-tail dancing grounds in Michigan. Of these, twenty-seven were located on cultivated lands, and sixty-eight were on wild lands. Although the majority of these contained no woody cover, 35 percent had such cover present, but rarely did this cover exceed 30 percent of the surface area.

Favored sites for both sharp-tails and pinnated grouse appeared to be low, mottled or sparse vegetation with good visibility, allowing for good footing and unrestricted movements. Elevated, rather than level or depressed sites, were preferred for both species; of sixty-five pinnated grouse and ninety-five sharp-tail display grounds, 47 percent were in elevated situations and only four were located in depressions.

In Wisconsin, Grange (1948) found that wild hay meadows and marshes were frequent display locations for pinnated grouse and sharp-tails, with sharp-tails exhibiting an apparently greater preference than pinnated grouse for wet marshes. A variety of other cover types was also found to be used by both species, including abandoned fields, cultivated fields and, less commonly, upland grassland, peat burns, and clover fields.

In Alberta, Rippin (1970) noted that of thirty-six display grounds studied by him, thirty-two were on open, dry, and elevated sites, three were on level ground, and one was on an elevation with heavy shrub cover. In the Nebraska sandhills Kobriger (1965) found that three-fourths of all prairie grouse display grounds studied were on wet, mowed sites. Similarly, Sisson (1970) reported that twenty-six of thirty-six sharp-tail dancing grounds in the Nebraska sandhills were within one-eighth mile of a windmill, where the vegetation was fairly low as a result of grazing and trampling of vegetation by cattle, and where visibility was good in all directions.

Nesting and Brooding Habitat Requirements

Ammann (1957) has provided a fairly detailed analysis of nesting requirements for sharp-tails in Michigan. He reported that they choose a wider variety of sites with respect to woody cover than do pinnated grouse, with site conditions varying from open to 75 percent shaded. Most nests were either protected by overhead cover or were within a few feet of such cover. Of twenty-nine nests found none was more than ten feet from brushy or woody cover. Of ten nests studied, six were in open aspen, three were in cut-over pines, and one was in an open marsh. These sites averaged 43 percent shrub cover, from three to six feet high, and 4 percent tree cover in excess of six feet tall. Associated shrubs were chokecherry, willow, and alder, and associated trees were aspen, spruce, and Juneberry. Of seven additional nests, four were located at the base of a small tree or bush, and there was one each in a hayfield, on an aspen-birch ridge, and in a heavy grass–sweet fern site.

Hamerstrom (1939) reported on cover sites for seventeen sharp-tail nests in Wisconsin. Of these, eight were at the edges of marshes, brush, or woods in brushy or woody (aspen, willow, etc.) cover. Three were in small openings of dense brush such as aspen or willow, two were in openings or edges of jack pine–scrub oak woods, two were in grass meadows, one was in a dry marsh, and one was in a mixture of scattered brush, trees, and grass. In this study as well as Ammann's, an apparent avoidance of cultivated areas for nest sites would seem to be present.

Since the males do not participate in nesting, they gradually move away from their display grounds to foraging and daytime resting sites that usually include brushy cover, aspen or willow thickets, or young conifer stands. In Utah, summer daytime resting places gradually change from weeds and grass during June and early July to shrubs and bushes in late July and August (Hart, Lee, and Low, 1952). For night roosting fairly open and upland cover with good ground cover is preferred by sharp-tails over marsh and bog vegetation (Ammann, 1957).

Brooding habitat requirements have been analyzed by Hamerstrom (1963) in the Wisconsin pine barrens and by Ammann (1957) for Michigan. Ammann concluded that the birds tend to favor somewhat more woody cover than that chosen for nest sites but in general remain in areas that do not exceed 50 percent shading by woody cover. Peterle (cited by Ammann) estimated a higher (70 percent) average over-all shading by woody cover, with shrubs covering 43 percent of the area and trees covering an average of 70 percent in locations where fifteen broods were observed.

Hamerstrom's observations on about 190 broods confirm the importance of opening in forested areas as brood habitat. Of his brood habitat records, about 80 percent were in open situations, 14 percent were in edge situations, and only 5 percent were more than fifty yards inside woody habitats. He concluded that brood cover should be basically grassland, with some shrubs and trees, but the taller the woody species present, the fewer there should be. Shrubs are more important than trees, since they provide not only cover but also food sources for chicks. Thus, berry-producing species such as blueberries, cherries, and Juneberries are valuable, as are catkin-bearing shrubs that can be used as a source of winter foods. Aspens and willows, although valuable as sources of winter buds, are most useful in small thickets and young trees. Hamerstrom stressed the importance of distinguishing the open, predominantly herbaceous brooding habitat from the fall and winter woody cover that is also critical to sharp-tail survival.

POPULATION DENSITY

Some of the best figures as to spring population densities for sharp-tailed grouse come from the work of Grange (1948). Using spring dancing ground counts and assuming a 55 percent ratio of males in the total populations, he calculated an estimate for 1941 of 235.2 acres per bird on 130,560 acres and 186.7 acres per bird on the same area in 1942. Considering only the occupied range in 1942, the average area per bird figure was calculated to be 138 acres. Ammann (cited by Edminster, 1954) reported spring densities on thirteen square miles of habitat on Dummond Island, Michigan, for a three-year period as averaging one bird per 45 acres, while the fall populations of sharp-tails on the island were approximately one bird per 18 acres of occupied range over a seven-year period. This island represents prime Michigan sharp-tail habitat, and these figures must be regarded as being unusually high densities which have not been maintained recently. Edminster (1954) summarizes a variety of other fall density estimates from various states that in general indicate that from 27 to 125 acres per bird in summer or fall are probably typical. One other high density figure has been reported for Saskatchewan, with Symington and Harper (1957) estimating late summer populations of between twenty-five and forty birds per square mile (16 to 25.6 acres per bird) in the Sand Hills area, where an ideal combination of native grasses, shrubs, and small trees occur.

FOOD AND FORAGING BEHAVIOR

Dependable and nutritious winter food sources are critical to the survival of all grouse, and the sharp-tail appears to be somewhat flexible in its winter diet in comparison with other grouse species. In central Wisconsin, paper birch (*Betula papyrifera*) buds and catkins are the primary winter diet, with aspen (*Populus tremuloides*) of secondary importance. Among shrubs, rose (*Rosa*) hips and hazel (*Corylus*) buds and catkins are important foods (Grange, 1948). In Ontario, the paper birch is also the primary winter food, supplemented by browse of willow, aspen, blueberry, and mountain ash (Snyder, 1935). In North Dakota, willow buds provide the most important single source of winter foods, but chokecherry, poplar, and rose hips are also major supplementary species (Aldous, 1943). During winter in Utah, sharp-tails move during periods of heavy snow into thickets of maple, chokecherry, and serviceberry, where they feed on the buds of these species. In the Nebraska sandhills the sharp-tailed grouse appears to be more efficient than the pinnated grouse in finding winter foods and surviving the severe weather conditions and is much more common and more extensively distributed through that region (Kobriger, 1965; Johnsgard and Wood, 1968).

Throughout the range of the species, the percentage of woody mast foods sharply decreases in spring as herbaceous plants become available after periods of thawing. Such plants include cultivated grain species, clover, alfalfa, and native annuals and perennials. Jones (1966) found that during the spring and summer months green materials comprised the bulk of the diet in Washington, with grass blades alone (especially *Poa secunda*) totaling half of the spring foods and three-fourths of the summer diet. Flower parts were the rest of the spring and summer foods, particularly those of dandelion (*Taraxacum*) and buttercup (*Ranunculus*). The importance of dandelion continued on into fall, when it seeds and grass leaves were the leading food sources. Apparently the sharp-tail relies to a lesser extent on animal sources of food during the summer than does the pinnated grouse (Jones, 1966), although Grange (1948) reported that grasshoppers are a major summer food, and Edminster (1954) estimated that from 10 to 20 percent of the adult summer food is of insect origin. Kobriger (1965) found that the juveniles had increased the amount of vegetable food in their diets to more than 90 percent; he reported that in Nebraska such important food plants included clovers, roses, cherry, and dandelion, the most important of which were favored by wetland mowing practices.

During fall, a diverse array of seeds and cultivated grains are taken in the diet, especially in agricultural areas. Otherwise the fruits of shrubs such as roses, snowberry, wolfberry, bearberry, blueberry, mountain ash, and poison ivy are taken, as well as seeds and green leaves of herbs, shrubs, and trees. Probably a superabundance of suitable foods is normally available during this time, and much local or yearly variation in foods taken might be expected to occur. Grange (1948) has pointed out that in general the sharp-tail closely resembles the ruffed grouse in its food cycle, and differences occur only because of the sharp-tail's stronger preference for more open habitats. Differences in foods taken are most pronounced in late summer and fall, but from late fall through spring they may be nearly identical. The primary differences noted between the sharp-tail and the pinnated grouse were that the pinnated uses a greater amount of grains and weeds and more generally depends on food sources associated with cultivation. Pinnated grouse may also feed to a somewhat larger extent on insects, especially grasshoppers, than do sharp-tails.

MOBILITY AND MOVEMENTS

Seasonal Movements

By far the most complete summary of sharp-tail movements is that of the Hamerstrom and Hamerstrom (1951), and the following account is based on their analysis of seasonal movements in this species. Evidence for a definite seasonal migration dates from fifty to one hundred years ago, when most or all of the original sharp-tail range was occupied. At that time, marked seasonal movements evidently did occur, but there is no clear evidence indicating migratory distances or even the directions involved. In areas of mountains or hills where woody cover occurred, an upward altitudinal migration apparently occurred, but few if any cases of a downward movement have been reported. Much of what has been interpreted as migration has consisted simply of movements to woody cover for the winter period, with distances of such movements gradually being reduced as the birds were driven out of their grassland habitats to woody edges, ravines, and similar brushy or woody situations. Thus, long-distance movements from prairies to wooded wintering habitats have in recent years been completely eliminated, although seasonal changes in habitat preferences still persist in local areas.

With the advent of agriculture, not only were the prairies made relatively unsuited for breeding grounds for sharp-tails but also the availability of fall and winter grain sources has influenced their movements. However, the sharp-tail has not been so strongly influenced by this food source as has the pinnated grouse, and is less likely to leave its brushy winter habitat to obtain grain than is the pinnated grouse. Where sharp-tails have simply incorporated grain into their winter diets they have thus altered their winter behavior very little, but in some areas the availability of grain throughout the winter has enabled the birds to winter in relatively open situations.

During the period of habitat shift from open to relatively brushy habitats, fall "packing" occurs, as coveys or broods gather into small flocks, which in turn form packs of up to several hundred birds. To a smaller extent, packing may occur in late winter during the reverse movement to breeding grounds.

The Hamerstroms presented banding data related to mobility for 167 sharp-tailed grouse banded in Wisconsin. Of the 162 birds for which the point of return was known, 81 percent were retaken within two miles of the point of banding. Only 12 percent had moved more than three miles, and only 10 percent were retaken more than five miles away. The longest distance away from the point of banding was twenty-one miles. Similarly, Aldous found that short-range movements were the rule, with the maximum distance for any return fifty-eight miles. Judging from comparable data on Wisconsin pinnated grouse, the relative over-all mobility of the two species would appear to be about the same. By transplanting sharp-tails and plotting their later recoveries, the transplanted birds were found in general to move farther than nontransplanted birds but to show no tendency to return to the point of banding. The maximum mobility of these transplanted birds was found to be between twenty-six and twenty-seven miles from the point of release.

The relative distances of movements of sharp-tails from their wintering quarters to spring display grounds doubtless vary greatly in different areas. Kobriger (1965) found that in the Nebraska sandhills the dispersal of 35 male sharp-tails from winter feeding stations to spring dancing grounds ranged from 0.2 to 3.3 miles, and averaged 0.9 mile. The majority of these birds moved from their wintering areas to the nearest dancing ground. However, this probably implies that the birds picked the suitable wintering area nearest their dancing ground rather than vice versa, since Evans (1969) found a high degree of fidelity of male sharp-tails to specific leks between successive years. Similarly, most nests are located within a mile of the nearest dancing ground (Hamerstrom, 1939; Hamerstrom and Hamerstrom, 1951).

Daily Movements and Home Ranges

The Hamerstroms (1951) reported that in the fall sharp-tails had a rather large covey range that totaled about one hundred to two hundred acres in extent with from three to six such coveys usually to be found in an area of one thousand to fifteen hundred acres. They estimated that the usual winter daily cruising radius was about one mile.

Kobriger (1965) tracked a sharp-tail male by radio telemetry through the summer months, during which it moved about 2.5 miles from its dancing ground. Similarly, a female was tracked from a dancing ground to a nest site 2 miles away. In the Nebraska sandhills sharp-tail display grounds average less than 1 mile apart, and it is thus probable that females may move considerably greater distances than this between a dancing ground and their selected nest sites.

REPRODUCTIVE BEHAVIOR

Territorial Establishment

Territorial establishment by sharp-tailed grouse probably occurs as early as the first fall of life. The Hamerstroms (1951) found that at least three of eighteen males seen on a dancing ground in North Dakota during late September were young birds. Likewise, Rippin (1970) found that although only adult males were among those trapped or shot on a display ground in late August, by late September and early October several juvenile males were also present. This regular fall period of display, which is also typical of pinnated grouse but not the sage grouse, may provide an important basis for the learning of traditional display sites by young birds. Rippin found that when he killed all of the males using a dancing ground during the spring, there was no usage of that display site the following fall, but on another area where he killed all but one of the displaying males, the lone bird formed a nucleus for display behavior with several other juvenile birds that following fall. Young probably begin trying to establish peripheral territorial areas their first fall of life, and these territories are then held again the following spring. Rippin reported that on two control dancing grounds (on which he did not experimentally remove any males), the percentage of immature males was 43 percent in 1968 and 37 percent in 1969. On his experimental grounds, he first mapped the relative territorial positions of the participating males; in each he recognized one or more centrally located males and approximately three outer rings of less dominant males defending peripheral territories. On one display ground which contained eighteen males, a marginal male originally defending a peripheral territory gradually established itself as a centrally located bird as Rippin progressively reduced the number of males on the dancing ground to five birds. When the ground was reduced to four participating males, no single bird was able to maintain a central dominant position. The clear result of his studies indicated that a strong centripetal tendency was present in all the males, with each attempting to attain and defend a relatively central territory.

When such display ground social structures are not disrupted by the death or removal of males, they exhibit a high degree of stability. Evans (1969) found that of ten males that were marked one spring, five returned to the same dancing ground the following spring, while the other five disappeared and apparently had died. The areas defended by the five returning males were virtually the same as those they had defended the previous spring, with a single minor exception. Hjorth (1970) analyzed Evans's data and concluded that on two grounds the average territorial size was about 90 square meters, ranging from 14 square meters in the central area to 170 square meters on the periphery. He also determined that the average territorial size for a Montana display ground was about 50 square meters, with the four central territories averaging 25 square meters.

The average sizes of display grounds, in terms of numbers of territorial males present, probably vary with population density. Ammann (1957) provides average numbers of birds of both sexes present on 10 different sharp-tail dancing grounds, which averaged 12.4 but ranged from 3 to 29 birds in different years and on different grounds. In the Nebraska sandhills, display grounds of both the sharp-tail and the pinnated grouse usually have an average of between 9 and 10 males (Johnsgard and Wood, 1968). Grange (1948) indicated that the average number of males on 14 sharp-tail grounds in Wisconsin was 6, while 7 pinnated grouse grounds averaged 7 males in attendance. In Utah, Hart, Lee, and Low (1952) reported the average number of birds present on 29 dancing grounds as 12, although as many as 50 were seen. Lumsden (1965) summarized data from several areas in Ontario that indicated from 2 to 24 males present on dancing grounds. In North Dakota the twelve-year average for 1,664 dancing grounds was 12.9 males (Johnson, 1964). It would seem that from 8 to 12 males represents a typical dancing ground for sharp-tailed grouse in most parts of their range.

Lumsden (1965) confirmed the observations of earlier persons working with pinnated grouse and sage grouse as to the reproductive advantage of holding central territories in sharp-tailed grouse dancing grounds. He reported that such central positions were held by socially dominant birds that readily achieved superiority in disputes with neighbors. These central territories were often smaller than peripheral ones, and Lumsden thought that normally only fairly old males could successfully hold such territories. On one display ground Lumsden noted that the dominant male performed 76 percent (thirteen) of the copulations or attempted copulations that were observed, which emphasizes the enormous selective value of occupying such central territories.

Territorial Advertisement and Defense

Lumsden (1965) has classified the social displays of the sharp-tails as those which serve aggressive functions, those which are concerned with courtship and mating, and those which are specifically associated with advertising the location of the display grounds. In addition, several signals serve as a predator warning system. Lumsden's account is unusually complete, and his terms and descriptions will be utilized here. More recently, Hjorth (1970) has made an equally detailed analysis; his comparable terms will be noted and a few divergent observations briefly mentioned.

Signals which serve primarily to advertise the location of the dancing ground and of specific males include the flutter jump and cackling calls. Both sexes perform cackling calls. Cackling by females is usually performed as they approach the dancing ground, and this stimulates strong responses by the males, especially flutter-jumping. Flutter-jumping was first described for the pinnated grouse, and it is virtually identical in both species. The male jumps into the air a few feet, sometimes uttering a *chilk* note as it takes off, flies a few feet forward, and lands again. In so doing, the male clearly advertises its own presence, as well as the location of the dancing ground as a whole. Cackling by males may occur between flutter jumps, or may be uttered by males when others are flutter-jumping.

A large number of male sharp-tail displays are primarily aggressive and serve to establish and maintain territories. Secondary functions no doubt include the attraction of females to the male and allow for sexual recognition. These primarily aggressive signals include several calls and postures. The calls may be called the *lock-a-lock*, "cooing," the "cork" call, and the *chilk* and *cha* calls. Lumsden regards the last two calls as being associated with courtship, since they are most often uttered when hens are present.

The *chilk* and *cha* notes are both loud, high-pitched notes that carry great distances. They are often uttered before or after flutter-jumping, and often during the "tail-rattling" display, and both may be uttered with great rapidity. They evidently grade into one another and probably serve similar functions.

The "cork" note is a squeaking sound resembling that produced by pulling a cork from a bottle and is only uttered during the tail-rattling display. It is most often heard when a female is near but may be elicited by another displaying male. A similarly aggressive call is called "whining," which consists of drawn out and repeated sing-song *kaaa—kaaaaa* notes. Such notes are usually associated with territorial defense and are often uttered by birds when facing one another.

The *lock-a-lock* call is a gobbling note that is produced by males when they are standing at rest. With head lowered slightly, a male may utter this call as he approaches his territory before dawn. It is not uttered in the presence of females and apparently serves only an aggressive function.

The "cooing" display is a combination of posturing ("oblique" posture of Hjorth, 1970) and sound production that is clearly homologous with the "booming" of pinnated grouse. As in that display, the tail is partially cocked, the esophagus is inflated, and the head is distinctly lowered ("bowing" of Hjorth, 1970), as a low-pitched cooing sound of one or two notes is uttered. However, the folded wings are not strongly lowered, and the throat skin is not as strongly distended as the pinnated grouse's is during booming. The neck skin color is usually pink to purple and thus is also different from that of the greater prairie chicken. Lumsden believes that cooing does not serve as a sexual signal but rather is evoked in aggressive situations, thus also differing functionally from the booming display.

Several postures or movements are also closely associated with territorial defense. These include an "upright advance" ("wide-necked upright" of Hjorth, 1970), which is an aggressive approach posture of a male during which the tail is cocked and the neck feathers are erected to expose the apteria. "Walking or running parallel" consists of two males' moving along their territorial boundaries while threatening one another, often while uttering the *lock-a-lock* call. During this display the head is usually held low, the eye-combs are enlarged, and the tail is cocked. During "ritual fighting" the birds face one another, often while squatting, and utter aggressive calls while periodically making short lunges toward the other bird. When not attacking, they usually hold their wings partly open and on the ground. During overt attacks the birds leap up into the air, flailing one another with their claws and beaks and sometimes striking with the wings. Between such attacks the birds watch each other intently, and Lumsden reports that "displacement sleeping" may occur when the attack intensity wanes to a certain point. Should a male attempt to withdraw from such an encounter, it typically lowers its tail, covers its neck skin, withdraws its eye-combs, and sleeks its feathers. These submissive patterns give the bird the appearance of a female and tend to inhibit attack by males. Lumsden reported that the sharp-tails he observed in Montana, but not those in Ontario, performed a shoulder-spot display when fighting and also just prior to copulation. This consists of exposing the white underwing coverts in the region of the elbow. The shoulder-spot display is a conspicuous feature of several grouse species, such as the pinnated grouse and in several seems to indicate fear or submission. However, Hjorth (1970) did not observe this display in Montana sharp-tails and I have not seen it in Nebraska. Recently Lumsden (1970) has reviewed the occurrence of this display in various grouse species and has concluded that in some species (such as black grouse and capercaillie) it serves as an aggressive signal function among males, while in females it indicates an expression of fear.

Much the most complex and interesting of the male displays is the "tail-rattling" or "dancing" display of sharp-tails. Lumsden considers this to be a courtship display, but it is also closely associated with territorial defense and proclamation. It consists of a highly ritualized series of rapid stepping movements, performed with the tail erect, wings outstretched, head held forward and rather low, and neck feathers erected to exhibit the bare purple skin. With the cocking of the tail the white undertail coverts become exposed and appear to be somewhat expanded for maximum visibility. In this rigid posture the male begins a series of very short and rapid stepping movements (eighteen to nineteen per second according to Hjorth, 1970), causing him to move forward in a generally curving direction ("aeroplane display" of Hjorth, 1970).

In synchrony with the stepping movements, the male also performs a strong lateral vibration of his tail, producing a clicking or rattling frictional sound which is a combination of these pattering sounds and the scraping noises of the overlapping tail feathers. Hjorth (1970) has recently found that during tail-rattling not only are the lateral rectrices alternately spread and shut, but the male also occasionally performs a rapid (0.08 second) symmetrical tail-spreading while breaking his stamping rhythm momentarily.

Not only are the foot and tail movements of the male a highly coordinated series of activities, but males tend to perform the tail-rattling display in highly synchronized fashion. Two or more closely adjacent males will start and stop their dislay almost simultaneously, and sometimes all of the males on a dancing ground will become silent simultaneously. At such times the birds appear to be highly attentive and sensitive to disturbance, whereas when they are all actively "dancing" they remain nearly oblivious to their surroundings.

When performing the tail-rattling display in the presence of a female, the male often alternates this display with a stationary posture Lumsden has called "posing." During this posture the male usually faces or nearly faces a female, with wings slightly spread and drooped and the eye-combs greatly enlarged. Soft crooning notes may also be uttered. Typically the male moves from this posture into a crouching or "nuptial bow" position before the female, in which he lowers his body to the ground, fully spreads his wings to the sides, and almost touches the ground with his bill ("prostrate" of Hjorth, 1970). The rear end of the bird is held high, so that the tail remains vertical, and in general the upper body surface and dorsal view of the tail appear to be presented to the female. In contrast to the comparable posture of the pinnated grouse, the male may perform several short and repeated bowing movements, while in the pinnated grouse the male typically remains prostrate and motionless before the female for several seconds. Although this display is normally performed by a male that is beside a female and not being bothered by rival males, Lumsden noted that he observed it as a precopulatory display in only one of nineteen copulation sequences.

Most copulations by sharp-tailed grouse occur before or approximately at the time of sunrise. Preliminary postures may include the nuptial bow, posing, or tail-rattling displays. The female squats in the usual manner and is immediately mounted by the male. Usually the hen runs forward rapidly immediately after copulation, then vigorously shakes her body and wing feathers. Following a successful copulation the hen often leaves the display ground within a few minutes, and there is no evidence to date that more than one copulation is needed to fertilize all of the eggs in a single clutch.

Vocal Signals

In addition to the calls already mentioned, Lumsden has described several other calls. In a situation of uneasiness or slight disturbance, a *yur* note with a downward inflection is uttered. In flight, a series of rapid calls *tuckle . . . tuckle . . . tuckle*, or *tuk . . . tuk . . . tuk*, are frequently uttered, and the same calls may be produced prior to flight.

One other vocal signal that serves as a courtship signal, or at least is produced only when hens are on the display ground, is the *pow* call. When courting a hen, males will utter this call several times in rapid succession. Most probably, as Lumsden has suggested already, it is homologous to the loud *whoop* call of greater prairie chickens.

Other Signals

Lumsden has described several predator-response postures of sharp-tailed grouse, which include an "upright alert" posture, in which the bird stands upright to its fullest extent with its feathers sleeked and crest raised. A "prostrate alert" is performed in a similar situation, but with the bird in a crouched and "frozen" posture. "Alarm strutting" may be performed as the bird walks around or away from a source of possible danger, in a stiff gait and with occasional tail flicks, which reveal the white outer tail feathers.

Nesting and Brooding Behavior

The female begins to make a nest scrape in a protected site at about the time she begins to visit the dancing grounds or possibly even before. Following successful mating, she leaves the dancing ground and probably will not return to it again, except in the event of renesting. The eggs are laid on an approximately daily basis, until the total clutch of about twelve eggs is produced (Hamerstrom, 1939; Ammann, 1957). The female typically begins incubation at about the time the last egg is laid, and the incubation period is twenty-three to twenty-four days.* Renesting attempts by females evidently do sometimes occur, but probably contribute no more than 10 percent of the offspring in an average season (Ammann, 1957).

Following hatching, the female leads the young away from the nest location fairly rapidly, and they particularly tend to move to fairly open areas where insects and green herbaceous foods are abundant (Hamerstrom, 1963). Although the young have been known to move as far as a quarter mile in a single day before fledging, it is probable that the summer brood territory is normally less than a half mile in diameter (Edminster, 1954). Young sharp-tails feed to a large extent on insects during their first few weeks; with grasshoppers, spiders, ants, and weevils all contributing to their diet, while leaves and berries are also important sources of foods

*W. Lemburg, 1970: personal communication.

(Grange, 1948). Chicks are able to fly to a very limited degree by the time they are ten days old, and from that time become increasingly independent of their mother. By the time they are six to eight weeks old, they are virtually fully independent, and broods begin to gradually break up and the young birds disperse, often fairly long distances.

EVOLUTIONARY RELATIONSHIPS

There can be little doubt that the nearest living relative of the sharp-tailed grouse is the pinnated grouse, and I agree with Short (1967) that they are obviously congeneric. Similarities in their downy young as well as in their adult plumage patterns bear this out, as well as the frequency of hybridization under natural conditions (Johnsgard and Wood, 1968). The two forms also share a number of common display patterns, such as "booming" and "cooing," "foot-stamping," the "nuptial bow," and "flutter-jumping." The sharp-tail's *pow* call no doubt is homologous to the *pwoik* of the pinnated grouse, and the whining and cackling calls of the two species are very similar. The sharp-tail's *lock-a-lock* aggressive call probably corresponds to the pinnated's *hoo-wuk;* I have heard a hybrid male utter an intermediate call sounding like *wuk-a-wuk'*. However, the lateral tail-rattling of the sharp-tails is replaced in the pinnated by symmetrical tail-fanning movements, the forward "dancing" is represented by foot-stamping almost in place, and cooing in the sharp-tail appears to have much less visual and acoustical importance than the homologous booming of the pinnated grouse.

Short (1967) suggests that the sharp-tailed grouse is probably closer to the ancestral prairie grouse type than is the pinnated, on the basis of its less specialized neck feathers (rudimentary pinnae) and reduced esophageal sacs. However, its tail feather structure is specialized for the tail-rattling display (Lumsden, 1968), and these differences largely reflect the relative importance of "booming" and "dancing" in the two species. I would suggest that both species have diverged equally from a common forest-dwelling ancestral type, the pinnated in a more easterly and southerly location (oak woodland or savanna habitat) and the sharp-tail in a more westerly and northerly location (grassland, coniferous forest edge habitat). There was probably little contact between these two forms until fairly recently, when human activities greatly altered the habitats of both species (Johnsgard and Wood, 1968).

19

Tree Quails: Long-tailed Tree Quail

Dendrortyx macroura (Jardine & Selby) 1828

OTHER VERNACULAR NAMES

CORDORNIZ Coluda; Gallina del Monte, Gallina de la Montaña, long-tailed partridge, long-tailed wood partridge, Perdiz Coluda, Perdiz del Volcán.

RANGE

Highlands of Mexico from Michoacán and Veracruz south to Oaxaca.

SUBSPECIES (*ex Check-list of the Birds of Mexico*)

D. m. macroura: Eastern long-tailed tree quail. Resident in the mountain forests of the Valley of Mexicó and the highlands of Veracruz.

D. m. griseipectus Nelson: Gray-breasted tree quail. Known only from the heavy oak forest of the Pacific slope of the Cordillera, in Mexico, Distrito Federal, and Morelos.

D. m. diversus Friedmann: Jalisco long-tailed tree quail. Resident in the highland forest of northwestern Jalisco.

D. m. striatus Nelson: Guerrero long-tailed tree quail. Resident above eight thousand feet in the highland forests from southern Jalisco to Michoacán and the cordillera of Guerrero.

D. m. oaxacae Nelson: Oaxacan long-tailed tree quail. Resident in the mountain forests of eastern Oaxaca from the Cerro San Felipe to Mount Zempoaltepec.

D. m. inesperatus Phillips. Resident in mountains near Chilpancingo, Guerrero. Recently described (1966) and not in Mexican *Check-list.*

MEASUREMENTS

Folded wing: Adults, both sexes, 141–69 mm (males average 5 mm longer than females).

Tail: Adults, both sexes, 119–75 mm (males average 11 mm longer than females).

IDENTIFICATION

Adults, 12–15 inches long. The sexes are alike in plumage. This is the largest and heaviest of North American quail, and it and its two close relatives are the only ones to have extremely long tails. This species differs from its two congeners in that it has a black throat and forehead and blackish feathers around the ears. A bushy, brown crest is present, and the upper back and chest are reddish brown edged with gray, while the lower back is mottled with olive brown, black, and tawny. The breast is gray, streaked with reddish brown, grading to olive on the sides and abdomen. The bill, legs, feet, and bare skin around the eyes are all red. (Modified from Leopold, 1959.)

FIELD MARKS

Rarely seen in the field, this species inhabits dense underbrush of mountain slopes in Mexico. It and related species have a long-legged and upright appearance but can compress the body and slip away unobserved (Dickey and van Rossem, 1938). More often heard than seen, its calls include grouse-like alarm notes (Schaldach, 1963). Its elaborate song is heard most often at dawn and is distinctive; it is a series of about five grunting, hooting notes that rise in volume and are followed by a loud, ringing *kor-EEE-oh,* repeated several times (Warner, 1959), or a series of *ko'-or-eee'* phrases spaced about one second apart, often in a chorus involving several birds.

Tree Quails: Bearded Tree Quail

Dendrortyx barbatus (Gould) 1846

OTHER VERNACULAR NAMES

Bearded wood partridge, Chiviscoyo.

RANGE

Resident in the mountain forests of the state of Veracruz, Mexico, northward to eastern San Luis Potosi and eastern Hidalgo.

SUBSPECIES

None described.

MEASUREMENTS

Folded wing: Adults, both sexes, 147–66 mm (males average 5 mm longer than females).

Tail: Adults, both sexes, 110–21 mm (males average 11 mm longer than females).

IDENTIFICATION

Adults, 9–10 inches long, sexes alike in appearance. Similar to the preceding species, but with a gray throat and chest, a buffy brown crown, and a cinnamon brown breast and abdomen.

FIELD MARKS

If seen, the buffy brown crown and gray chin and throat region will readily separate this species from the long-tailed tree quail, in the few areas where both occur (Orizaba and Cofre de Perote in Puebla and Veracruz). Its calls are similar to those of the long-tailed tree quail.

Tree Quails: Buffy-crowned Tree Quail

Dendrortyx leucophrys (Gould) 1844

OTHER VERNACULAR NAMES

Gallina de Monte, highland wood partridge, long-tailed partridge, white-throated wood partridge.

RANGE

Highlands of Chiapas, Guatemala, Honduras, Nicaragua, and Costa Rica. Few specimens are known from Chiapas, but the species evidently occurs both in the Sierra Madre of Chiapas and in the interior montane forests north of the Río Grijalva (see distribution section).

MEXICAN SUBSPECIES

D. l. leucophrys: Guatemalan buffy-crowned tree quail. Resident in the moist mountain forests of Chiapas. Another race (*D. l. nicaraguae*) has been described from the Pacific Cordillera of adjoining Guatemala, but apparently is not valid (Baepler, 1962).

MEASUREMENTS

Folded wing: Adults, both sexes, 130–55 mm (males average 14 mm longer than females).

Tail: Adults, both sexes, 108–43 mm (males average 16 mm longer than females).

IDENTIFICATION

Adults, 12–14 inches long, sexes alike in appearance. Differs from the two other species of tree quails in the pale buffy forehead and the white eye-stripe, chin, and throat. The lower throat and breast are also more grayish than is true of the bearded tree quail, and the tail is longer than in that species but shorter than in the long-tailed tree quail.

FIELD MARKS

Like the other tree quails, this species is rarely seen, but is usually detected by its repeated ringing calls, which are loud and rollicking, consisting of four syllables (Edwards, 1968). If the bird is seen, the long tail, red legs, and relatively slim body will identify it as a tree quail. When walking on the ground it often holds its tail at a slightly cocked angle, with the feathers somewhat compressed and vaulted, reminiscent of bantam chickens.* This is the only species of tree quail known to occur in Chiapas and unlike the other two species has a black bill rather than a bright reddish bill color, as well as a white forehead.

AGE AND SEX CRITERIA (All Species)

Females apparently are not readily separable from adult males by external characteristics, but they do tend to have shorter tails (see measurements).

Immatures evidently have the outer two primaries frayed (true of at least *D. leucophrys*) and have buffy tips on the upper greater primary coverts (also in *leucophrys*). Petrides (1942) indicates that age of *D. macroura* can also be determined by conventional methods.

Juveniles of at least the long-tailed and buffy-crowned tree quails have white shaft-streaks in the breast, belly, and back feathers, expanding to form large V-shaped markings or broad white bars at the ends of the feathers.

Downy young of the bearded tree quail (illustrated in color plate 110) are representative of the genus. The corresponding plumage of the buffy-crowned tree quail is apparently undescribed and the only description of the long-tailed tree quail available is that of Warner (1959), which was based on somewhat older birds starting to assume the juvenal plumage. Both species are a nearly uniform auburn brown on the back, with no darker or lighter streaking evident and are fairly bright yellow below, particularly on the throat. The crown is dark auburn in both, and the face is yellowish with a large dark brown ear-patch and a smaller loreal stripe. They most resemble spotted wood quail of the same age, the difference being the latter's

*Miguel Alvarez del Toro, 1970: personal communication.

duller and more olive-colored underparts and more reddish face, especially above the eyes. Two recently hatched specimens of the long-tailed tree quail in the United States National Museum are comparable in age and very similar in appearance to the downy bearded tree quail illustrated here.

Two quarter-grown specimens of the buffy-crowned tree quail in the Field Museum of Natural History still have down-covered heads and exhibit a very similar pattern, with a rusty brown crown, a brown ear-patch, yellow to buffy cheeks and superciliary stripes, and a yellow throat.

The downy specimen of the bearded tree quail shown in the color plate is from the James Ford Bell Museum of Natural History, Minneapolis, Minnesota.

DISTRIBUTION

The distribution of the three species of tree quails is largely but not entirely complementary, with a limited degree of overlapping in eastern Puebla and Veracruz, where the long-tailed quail and bearded tree quail occur together on Pico de Orizaba and Cofre de Perote (Leopold, 1959). Of the three, the long-tailed tree quail has the largest range in Mexico and occurs in cloud forests on most of the mountain ranges north of the Isthmus of Tehuantepec northward to northwestern Jalisco and the vicinity of Orizaba, Veracruz. Further northern extension on the western part of the range is presumably blocked by a break in the mountains (to about three thousand feet) at 21 degrees north latitude near Guadalajara, but no equivalent barrier blocks possible northward extension in Veracruz. Perhaps, however, competition with the bearded tree quail in this area has prevented such range expansion.

The bearded tree quail occupies a comparable cloud forest habitat of the Sierra Madre Oriental from San Luis Potosí southward through eastern Hidalgo and eastern Puebla to Veracruz, at the eastern end of the Sierra Volcanica Transversal. Its range may extend somewhat farther south than Orizaba, but in any case probably occurs no farther than central Veracruz, where there are apparently breaks in the cloud forest (Martin, 1955). In southern Oaxaca the long-tailed tree quail likewise reaches the southern limits of its range, probably near La Cima (Binford, 1968).

The buffy-crowned tree quail occurs only south of the Isthmus of Tehuantepec, and Leopold (1959) indicates its range as including only the Sierra Madre de Chiapas of extreme southern Chiapas. However, Miguel Alvarez del Toro informed me* that not only has he observed the birds in cloud

*Miguel Alvarez del Toro, 1970: personal communication.

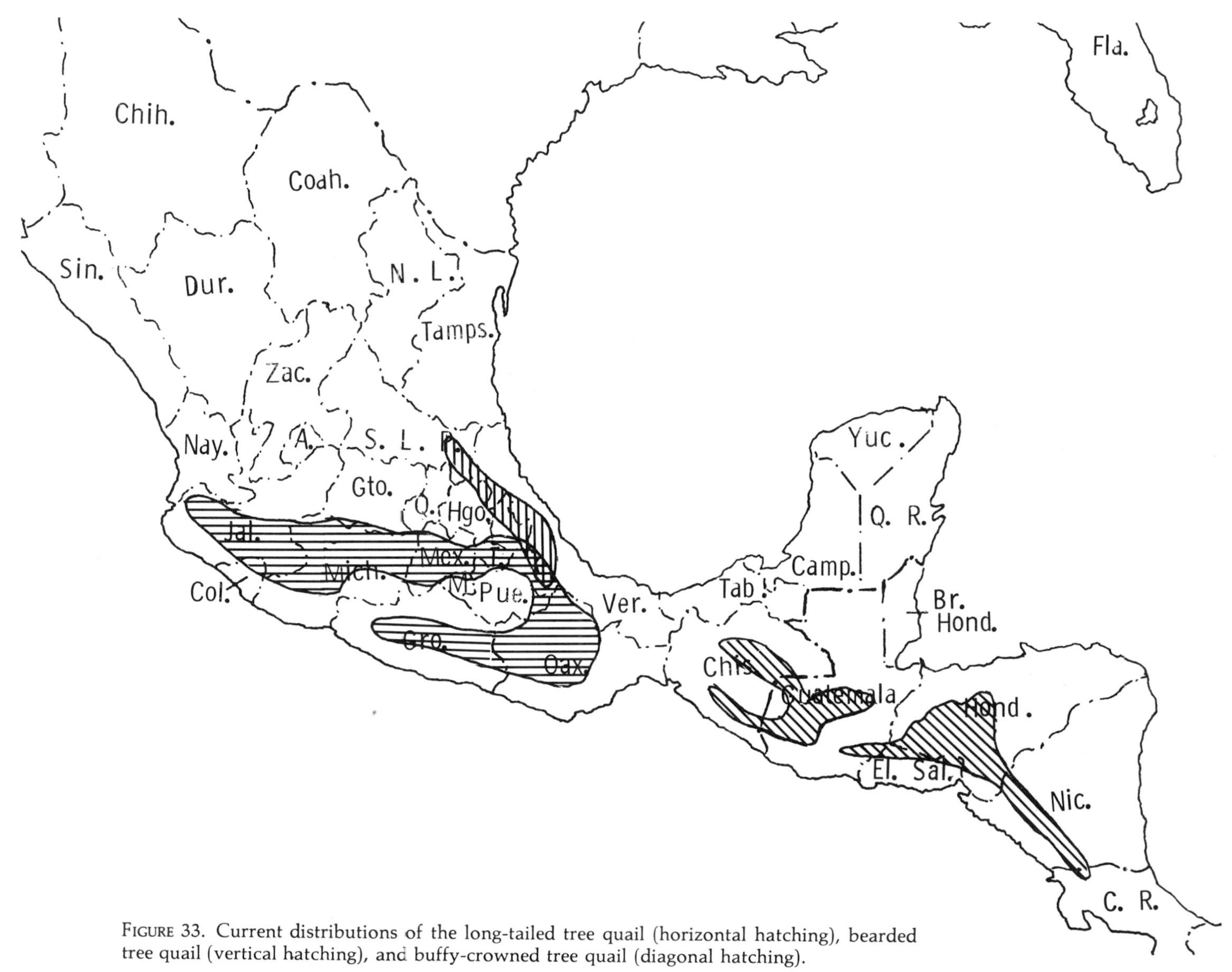

FIGURE 33. Current distributions of the long-tailed tree quail (horizontal hatching), bearded tree quail (vertical hatching), and buffy-crowned tree quail (diagonal hatching).

forests of that area (above Mapastepec, at two thousand meters elevation), but also in pine forests near Jitotól, north of Tuxtla Gutierrez, on the Gulf drainage. He had also heard of a specimen's being brought into a mission school at Pueblo Nuevo Solistahuacán, not far from Jitotól. An extensive area of cloud forest occurs between this village and Tapilula, and at Tapilula I was told by a well-informed local resident who keeps birds that tree quails are not uncommon in the nearby forests and that he had sometimes bought young birds that were brought in to him. This area would probably represent the extreme northern limits of the species' total range.

HABITAT AND POPULATION DENSITY

The preferred habitat of all three tree quails consists of cool, moist montane forest, particularly cloud forests. However, they also occur in the moister pine or pine-oak forests that are usually adjacent to the cloud forests of Mexico.

The habitat of the buffy-crowned tree quail south of Mexico has been variously described by ornithologists. Dickey and van Rossem (1938) indicate that in El Salvador the bird is found from the upper limit of the arid lower tropical zone to an elevation of at least 8,000 feet in the humid upper tropical zone but is most common in the arid upper tropical oak association. In Honduras, Monroe (1968) reported that it occurs at about 1,000 meters in pine, oak, and cloud forest but prefers drier habitats. In Costa Rica the species is said to occur in thick growth and brush and possibly also in grassy, parklike montane pastures, but Slud (1964) only personally observed it in Honduras cloud forest at 6,000 feet. In Guatemala, Griscom (1932) reported hearing it frequently in mountain undergrowth above 3,000 feet, while Saunders, Halloway, and Handley (1950) noted that it occurs at from 2,000 to 8,000 feet in second growth and heavy forests, with one reported occurrence at 1,000 feet. Wetmore (1941) found the species in dense "rain forest" (probably cloud forest, since the elevation was nearly 9,500 feet) on the Sierra Santa Elena, Guatemala. Alvarez del Toro informed me* that near Mapastepec, Chiapas, he found it in cloud forest in association with three other rare species, the horned guan (*Oreophaisis derbianus*), the black chachalaca (*Penelopina nigra*), and the quetzal (*Pharomachrus mocinno*). The only estimate of density that I am aware of is for an area of mature pine-oak forest near San Cristobál de las Casas in Chiapas, at an elevation of 7,700 feet, where the breeding density was estimated at one pair in a fifteen-acre study area (*Audubon Field Notes*, 1959, 13:478).

*Miguel Alvarez del Toro, 1970: personal communication.

Little has been written of the habitat of the bearded tree quail. It has long been known to occur in the cloud forests near Jalapa, Veracruz (Gould, 1850). Edwards (1968) lists the species among those of Xicotepec de Juarez, Puebla, where remnant cloud forests still occur at an elevation of about four thousand feet. He also lists it and associated moist-forest species as occurring at elevations of about five thousand feet near Teziutlán, Puebla.

I inquired in 1970 as to the occurrence of the bearded tree quail near Xilitla (elevation 2,300 feet) and was told by residents that it is still fairly common in forest remnants. Stopping at various villages southwest of Tamazunchale to Jacala, Hidalgo, we were also assured by natives that the species was to be found in moist forests in that area of eastern Hidalgo. Near Puerto El Rayo (elevation 5,500 feet) I obtained five specimens of this species, which are perhaps the first obtained from that Mexican state.

The long-tailed tree quail is likewise a species which inhabits cloud forest and adjacent vegetational zones. Schaldach (1963) reports that it occurs in both the arid and humid pine-oak forests of the Volcanes de Colima, and during winter it occasionally is found as low as the lower edge of the arid pine-oak forest, but its normal habitat is in the higher cloud forests. Rowley (1966) reported that it occurred throughout the year in cloud forests near La Cima, Oaxaca. A study of the breeding birds of this vicinity, at an elevation of 6,000 feet, provides the only estimate of population density known to me, namely two pairs per one hundred acres (*Audubon Field Notes*, 1965; 19:598). Binford (1968) indicates that in Oaxaca the species occurs from 5,800 to 9,000 feet, in humid pine-oak and cloud forest habitats. Edwards and Martin (1955) and Edwards (1968) report that the long-tailed tree quail occurs in fir forest and, less commonly, pine-oak south of Lake Pátzcuaro, Michoacán, at an elevation of almost 9,000 feet on Cerro Moluca. Warner (1959) reported that they are found in the least disturbed humid fir-pine-oak forests between 2,800 and 3,300 meters (9,000–10,600 feet) in the Zempoala lagunas south of Mexico City.

FOOD AND FORAGING BEHAVIOR

Little is known of the food requirements of the tree quails. Dickey and van Rossem (1938) list seeds and flower buds among the stomach contents of a specimen of the buffy-crowned tree quail taken in El Salvador. Warner (1959) stated that flowers, flower buds, and small fruits are taken from arboreal perches by the long-tailed tree quail, although it also performs much scratching in the leafy litter. Seeds, vegetable matter, and arthropod remains were also found in specimens he examined. Leopold (1959) noted that a specimen of the long-tailed tree quail that he obtained had a crop full of legume seeds, mostly of tick trefoils (*Desmodium*). Wetmore (1941) noted that the crops of two buffy-crowned tree quails that he shot contained small drupes.

We were told by natives in eastern Hidalgo that bearded tree quail almost never venture out of the dense forest but do visit fields when the black beans are ripening. Of the captive specimens which I had, I found that they not only liked such seeds as black beans and whole corn (both soaked for several hours) but in particular relished soft fruits, such as grapes and bananas. The heavy bills of the tree quails are effective in tearing fruits apart, and they can also handle relatively large seeds such as beans.

MOBILITY AND MOVEMENTS

Virtually nothing is known of possible movements in these birds, but in all likelihood they are virtually sedentary in their mountain-forest habitats. Schaldach (1963) did mention a possible movement to lower elevations during the dry winter season.

Except perhaps at night, tree quails are largely to be found on the ground, and can move through the underbrush with amazing agility. Dickey and van Rossem (1938) stated of the buffy-crowned tree quail: "It has a 'long-legged' appearance with erect posture when unobserved, but on the least alarm will flatten out and dart away through the brush with rapidity and silence. The body is compressed laterally to a point equalled only by some of the rails, and is thus well adapted for slipping through the close growing stems of its usual habitat." Of the same species, Saunders, Halloway, and Handley (1950) noted that it was difficult to flush in short cover, since the birds would quietly run away. In heavy cover they could be flushed more easily but provided only the briefest targets before being lost to view. After landing they apparently continued by running farther, since those that were chased were not flushed again. In the presence of dogs, however, they are more prone to fly into trees, from which they can more readily be shot (Schaldach, 1963).

SOCIAL AND REPRODUCTIVE BEHAVIOR

Except during the breeding season, these birds are usually to be found in small coveys. Miguel Alvarez del Toro informed me* that in his experience the buffy-crowned tree quail was usually found in groups of four to six birds, but as many as about a dozen birds have been seen in a group. Dickey

*Miguel Alvarez del Toro, 1970: personal communication.

and van Rossem (1938) indicated that the birds usually move in small flocks, which break up as the breeding season approaches. These authors located a roosting tree, which was some ten feet higher than the surrounding ones. Every evening a considerable number of birds were known to converge on this tree, since the calls of the entire area became concentrated there. Calling typically occurs both at dawn and again at dusk, in a manner similar to chachalacas (Warner, 1959; Dickey and van Rossem, 1938).

The timing of the breeding season in Mexico is still not entirely clear. Schaldach (1963) observed that nesting of the long-tailed tree quail occurred in Jalisco in June and July, with the first young birds being seen in mid-June. Warner (1959) judged that nesting begins in late April or early May, and he found a nest with fresh eggs on July 1. He also obtained two young birds in early June in the same general area. Rowley (1966) reported finding two nests of this species, the first of them in mid-April, while the date of the second was not indicated.

I was told by natives in eastern Hidalgo that the bearded tree quail in that region nests in May and June, and a source in Tapilula, Chiapas, indicated that buffy-crowned tree quail breed during the rainy period of June and July. Downy chicks of the bearded tree quail have been taken near Xilitla, San Luis Potosí, in June (Lowery and Newman, 1951), and older young of the long-tailed tree quail have been collected in August, September, and December (Leopold, 1959), suggesting a fairly long breeding season for at least that species. Two recently hatched (remiges just beginning to appear) downy young of this species in the United States National Museum were collected in late May at Omilteme, Guerrero.

The breeding season of the buffy-crowned tree quail in Mexico can only be guessed at, since no eggs or young birds have yet been collected there. A female collected during May in Guatemala exhibited a brood patch (Baepler, 1962), and the Field Museum has five quarter- to half-grown young from Honduras, all of which were collected between April and July, suggesting that this species breeds at the same time as do the other two.

The only nests of this species which have so far been found are those of the long-tailed tree quail. Warner (1959) located a nest in a semiopen pine-fir forest which contained a dense growth of shrubs and many young trees, especially firs. The nest was at an elevation of 2,900 meters, on a very steep slope in a brush tangle. Dead branches in this brush tangle jutted out over a two-foot-high rock face, so that a sloping roof for a cavity three or four feet long and two feet wide at the ground was formed. Dead twigs, branches, and leaves had formed a mat on the branches above, making a light-impervious roof. Only a single opening about six inches wide led to the nest, from which two trails diverged into the forest. The nest, about twelve inches from the opening, was a shallow depression lined with fine grasses. Six eggs were present about twenty-five hours after four eggs had been counted the previous day.

Two additional nests of this species were found by Rowley (1966). One consisted of a few dead leaves and needles in a slight depression at the base of a small shrub, while the other was located amid a rock outcrop in a crop clearing. Both were poorly concealed, and each contained four slightly incubated eggs. The eggs of this species are pale buffy to cream-colored, with small light brown to reddish brown spots, according to these authors. Schaldach (1963) reported seeing young numbering five and seven following their presumed mothers, suggesting that up to seven eggs are normal clutch sizes. Dickey and van Rossem (1938) were told by natives that the buffy-crowned tree quail nests on the ground and lays four or five eggs. I was told by a native in eastern Hidalgo who was keeping captive bearded tree quails that they nested in litter on the ground, laying three or four eggs. A captive pair that I exported from Mexico and placed in the care of Frank E. Strange constructed a fairly simple nest of palm leaves around a depression they had dug in the corner of their cage. The female laid a total of sixteen eggs (which were removed as they were laid), at intervals of from one to eighteen days. Five of these eggs hatched after incubation periods of from twenty-eight to thirty days.

The eggs of the buffy-crowned tree quail apparently average slightly smaller than those of the long-tailed tree quail (44 by 33 mm compared with 49.2 by 33.5 mm) and are reddish buff, with spots and blotches of reddish brown (Leopold, 1959). The eggs laid by the bearded tree quail pair were all a uniform dull white color, and averaged 46.6 by 31 mm (five eggs).

Vocal Signals

The vocalizations of the tree quails are perhaps the most impressive of all the New World quails. Rowley (1966) reported hearing the long-tailed tree quail singing particularly in late evening but also in early morning hours, throughout the year but especially during the spring. Warner (1959) indicated that singing of this species was first heard in February and lasted until July, but singing occurred only at dawn and dusk and was rare even at those times. Griscom (1932), quoting Anthony, indicated that the buffy-crowned tree quail calls at all hours, and the Guatemalan native name *guachoque* or *guachoco* is derived from its call. In Costa Rica the species is known as the *chirascuá* (Slud, 1964), apparently for the same reason.

Warner (1959) has provided a description of the long-tailed tree quail's typical morning and evening calls. He indicated that a series of soft, gutteral hooting sounds precedes the louder calls, but can only be heard at close distances. These preliminary notes sound like *whoop, whOOp, whOOOp, WHOOP*, and are followed by a loud, ringing *koor-EEE-oh, koor-EEE-oh, koor-EEE-oh, koor-EEE-oh*. Schaldach (1963) noted that when a bird is treed by a dog, it typically utters grouse-like alarm notes.

Tape recordings made by L. I. Davis in Guerrero and filed in the Laboratory of Ornithology's Library of Natural Sounds are from birds of unknown sex. The typical calls are piercing screams, the phrases averaging slightly under one second apart and sounding to me like *ko'-or-eee'*, in a series of as many as fourteen phrases. From two to four *whoop* notes may precede such a series but also may serve to connect two series of phrases. On this tape two birds sometimes sing simultaneously, but not in an obviously structured duet.

Little has been written on the calls of the bearded tree quail, but I obtained several recordings from captive birds while in Mexico. Both sexes utter loud, piercing distress calls when held in the hand. These down-slurred whistles often alternate with or teminate in rattling calls that have strongly pulsed characteristics. When disturbed, the same rattling calls are usually used, and they sometimes also utter a very faint, rising whistle when they appear to be agitated.

To determine the effects of separation, I removed two females from their mates and placed them outside about twenty feet away, then hid myself. Within fifteen minutes one of the females emitted some very soft notes, which were immediately answered by one of the males. This rapidly built up into an alternated call-and-answer series of notes, and the other pair of birds soon joined in, making a terrific din. The separation call of the male was a three- or four-syllable sound, which although seemingly pulsed, actually consisted of continuous sound energy. It sounded like *ko-orr-EE-EE* or *ko-or-EEE*, with the last syllable or two of somewhat higher pitch and amplitude. The native name *Chiv-is-coy'-o* no doubt refers to this call. The sound is a nearly pure whistle, with most of the energy at a frequency of about 2,000 Hz. The females' answering note was a somewhat weaker and slightly higher-pitched sound, centering slightly above 2,000 Hz. It lasted about one second, and sounded like *ko-or-ee-ee-ee-eee*, with a varying number of *eee* syllables, and with most of the notes of about the same frequency and amplitude. This separation chorus lasted for at least ten minutes, and when I appeared again to stop the recorder the males continued to pace their cages and call loudly, in spite of the termination of the females' calls.

The morning and evening chorus of these captive birds was heard on only a few occasions, when they were placed in unusually quiet and undisturbed surroundings. Because of the massed singing by most or all of the five birds, it was almost impossible to get a clear recording of the individual calls, which generally sounded much like the separation calls. The whistle notes were, however, preceded by one or more preliminary softer calls that no doubt correspond to those described by Warner for the long-tailed tree quail, and thus the complete call sounded something like *whoop, whoop, KO-OR-EE.* Calling started before sunrise and terminated when it became fairly light outside, usually within fifteen or twenty minutes. It is clear that females as well as males participate in these choruses, which probably function to allow individuals or flocks to locate one another for evening roosting or to announce flock locations at dawn.

EVOLUTIONARY RELATIONSHIPS

Holman (1961) judged on osteological evidence that the genus *Dendrortyx* is the most generalized of the entire group of New World quail, and this would seem to be an attractive idea. The view that the New World quails evolved from a semiarboreal offshoot of the early cracid stock in a forested semitropical environment would fit very well with the present zoogeographic relationships of these birds and also with similarities in the behavior patterns of these two galliform groups. The Cracidae appear to be largely fruit-eaters but do also scratch about in litter for other vegetable matter and arthropods. Unlike the tree quails, they nest largely in bushes or trees but do sometimes nest on the ground. They also have fairly small clutches, and the young may be attended by both parents or by the female alone. The well-developed tails and the importance of vocalizations of cracids are further similarities with the tree quails, but probably reflect in part similar niche adaptations to a common arboreal existence in heavily vegetated habitats.

The nearest relatives of the tree quails appear to be the forest-dwelling wood quails of the genus *Odontophorus,* which differ from them primarily in their adaptations for a more highly terrestrial existence. Similarly, the plains- and desert-adapted species of New World quail might be derived from a tree quail ancestral type, although less directly. The barred quail seems to have a definite evolutionary affinity with the tree quails; indeed, Holman (1961) regards *Philortyx* as the possible nearest living relative of this group. However, *Philortyx* exhibits fewer ecological and general behavioral similarities to the tree quails than do the species of *Odontophorus,* and both downy and adult plumage patterns of the tree quails and wood quails are very similar.

20

Barred Quail

Philortyx fasciatus (Gould) 1844

OTHER VERNACULAR NAMES

BANDED quail, Chorrunda, Codorniz listada.

RANGE

Resident in semiarid tropical scrub of the Pacific slope from southwestern Jalisco to southeastern Guerrero, and inland to Morelos and Puebla, Mexico.

SUBSPECIES

None described.

MEASUREMENTS

Folded wing: Adults, both sexes, 94–104 mm (no consistent differences in the sexes).

Tail: Adults, both sexes, 58–68 mm (no apparent sexual differences).

IDENTIFICATION

Adults, 8–8.5 inches. The sexes are almost identical in appearance. This small quail of the arid parts of western Mexico has a brownish to grayish white head with a nearly straight crest of several feathers that are barred with black and brown. The body is generally grayish brown to gray, with the flanks and breast strongly barred vertically with brownish black and white.

FIELD MARKS

The heavy vertical barring of the flanks, and the generally grayish barred coloration throughout are distinctive field marks, as is the relatively straight crest of the same coloration. The elegant quail is somewhat similar to the barred quail and their ranges may partially overlap, but the elegant quail has spotted flanks and a larger, more differentiated crest.

AGE AND SEX CRITERIA

Females are difficult to distinguish from males, but on the basis of a few captive birds they apparently have a more brownish rather than grayish head and neck than do males, and the crest of females is shorter and less recurved than in males. However, examination of skins does not show consistent separation by these criteria. Judging from twenty-nine specimens, adult individuals with maximum crest length of 25 mm or longer are males, while those with crests of under 21 mm are females, but crests of intermediate length occur on both sexes.

Immatures have their outer two primaries relatively pointed and frayed and have buffy-tipped upper primary coverts (Leopold, 1959). Up until 4½ months of age, black feathers persist on the cheeks, throat, chin, and forehead.*

Juveniles acquire black facial feathers which (in captive birds) first appear at sixty-three days of age.* Prior to that time their throats are whitish and white shaft-streaks are evident in their body feathers.

Downy young of this species (illustrated in color plate 110) have not been described heretofore, but three specimens of known age (ranging from two days to three weeks old) in the Los Angeles Museum provide an excellent basis for description. In all, the throat is bright mustard yellow, darkening to olive yellow on the breast, sides, and abdomen, and to dull

*F. E. Strange, 1970: personal communication.

brown on the legs and under the tail (this dull brown extends considerably farther forward in the youngest bird than in the older ones, in which it is progressively less conspicuous). The back color is a dull chestnut, nearest that of *Colinus* but lacking any indication of darker or lighter streaking, being similar in this regard to *Dendrortyx* downies. The crown is a brighter chestnut than the back and is indistinctly margined with blackish, again similar to that of *Colinus*. No crest is evident, even on the oldest specimen. A conspicuous ochre-tinged superciliary stripe and subocular patch are present, between which a large brown auricular mark occurs, somewhat larger than in the mountain quail and similar in size and shape to that occurring in downy spotted wood quail. Juvenal primaries approximately 1 cm long are already present in the intermediate-aged (about one week old) specimen, and scapulars with conspicuous buffy shaft-streaks that widen at their tips, are starting to appear in the three-week-old specimen. In summary, the *Philortyx* natal pattern provides an excellent transition point between the relatively unpatterned and presumably primitive condition of *Dendrortyx*, and the highly patterned condition existing in *Oreortyx*, *Colinus*, and *Callipepla*.

DISTRIBUTION

Although the barred quail is often said to be limited to the "highlands" of western Mexico, it is in general much more typical of low and arid country, such as rain-shadow areas and interior river basins. Leopold's range map (1959) illustrates this situation, and he mentions that although the birds may occur at elevations of up to 5,000 feet or more, the densities are greater at lower elevations. To point out this altitudinal relationship more clearly, a more detailed range map has been prepared, with elevations above the 1,600-meter (5,200-foot) contour level indicated by shading. The predominance of records at lower elevations is quite clearly apparent in this map, as is the relationship of the species' distribution to the Río Balsas, Río Tepalcatepec and Río Armería drainage basins. Except for these major river systems, the species is largely limited to coastal regions, where it is abundant in thorn forest or tropical scrub habitats.

The species extends into Jalisco at the northern edge of its range (Schaldach, 1963) and to the south extends virtually to the Guerrero-Oaxaca border. I know of no specimens from Oaxaca, nor does Binford (1968) list the species for that state, but potential coastal scrub habitat does occur at the western limit of the state. At Copala, Guerrero, I was told by local residents that both barred quail and bobwhites occur (by showing them live specimens of each and asking them to comment on them),

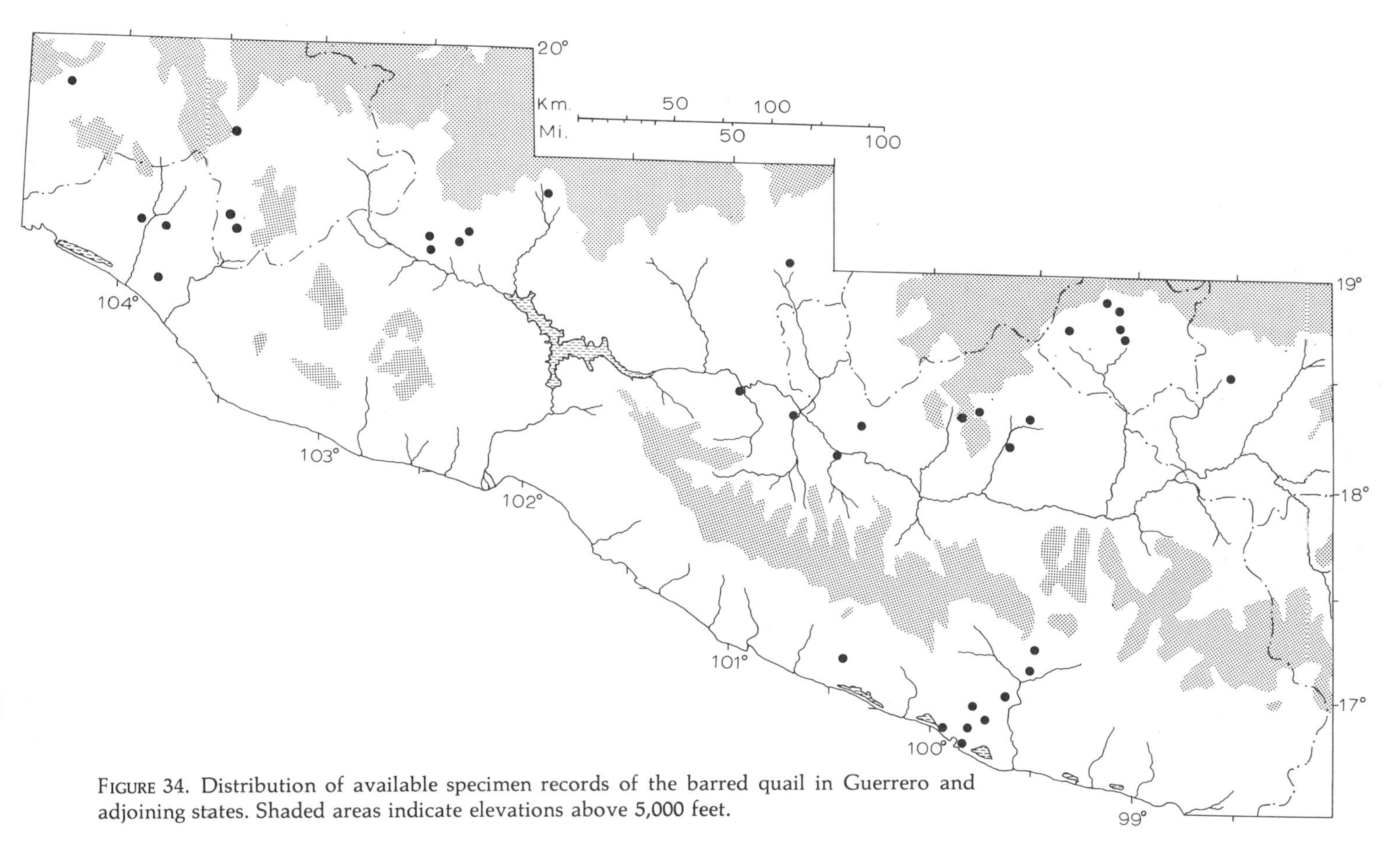

FIGURE 34. Distribution of available specimen records of the barred quail in Guerrero and adjoining states. Shaded areas indicate elevations above 5,000 feet.

however, persons from Pinotepa Nacional, Oaxaca, who were asked to identify these birds were not familiar with the barred quail.

I have seen no specimens of barred quail from the state of Mexico, but it is listed as occurring there by Friedmann, Griscom, and Moore (1950).

HABITAT AND DENSITY

Schaldach (1963) describes the preferred habitat of this species as open lowland thorn forest near cultivated fields, but also occurring as high as 5,000 feet on the Volcanes de Colima, Jalisco. Leopold (1959) noted that the birds were most abundant in weed-bordered agricultural areas in the valleys and alluvial flats of the lower Río Balsas basin. At Emiliana Zapata, near Cuernavaca, I found barred quail in an area that consisted of irrigated sugar cane and rice fields alternating with thorny brush thickets, but the birds were far more abundant near Acapulco, where they inhabited ungrazed or lightly grazed coconut plantations that contained brush thickets of mimosa and staghorn acacia that largely covered depressions or excavations where the moisture availability was favorable. In these nearly impenetrable thorn thickets the birds found shade and protection, and weedy herbs no doubt provided abundant food in the form of seeds and green leaves.

Leopold (1959) estimated that in favorable situations the density of barred quail probably exceeded one per acre, and coveys were spaced as closely as 150 to 200 yards apart. I flushed several coveys less than 100 yards from one another and would agree that at least one bird per acre would be a reasonable estimate of density in ungrazed or lightly grazed plantations near Acapulco. Barred quail appeared to be more common near waterways, but this probably reflected the increase in woody or brushy cover, rather than any need for drinking water.

FOOD AND FORAGING

Leopold (1959) examined a considerable number of these birds and found a wide variety of weed seeds, especially those of legumes such as *Desmodium* and *Crotalaria*. He also found a variety of seeds from various weedy herbs, such as sunflowers, thistles, corn cockle, and doveweed (*Croton*). He noted that beans and sesame seeds were also present in the crops. I found that a captive specimen preferred smaller seeds than those taken by bobwhites in the same cage, and it particularly liked sesame seeds. Barred quail also probably consume some insects, judging from observations of Blake and Hanson (1942).

It is doubtful that the birds do any arboreal foraging, but they often, if not regularly, roost in trees. On several occasions I flushed small coveys out of trees in early morning hours, and once a flock of about ten birds flew into some trees in late afternoon after being flushed. Leopold (1959), however, noted that they may also roost in open low grass or weeds.

MOBILITY AND MOVEMENTS

Judging from my limited observations, the barred quail are highly sedentary. Most flights made by the birds were relatively short, under one hundred yards in length, and it was possible to return to the same area a day or more after flushing a covey and be confident of finding it not far from the earlier point of observation. The birds seemed highly reluctant to run, probably because of the irregular distribution of thorny bushes, and once hidden in such shrubs they could scarcely be flushed without thrashing the shrubbery with sticks, throwing objects such as coconuts into the middle of the brush, and making a considerable amount of noise.

Davis (1944) noted that in areas of sparse cover the quail tended to seek cover by flying a short distance, landing, and then running. More typically, they flew to the nearest dense brush and hid among small trees. When flushed from one tree they would simply fly to another.

SOCIAL AND REPRODUCTIVE BEHAVIOR

The barred quail is highly social, and rarely will single birds remain alone when separated from a covey. Indeed, I have sometimes seen a single bird flush from a brush patch in which a covey was hiding, hover in the air, and drop back down again in the brush if its flockmates did not take off. When flushing the birds utter a squealing *pee-urr* sound that is highly distinctive and is the basis for the local name *chillona*, or screamer, in the area around Cuernavaca. The other local name, *chorrunda*, is apparently onomatopoetic.

Leopold noted that the average size of eighteen coveys he observed was twelve birds and ranged from five to twenty. These were apparently fall and winter coveys, containing perhaps 50 percent young birds, and were considerably larger than the coveys I saw in June, which consisted of mature birds that had not yet separated into pairs for breeding. The average of seventeen coveys (some of which were doubtless repeats) was 5.8 birds, and the range was from 3 to 10 birds. This reduction of average covey size by at least half prior to breeding would suggest a fairly high annual mortality rate for the species, certainly in excess of 50 percent.

It is clear that the breeding season of barred quail is unusually late, and certainly later than that of bobwhites breeding in the same area. I observed thirty bobwhites in the vicinity of Acapulco, all but eight of which were already paired, whereas none of the approximately one hundred barred quail I observed there between June 17 and June 25 were yet paired. Blake and Hanson (1942) collected a female barred quail with a hard-shelled egg in August in Michoacán, and Davis (1944) collected birds in breeding condition near Acapulco in early August. We were likewise told by natives near Cuernavaca that this species nests in August and lays fourteen to sixteen white eggs. None have been found in the wild by ornithologists to my knowledge, and Leopold (1959) had no information on nests or eggs.

Although the timing of egg-laying by captive birds cannot be regarded as typical of wild ones, it is interesting that the captive barred quail of F. E. Strange, Redondo Beach, California, have shown such a delayed breeding period. In 1967, the first year he had the wild-caught birds, they laid three eggs in early August. In 1968 they laid seven eggs in sixteen days, starting July 30. In 1969 they also laid seven eggs in sixteen days, starting July 26. In 1970 the laying began about a month earlier (July 1) than in the previous years, with seven eggs being laid. A second pair laid five eggs between August 25 and September 14, 1970. The eggs are entirely white, with five having measurements of 23–24.5 mm by 30–31 mm, averaging 23.7 by 30.2 mm. The eggs were laid in a simple nest in grass, which was slightly roofed over. Sixteen of these eggs had an average incubation period of 22.6 days, ranging from 21 to 23.

The role of the male in caring for the nest and young has not yet been determined, but the highly social nature of this species would lead one to expect that both sexes might attend the brood.

Vocal Signals

The characteristic squealing *pee'-urr* notes, which are uttered with a downward slur, can be heard nearly every time a covey is flushed. When individual birds are separated from the covey after flushing, they regain contact by uttering soft *cheep-cheep'* whistled calls. On one occasion I saw a bird that had been separated after flushing stand near the edge of an acacia thicket with crest erect and utter a fairly loud *ca-ut'-la* call over twenty times in fairly rapid succession, which finally terminated with two whistled *pee'-urr* notes. This three-noted call, which I heard on only a few occasions, would seem to be the basis for the Spanish name *chorrunda*. It may well correspond to the *pay-cos* call of the scaled quail and the comparable *chi-ca-go* call of the California quail, for it has very similar sound characteristics and may fulfill a comparable separation call function. Zimmerman and Harry (1951) heard *pip-pip-pip* notes uttered by birds fleeing on foot and described the flight call as *pee-pee-pee-eeee.*

The single bird which I have kept in captivity has produced several calls, but their functions are uncertain. Judging from crest length and the appearance of the head plumage, the bird seems to be a male. Often in late afternoon or early morning it will utter a long series of *ca* or *cow* notes, like the first note in the *ca-ut'-la* call described above. These series may consist of as few as ten or as many as twenty-six individual notes, all uttered at about the same amplitude and frequency, but increasing slightly in cadence. One recorded series of thirteen notes lasts seven seconds, while another of fifteen notes lasts eight seconds. The male could be stimulated to utter this call by our playing it back to him, and this, plus the conditions under which it was normally uttered, made me believe that it serves as a location-announcement call.

I was never able to elicit a typical hand-held distress call from this bird, but sometimes while being held or otherwise disturbed it frequently utters a low, rattling note, quite similar to that produced by the bearded tree quail under the same circumstances. However, a juvenile that I once handled produced a long series of typical quail distress notes.

The general sounds of the non-whistled *ca* and *ca-ut'-la* calls of the barred quail are surprisingly similar to calls such as the *pay-cos* of the scaled quail, and I heard a few calls of interest that were produced by some barred × scaled quail hybrids, reared by Alvaro Aragon at the Estación de la Fauna, Centro de Investigaciones Basicas, Progreso, Morelos. These birds, when disturbed, produced a series of low rattling or "chittering" notes, apparently comparable to the barred quail's rattling notes mentioned above and the repeated *pit* sounds of disturbed scaled quail. As I left the cage, one of the males uttered five rapid nasal notes with a slight head-throw as each note was uttered, doubtless representing the "head-throw" call of the scaled quail. One hybrid female was held in the hand, but did not utter a distress call. These twelve hybrids are the only ones that have yet been reported involving the barred quail. Since none of the females, then two years old, had yet laid any eggs, it may be presumed that they are completely infertile.

EVOLUTIONARY RELATIONSHIPS

The barred quail seems to represent an important "transition" species, with some traits, primarily skeletal, that appear to be primitive and suggest affinities with the tree quail group. In its social behavior and its vocalizations, however, it appears to be closer to the *Callipepla* group of arid-adapted terrestrial quails, which it also resembles in its ecology. There would seem to be no good reason for merging *Philortyx* with *Callipepla*, as Delacour (1961) has suggested doing; if anything the species may be more closely related to *Oreortyx* or *Colinus* than to *Callipepla*, but in neither case does the relationship appear to be particularly close.

The adult plumage pattern of the barred quail is probably more like that of the scaled quail than any other species, and the plumages of the hybrid barred × scaled quail mentioned earlier emphasize the great similarities in adult plumage patterns. The appearance of a black throat in the juvenal plumage is unique to the barred quail and presumably represents an ancestral trait that has been suppressed in adults. Black throats are, of course, found in many New World quails, such as the long-tailed tree quail and adult males of Gambel and California quails, as well as in some races of bobwhites.

21

Mountain Quail

Oreortyx pictus (Douglas) 1829

OTHER VERNACULAR NAMES

CODORNIZ de Montana, mountain partridge, painted quail, plumed quail, San Pedro quail.

RANGE

Resident in the western United States from southern Washington and southwestern Idaho east to Nevada and south to Baja California. Also introduced in western Washington and western British Columbia (Vancouver Island). Introduced but of uncertain status in western Colorado.

SUBSPECIES (*ex A.O.U. Check-list*)

O. p. pictus (Douglas): Sierra mountain quail. Resident in mountain regions of extreme western Nevada west to the west side of the Cascade Range in southern Washington and south to the Sierra Nevada and inner Coast ranges of California.

O. p. palmeri Oberholser: Coast mountain quail. Resident from southwestern Washington south through western Oregon to northwestern San

Luis Obispo County, California. Also in southern Vancouver Island, British Columbia.

O. p. confinis Anthony: San Pedro mountain quail. Resident in lower California in the Sierra Juarez and Sierra San Pedro Martir.

O. p. eremophila van Rossem: Desert mountain quail. Resident in the mountains of southern and west central California in the Sierra Nevada south to the Baja California boundary and somewhat beyond and in extreme southwestern Nevada.

O. p. russelli Miller: Pallid mountain quail. Resident in the Little San Bernadino Mountains in Riverside and San Bernadino counties, California.

MEASUREMENTS

Folded wing: Adults, both sexes, 125–40 mm (males average 2 mm longer than females).

Tail: Adults, both sexes, 69–92 mm (males average 4 mm longer than females).

IDENTIFICATION

Adults, 10.6–11.5 inches long. The sexes are very similar in appearance. This relatively large western quail differs from all others in that both sexes have straight, narrow, and blackish crests composed of only two feathers, which appear with the juvenal plumage. The throat is chestnut, edged with black, and this is separated from the slate gray chest, neck, and head by a white line. Otherwise the birds are plain olive gray on the back, wings, and tail. The flanks are a rich, dark brown, with conspicuous vertically oriented black and white bars.

FIELD MARKS

The slender plumes and boldly patterned flanks will serve to identify mountain quail without difficulty. The California quail may occur in the same areas but has a shorter, curved crest of "teardrop" shape and dull brown flanks that are narrowly streaked with white. A loud, clear, whistled *quee-ark* or *plu-ark* is the advertising call of the male during spring.

AGE AND SEX CRITERIA

Females have slightly shorter plumes than males (average of twelve is

58 mm with a maximum of 66 mm, as compared to a minimum of 66 mm and an average of 72 mm for twelve males) which are also browner. Ormiston (1966) reported that nine adult females averaged 62.1 mm and ten males averaged 85.3 mm in crest length, but he did not find this difference statistically significant. McLean (1930) reported that in addition to having a longer crest, the male is more brightly colored beneath and the gray of the hind-neck is more sharply defined than is that of the female. Schlotthauer (1967) likewise noted that in females the brown back color extends to the top of the head, while in males the back of the neck is grayish blue. F. E. Strange* believes that the neck color is the most reliable criterion but has limited use with dark coastal birds.

Immatures have buff-tipped greater upper primary coverts, as compared with the uniformly gray coverts in adults (van Rossem, 1925). The two outer primaries are more pointed and frayed than the inner ones.

Juveniles have dull fuscous crest feathers (under 60 mm) of which the terminal third is banded with tawny drab (Ridgway and Friedmann, 1946) and have whitish chins surrounded by dark gray throats.

Downy young (illustrated in color plate 110) of this species are quite distinct from *Callipepla* downies and approach *Colinus* in some respects. Besides being slightly larger than any of these, mountain quail downies exhibit more whitish tones, especially on the sides of the head and body, and particularly just below the chestnut crown. The black-bordered chestnut color is also present on the back as a middorsal stripe, which in *Callipepla* is a pale buff or dull mummy brown. A second blackish stripe, separated from the middorsal stripe by a white line, occurs above the legs, and black is also evident on the upper neck region. There is a large blackish mark extending from the rear of the eye to the ear region, where it expands considerably in size.

DISTRIBUTION AND HABITAT

The mountain quail is perhaps the most temperate-adapted of any species, inasmuch as it is the only United States quail species that barely extends its range into Mexico, and thus is limited to the extreme northern part of the Baja peninsula. Like the montane tree quails, it is larger than the species of the arid lowlands, although it does not quite reach the body size of *Dendrortyx*. Nevertheless, it occupies a comparable climatic zone, being found in dense brush, in coniferous forests, around the edges of mountain

*F. E. Strange, 1970: personal communication.

FIGURE 35. Current distributions of the mountain quail (shaded) and barred quail (hatched).

meadows, and sometimes on fairly high crests (Leopold, 1959). During the breeding season the vertical distribution of *pictus* and *eremophila* in California is from about 1,500 or 2,000 feet to 9,500 or 10,000 feet, although the coastal form *palmeri* occurs only up to 5,600 feet (Grinnell and Miller, 1944). The habitats of these three subspecies in California include brushy mountainsides, particularly those covered with chaparral vegetation, such as manzanita, snowbush, chinquapin, and similar broad-leafed hardwoods. Coniferous forest edges, open forests, or forests disturbed by logging or fires provide additional habitat for this species. The desert mountain quail extends its breeding range into sage, piñon and juniper vegetation where water is available locally (Grinnell and Miller, 1944), and the vertical range of mountain quail in the Sierra Nevadas extends lower on desert-facing slopes than on moister ones (Sumner and Dixon, 1953). Sumner and Dixon indicate that brushy areas of California black oak (*Quercus kelloggii*) and ponderosa pine (*Pinus ponderosa*) are favored breeding habitats, while the lower blue oak (*Q. douglasii*) zone is used in winter. McLean (1930) stated that the Sierra form of mountain quail is most often associated with white-leafed and mariposa manzanita (*Arctostaphylos vicida* and *A. mariposa*), often dropping down in winter to the chamise or greasewood (*Adenostoma fasciculatum*) zone. However, the coastal form is generally found in the dense undergrowth of the redwood (*Sequoia sempervirens*) belt.

In Washington, where the species was introduced in the late 1800s (apparently primarily from *palmeri* stock), it inhabits brushy burns and clearings, brushy canyon thickets, and areas near farms and woodland borders (Jewett et al., 1953). In Oregon the coastal race *palmeri* likewise inhabits cutover lands and edges of clearings in the humid forest zone, while the interior race *picta* is found in more open country (Gabrielson and Jewett, 1940).

The species was introduced into British Columbia in the 1870s and 1880s and currently persists only on Vancouver Island, where it is sometimes fairly common (Guiguet, 1955). In western Idaho the mountain quail may or may not be native, but it occurs along the lower parts of several river systems, including the Snake, Boise, Clearwater, and Salmon (Ormiston, 1966). It also occurs sparsely in the northern and western parts of Nevada (Gullion and Christensen, 1957), possibly also representing introduced stock. Beginning in 1965, a series of releases of mountain quail were made at the western edge of the Uncompahgre Plateau, Mesa County, Colorado (Colorado Outdoors, 15[6]:1, 1969). Subsequent sightings of the birds have indicated considerable survival, and a possible establishment of the species has been attained.

POPULATION DENSITY

Few estimates of population densities of mountain quail have been made. Edminster (1954) cited California research indicating an early spring density of one bird per three acres following a winter of high survival, and near water densities of up to one bird per two acres occurred.

In the fall, in areas where the average covey size is relatively high (eleven birds), the late summer and fall density of birds may reach one bird per five acres (*P. R. Quarterly*, April, 1950, p. 136).

HABITAT REQUIREMENTS

Winter habitat of the mountain quail typically consists of mixed brush and herbs, with the brushy species including such plants as chamise, Fremont silk-tassel, manzanita, scrub oaks, and other species (Edminster, 1954). Edminster judged that snow cover was not usually important in winter survival, since the bird can use shrubs and trees for sources of food when herbaceous vegetation is covered. Snow may, however, be important in the northern parts of the range or set an upper altitudinal limit for winter survival in mountainous country. In a winter of unusually cold weather and heavy snowfall, no noticeable decrease in wintering quail was seen in two California study areas (*P. R. Quarterly*, July, 1949, p. 307).

In spring, the birds return to their breeding habitats and seek out suitable nesting areas. Edminster (1954) indicated that the birds prefer moderately open brush and tree cover on slopes. Woody cover shading from one-quarter to one-half of the ground was regarded as being best for nesting and roosting. Where the mountain quail nests in desert habitats, it is often associated with such woody plants as juniper (*Juniperus*), thornbush (*Lycium*), black brush (*Coleogyne*), and desert apricot (*Prunus*), and cover apparently is not a limiting factor (*P. R. Quarterly*, October, 1948, p. 408). Hillsides of at least a twenty-degree slope are used by birds to escape by running uphill, and such slopes serve purposes similar to that provided by plant cover. In desert areas the availability and distribution of water is probably important; the birds are apparently restricted to remaining no more than a mile from water (*P. R. Quarterly*, January, 1948, p. 11).

Nesting cover in various parts of the California range varies greatly as to plant species, but most such cover contains large shrubs, trees, or both, usually in dense growth formation. Mixtures of trees and shrubs may be more valuable than either alone, perhaps because of decreased density in the shrub layer. Small trees are more useful than are large ones for roosting, and the mast from trees such as ponderosa pine, firs, and oaks provides important food. Roads in unusually dense cover may provide useful clearings where dusting may occur and young birds can dry out and warm up early in the morning (*P. R. Quarterly*, October, 1949, p. 459). Nesting range may possibly be selected on the basis of abundant green plant food, which may occur on flatlands adjacent to wooded hills (*P. R. Quarterly*, October, 1948, p. 408).

In the central Sierra Nevadas, nesting occurs both in the foothill chapparal belt and also at high elevations near timberline. The foothill nesting population is composed of a sedentary population, whereas the timberline nesting population moves upward every year from the foothills through a heavily vegetated forest zone where few quail nest. These birds nesting in higher elevations evidently are much more dependent on available free water than are the foothill nesters; their nests are usually no more than a few hundred yards from it, and they frequently visit watering places. However, the foothill residents may nest more than a mile from water and not visit watering places until after the young are hatched (*P. R. Quarterly*, October, 1949, p. 459).

Since chicks require water soon after hatching, its availability is an important aspect of brooding cover. Insects and succulent green vegetation are also likely to be abundant near water, as well as shady cover and safe roosting places. Miller and Stebbins (1964) never found adults more than a mile, or young more than half a mile, from water in the Joshua Tree National Monument, and usually they were much closer. They also knew of no nesting success except near springs. Edminster (1954) judged that few broods were raised more than a quarter mile from a source of water. Ormiston (1966) likewise considered that free water was an essential part of mountain quail habitat in Idaho.

Fall habitat needs of the mountain quail include suitable food sources. Edminster (1954) noted that oak-pine stands provide important mast sources, on which the birds feed until the weather forces them to lower elevations. In the western Sierra Nevada range the birds were found in stands of ponderosa pine, California black oak, and mountain misery (*Chamaebatia*) during September, and in early October they were seen in a variety of associations of mixed conifers, oak, and chapparal species where water was commonly present (*P. R. Quarterly*, April, 1950, p. 136).

FOOD AND FORAGING BEHAVIOR

Most of the limited data on mountain quail foods comes from fall collections, such as the analysis by Yocom and Harris (1952). Of thirty-three quail they studied in Washington, smooth sumac (*Rhus*) fruits and seeds comprised nearly a quarter of the diet. Other important sources of fruit food include hackberry (*Celtis*), serviceberry (*Amelanchier*), grape (*Vitis*), gooseberry (*Ribes*), manzanitas, nightshade (*Solanum*), elder (*Sambucus*), Christmasberry (*Photinia*) and snowberry (*Symphoricarpos*). Seeds of trees including those of various pines, Douglas fir (*Pseudotsuga*), and black locust (*Robinia*) are used, as well as acorns, and a host of legume and other weed seeds are also eaten (Edminster, 1954). Tubers and roots are also used to some extent for fall foods and may comprise about 10 percent of the early fall diet but are not eaten much at other times of the year.

Winter foods of the mountain quail consist of acorns and seeds of a diverse array (Martin, Zim, and Nelson, 1951). In addition to acorn meats, pine seeds and greens may also be taken in fall and winter (*P. R. Quarterly*, April, 1948, p. 165).

As greens become available in late winter and spring, they are heavily utilized and may make up from 25 to 40 percent of their diet. Leaves and, later on, buds and flowers are used through the summer, and collectively comprise about a quarter of the annual diet. The yearly average of food from animal sources is only about 3 to 5 percent, with fruit, mast, and seeds making up most of the remainder of the total food intake (Edminster, 1954).

Judd (1905a) provided an analysis of foods from the crops of twenty-three mountain quail collected in California, of which only 3 percent by volume came from animal sources. Grain comprised 18 percent, seeds of legumes, weeds, and grasses totaled 47 percent, fruit comprised 8 percent, and miscellaneous vegetation made up the remaining 24 percent.

The most complete study on mountain quail foods so far available is that of Ormiston (1966), which was based on forty-eight adult samples collected from spring to fall, and twelve samples from young birds. During spring, two early-maturing annual herbs, chickweed (*Holosteum*) and microsteris (*Microsteris*), were the most important foods, with the birds consuming the developing seed heads. Chickweed and blue-eyed Mary (*Collinsia*) seed heads were found in May samples, and barley (*Hordeum*) occurred in large quantities in one May sample. Underground bulblets of fringecup (*Lithophragma*) were found in May samples and evidently became increasingly important in late summer and early fall, when they made up nearly half of the sample volumes. Seeds of grasses, hawthorn (*Crataegus*), pines, and sweet clover (*Melilotus*) were also important fall food sources. Large weedy species such as thistles (*Cirsium*), ragweed (*Ambrosia*), and teasel (*Dipsacus*) provided important fall seed sources as well.

Foods of young mountain quail collected by Ormiston contained only

7.5 percent animal matter, and Lahnum (1944) reported that 20 percent of the food contents of ten young quail was of insect origin, so it would seem that a surprisingly small amount of the food taken by young quail is of animal matter. Flower heads of chickweed and miner's lettuce (*Claytonia*) were the major foods of chicks under a week old, while older chicks began to consume fringecup bulblets, and seeds of miner's lettuce and various woody plant species. By the time the chicks were eight weeks old they fed largely on the dry seeds of various herbaceous species and also continued to feed on fringecup.

By fall, with the ripening of the acorn crop in California, the birds once again begin to concentrate on it. Miller and Stebbins (1964) described how unripe acorns are shelled by the mountain quail. At the green base, where the shell is still soft, the bird opens a hole and tears or cuts away enough of the rest of the covering to extract the meat. Quite possibly the birds pull such green acorns from the trees before they would normally fall to the ground.

MOBILITY AND MOVEMENTS

The unique vertical migration of the mountain quail is no doubt a reflection of the fact that it breeds at higher elevations and in an associated cooler climate than do any of the other North American quail species found in the United States. The migrational movements are fairly leisurely and are normally undertaken on foot, although the birds will sometimes fly across canyons (Leopold, 1959). In the west Sierra Nevada slope the total migratory movement may be twenty miles or more (*P. R. Quarterly*, January, 1951, p. 9).

While in the wintering habitat, daily movements are not great; one study indicated that the maximum was about one thousand yards per day, and the minimum about four hundred yards, as the birds moved from roosting and loafing areas under scrub oaks to forage in low brush (*P. R. Quarterly*, January, 1948, p. 11).

By late February, movement back to the breeding areas begins, with the coveys remaining intact until the nesting range is reached. At this time the males become intolerant of one another, and dispersion of pairs occurs.

Ormiston (1966) found that during the summer daily movements were limited and did not exceed half a mile unless the birds were disturbed. In his study area in Idaho he found little evidence of major seasonal movements, with marked birds remaining within a one-square-mile area at all seasons. The longest move recorded for any marked individual was about one mile, including a 700-foot movement upslope. Sumner and Dixon

(1953) observed surprisingly long flights of about half a mile by disturbed birds, while Miller and Stebbins (1964) saw a bird fly 150 yards upslope at a twenty-five-degree angle.

There is little movement in the summer during brood-rearing; Ormiston (1966) noted that when birds were young, coveys remained in a two- or three-acre area for several days at a time. However, there is a movement toward areas of available water. In late July of 1947 a concentration of several thousand mountain quail occurred at Jackass Spring, Panamint Mountains, Inyo County, California. A similar but smaller concentration occurred at various springs in Joshua Tree National Monument the same month, with a minimum of 730 birds at twelve watering points, or an average of 60.8 birds per spring. When a small amount of rain fell in August, the birds immediately left the springs and were found later two or three miles from water, feeding on new plant growth produced by the rain. Banded birds were seen from one to five miles away from the point of banding during August and September (*P. R. Quarterly*, January, 1948, p. 11). In succeeding years, birds may return to the same water hole. Of seventeen banded birds observed at watering holes a year after banding, most were at the same water hole and none was more than a mile away from the point of banding. Only about 10 percent of the birds banded one summer were seen the following one (*P. R. Quarterly*, October, 1948, p. 408).

Mountain quail probably need to visit water sources only once a day and can hold up to 12 cc of water in their crops (*P. R. Quarterly*, January, 1948, p. 11). In the Jackass Spring area such watering usually occurred after 10:00 A.M., and most usage was near noon. However, in other areas, the birds were seen to come in at all hours of the day but especially during early morning. Ormiston (1966) noted that coveys were usually found near streams between 8:00 and 10:00 A.M., and after they finished drinking they fed, dusted, and finally moved to heavy cover to spend the hottest part of the day. A second period of feeding occurred from late afternoon until just before dark, when the birds went to roost in heavy cover, probably on the ground.

Miller and Stebbins (1964) reported a similar late afternoon visit to water holes during late summer. The birds would arrive on foot in coveys of six to twenty, walking single file, and approach the spring with great caution. When frightened the birds invariably move uphill and prefer to run rather than to fly unless the cover is unusually open.

In the Sierras the movement back down the mountains toward the winter habitat starts in late August or early September, and by the first of October the birds are usually gone from elevations above five-thousand feet, regardless of the weather conditions that may be prevailing (Bent, 1932).

SOCIAL AND REPRODUCTIVE BEHAVIOR

As in all New World quail, the covey forms the basis for the social group for nearly the entire year. Except where drought conditions cause groupings, most coveys are probably basically family groups. In the Sierra Nevada, covey size has been reported to average 7 birds, and in the San Gabriel Mountains 5 birds represented an average covey size (*P. R. Quarterly*, April, 1950, p. 136). The average of twenty-one coveys from late summer through winter at Joshua Tree National Monument was 9.1 birds and ranged from 3 to 20 (Miller and Stebbins, 1964). Coveys consisting of family groups would be expected to average a pair and up to perhaps as many as 10 young, probably averaging about 5 in well-grown broods, assuming a 50 percent brooding loss. Unsuccessful adults probably join such family groups, thus increasing their numbers. In unusually dry years, little or no nesting occurs, and at such times fairly large coveys consisting entirely of adults may be seen in early summer (Leopold, 1959).

In California the mating season begins in March at low elevations or early April higher in the mountains, and mate selection occurs while the birds are still in coveys (McLean, 1930). The onset of mating may be recognized by the location call of unmated males, which is usually uttered from a prominent stump, rock, or branch in a break in the woody cover. This call, a clear whistle that drops slightly in pitch toward the end, and sounds like *quee-ark*, *kyork*, *queerk* or *plu-ark*, can sometimes be heard for three-quarters of a mile (McLean, 1930). Grinnell and Storer (1924) indicate an average interval between calls of about 6 or 7 seconds, and a recorded series in the Cornell University Laboratory of Ornithology's Library of Natural Sounds averaged 8.5 seconds apart over a 6.7-minute period. The head is quickly thrust upward and thrown back and the crest suddenly erected as each call is uttered. Although the call, or a whistled imitation of it, may stimulate other males to respond (Dawson, 1923), it should not be regarded as a territorial proclamation signal. Rather, as in the other quails, it simply represents the announcement of the location of an unmated male, to which available females might be attracted.

As in the other quail, pair formation is probably a fairly simple process, but it has not been described adequately and I have not personally observed it. The strong similarity in the sexes would suggest that sexual recognition in this species may be more difficult than in the genera *Callipepla* or *Colinus*, and one might expect that initial male responses to females would be largely aggressive. The striking flank markings would suggest that lateral displays are important visual signals, and a male hybrid mountain × California quail in my collection has a strongly developed frontal display (without wing-spreading) that exhibits the throat markings very well.

In one California study, male crowing was first heard on February 20, and the first pair seen February 26. By March 6, a total of seven pairs had been located, but some coveys were still present. These all broke up by the end of March (*P. R. Quarterly*, July, 1949, p. 307). As males become antagonistic toward one another, the population spreads out, with a nesting pair occupying from five to fifty acres (*P. R. Quarterly*, January, 1950, p. 10). In the Joshua Tree area, April is the period of nesting (Miller and Stebbins, 1964), with a probable average hatching date in 1948 of May 7 (*P. R. Quarterly*, October, 1948, p. 408). However, in the central Sierras, nesting is from mid-June to mid-July (*P. R. Quarterly*, January, 1948, p. 10). The average clutch size of eleven nests was 10 eggs in one study in the Sierras (*P. R. Quarterly*, January, 1948, p. 10). Grinnell, Bryant, and Storer (1918) summarized literature references on clutch sizes of this species and added their own observations. If two clutches of 19 and 22 eggs are excluded as being the probable result of two females, the average clutch size for twenty-nine clutches would be 8.7 eggs. A few of the smaller clutch records were probably of incomplete clutches, thus 9 to 10 eggs would seem to be a typical clutch size for mountain quail.

Nests are usually well concealed, often being placed under fallen pine branches, amid weeds or shrubs at the base of large trees, beside large rocks in the shade of shrubs, or in masses of shrubby vegetation (Bent, 1932). Nests are usually located near paths or roads and are probably always within a few hundred yards of water. The incubation period is twenty-four to twenty-five days.* The male takes an active role in nest and brood defense and will perform distraction displays such as injury feigning (Bendire, 1892). Males also regularly exhibit brood patches (Miller and Stebbins, 1964), indicating that they might assist with incubation, particularly if the female should be killed. One California study indicated that most broods were led by a single adult, which might be of either sex (*P. R. Quarterly*, October, 1948, p. 408), but broods tended by both adults averaged larger than those with only one present.

So far, there is no evidence that two broods are ever normally raised in mountain quail, although unsuccessful pairs will often make a second or even third attempt to nest (Leopold, 1959). One California study indicated that eight of fourteen nests under observation were successful, and the hatching success of the eggs in successful nests was 95.8 percent (*P. R. Quarterly*, January, 1948, p. 10).

Vocal Signals

The unmated male announcement call is undoubtedly the best-known

*F. Strange, 1970: personal communication.

of the mountain quail vocalizations. Miller and Stebbins (1964) noted that the male's whistled call may also occasionally be heard in October from birds in flocks, which might be a reflection of a fall resurgence of sexual activity. An important covey-maintenance call is the assembly or rally call, used to reunite separated birds. This is a loud *cle-cle-cle* or *kow-kow-kow* series of notes (Miller and Stebbins, 1964; McLean, 1930), which are quite distinctly different from the brief assembly calls of *Callipepla* or *Colinus* and more closely approach the repeated call notes of *Philortyx*. The alarm note is a *scree* (Miller and Stebbins, 1964), or a shrill *t-t-t-r-r-r-r-rt* (Haskin, in Bent, 1932), rapidly delivered in a sharp crescendo and accented like a barnyard fowl's cackle.

A variety of other calls have been described as associated with enemy avoidance. The male is said to utter a shrill *quaih-quaih* while performing distraction displays (Bendire, 1892). The call of the female with young that stimulates them to "freeze" is a nasal *kee'-err* and a hen-like *kut, kut, kut,* while a low *whew, whew, whew* is uttered as they rush for cover (Hoffman, 1927). The hand-held distress call of both sexes is a loud, repeated *psieuw*.

EVOLUTIONARY RELATIONSHIPS

Holman (1961) regarded the scaled quail as the nearest relative of the mountain quail, with somewhat lesser affinities to the other crested quails ("*Lophortyx*") and to the bobwhites. Certainly the occurrence of wild hybrids between the mountain and California quail would suggest a moderately close relationship between these two species, but I would suggest that *Oreortyx* was derived from a pre-*Callipepla* type prior to the separation of gene pools into the currently extant species. It would seem likely that *Oreortyx* developed in the mountains of southwestern North America in a semiarid woodland or chapparal habitat after being isolated from stock adapted to more arid habitat such as that of the Gambel quail. Apparently the mountain quail had a considerably more widespread distribution in pre-Columbian times, since its remains have been found in cave deposits of New Mexico (Howard and Miller, 1933).

22

Scaled Quail

Callipepla squamata (Vigors) 1830

OTHER VERNACULAR NAMES

BLUE racer quail, blue quail, Cordorniz Azul, Codorniz Escamosa, cottontop quail, Mexican quail, scaled partridge, top-knot quail, Zollin.

RANGE

From southern Arizona, northern New Mexico, eastern Colorado, and southwestern Kansas south to central Mexico. Introduced into central Washington and eastern Nevada.

SUBSPECIES (*ex. A.O.U. Check-list*)

C. s. squamata: Mexican scaled quail. Resident in Mexico from northern Sonora and Tamualipas south to the Valley of Mexico.

C. s. pallida Brewster: Arizona scaled quail. Resident from northern Sonora and Chihuahua north to Arizona, New Mexico, Colorado, Kansas, Oklahoma, and western Texas; introduced into central Washington (Yakima and Grant counties) and Nevada (Elko, Nye, and White Pine counties).

C. s. castanogastris Brewster: Chestnut-bellied scaled quail. Resident in southern Texas south through Tamaulipas, Nuevo Leon, and eastern Coahuila, Mexico.

MEASUREMENTS

Folded wing: Adults, both sexes, 109–21 mm (males average 2 mm longer than females).

Tail: Adults, both sexes, 75–90 mm (males average 2 mm longer than females).

IDENTIFICATION

Adults, 10–12 inches long. The sexes are very similar in plumage. Scaled quail have a predominantly bluish gray coloration (thus "blue quail"), and are extensively marked on the back, breast, and abdomen with blackish "scaly" markings. The crest is bushy, varying in color from buff in females to more whitish in males. Otherwise, the head is light grayish brown, the lower back, wings, and tail are brownish gray to gray, and the flanks are grayish to brownish with lighter shaft markings. Males of one race (*castanogastris*) have chestnut coloration on the abdomen similar to that of male California quail.

FIELD MARKS

The "cottontop" crest is often visible from some distance, and the generally grayish coloration of the bird sets it apart from all other quail in the arid habitats where they occur. They are usually reluctant to fly, preferring to run rather than remain hidden. The distinctive *pey-cos* location calls (stronger in males) will often reveal the presence of scaled quail in an area.

AGE AND SEX CRITERIA

Females may be distinguished from adult males by their less conspicuous crests (males' crests average 40.6 mm, females' 36.8 mm) and by the dark brown shaft-streaks on the sides of the face and the throat, as compared with the unstreaked pearly gray to white coloration of the male in this area (Wallmo, 1956a).

Immatures of both sexes have buff-tipped greater upper primary coverts associated with the first seven primaries.

Juveniles have poorly developed crests, central tail feathers with much cross-barring of darker and whitish coloration (Ridgway and Friedmann, 1946), and whitish shaft-streaks on the upper parts. They are quite similar to juvenile California quail but are paler and more streaked, and they are grayer below mottled with dull white (Dwight, 1900).

Downy young (illustrated in color plate 110) differ from those of elegant quail by their considerably paler lower back and upper leg coloration and from California and Gambel quail young by their grayer over-all body tone, with yellow or cinnamon-buff tints limited mostly to the head area. The two pale lines delimiting the darker middorsal stripe in scaled quail downies are nearly white rather than being buffy or cinnamon as in Gambel and California quail.

DISTRIBUTION AND HABITAT

The geographic distribution of the scaled quail more or less conforms to the Chihuahuan desert and adjacent desert grasslands, just as the Gambel quail's distribution centers on the Sonoran desert. The southern limit of the Chihuahuan desert extends approximately to the southern limits of San Luis Potosí (Leopold, 1959; Jaeger, 1957), whereas the scaled quail is common as far south as Hidalgo in locally arid habitats lying in the rain shadow of the Sierra Madre Oriental. This area represents the southern limit of natural mesquite (*Prosopsis*) grassland, but Leopold (1959) believes that the apparently recent extension of the scaled quail's range farther southward to the Valley of Mexico has been brought about by clearing of the pine-oak forest, overgrazing, and agriculture with the resulting formation of a secondary desert habitat. Leopold reported that in Mexico the bird thrives best where there is a combination of annual weeds, some shrubby or spiny ground cover, and available surface water. The natural desert habitats best provide this combination of characteristics, and the secondary deserts mentioned above, as well as the more extreme creosote bush deserts, support only relatively low populations. Dixon (1959) points out that the scaled quail was noted in all of four different studies of Chihuahuan desert birds and also occurred in a study of Tamaulipan thorn scrub habitat in south central Texas.

In Texas the scaled quail occurs in the Panhandle and trans-Pecos area eastward to the western parts of the Edwards Plateau and southeastward locally to McMullen and Hidalgo counties (Peterson, 1960). Its range is largely complementary to that of the bobwhite (McCabe, 1954), although a slight amount of range overlap does occur. Hamilton (1962) noted that the scaled quail is typically found in mesquite or juniper savanna habitats, while the

FIGURE 36. Current distribution of the scaled quail.

bobwhite typically occurs in scrub oak woodland, riparian woodland, or juniper-oak woodland. Scaled quail in Texas prefer calcareous soils having a combination of grass and brush and cannot survive where heavy woody cover is lacking (Principal game birds and mammals of Texas, 1945). During the breeding season the Arizona race of scaled quail is also found on open mesquite grassland and farming land, while the chestnut-bellied scaled quail prefers open prickly-pear cactus (*Opuntia*) flats. The winter habitats are around ranches, creek bottoms, and canyons in the case of the Arizona race, while the chestnut-bellied race prefers gravelly hills covered with black brush (*Acacia*).

In Oklahoma the scaled quail is common only in Cimarron County but also occurs less abundantly in sixteen other western Oklahoma counties. Of seventy reports of scaled quail occurrence as to habitat type in Oklahoma, 47 percent were in sand sagebrush (*A. filifolia*) habitats, 21 percent in short grass–high plains habitat, 13 percent on mesquite grassland, 10 percent on mixed-grass prairies, and the remaining 9 percent on shin oak, post oak–black oak, and tall-grass prairies (Schemnitz, 1959).

The range of the scaled quail in Kansas is extremely limited, and it is found locally south and west of Pawnee County in the southwestern part of the state (Johnston, 1964). It occurs in roughly the same areas as the lesser prairie chicken, namely where sandy soils occur along the Cimarron and Arkansas rivers and a combination of grasses and sagebrush predominate (Baker, 1953).

In Colorado the species extends along the Arkansas and Cimarron river basins from the Kansas and Oklahoma borders on the east and the New Mexico border on the south, westward to the foothills of the front ranges of the Rocky Mountains (Hoffman, 1965). Its altitudinal range in the state is mainly from 3,400 feet to 7,000 feet, but it is found as high as 8,000 feet. Based on observed quail usage, the most important habitat type in eastern Colorado is the sand sagebrush community on sandy soils, which in Hoffman's study accounted for over 40 percent of the quail observed. The second most important habitat type is dense cholla cactus and/or yucca grassland, a shortgrass community in which through grazing the cactus or yucca have developed into thick stands. The third most important habitat type is the piñon pine (*Pinus edulus*) and juniper (*Juniperus*) woodland community, which is typical of stony soils and rocky outcrops. All other natural and agriculturally modified habitats are of considerably less value to scaled quail, judging from numbers observed (Hoffman, 1965).

In New Mexico the scaled quail extends over most of the nonforested areas of the state up to an elevation of at least 6,990 feet, and its range is largely coextensive with those of mesquite, blue chaparral (*Condalia*), and cholla cactus (Ligon, 1961).

In Arizona the scaled quail occurs only in the southeastern part of the state, where it is associated with grassland vegetation and is replaced by the Gambel quail whenever the grasses are replaced by mesquite and cholla cactus as a result of overgrazing (Phillips, Marshall, and Monson, 1964). As a result, its range in that state may have decreased considerably in recent decades.

In central Washington state the species has been introduced and is well established in Yakima County and also in the eroded basalt scablands below the potholes of Grant County, where the birds are fairly common in the dense sagebrush (*Artemisia tridentata*) and grass habitats. A similar sage–shad scale habitat is used by the birds in Nevada, where they have been introduced in several eastern counties and now appear to be well established (Tsukamota, 1970).

POPULATION DENSITY

Densities of this species probably vary greatly in different habitats, and even in the same habitats during different years. In southern Texas, concentrations of about 1 bird per acre were reported on areas as large as 200,000 acres during 1940 and 1941 (Principal game birds and mammals of Texas, 1945). By comparison, at the northern edge of its range in Colorado, Figge (1946) reported a winter population of 333 scaled quail on 8,960 acres, or 1 bird per 27 acres. More recent studies by Hoffman (1965) indicate lower scaled quail populations averaging only about 10 birds per square mile, or 1 per 64 acres. Winter covey counts by Schemnitz (1961) in Oklahoma indicated that the population density on an over-all acreage basis on his study area was 1 quail per 12.9 acres, but if occupied ranges only were considered, the density was 0.84 acres per bird. By the same consideration of using occupied range only, Wallmo (1956b) found an average winter density of 1 quail per 10.1 acres. These figures simply point out the great locational and probably yearly differences to be expected in quail populations occupying desert or other habitats that are often marginal for survival.

HABITAT REQUIREMENTS

Habitat usage and requirements of the scaled quail have been well analyzed by Schemnitz (1961), whose work provides the basis for the following summary. During winter, quail fed in soapweed (*Yucca*) or soapweed–sand sage pastures, weed patches, or grain stubble fields during the early morning, then moved to resting cover, often consisting of man-made structures or piles of brush. Escape cover consisted of soapweed or soapweed–sand sage–grassland habitat or heavier cover, depending on degree of disturbance. Man-made structures not only served as protective shelter but also were usually associated with food plants in the form of weedy herbaceous plants. Midday periods were spent in the shade of tree cactus (*Opuntia*) plants. Wallmo (1956b, 1957) emphasized the importance of midday shade and loafing cover and noted that night roosting cover must not be so dense or thick that it prevents easy movements by the birds. Schemnitz (1964) also pointed out that scaled quail cover should provide overhead protection but opportunities for ground-level movement, since the species typically runs when disturbed. In contrast, the bobwhite, which more often "freezes" when disturbed, inhabits heavier woodland and brush habitats.

During the spring the birds moved from the heavier cover associated with winter areas to less dense cover, perhaps because of a seasonally lower hawk population. Soapweed and sand sage continued to be used for resting purposes, along with annual forbs and grasses. The nesting cover (based on fifty nests) consisted of a variety of forb or shrub cover types, with two-thirds of the nests being found under dead Russian thistle (*Salsola*), machinery and junk, or mixed forbs and soapweed. Similar nest-site requirements were suggested by Russell (1932), who found sixteen of twenty-three New Mexican nests in Russian thistle, forbs, soapweed, Johnson grass (*Sorghum*), or under overhanging rocks. Schemnitz (1964) found that grassy situations provided nesting cover for only three of the fifty nests. During the summer, the birds studied by Schemnitz foraged in fairly exposed grassland areas and loafed under soapweed clumps, where dry sandy soil was usually available for dusting.

Considering usage by life-form of the habitat, Schemnitz found that the habitats dominated by shrubs three to twenty feet high contributed the majority (54 percent) of more than two thousand flush observations of scaled quail, with man-made cover providing about 30 percent, and the remaining 17 percent more or less equally divided among forb clumps, cropland, and open grassland. In piñon-juniper ranges, skunkbush (*Rhus*), tree cactus, and dense soapweed provided favored shrub cover types, in short-grass habitats skunkbush was used most heavily, and on sand sage habitats a combination of dense soapweed and sand sage represented the major shrub cover type used by scaled quail. Skunkbush and man-made structures are used throughout the year by scaled quail for cover, and where they are available they received a total usage that was far in excess of their relative availability on the habitat. On the other hand, croplands and open grasslands were used much less frequently than their availability might have suggested.

The importance of available water as a habitat requirement for scaled quail is somewhat controversial. Wallmo (1956b, 1957) questioned its importance, and noted that he had observed coveys from three to seven miles from water during his studies. However, Schemnitz (1961) never observed quail farther than one and a quarter miles from water and furthermore found that they were distributed closer to water sources than a random distribution pattern would dictate. However, food or cover distributions might also be positively correlated with water distribution, and thus a direct relationship between the occurrence of water and quail cannot be positively stated. The water requirements of the scaled quail have not been as intensively studied as those of other southwestern quail, but some early observations (Vorhies, 1929) suggest that the birds can survive well without free water.

FOOD AND FORAGING BEHAVIOR

Apparently the usage of insect food by the scaled quail varies considerably in different areas or years, with some studies (Martin, Zim, and Nelson, 1951; Principal game birds and mammals of Texas, 1945; Bailey, 1928) indicating that up to 30 percent of the total food may be of this source, while other persons (Wallmo, 1956b; Kelso, 1937; Schemnitz, 1961) indicate that 7 percent or less of the food may be of animal origin.

Studies in Texas (Principal game birds and mammals of Texas, 1945) indicate that in the plains area of northwestern Texas weed and grass seeds are eaten extensively, while the chestnut-bellied scaled quail of south Texas relies heavily on seeds of woody plants (Lehmann and Ward, 1941). The two most important of these seed sources are elbowbrush (*Forestiera*) and cat's-claw (*Acacia*). Similarly in the trans-Pecos area the Mexican huisache (*Acacia*) is an important food, and on the lower plains and panhandle areas the seeds of mesquite and hackberry (*Celtis*) are relatively frequently taken. Mesquite is also used by birds on the Edwards Plateau, together with the seeds of sennabeans (*Vigna*) and weedy herbs (*Amaranthus* and *Solanum*).

The study by Schemnitz (1961) provides comparable information on scaled quail food usage in piñon-juniper and sand sage–grassland communities. In this area tree fruits are of minor importance, and of the twenty leading foods, thirteen were seeds of annual and perennial forbs, two were agricultural grains, two were insects, and the remaining three were grass seeds, tree fruits, and leafy materials. A variety of weedy forbs, such as pigweed (*Amaranthus*), Russian thistle, sunflower (*Helianthus*), and ragweed (*Ambrosia*) made up the majority of winter foods. Sorghum grain was the only distinctly preferred food among the cultivated grains, and grass seeds were likewise little utilized. In contrast to the Gambel quail, for which herbaceous legumes are a staple food source, only one species (*Psoralea*) was found to be an important food in Oklahoma. However, leguminous forbs such as lupines (*Lupinus*), locoweed (*Astragalus*), and deervetches (*Lotus*) have been reported in Texas foods. Schemnitz found a surprising diversity of foods consumed, with up to as many as twenty-four food types in one crop, which he considered a desirable foraging adaptation and one that might help support a relatively high bird population.

Schemnitz noted that scaled quail typically foraged from daybreak to about 10:00 A.M., and again from about 4:00 P.M. to dark, varying somewhat with the season and the temperature. Although the birds sometimes foraged during rain, they usually did not feed during snowstorms but waited until the snow had ceased falling. When the snow was fairly deep the birds perched in trees up to twenty-five feet above the ground, where they could reach the seeds of hackberry, skunkbush, and juniper.

MOBILITY AND MOVEMENTS

The only major study of scaled quail home ranges and movements to date is that of Schemnitz (1961), which will be the basis for the following discussion. In the winter, scaled quail gather in fairly large flocks that may number up to 100 or more birds. By marking individual birds, Schemnitz estimated that the average size of a winter home range in 1954–55 was 52.3 acres, but ten such home ranges varied from 24 to 84 acres. During the following winter the average estimated home range was slightly larger (69.5 acres) for the same home ranges, and all ten of the home ranges studied the previous year were again occupied. These winter coveys averaged about 30 birds during the two winter periods, ranging from 7 to 150, and generally larger coveys were present in the sand sage–grassland habitats than in short-grass or piñon-juniper habitats. The maximum diameter of a winter home range found by Schemnitz was 1 mile, or less than an estimated 1.5-mile cruising radius reported by Figge (1946) for Colorado birds, and the 0.75-mile ranging distance from winter roosting sites estimated by Russell (1932) for New Mexico. Wallmo (1956b) found that winter coveys had ranges averaging about 450 acres and restricted their daily movements to areas within 160 acres.

Schemnitz found only a limited amount of cover shifting among the winter coveys, a situation reported earlier by Wallmo (1956b). However, winter home ranges generally overlapped only slightly or not at all, and thus opportunities for covey mixing were rather limited.

Long-tailed Tree Quail, Male

C. G. Pritchard

Bearded Tree Quail, Male

J. P. O'Neill

Mountain Quail

D. Landau, courtesy Colo. Game, Fish & Parks Dept.

↑ Barred Quail and Scaled Quail, Males

↓ Elegant Quail, Male

Ken Fink

Gambel Quail

D. Landau, courtesy Colo. Game, Fish & Parks Dept.

↑ Scaled Quail

↓ Gambel Quail, Male and Female

Hybrid Scaled x Gambel Quail

L. A. Fuertes, courtesy Cornell Laboratory of Ornithology

Raphael Payne

↑ California Quail, Male

↓ Bobwhite Quail, Males and Female

Ken Fink

Spotted Wood Quail, Male (Red Phase) and Female (Olive Phase)

J. P. O'Neill

Ken Fink

↑ Masked Bobwhite, Male

↓ Black-throated Bobwhite, Male

Buffy-crowned Tree Quail (above) and Singing Quail (below)

C. G. Pritchard

↑ Mearns Harlequin Quail, Male

↓ Mearns Harlequin Quail, Female

Salle Harlequin Quail, Male and Female

C. G. Pritchard

↑ Chukar Partridge, Male

↓ Gray Partridge, Male

Gray Partridge, Male

D. Landau, courtesy Colo. Game, Fish & Parks Dept.

C.G.Pritchard

Winter home ranges were not distinct from, but rather part of, the larger summer home ranges. The summer home ranges of three coveys studied by Schemnitz were 720, 1,220, and 2,180 acres, but within these larger areas individual pairs probably occupied fairly small home ranges. Studies of individual birds marked on their winter ranges and seen again during the summer indicated movements of from as little as none to as much as 2.75 miles from the winter range. In the case of three pairs, the birds returned with their brood to the winter home range occupied the year previously.

Although scaled quail are not generally considered highly mobile, one documented case of apparent mass dispersal during late fall and winter has been established. Campbell and Harris (1965), while banding over two-thousand birds during the years 1960 and 1964, found that during the late part of 1961 and early 1962 a substantial population dispersal occurred. This dipersal involved both sexes and adult as well as immature birds. Thirteen banded birds were known to have moved at least ten miles or more, and a maximum movement of sixty miles was found for one subadult male. The movements did not have any clear directional tendencies and probably should be interpreted as population dispersal rather than possible migration.

SOCIAL AND REPRODUCTIVE BEHAVIOR

The fairly large winter coveys of scaled quail remain intact until the males begin to come into reproductive condition, and the combination of increasing male aggression toward other males and the separation of paired birds from the coveys gradually cause the dissolution. Schemnitz (1961) noted that in Oklahoma this breakup of winter coveys began to occur shortly after the period from March 1 to April 15, which was marked by male fighting and intolerance among mated pairs. He reported the first *whock* call of unmated males on April 13, and the earliest copulation that he observed was on April 5. Nests, however, were not found until early May, a rather surprisingly late date for a desert-nesting bird. Likewise, Leopold (1959) reported that in Mexico most nesting occurs from June through August and pointed out that it is during this time that the summer rains usually fall, resulting in an abundance of water, insects, and succulent foods. This long nesting period, which extends into September or even October as far north as Oklahoma, no doubt is an adaptation to allow nesting during the most favorable period or possible renesting attempts if initial efforts are unsuccessful.

Nests are usually located under shrubs or some other protected and shady site, and a fairly large clutch is typical. Wallmo (1956b) estimated that

←
Row 1: Mountain, barred, and bearded tree quails. Row 2: Scaled, elegant, and Gambel quails. Row 3: California quail, black-throated bobwhite, and bobwhite. Row 4: Spotted wood, singing, and harlequin quails.

fourteen eggs is an average clutch size based on personal observations and literature sources, and Schemnitz (1961) reported a similar average clutch size of 12.7 eggs. Male scaled quail evidently share in incubation less regularly than do bobwhite males; Schemnitz noted only one definite case and the presence of a second bird in the vicinity of the nest for only six of fifty nest locations. Incubation requires from twenty-two to twenty-three days, although a twenty-one-day incubation period has been commonly estimated.

There is still no clear evidence that males normally take over the care of the first brood, which would enable the female to begin a second one, although this possibility should not be dismissed. Wallmo (1956b) reported one such case in which the male raised the first brood while the female began laying again. The available data (summarized by Schemnitz, 1961) indicate a low average hatching success of scaled quail, generally under 20 percent, and this together with a high adult mortality rate would suggest that persistent renesting or possibly double brooding would be the only way that populations might be maintained. Average brood sizes in Oklahoma were apparently fairly high (7.8 to 11.5 young), but the percentage of adults without broods ranged from 38 to 70 percent during the three years of Schemnitz's study. Similarly, Hoffman (1965) reported an over-all average brood size of 8.7 young for a six-year period, and an average young to adult ratio of 2.8 to 1 during the same period based on these brood counts. A very similar juvenile to adult ratio of 2.86 to 1 (74.1 percent juvenile) was reported for fall hunter samples by Schemnitz. This would suggest that each adult pair must have averaged between five and six young that were raised to the November to January hunting season, which could hardly be possible if roughly 50 percent of the adults were unsuccessful nesters and only a single brood was raised by successful breeders.

During extremely dry summers little or no successful nesting occurs in quail, and the birds may not even attempt to nest. Leopold (1959) attributed this to a possible weakening of the adults because of the resultant poor diet, a reduced hatching success of eggs because of the lack of moisture, or reduced food and water supplies for the developing chicks and consequent high chick mortality.

As the chicks mature, the broods gradually become organized into larger covey units. During trend-route counts from July to early September in Colorado, the covey sizes seen averaged about 11 to 17 birds (Hoffman, 1965). Later area-covey counts made from mid-November to the early winter period provided yearly average covey sizes of 17 to 23 birds, suggesting a gradual merging of broods in late fall to form the fairly large winter coveys that are typical of this species. Wallmo (1956b) noted that seven fall coveys averaged 38.7 birds, while by spring the average size of twelve coveys observed during two different years had been reduced to 18.8 and 21.7 birds.

Vocal Signals

Surprisingly little has been written on the vocalizations of the scaled quail. The best-known call is the separation call, used by individuals separated from their covey as well as by both sexes when visually separated from their mates. This is a two-syllable, nasal call *pe-cos'* or *pey-cos'*, with both syllables having the same, uniform pitch, although the second syllable is of longer duration and somewhat greater amplitude. The two syllables have sharp starting points that are two-fifths of a second apart, and the call is repeated several times at intervals of about one second. Males which are unmated will respond to the playback of female *pey-cos* calls by approaching the recorder during the breeding season (Levy, Levy, and Bishop, 1966), which provides a census method for male populations. It is not yet established whether mated males can differentially distinguish the separation calls of their mates from those of other females, as is known to occur in Gambel and California quails.

The announcement call of an unmated male is a single-note, slightly nasal whistle, which Schemnitz (1961) described as a *whock* whistle and Wallmo (1956b) called a *squawk* or *kwook*. This is usually uttered from a conspicuous calling point, and is probably uttered during the entire period that unmated males are in reproductive condition, as has been proven for the corresponding call in Gambel quail. Wallmo (1956b) heard it only in males, probably only unmated ones.

Wallmo (1956b), who described the separation or "gathering" call as a *chin-tang'* or *chuk-ching'*, indicated that the group alarm note is similar, but more excited and more rapid, sounding like *chink-chank'-a*. Bendire (1892) also indicated the same similarity in these two calls. When birds were removed from traps they sometimes uttered a fright call, *tsing*. This call is very much like the down-slurred distress calls of other New World quails, as a comparison of sonagrams will readily show.

So far only a single type of male-to-male aggressive call has been noted in my laboratory. When confronted with other males (or a mirror), paired males utter a strong series of nasal calls, each of which is associated with a rapid and vigorous head-throw, with the bill being raised to the vertical and the head drawn well backward. Up to seven or more of these are given in rapid sequence, at intervals approximately one-half second apart. The female also uncommonly performs a version, weaker both in relative movement and sound amplitude, of the same display under conditions of disturbance, but this does not occur with predictable regularity as it does in males. In both the releasing situation and its sound characteristics the "head-throw" call is clearly homologous to the *squill* of the California quail and the *meah* of the Gambel quail, and male hybrids of the scaled quail and each of these species regularly perform intermediate calls and postures in this situation. Strangely, the scaled quail apparently lacks, or at most has very poorly developed, any aggressive calls that correspond to the *wit-wit* and *wit-WUT* calls of these two species, thus the scaled quail's head-throws are neither preceded by nor alternated with other threat calls, as is the typical situation in the Gambel and California quails. Likewise the scaled quail apparently almost lacks the typically repeated soft *chip* sounds made by these species in situations of mild alarm, with the head-throw call or a variant of it serving to keep the covey together as they retreat through the brush. Daniel Hatch* noted that about a third of the birds he heard calling in this situation uttered the head-throw call (males?), another third produced *chip'* and *chip-eee'* calls, and the remainder uttered only a *chip-eee'* note. Bendire (1892) described this call as a *chip-churr* sound. He also noted that when chased by a hawk the birds uttered a gutteral *oom-oom-oom;* I have not had an opportunity to hear the response of this species to avian predators.

Laboratory-produced hybrids between the scaled quail and the bobwhite produce a call that is intermediate between the *pey-cos* and the *hoy, hoy-poo* complex when placed in a situation that would elicit separation calls. The male call that is uttered in male-to-male aggressive situations lacks a definite head-throw component, but acoustically appears to be intermediate between the head-throw call and the bobwhite's caterwaul call.

The total adult vocal repertoire of the scaled quail is thus a surprisingly limited one, that includes an unmated male announcement call, a separation call used by both sexes, an agonistic call that is largely but not entirely typical of males, an alarm *chip* note that is probably used by both sexes, an avian predator call, and a distress call. Wallmo (1956b) mentioned hearing various "conversational" or contact notes that might be added to this list, and doubtless one or more parental calls also occur. I have not heard calling by either sex during copulation, and the tidbitting display of males to females is likewise silent. It would thus seem unlikely that more than ten call-types are present in the scaled quail, or far fewer than have been found to occur in the bobwhite.

*Daniel Hatch, 1971: personal communication.

EVOLUTIONARY RELATIONSHIPS

Even if *Lophortyx* were not merged with *Callipepla* there could be no question that the elegant, Gambel, and California quails are the nearest relatives of the scaled quail and the lack of a distinctively colored and elongated crest in this species is of no taxonomic significance beyond the species level. It is difficult to judge with which of these three species the scaled quail has the greatest affinities, but the elegant quail bears an interesting allopatric relationship to the scaled quail, and one might readily imagine that speciation occurred following isolation from a common ancestral type by the Sierra Madre Occidental mountains. Both species are desert-adapted and dependent on the presence of shrubby or brushy vegetation in relatively scattered (for the scaled quail) or continuous (for the elegant quail) groupings. Both also have strong similarities in their vocalizations, their downy young, and their general plumage patterns; although differences in adult plumages do occur they are not any greater than between those of the scaled and the California or Gambel quails. However, the only known hybrids between the scaled and elegant quail have apparently been sterile (Banks and Walker, 1964), whereas at least a limited degree of hybrid fertility exists between the scaled quail and both the Gambel and California quails.

There is apparently also a partial sterility barrier between the scaled quail and both the barred quail and the bobwhite quail, with female hybrids representing these crosses apparently either laying no eggs (scaled × barred) or laying subnormally small ones (scaled × bobwhite). One might presume therefore that the scaled quail does not provide a definite "link" between the crested quails (*"Lophortyx"*) and *Colinus*, nor between these species and *Philortyx*. For these reasons, and the very weak morphological criteria for separating *Callipepla* from *Lophortyx*, it seems most reasonable to consider the scaled quail and the three crested quails as a close-knit evolutionary unit.

23

Elegant Quail

Callipepla douglasii (Vigors) 1829

OTHER VERNACULAR NAMES

BENSON quail, Codorniz Gris, crested quail, Douglas quail, Lesson quail, Yaqui quail.

RANGE

Western Mexico from Sonora and Chihuahua to Nayarit and Jalisco.

SUBSPECIES (*ex Check-list of the Birds of Mexico*)

C. d. douglasii: Douglas elegant quail. Resident in extreme southern Sonora, south through Sinaloa and northwestern Durango (properly called *elegans*, according to van Rossem, 1945).

C. d. bensoni (Ridgway): Benson elegant quail. Resident in Sonora, from close to its northern boundary to Guaymas and San Javier.

C. d. teres (Friedmann): Jalisco elegant quail. Resident in northwestern Jalisco, but not extending to Colima (Schaldach, 1963). (Properly called *douglasii*, according to van Rossem, 1945).

C. d. impedita (Friedmann): Nayarit elegant quail. Resident in Nayarit.

C. d. languens (Friedmann): Chihuahua elegant quail. Known only from western Chihuahua (of doubtful validity according to van Rossem, 1945).

MEASUREMENTS

Folded wing: Adults, both sexes, 98–115 mm (males average 3 mm longer than females).

Tail: Adults, both sexes, 65–94 mm (males average 4 mm longer than females).

IDENTIFICATION

Adults, 9–10 inches long. The sexes are somewhat different in appearance. The head coloration is mostly brown (females) or gray (males), streaked or spotted with black, and with a straight, pointed crest of graduated feathers that are orange-buff (males) or mottled brown (females). The upper portions of the back, wings, and tail are uniformly gray (males) or mottled with grays and browns (females). The underparts from the breast to the abdomen are grayish or brownish, with paler spots that are generally rounded and increase in size posteriorly.

FIELD MARKS

The pale, rounded flank spotting of both sexes sets this species apart from all other quails, and the fairly straight crest that narrows toward the tip rather than being recurved forward and enlarged toward the tip distinguishes the elegant quail from its near relatives. It inhabits arid desert in northwestern Mexico and in some areas is found in company with Gambel quail. The location call is a loud, two-note rasping sound which has the cadence of the scaled quail's *pey-cos* call, and the unmated males' call is likewise very similar to the corresponding call of the scaled quail.

AGE AND SEX CRITERIA

Females have shorter crests than do adult males (average of ten is 28 mm compared with 39 mm in males), and these crest feathers are dark brownish, spotted or barred, rather than orange cinnamon.

Immatures (presumably) have buffy-tipped greater upper primary coverts and pointed outer primaries.

Juveniles resemble females but are generally darker, more rufescent above, and the breast and abdomen feathers have barring rather than round white spots (Ridgway and Friedmann, 1946). No doubt shaft-streaks are also present dorsally.

Downy young (illustrated in color plate 110) of the elegant quail have an appearance quite similar to that of scaled quail young, but whereas the downy scaled quail have a brownish buff spinal stripe isolated by two narrow black lines, elegant quail downies have a less contrasting dorsal stripe, consisting of a spinal stripe of mummy brown that becomes darker toward the sides. Also, below the pale line that separates the dorsal stripe, a second dark brown area occurs, which is bisected by a second pair of narrow buffy lines. The relatively dark lower back and tail coloration also distinguishes elegant quail downies from those of California and Gambel quails.

DISTRIBUTION AND HABITAT

The range map prepared by Leopold (1959) provides an accurate indication of the elegant quail's range, except that the southern tip of the range is probably in Jalisco, rather than extending to Colima (Schaldach, 1963).

The northern limits of the elegant quail's range are in northern Sonora. Van Rossem (1945) lists the northernmost records as being Opodepe and eighteen miles north of Cumpas (near Nacozari, or about sixty miles south of the United States border). Recently there have been several sight records for this species as Nogales, Arizona (*Audubon Field Notes*, 18:476, 527, 1964; 23:391, 1969). After inquiring about these records to William Harrison, Nogales, I received this reply:

> Beginning in the spring of 1964 elegant quail occurred around the Nogales area, especially west of town on the road to St. Joseph's Hospital. They were seen repeatedly at several good locations near water spots. A local doctor (M.D.) who travelled the road daily and who is very much interested in birds(and quite familiar with Mexican species particularly) reported seeing adults with young during the late summer of 1965 (and at other times?). Finally we talked with the operator of a used car lot on the road to the hospital who told us that the quail had originally escaped from a pen across the line in Nogales, Mexico (just one-half mile distant). I suspect this may be the case, for I have not seen elegant quail south of Nogales, a territory I have worked extensively during the last ten years. I understand that the species almost reaches the border at Douglas and Sasabe, but both of these areas are slightly lower in elevation than Nogales, and I know that the bird prefers thorn scrub.*

*William Harrison, 1970: personal communication.

Most of van Rossem's plotted specimen records are along interior river systems including the Sonora, Moctezuma, Yaqui, and Mayo. He earlier (1931) reported that the elegant quail is rare on the coastal plain, and common in broken, lower hill country. The only coastal record for Sonora he obtained was for Guaymas, where he reported a mated pair of male Gambel and female Benson elegant quail.

Alden (1969) indicates that the elegant quail is common in tropical deciduous forest along the road to Alamos, Sonora, and is also common in the valleys of the Fuerte and Culiacán rivers, Sinaloa. Edwards (1968) reported that the bird occurs near the coast at Mazatlán, Sinaloa, and San Blas, Nayarit, as well as farther inland near Acaponeta and Tepic, Nayarit.

The elegant and Gambel quails are widely sympatric in Sonora and northern Sinaloa but apparently exhibit ecological differences that reduce contact between them. Alden (1969) reported that whereas the elegant quail is common in the Río Fuerte valley, Sinaloa, the Gambel quail is only occasionally found there. It would seem that the Gambel quail is more highly desert-adapted, being found most commonly in desert or mesquite grasslands, while the elegant quail is primarily a bird of the thorn forest foothills and scrub thickets of river valleys. Leopold (1959) indicated that dense second growth of tropical forest is a favored habitat but that open fields and pastures are avoided. In such areas of heavy brush as Nayarit he estimated the population to achieve pockets of more than a bird per acre. The only other estimate of population density of which I am aware is a breeding bird census in Sinaloa thorn forest (*Audubon Field Notes*, 22:686, 1968), where the population of elegant quail was placed at 1.5 territorial males on 22.5 acres.

FOOD AND FORAGING BEHAVIOR

Little has been written on the food of this species. Leopold (1959) reported an assortment of weed seeds, fruits, and insects in the samples he examined, the weed seeds including a predominance of unidentified legumes. He reported that the birds scratch vigorously while they are foraging, and after filling their crops they loaf and dust along roads and trails.

MOBILITY AND MOVEMENTS

It would seem that the elegant quail is not highly mobile. Leopold (1959) reported that when frightened the birds would fly or jump into dense *monte* brush and "freeze," returning to the ground only long after the end of the disturbance. Rarely was he able to force a covey into flight. At night they return to the brush for roosting, sleeping in vines or bushes a few feet above the ground. In the mornings they begin foraging again under the brush or along the edges of clearings. Like other quail, they probably move to water at least once a day when it is available but may utilize succulent green vegetation whenever free water is not within their normal range of mobility.

SOCIAL AND REPRODUCTIVE BEHAVIOR

Leopold (1959) found that in Nayarit coveys, which had averaged six to twenty birds, began to break up in mid-April, and males were then crowing from low perches. He estimated, on the basis of gonad development in birds he collected, that nesting would probably begin in early May. In southern Sinaloa nesting probably occurs in April and May, and the nests are reportedly on the ground, with eight to twelve eggs normally, but up to twenty sometimes found (Miller, 1905).

The eggs of the elegant quail are pure white and, if those I have seen are typical, are somewhat more elongated than is typical of related species. The average of ten eggs laid by birds in my collection was 33.9 x 23.9 mm, with ranges of 31 to 35 mm and 23 to 24 mm. In captivity at least the maximum rate of laying is about one egg per day; one female laid seven eggs in an eight-day period, followed by four more eggs during the next twenty-six days. Over a several week period, if the eggs are removed every day, the rate of laying is approximately two days per egg. Thus, one female laid twenty-four eggs in fifty-three days, followed by three more a month later. The incubation period is the same as in related species of *Callipepla*, namely twenty-two or sometimes twenty-three days.

Vocal Signals

Virtually nothing has been written on the vocalizations of this species. The birds in my collection have called very little, thus only a few general comments are possible. The male call that apparently serves as a separation call is a somewhat nasal two-note *ca-cow'* that usually occurs in groups of two to five calls (average of seven is three), separated by intervals of about one-fourth second. The midpoints of the first and second syllables of the call are almost exactly one-half second apart, as is also true of the scaled quail's *pey-cos* call. Unlike the scaled quail's call and in common with the *chi-ca-go* call of the California and Gambel quails, the frequency of the call is not constant, but rises and falls during each note. In cadence and frequency characteristics, the call thus is intermediate between the two call-types. The advertising call of unmated males is a sharp, nasal whistle that corresponds to the unmated male *cow* call of the California and Gambel quails and the comparable *wock* whistle of the scaled quail, which it very closely resembles when compared sonagraphically.

When disturbed both sexes utter a chipping or clucking *chip-chip'*, and at least the male often utters a sharp whistled *wheet'* when alarmed. When held in the hand, both sexes produce sharply down-slurred distress whistles, usually repeated in a long series for ten or more seconds.

EVOLUTIONARY RELATIONSHIPS

Holman (1961) indicated that the elegant, Gambel, and California quails showed greater interspecific differences from one another than did the Recent species of *Colinus*. His data suggest a slightly greater affinity of the elegant quail with the Gambel quail than with the California quail (sharing ten as compared to eight of thirty osteological characters), but collectively the characters of the group are very much like those of *Colinus* and the scaled quail. The adult plumage pattern of this species is quite distinct from all of the other three species of *Callipepla*, whereas the downy plumages of all four species are relatively similar. Largely on zoogeographic grounds and general similarities in vocalizations, I would be inclined to believe that the scaled quail represents the nearest living relative of the elegant quail.

24

Gambel Quail

Callipepla gambelii (Gambel) 1843
(Lophortyx gambelii in *A.O.U. Check-list)*

OTHER VERNACULAR NAMES

A*RIZONA* quail, Codorniz de Gambel, desert quail, Olanthe quail.

RANGE

From southern Nevada, southern Utah, and western Colorado south to northeastern Baja California, central Sonora, northwestern Chihuahua, and western Texas.

SUBSPECIES (*ex A.O.U. Check-list* and *Check-list of the Birds of Mexico*)

C. g. gambelii: Southwestern Gambel quail. Resident from southern Utah and southern Nevada south to the Colorado and Mojave deserts, northeastern Baja California, and introduced in north central Idaho.

C. g. fulvipectus Nelson: Fulvous-breasted Gambel quail. Resident in north central to southwestern Sonora, and probably north to southeastern Arizona and southwestern New Mexico.

C. g. pembertoni (van Rossem): Tiburon Island Gambel quail. Resident on Tiburon Island, Gulf of California.

C. g. sana (Mearns): Colorado Gambel quail. Resident in western Colorado in the drainage areas of the Uncompahgre, Gunnison, and Rio Grande rivers.

C. g. ignoscens (Friedmann): Texas Gambel quail. Resident of desert areas in southern New Mexico and extreme western Texas.

C. g. stephensi (Phillips): Recently described (1959). Resident in Sonora, near the Sinaloa border (not yet verified).

C. g. friedmanni (Moore): Sinaloa Gambel quail. Resident in coastal Sinaloa from Río Fuerte south to Río Culiacán.

MEASUREMENTS

Folded wing: Adults, both sexes, 105–22 mm (males average 2 mm longer than females).

Tail: Adults, both sexes, 83–107 mm (males average 5 mm longer than females).

IDENTIFICATION

Adults, 9.5–11 inches long. The sexes are different in appearance. This southwestern quail has a blackish, forward-tilting and teardrop-shaped crest as in the California quail but completely lacks the scaly patterning of the underparts typical of the latter. Only on the back of the neck of males is some scaly patterning evident, but this is ill-defined. Male Gambel quail also have a black forehead and reddish-brown crown coloration, and both sexes have more rufescent brown flank coloration than occurs in the California quail. Otherwise the birds are generally grayish brown to brown on the upperparts and tail and have buffy underparts that may be streaked with brown (females) or have an extensive black area on the abdomen (males). Males also have the characteristic black throat pattern that is lacking in females.

FIELD MARKS

Generally limited to desert regions of the southwest, Gambel quail can be identified in the field by the combination of "teardrop" crests and unscaled underparts. The rich reddish-brown flanks of both sexes are visible at considerable distances, and at close range the reddish crown color of males and the black mottling of their underparts may be evident. This species' calls are similar to those of the California quail, but are less metallic and more nasal. The distinctive location call consists of occasionally repeated *chi-ca-go-go* notes (occasionally California quail will also add a fourth syllable to their location call).

AGE AND SEX CRITERIA

Females have dark brown rather than black crests and lack black throats.

Immatures have mostly buff-tipped greater upper primary coverts which are carried for the first year (Leopold, 1939). The outer two primaries may be somewhat more pointed and frayed than the inner ones in immature birds.

Juveniles resemble females but have dull brown crests and broad bands of pale cinnamon buff above the eyes. They are very similar to California quail of this age except that the nape feathers lack dusky borders and are uniformly gray with more distinct shaft-streaks (Dwight, 1900).

Downy young (illustrated in color plate 110) of this species cannot be easily distinguished from California quail of the same age, but are perhaps in general slightly paler and less yellowish in tone overall. The pale spinal stripe is somewhat tinged with darker streaks in the Gambel quail, while in the California quail it is a slightly brighter buff. Furthermore, the downy California quail generally has less sepia brown (that is, more buff) coloration on the forewing than do the Gambel and scaled quails.

DISTRIBUTION AND HABITAT

A detailed analysis of the range and habitat of the Gambel quail has been made by Gullion (1960), from which the accompanying range map is largely derived. No major changes in ranges have occurred since that time, and his review of the species' distribution by states cannot be improved upon. He found that the species is found in three major climatic and habitat types. One of these is the mesquite (*Prosopsis*), saltbush (*Atriplex*), tamarisk (*Tamarix*), and desert thorn (*Lycium*) shrub associations of desert valleys from Texas west to southern California, Nevada, Utah, and northern Mexico. These areas have similar altitudinal ranges, low annual precipitation totals, and mild winter temperatures.

Especially in the western part of the species' range, it also occurs in upland desert habitats, particularly where a fairly uniform desert vegetation is dominated by cat's-claw (*Acacia*), creosote bush (*Larrea*), desert thorn, skunkbush (*Rhus*), yuccas (*Yucca*), burroweed (*Franseria*), and prickly pear (*Opuntia*). This habitat type occurs on the Mohave desert areas of Arizona, California, and Nevada and to a reduced extent in southwestern

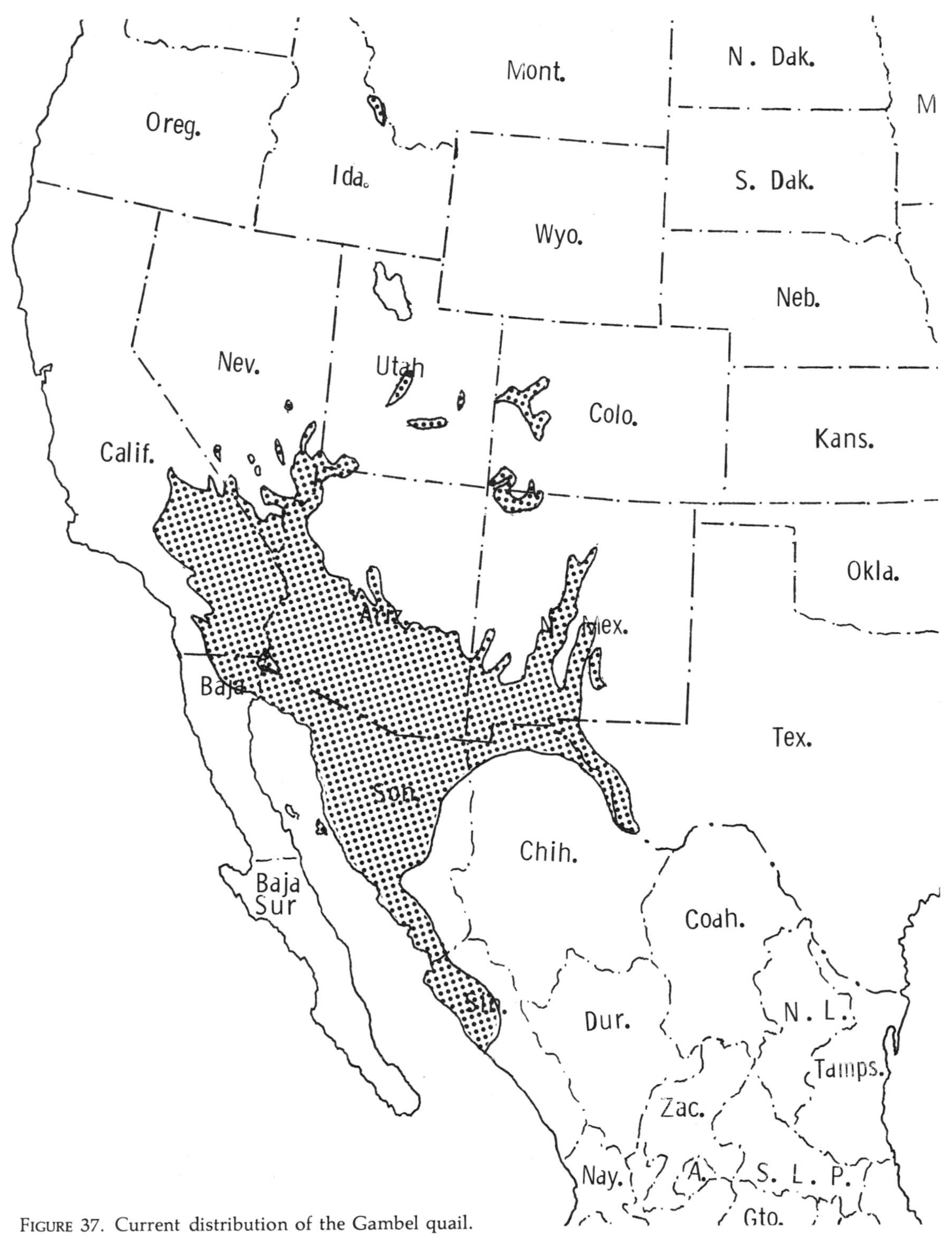

FIGURE 37. Current distribution of the Gambel quail.

New Mexico and Utah. The altitudinal range is from three thousand to forty-five hundred feet, and winter temperatures average considerably above freezing. Although precipitation averages more than in the valley habitats, it is still only from about 3 to 9½ inches. These birds exhibit greater population fluctuations than is typical of lowland habitats, depending on annual productivity. Winter precipitation variation is one of the most important factors regulating these population changes.

In addition to these two warm desert habitats, the species also occurs in the Colorado River basin areas of New Mexico, Colorado, and Utah and as an isolated population in Idaho, all of which are subjected to considerably colder temperatures. The vegetation here is essentially that of the Great Basin desert, with such shrubs as greasewood (*Sarcobatus*), rabbit brush (*Chrysothamnus*), skunkbush, saltbush, and sagebrush (*Artemisia*) being almost universally present. These habitats and climates are marginal for the Gambel quail, and at least in some areas the presence of food in the form of agricultural crops such as alfalfa may be critical for survival. Gullion also suggested that such populations are marginal where snowfall exceeds twenty inches or where at least an inch of snow is on the ground for more than about forty days a year. Where the northern population survives best, the winter precipitation totals are normally quite low, usually well below 50 percent of the total annual precipitation.

POPULATION DENSITY

Breeding populations of the Gambel quail have not been intensively studied as to population densities. Hensley (1954), in studying the birds of desert habitats in Arizona, estimated that the average number of breeding quail pairs per 100 acres on 210 acres of study areas was six, or one pair per 16.6 acres. However, based on one study area of 70 acres, he had an estimated maximum population of twelve pairs per 100 acres, or one pair per 8 acres.

In a study of the breeding bird population of a cholla cactus (*Opuntia*), palo verde (*Cercidium*), and saguaro (*Cereus*) desert community in Arizona, an estimate of 20 territorial male quail per 100 acres has been made (*Audubon Field Notes*, 19:610–611, 1965), or presumably one pair per 5 acres. Also, Hensley (1954) reported that four pairs of Gambel quail occupied a mountain canyon study area measuring twenty-five by eight hundred yards (4.1 acres), suggesting that under favorable conditions a population density of at least one bird per acre may sometimes occur. Gullion (1962) reported that an estimated total of 472 quail were present on a 777-acre study area in Nevada, or 1.6 acres per bird. This total apparently referred to a late winter population.

HABITAT REQUIREMENTS

Gullion (1960) has suggested several biotic and physical environmental features that may represent limiting factors for Gambel quail. Soils having good populations are residual soils of decomposed granite in the uplands of Nevada; such soils support a relatively luxuriant and diversified vegetation. Transported soils of river bottoms also support luxuriant shrub growth and high quail populations. Populations are also highest where January temperatures do not drop below forty degrees F., and, as mentioned earlier, winter snow cover is probably an important limiting factor in northern marginal populations. However, the Colorado race of Gambel quail is known to survive winter temperatures as low as eight degrees below zero in New Mexico, the Texas race of Gambel quail occurs in areas having minimum winter temperatures of five degrees below zero, and in Utah and Idaho the introduced race *gambelii* have survived temperatures of approximately forty degrees below zero.

Although lowland populations of Gambel quail depend on subsurface moisture that may originate several hundred miles away, upland populations evidently require a winter precipitation of more than five inches (Gullion, 1960). This of course is not a reflection of drinking water needs but of the effects of the precipitation on vegetational growth. Swank and Gallizioli (1954) considered December to April in Arizona to be the most critical months for precipitation, and Gullion (1960) correlated quail populations with the precipitation totals of the preceding October to March. Apparently winter germination and growth of green plants is vital to the breeding success of this species, possibly because of its effect on vitamin A storage in potential breeding birds (Hungerford, 1964). Raitt and Ohmart (1968) reported that in New Mexico the fall productivity index based on age ratios was closely correlated with amounts of precipitation during the preceding May and June rather than those of the previous fall, winter, or early spring, indicating a lack of strict dependency on such winter rainfall. They suggested that the effects of irrigation or a winter climate that permits an accumulation of soil moisture might account for this apparent difference in climatic correlation.

The importance of free water for drinking purposes by Gambel quail is not completely clear. Gullion (1960) believed that where a combination of high humidity and fleshy plants occurs, the birds can live an entire lifetime without drinking water. Hungerford (1960) concluded that water catchments were nonessential in southern Arizona, where moist succulent plant foods are normally available. However, on desert uplands, such as in Nevada, there may be a critical period for moisture from about mid-June to mid-July, when succulent spring annuals have dried up and summer thunderstorms have not yet occurred. During such times, if succulent plants are not available, artificial watering structures may be quite important to the species (Gullion, 1960). Miller and Stebbins (1964) report that in Joshua Tree National Monument the Gambel quail occurs primarily in the vicinity of springs, and the greatest distance from water which they have recorded for this species was one and a half miles at a time when succulent vegetation was widespread. Most coveys probably stay within a mile of water when it is needed.

Nesting cover requirements for the Gambel quail are simple, consisting of desert shrubs or trees, with the primary requirement apparently being a source of shade from the midday sun (Bent, 1932). Brooding requirements no doubt include brushy escape cover, shade for resting, and foraging sites where insects and small green plant growth is readily available. Grit sources and dusting locations are readily available in desert habitats.

FOOD AND FORAGING BEHAVIOR

In common with the California quail, the Gambel quail relies very little on animal sources of food, adults taking perhaps as little as 0.5 percent of their annual food from this source (Judd, 1905a), with a maximum usage of 12 or 13 percent during spring and summer (Martin et al., 1951). Otherwise, the birds rely predominantly on the foliage and seeds of a large array of plants.

Judd's analysis (1905a) of twenty-eight food samples from Arizona and Utah indicated that virtually no fruit material is consumed and only a very small amount of cultivated grains (3.9 percent of annual total). Rather, leafy materials, mainly legumes, and seeds of a variety of species made up over 95 percent of the total sample, with these two food categories totaling 31.9 and 63.7 percent by volume, respectively. Legume seeds alone made up 21.2 percent of the total food material, especially of those of alfalfa and bur clover (*Medicago* spp.). Gullion (1960, 1966) noted that at least ninety-one species of plants are consumed by Gambel quail in southern Nevada, but the availability of species representing only three groups, namely deervetch (*Lotus* spp.), filaree (*Erodium*), and a few herbaceous legumes (*Astragalus* and *Lupinus*) determines the abundance of Gambel quail in this area.

Hungerford (1962) examined the seasonal variations in food consumed by Gambel quail in southeastern Arizona, based on the study of 221 samples. He found that various legumes (*Lotus, Lupinus, Mimosa, Prosopsis*) were the most important food sources, with their leaves, flowers, and seeds all being consumed. Filaree seeds and flowers were a highly preferred food source as well. On a yearly basis, seeds made up 60.7 percent of the diet and were important foods throughout the year. Considering only life-form of food sources, forbs were most important, making up 54.2 percent of the annual diet, shrubs were second, totaling 31.8 percent, and grasses, animal foods, and unknown plants made up the remaining amount. During spring, a high 1:1 ratio of succulent to nonsucculent plants was present, while during fall and winter this ratio dropped to about 1:2. Apparently these succulent food sources, during dry periods or in areas where free water is not normally available, provide important sources of moisture and are highly important aspects of the quail's ecology.

A study by Campbell (1957) on the fall foods of the Gambel quail in New Mexico provides an additional index of the diverse food usage of this species. Of fifty-seven crops studied, all had seeds and/or fruits present, and collectively eighty-seven plant species representing twenty-seven different families were present in the crops. However, foods representing twenty-two species of plants accounted for more than 90 percent of the sample volume, including five species of legumes, four composites, four grasses, and three chenopods. Campbell concluded that the Gambel quail's flexibility in foraging behavior in utilizing so many different food sources helped to explain its success in agricultural areas, where the vegetational complex is quite different from that prevailing in undisturbed desert habitats.

MOBILITY AND MOVEMENTS

The movements and social organization of Gambel quail coveys has been studied by Gullion (1962) in Nevada, on a 777-acre area of thorn shrub vegetation. A total of twenty-four coveys were present on the area, ranging from 3 to 40 birds and averaging 12.5. An estimated total of 472 birds were present, of which 217 banded birds were used to establish covey organization and movements. There were three major areas of use on the study area, with some overlapping of home ranges. The home ranges of ten coveys varied from 19 to 95 acres, averaging about 35.7 acres per covey. No clear correlation occurred between covey size and size of home range, with the largest covey (22 birds) having a 95-acre range, the second largest (21 birds), a 37-acre range, and a still smaller covey had an intermediate range.

During the winter, covey movements appeared to be erratic. From late

December to the following April, the ten coveys ranged over areas with diameters of from 1,500 to 4,200 feet, averaging 2,340 feet. One covey of twenty-two birds consisted of at least four subgroups and moved about over a sixty-three-acre area, then all moved into a new area 2,200 feet away. After staying in the new area for at least ten days, the covey disappeared from the study area, with a few of the birds eventually returning to the location where they were originally trapped.

Seasonal variations in covey movements were considerable and were influenced by the age composition of the coveys, with coveys composed of adults moving considerably farther than did brood coveys. During the winter period of December through late January, five adult coveys moved an average of 103 feet per day, while thirteen brood coveys averaged 63 feet per day. The movements increased in late January and early February, with average daily movements of 264 feet for adults and 131 feet for broods. During late March and early April there was a considerable prenesting shuffle, with coveys actively moving about, and the five adult coveys averaged 1,029 feet per day during this time. However, after about the first week of April, most of the coveys became sedentary, with the exception of a few new arrivals on the study area.

Individual movements of three birds during periods between late morning and midafternoon ranged from 400 to 1,250 feet, while the movements of forty-two banded birds over twenty-four-hour periods averaged 755 feet but were as much as 2,800 feet. One male moved at least 2,400 feet in a forty-eight-hour period and another male at least 3,800 feet in ninety-six hours. Another male moved 4.7 miles between April and November, while a fourth male moved 5 or 6 miles between late April and October. The longest recorded movement was of an adult female, which moved 6.5 miles from its banding site in somewhat over two years, and was at least four and one-half years old when it was killed.

No definite fall dispersal pattern for single quail could be established, but a spring dispersal pattern was clearly evident. This dispersal, which consisted of covey-shifting, was performed mostly by young males, plus a few young females. Although the evidence was not clear, major dispersals over long distances probably involved entire coveys rather than individual birds.

SOCIAL AND REPRODUCTIVE BEHAVIOR

Gullion's (1962) study indicated that coveys of Gambel quail consist basically of family units of 5 to 7 birds or their aggregates (9 to 13; 17 to 22). Winter coveys might consist either of such combined broods or of varying numbers of nonbreeder adults. Although some overlapping of home ranges of coveys does occur, there is considerable covey fidelity, with little of the covey exchange that has been reported for other species of quail. Such covey exchange that Gullion found (20 of 217 birds) occurred mostly during the prenesting shuffle, with only 5 birds shifting during the earlier winter period.

The study by Raitt and Ohmart (1966) in southern New Mexico provides one of the best analyses of seasonal variations in social behavior that is available for the Gambel quail. During late winter, pair formation and increased hostility among males begins to cause the dissolution of coveys, which in New Mexico begins in March. The process of pair formation is a subtle one, which apparently occurs over a prolonged period of contact. Raitt and Ohmart thought that chases of females by males, during which they uttered explosive high-pitched notes together with longer and lower-pitched, softer notes, might be associated with pair formation under natural conditions. Such chases rarely if ever occur in captive birds which have been held in pairs through the prebreeding period, but if a female is introduced to a lone male in breeding condition strong chases of this type will immediately occur, and care must be taken that the female is not killed by the male. Thus, it would seem that initial male-to-female responses are not greatly different from male-to-male behavior, except that the female attempts to escape and performs submissive responses such as huddling that usually serve to break off attacks by the male. I have not seen strong wing-drooping during such display in the Gambel quail, but evidently it does occur. Gorsuch (1934) described such an encounter as follows:

"One day, while observing a whistling cock that was known to have used the same bush from which to call for over three weeks, a clucking sound was heard from down the wash and shortly a hen appeared. Immediately the cock sighted her his notes became fewer and shorter, and when she was within thirty feet of his perch he became greatly excited, jumping about the bush as if much disturbed, and talking to her meanwhile in a variety of notes. When she approached to within fifteen feet he . . . leaped to the ground and slowly but eagerly advanced to her. After walking around the hen in short circles several times, expanding his chest and trailing his wings in display they engaged in low-voiced conversation and wandered slowly away; it was definitely known that no nest existed within 200 yards of this whistler's post."

When males are chasing males, fighting may occur; this should not be regarded as territorial defense but only as a means of establishment of social dominance. Such attacks consist of rapid pecking movements and short vertical flights as each bird tries to get above the other bird and peck its skull. After a few such attacks, one bird usually makes a quick retreat and in a small cage may be caught by the dominant bird, whereupon its back, nape, and skull may be seriously damaged by pecking.

As the coveys are breaking up and strong pair bonds are forming, *cow* calling by unmated males begins. In New Mexico this may occur as early as mid-March, but reaches a high level in April and May, declining in June and terminating completely in late July or early August. Its duration thus does not conform to the period of pair formation, and a census of calling males should obviously not be regarded as a census of pairs in the area. Rather, its cycle generally follows the testis activity cycle, and it is thus a reflection of male sexual tendencies of unpaired birds. Probably no *cow* calling occurs in mated males, according to Raitt and Ohmart, and the study of Ellis and Stokes (1966) confirmed this opinion. These authors indicate that the call, which they refer to as the *kaa*-call, is usually given from an exposed perch and has a function analogous to the advertising song of passerine species. During the call, the male stands in an erect posture, with his abdominal patch wholly visible and the crest held vertically erect.

Gambel quail are strongly monogamous. The gonadal activity cycle of the female lags about two weeks behind that of the male, and in New Mexico laying begins in late April. Gorsuch (1934) indicated that a depressed area about one and one-half inches deep and five to seven inches in diameter is scratched out and variably lined. The first egg is deposited shortly, and the remaining eggs are deposited daily thereafter, with lags of one to three hours on each succeeding day. After four to six eggs, a day is skipped, and the cycle begins again. After about three such cycles of four to six eggs, the clutch is complete. Gorsuch found clutches of up to nineteen eggs at forty-four nest sites, but twenty-nine of the nests had from ten to sixteen eggs present; thus, twelve to fourteen must be regarded as a typical clutch.

Incubation is performed by the female alone, with the male usually sitting at a perch some forty to eighty feet away. When the nest is approached by an intruder, the male typically performs a "broken wing" distraction display (Gorsuch, 1934). Incubation usually requires from twenty-one to twenty-three days, with pipping usually occurring on the twenty-third day. Gorsuch estimated that about ten days might be needed for nest selection and construction, thirty-eight to forty-two days for egg-laying and incubation, and nearly three months is required for raising the brood to an independent state. Thus two broods cannot be raised successively by a single pair even with the long nesting season typical of the southwestern desert. However, during highly favorable nesting seasons supplementary nestings may be achieved by two different methods. The males may take over the care of the brood, leaving the female free to begin a second clutch, or, more commonly, the chicks may be "weaned" when about a month old and left in the care of older birds of the area, thus allowing the pair to start a second clutch (Gullion, 1956a). In one desert area where such double-brooding occurred, the average number of chicks per adult pair was fifteen, whereas on the valley habitats where double-brooding did not occur the average number of chicks per adult pair was ten.

When the young are hatched, the family leaves the nest-site and does not return again. Brooding by the female occurs on shady and well-sheltered areas, while the male typically "stands guard." As the brood moves, the male usually takes the lead, with the chicks following and the hen bringing up the rear. Males leading young chicks regularly perform distraction displays, while the hen and young "freeze," or both adults may fly off as the young remain in place (Gorsuch, 1934). Like all young galliforms, the chicks feed almost exclusively on insect life during the earliest part of their life but soon begin to take leaves and other succulent vegetation and within a few months are consuming about 90 percent vegetable materials (Gorsuch, 1934).

Vocal Signals

The most complete analysis of vocalizations of the Gambel quail is that of Ellis and Stokes (1966), which will be followed here. They grouped the species'calls into those associated with group activity, with feeding relationships, with responses to enemies, and with agonistic and sexual phases of reproductive behavior.

Calls important in integrating covey activity are the basic contact *took*! note, a conversational *ut*-growl, and the location call. The contact note is uttered by both sexes and carries only a short distance. It occurs at all times of the day, but is especially associated with foraging. A similar call, the *ut*-growl, is the same note with an added trill and is especially prevalent when the birds find food or water after being deprived of them.

The location or separation call is a four-noted *ka-KAA-ka-ka* (also interpreted as *cow-COW-cow-cow* or *chi-CA-go-go*) and is produced by birds when separated from their mate or the covey. Both sexes produce the same call, but sufficient individual variation occurs in the call (which is the most acoustically complex as to cadence and amplitude characteristics) that individual recognition is typical. Visually isolated birds keep in contact by use of this call, and males can distinguish the location call of their mates from those of other females.

No specific food calls were noted by Ellis and Stokes, nor have I heard any. Evidently paired males do show or pick up food particles in front of their females, a display ("tidbitting") that is widespread in galliform birds, but Ellis and Stokes did not notice any associated calling. However, Prososki (1970) did hear calling in this situation.

Several calls are associated with responses to enemies. The most typical alarm note of Gambel quail, as well as other *Callipepla* species, is a repeated *chip-chip-chip* as the birds investigate any disturbance during moderate alarm or curiosity. When thoroughly frightened and rushing for cover, a bird utters a raucous *squawk* followed by a series of *chip* notes, or the two kinds of calls may be alternated. The *squawk* note is both louder and more prolonged than the *chip* sounds, but they probably intergrade with one another. During times when the birds are being held in one's hand, they usually utter loud, down-slurred distress *kee-OW*! notes, repeated almost indefinitely at intervals of about one-half second. Both sexes use the call, but individuals vary in the ease with which the call can be elicited from them.

The reproductive phase of sexual behavior has several associated calls. One of the most important of these is the *kaa* or *cow* call, already discussed in the section on social and reproductive behavior. Another is the location or separation call, *ka-KAA-ka-ka*, uttered by members of a pair whenever they are visually separated. Ellis and Stokes noted that during copulation at least the female, and probably also the male, uttered a series of short squealing calls. When an unpaired male is displaying toward a female, he faces her and utters a series of *wit-WUT* aggressive notes that are the same as occur when two males are threatening one another. At this time the head is bobbed somewhat, causing the erect plumes to vibrate, and the bird stands in an erect posture.

During aggressive encounters between two males, the same *wit-WUT* call is uttered, often alternated with pecking movements or actual attacks. In such situations the calling may be almost continuous as the birds face one another, seemingly unwilling to attack or retreat. After a varying number of such threats and attacks, one of the birds typically utters a cat-like *meah* call, at the same time lifting his beak almost to a vertical position. This call no doubt is homologous to the *squill* of the California quail but is both more prolonged and much slower in the associated head movements. This call usually stimulates the other male to respond in the same fashion and generally leads to a termination of the encounter.

Observations on the vocalizations of a male hybrid bobwhite × Gambel quail (Prososki, 1970) allows for the establishment of some probable vocalization homologies between these genera. The announcement call of the unpaired male bobwhite is a whistled *bob-white!* (Stokes, 1967). The hybrid's call was a similar two-note call, but the two notes were virtually identical in volume and frequency characteristics, sounding something like *cow-COW!*

The separation call of the male hybrid was apparently the same call as the male's announcement call, whereas in the bobwhite two calls (*hoy-poo* and *hoy*) serve this purpose. The calls are also used in agonistic situations by male bobwhites.

Two calls were produced in agonistic situations by the hybrid male, a two-noted *porquoi* and a growling *ker-ra-wa* call. Typically he would begin with a number of *ker-ra-wa* calls, followed by several *porquoi* notes. The *ker-ra-wa* calls sonagraphically most resemble the *hoy-poo* calls of the bobwhite, while the second note of the *porquoi* approached the *meah* in its acoustic characteristics. No sounds resembling the Gambel quail's *wit-WUT* call were produced.

The hybrids also produced chipping alarm calls, hand-held distress calls, contact calls, tidbitting calls, and copulation calls, all of which were comparable to those of both parental species, since interspecific differences are generally not great in these calls.

It is of interest that in this group of quails the male call that is used to announce the location of unmated males (thus also communicating information on species, sex, and reproductive state) is a simple, one-syllable note in at least three species (Gambel, scaled, and California quails). However, the call used by both sexes to announce the location of a bird separated from its mate and serving both for individual recognition and for homing purposes consists of two notes (in elegant and scaled quail), three (in California quail) or four (in Gambel quail), varying in cadence, pitch, and loudness but all having similar harmonic characteristics. In the Gambel and California quails the male announcement call is, in effect, a single note "excerpt" from the longer location call, while in the scaled quail the male's announcement note more closely approaches a pure whistle. This distinction between a harmonic-rich location call and a nearly harmonic-free whistle for a male announcement call is even greater in the bobwhite. The bobwhite also seemingly has a greater number of agonistic calls than do the species of *Callipepla*, and in general its acoustic communication system appears to be more complex.

The Gambel quail apparently has two basic male agonistic calls, one of which (the *wit-WUT*) is used during sexual display toward females and aggressive encounters with other males, and the other (the *meah*) which is used only toward other males and apparently serves to break off aggressive encounters. Similarly the California quail has two calls, the *wip-wip*, which serves the same function as the Gambel's *wit-WUT*, and the *squill*, which occurs during high-intensity male-to-male threat. In contrast, the scaled quail seems to lack a call comparable to the *wip-wip* or *wit-WUT*, and the head-throw call is performed by both sexes in agonistic situations, although it is used predominantly by males. Again, the bobwhite is the most complex in its agonistic vocabulary. Both sexes use the *hoy* and *hoy-poo* calls in agonistic situations, and two additional calls, the *squee* and "caterwaul," are largely but not entirely characteristic of the males (Stokes, 1967). The *hoy, hoy-poo,* and "caterwaul" calls seem to represent one intergrading motivational complex, while the *squee* call has a different seasonal and contextual occurrence. Thus a certain vocal duality is present, but it is difficult to judge possible homologies in these calls. One might only imagine that the evolutionary trend has been from a situation (as in the scaled quail) where both sexes perform a common call in an agonistic situation to one (as in Gambel and California quail) where the male has separate vocal signals for male-to-male situations and male-to-female situations, and finally (as in bobwhite), to a condition where both sexes have a complex intergrading series of calls associated with varying agonistic situations.

Ellis and Stokes (1966) list a total of ten call-types for the Gambel quail, of which at least seven are common to both sexes, two occur only in males, and one (the copulation call) occurs in the female and possibly also the male. Stokes's analysis (1967) of the bobwhite's vocalizations indicated a considerably larger number of vocalizations, but the intergrading qualities of many of the calls make a strict numerical comparison impossible.

EVOLUTIONARY RELATIONSHIPS

The close similarities in downy and adult plumage patterns, as well as strong behavioral similarities, clearly indicate that the Gambel and California quail are close relatives. The ecological differences between the two species prevent extensive sympatry, but where limited contact does occur hybridization has been found (Miller and Stebbins, 1964). It would seem reasonable that the Sierra Nevada range might have provided an effective geographic barrier that allowed speciation to develop to the point that now exists and has still virtually prevented any extensive population overlap, partly because of the major climatic differences prevailing on the two slopes of this range. It also seems possible to assume that the common ancestral type may have had a range in the southern part of the continent similar to that now occupied by the Gambel quail, and that as the ancestral California quail adapted to the moderate climate of interior California it gradually extended its range northward into the coastal portions of the Pacific northwest.

25

California Quail

Callipepla californica (Shaw) 1798
(*Lophortyx californicus* in *A.O.U. Check-list*)

OTHER VERNACULAR NAMES

CALIFORNIA partridge, Catalina quail, Codorniz Californiana, crested quail, San Lucas quail, San Quintin quail, topknot quail, valley quail.

RANGE

From southern Oregon and western Nevada south to the tip of Baja California. Introduced into southern British Columbia, Washington, Idaho, northern Oregon, and Utah.

SUBSPECIES (*ex A.O.U. Check-list*)

C. c. californica: Valley California quail. Resident from northern Oregon and western Nevada south to southern California and Los Coronados Islands of Baja California. Introduced in eastern Washington, central British Columbia, western Idaho, Oregon, Utah, and Colorado.

C. c. catalinensis (Grinnell): Catalina Island California quail. Resident on Santa Catalina Island and introduced on Santa Rosa and Santa Cruz islands, southern California.

C. c. plumbea (Grinnell): San Quintin California quail. Resident from San Diego County, California, through northwestern Baja California, Mexico.

C. c. achrustera (Peters): San Lucas California quail. Resident in southern Baja California, Mexico.

C. c. canfieldae (van Rossem): Inyo California quail. Resident in Owens River valley in east central California.

C. c. orecta (Oberholser): Great Basin California quail. Resident in the Warner Valley, southeastern Oregon.

C. c. decoloratus (van Rossem): Baja California quail. Resident in Baja California from 30° north latitude to about 25° north latitude.

C. c. brunnescens Ridgway: Coastal California quail. Resident in the humid coastal area of California from near the Oregon boundary south to southern Santa Cruz County. Introduced on Vancouver Island, British Columbia.

MEASUREMENTS

Folded wing: Adults, both sexes, 105–19 mm (males average 5 mm longer than females).

Tail: Adults, both sexes, 79–119 mm (males average 4 mm longer than females.

IDENTIFICATION

Adults, 9.5–11 inches long. The sexes are different in appearance. This widespread quail of the western foothills resembles the Gambel quail inasmuch as both sexes have forward-tilting, blackish crests that are enlarged terminally into a "comma" or "teardrop" shape. Both sexes also have clear bluish gray to gray chests that become buffy toward the abdomen and have darker "scaly" markings reminiscent of scaled quail. The flanks are brownish gray with lighter shaft-streaks, and the upperparts are generally gray to brownish gray, intricately marked with darker scaly markings. Males have black throats and a chestnut-tinged abdomen and are chocolate brown behind the plume, while the area in front of the eyes and above the bill is whitish.

FIELD MARKS

The combination of a "teardrop" crest and scaly markings on the lower breast and abdomen is distinctive for both sexes. Males of this species may be distinguished from the very similar Gambel quail by the combination of a whitish rather than blackish forehead, no black abdomen patch, and dull brown rather than chestnut brown flank and crown coloration. A three-note *chi-ca-go* call serves as a location call for both sexes.

AGE AND SEX CRITERIA

Females have dark brown rather than black crests and lack black throats.

Immatures have buff-tipped upper greater primary coverts which are carried for the first year (Sumner, 1935; Leopold, 1939), and the outer two primaries are relatively pointed and frayed. Maximum width (but not length) of the bursa of Fabricius may be used as an accurate indication of immaturity through December (Lewin, 1963).

Juveniles resemble females but have forehead feathers with indistinct pale grayish terminal spots and have shorter and lighter crests (Ridgway and Friedmann, 1946). See Gambel quail account.

Downy young (illustrated in color plate 110) are very difficult to distinguish from young Gambel quail (see that species' account), but they can be recognized from downy scaled quail by their less grayish white and more yellowish body tones, and by the fact that the pale spinal stripe in the California quail is cinnamon-buff rather than a dirty brownish buff. This species is considerably lighter and more yellowish on the lower back and tail than downy elegant quail.

DISTRIBUTION AND HABITAT

The California quail exhibits a rather complex distribution pattern that extends along the western coast of North America for about two thousand miles, from the southern tip of Baja California, Mexico, to the southern part of Vancouver Island, British Columbia. Along this entire range its coastal distribution is almost unbroken except for forested areas associated with the Coast and Olympic ranges. The climatic and precipitation variations along this coastal strip are considerable, ranging from hot scrub desert along much of Baja California, through a mild Mediterranean climate associated with chaparral vegetation in southern California and a cool, wet coastal forest (where the bird occurs in edge and successional vegetation stages) from central California northward to Puget Sound. In the interior of these coastal states, as well as in Nevada, Idaho, and Utah, the species also occurs in valleys and rain-shadow areas dominated by grasslands or semidesert sagebrush shrub, although many of these interior populations are introduced ones.

FIGURE 38. Current distributions of the California quail (shaded) and elegant quail (hatched).

In Mexico, Leopold (1959) reported that the highest populations are found in chaparral vegetation along the northwestern Baja coast and foothills and in scrubby tropical forest and brushland at the tip of the San Lucas Cape, but they also occur in desert washes wherever there is a combination of brushy cover and water available.

In California several races occur, but all are associated with brushy vegetation in combination with more open weedy or grassy habitats and available water supplies. Heavy forest and dense chaparral is avoided even by the coastal race, although dense-foliaged trees may be used for night roosting. The exact vegetational composition is probably not so important as life-form characteristics of the dominant vegetation, namely an interspersion of brush and more open vegetational types (Grinnell and Miller, 1944).

In Oregon the species was probably originally confined to the counties bordering California (*californica*) and Nevada (*orecta*), but trapping and transplanting activities have spread the bird's range to most of eastern Oregon and many western Oregon counties, with consequent mixing of subspecies stocks (Masson and Mace, 1962). The highest populations occur in the Columbia basin and in central and southeastern Oregon, in dry, semidesert vegetation.

The Washington population of California quail is likewise largely or entirely an introduced one, of uncertain subspecific designation. Its preferred habitat is thickets, brushy tracts, logged areas, and burned over districts, and although sometimes seen in second-growth timber it avoids heavy woods (Jewett et al., 1953).

In Canada the California quail is generally limited to one small introduced population on the southern part of Vancouver Island and another in the Okanagan Valley (Godfrey, 1966; Lewin, 1965). More is known of the Okanagan and Similkameen valley populations than the island population, and Lewin reported that about 390 square miles of these river valleys are occupied by an estimated population of about 250,000 quail. The quail are associated with orchards and irrigated areas and are generally found below two thousand feet elevation. A few also occur in native vegetation consisting of scattered thickets of aspen (*Populus*), rose (*Rosa*), Saskatoon berry (*Amelanchier*), and chokecherry (*Prunus*), but they do not extend into the higher coniferous woods (Lewin, 1965).

In Idaho the species occurs locally along watercourses of the Snake River valley from near the middle of the state to the Oregon line, and a limited population also occurs along the Snake and Clearwater rivers in northern Idaho and perhaps in the Clarkia and upper St. Joe river valleys as well (Upland Game Birds of Idaho).

In Nevada the range of the possibly originally native California quail has been greatly affected by release programs, but the birds are usually associated with rose and willow thickets along streams, where cover and water are both available. In western Nevada the heaviest populations occur in agricultural areas, but the birds are found wherever springs exist. In eastern Nevada their distribution is limited and spotty (Gullion and Christensen, 1957).

In Utah the species was first introduced over a century ago and thus is now found in scattered areas around the state, but it is primarily limited to semiarid foothills and valleys, especially along streams (Rawley and Bailey, 1964). An introduced population once occurred in north central Colorado, but now is wholly extirpated.* Recent attempts at establishing the species in Arizona may have been successful in the vicinity of the Little Colorado River near Springerville, but it is too early to be certain of this.

POPULATION DENSITY

Population densities doubtless vary considerably in this species according to habitat quality. Emlen (1939) reported on a "low density" winter population that contained 113 birds on a study area that represented a density of 1 bird per 7 acres. However, if only the occupied home ranges of the birds were considered, the four coveys' total occupied area was 93 acres, or 0.9 acres per bird. Raitt and Genelly (1964) reported on a population that also contained four winter coveys on approximately 100 acres. Over an eight-year period this area had fall populations ranging from 25 to 140 birds and averaging 101 birds, or 1 bird per acre. Since the average fall age ratio was 1.47 juveniles per adult, the average spring breeding population (ignoring spring to fall adult mortality) must have been at least 41 adults. Thus a spring breeding density of approximately 1 bird per 2 acres would seem probable. These figures are in general agreement with those of Glading (1941), who recorded late winter densities on a study area in central California that varied over a six-year period from 1.7 to 3.9 acres per bird.

Maximum population densities that have been noted for the species are some reported on a private hunting club property where artificial feeding and predator control measures were used, and fall populations of up to 4.8 birds per acre were attained (Glading, Selleck, and Ross, 1945).

HABITAT REQUIREMENTS

A fairly detailed analysis of habitat needs of the California quail has

*Glenn Rogers, 1970: personal communication.

been made by Emlen and Glading (1945). They classified quail habitat into four general types, desert, range land, dry farming land, and irrigated land, of which the range land is most extensive and most important to the species. Within these general categories, the basic habitat requirements of food, water, escape cover, roosting cover, nesting cover, and loafing cover are variably available. Irrigated lands provide water but may be limited in the various cover types, especially for roosting, nesting, and loafing. Dry-land farming areas are even less suitable, since they may lack available water in addition to escape cover or other cover types. Deserts usually provide both food and cover sources, and if water is locally available, they may support moderately large quail populations. Range lands vary greatly in quality of habitat, but the best offer available water, seed-producing herbaceous plants, and moderately open brushy cover that will serve for escape, nesting, roosting, and loafing.

Edminster (1954) has analyzed the aspects of cover that are most desirable for quail usage. Nesting cover is usually herbaceous rather than brushy, in a moderately open situation. Roosting cover is provided by tall shrubs or trees, with evergreen species being preferred for winter cover. Escape cover consists of dense growths of shrubs, vines, or herbaceous growth into which the birds can readily run when frightened. Feeding cover is usually not limiting, since the birds consume a large variety of seeds, but leguminous plants are preferred both for seeds and their leafy growth, perhaps because of their nitrogen content. Loafing cover consists of shady places under shrubs or trees, where relief from the midday sun is available and dry dust as well as grit may be readily available. The California quail depends more on available water or succulent plant material than does the Gambel quail, but it is more drought tolerant than the bobwhite (McNabb, 1969). Probably as long as insects and succulent vegetation are available the bird can survive indefinitely without surface water, and moderately saline water sources (but not sea water) can also be utilized (Bartholomew and MacMillen, 1961).

FOOD AND FORAGING BEHAVIOR

The animal portion of the diet of California quail is relatively small and even during summer probably contributes no more than 5 percent of the diet of adults (Martin, Zim, and Nelson, 1951; Edminster, 1954). Otherwise, nearly the entire remainder of the diet consists of herbaceous leafy materials and seeds, with grains and fruits playing a very subsidiary role in most areas.

Edminster (1954) summarized much of the early food studies of California quail and concluded that the most important food sources were legumes (25 to 35 percent of all foods taken) and annual weeds (20 to 60 percent), followed by grasses (10 to 25 percent) and the fruits and leaves of woody plants (3 to 5 percent). Of the important legumes, bur clover (*Medicago*), lupines (*Lupinus*), deervetches (*Lotus*), clover (*Trifolium*), acacias (*Acacia*) and vetches (*Vicia*) are major food sources, especially their seeds. The leaves and seeds of filaree (*Erodium*) and the seeds of turkey mullein (*Eremocarpus*) are important food sources among the weedy herbs (Edminster, 1954; Martin, Zim, and Nelson, 1951).

Two more recent California studies confirm these earlier conclusions as to the significance of legumes for this species. Shields and Duncan (1966) found that during the fall and winter, seeds comprised over 80 percent of the bird's diet, with four species of legumes (*Lotus*, *Lupinus*, and *Trifolium*) alone making up 60 percent of the sample volume. With the start of the winter precipitation, the intake of leaves increased from 6 percent of the diet in November to 41 percent in January, with the leaves of forbs, clover, and grasses all being utilized. The importance of legumes was also pointed out by the study of Duncan (1968), who compared the foods taken during fall in burned and unburned rangeland. Relatively little difference in the two habitat types was found, with seeds from five species of *Lotus*, *Lupinus*, and *Trifolium* again making up from 66 percent of the early fall diet in unburned areas to 80 percent of the diet in burned areas. Among non-legumes, filaree and turkey mullein were important seed sources.

Food studies from areas outside the California quail's native range are more limited and suggestive of greater dependence on nonnatural food sources. In Nevada a considerable utilization of grain crops, such as wheat, barley, and corn, as well as the legumes alfalfa and sweet clover, is indicated by Martin, Zim, and Nelson (1951). In eastern Washington, Crispens (1960b) found that wheat seeds were the most important source of food throughout the year. Seeds of various weedy species, such as pigweed (*Chenopodium*), teasel (*Dipsacus*), and locust (*Robinia*) were selectively utilized, and both sunflower (*Helianthus*) and Russian thistle (*Salsola*) were highly preferred food sources. Surprisingly, legumes were found in very limited quantities among these samples.

The general lesson to be obtained from these studies is that the need for brushy habitat by the California quail is largely a reflection of its protective cover requirements, while most of its food sources come from herbaceous forbs, particularly legumes.

MOBILITY AND MOVEMENTS

Emlen's study (1939) of California quail movements is still the most complete and will be summarized here. During the winter, the birds occupied home ranges roughly comparable to the size of the covey, with four coveys of twenty-one to forty-six birds using home ranges of seventeen to forty-five acres. These covey locations were associated with the distribution of brushy cover such as shrubs, perennial weeds, and vineyards. Each covey tended to feed together but sometimes broke up into smaller feeding units. Usually the birds of a covey roosted together but sometimes used two or three roosting sites. The coveys were separated by distances of from 350 yards to a half a mile, and contacts between coveys were thus infrequent. However, during such intercovey contacts, a "social barrier" between members of the two groups existed, which virtually prevented any covey shifting. Winter movements were very restricted, with rarely more than a fourth or at most a half of the covey's home range being used during any single day. Over a period of time, however, the birds would feed in different parts of the covey's home range.

Beginning in late February, coveys began to break up as pairs and unmated males began to break away from the group and apparently moved into more open farm land that was not suitable for winter use because of its limited cover. About half of sixty-seven marked birds separated from their coveys by the first of April, and the birds which left were predominantly males. At least one male moved a mile and a half before the nesting season. Further, younger males were evidently more inclined to leave the covey than older ones, since fourteen of the twenty-one males that disappeared were young. Only one of the twenty-one young males remained to nest on its winter territory, while seven of eighteen older males did so. Likewise, the young females tended to leave the winter range, while the adult hens all remained in the covey. By the middle of April the covey was composed of a nearly balanced ratio of the sexes and apparently consisted largely of older and mated birds. The second phase of covey breakdown was caused when these birds dispersed for nesting. Only a few nonnesting or late nesting birds remained around the winter roosting sites.

Movements during the summer were highly restricted and were largely limited to those of unmated males. These birds began to *cow* call in late April with the start of the nesting period and would attempt to approach females of mated pairs. Of eight such birds, four established "crowing territories" near the nest of an established pair, while the others assumed a more nomadic existence, sometimes covering a mile in a single day. Later, Genelly (1955) discovered that most such territories are held by old males, while the first-year males are principally nomadic. On the other hand, mated pairs limited their daily moves during egg-laying to from twelve to twenty-five acres while foraging, and returned at night to a roosting site, sometimes held in common with a neighboring pair. When incubation began, movements were even more limited, to about three to ten acres around the nest.

Many nesting attempts were unsuccessful, and losses of a member of the pair caused some shuffling. If a mated male was lost, the female soon mated with one of the unpaired "crowers" near the nest or became foster parent of an available brood. When males lost their hens they started crowing within a day, either at the same place or at distances from one-fourth to one and one-half miles away from the original nesting location.

With the hatching of young, the re-formation of coveys began, with broods forming covey nuclei. By the middle of August, nine such covey nuclei had been established, and these attracted individual nonbreeders or unsuccessful breeders, so that the covey sizes gradually grew. Brood mobility was very low during the first few weeks of life, probably being limited to a few acres, but they ranged up to ten or twenty acres by the end of the first month. Some older broods moved considerable distances when their brooding cover was destroyed, with one brood of ten-week-old chicks moving a mile from its point of hatching. However, most broods remained close enough to the nest site that they wintered on the covey home range nearest their place of hatching. Although little interbrood shifting occurred in very young broods, this increased after the young were three or four weeks old, and the adults would tolerate the presence of other chicks of the same age. Contacts became more frequent when the chicks were somewhat older, and soon mergers of broods occurred, with nine broods gradually being incorporated into six subcoveys.

The subcoveys retained their identities until late November, when they condensed into four coveys that exhibited ranges nearly identical to those held the previous winter. Eight of twelve marked birds returned to the winter range held the previous year, while four occupied new winter ranges, but in all probability less than half of the total number of adults returned to their previous winter ranges.

A more recent study by Genelly (1955) supported Emlen's view that the dominant, nesting territory-holding males are usually older birds, while the nomadic and unmated ones are primarily young birds. It would seem probable, therefore, that population dispersion and range extension would be primarily the result of movements by young birds, especially males. Lewin (1965) mentions a report of a male being seen during midsummer some twenty-two miles north of regularly inhabited range. Also, when birds have been released into new areas considerable movement sometimes occurs; Richardson (1941) noted several such movements in excess of twenty miles and one extreme case of a ninety-five-mile movement.

On the basis of movements of recaptured birds at various trap sites,

Raitt and Genelly (1964) obtained an index of relative mobility, which suggested that summer and winter movements are least, while spring and fall movements are more extensive, particularly during April and May. These observations tend to support Emlen's views that a good deal of individual movement occurs in spring, especially among males. Although fall mobility is also moderate, there is little interchange of covey members at this time, thus a "spring shuffle" rather than a "fall shuffle" may tend to bring about population mixing.

SOCIAL AND REPRODUCTIVE BEHAVIOR

The covey is the social unit of the California quail from late fall until early spring. Emlen (1939) and, later, Howard and Emlen (1942) have pointed out quite clearly that in the California quail the covey is a relatively closed social unit, with little opportunity for intercovey mixing. This mixing is reduced or prevented during late winter and spring by attacks on outsiders by resident birds of the same sex; such established covey members always socially dominate aliens that are introduced into a covey. However, Howard and Emlen emphasized that this aggressive behavior should not be considered territorial defense by covey members but rather a form of social dominance associated with confidence related to the residents' knowledge of the local range. Territorial behavior in the sense of a defended area does not occur in coveys or mated pairs of this species (or probably any New World quail); only some unmated "crower" males exhibit anything like proprietary behavior toward a specific piece of habitat.

The process of covey breakup and pairing has been well studied in this species, first by Emlen and later by Raitt (1960) and Genelly (1955). Perhaps because older males begin their reproductive development somewhat sooner than younger ones, pairing that occurs prior to covey breakup involves primarily older males, which mate with both adult and first-year females. Such pairing probably begins in late February or early March, and during early stages of pair formation some shifting about of partners may occur. Most pairing occurs before the testes are much enlarged (Anthony, 1970), thus pair formation does not necessarily involve copulation or other strong sexual behavior patterns on the part of the pair, although copulation attempts may occur. Genelly (1955) felt that an initial stage of "acquaintanceship" might be required, during which individual recognition develops. No striking displays need occur in association with pair formation (Raitt, 1960), and only rarely is the "rush" display of males seen. Genelly (1955) mentioned seeing it only when females were placed in traps, and I have seen it only when a female was introduced without prior contact into the cage of an unmated male. The display consists of several low notes followed by an extension of the neck and a lowering of the head, a fluffing of body feathers, a raising and spreading of the tail, and a slight extension and marked drooping of the wings, so that the primary tips touch the ground. In this posture the male approaches the female in a series of short rushes, from which the hen typically flees. The highly aggressive origin of the display may be seen from the similarity of it to threat postures assumed toward other males and the actual pecking attack that the male may perform on the female if she is unable to flee. In short, the display appears to be a strong assertion of dominance, and probably only the submission behavior of the female and her lack of male plumage features normally inhibits overt attack.

As the males and females of incipient pairs begin to remain with one another an increasing amount of time, male-to-male aggression also increases. This probably largely involves a chasing of other males from the vicinity of the mate, and an eventual exclusion of such unmated males from the covey. Since the sex ratio of spring coveys always has an excess of males, a forcible exclusion of surplus males is the only way that the covey can remain intact and persist as an integrated social unit. Raitt (1960) noted three major forms of hostile behavior: side-by-side nudging, chasing, and overt fighting. Nudging is the least aggressive of the three, and sometimes occurs among members of a pair or between adults and young, with the dominant bird pushing the other to one side as they both jostle for a common food source. Chasing consists of a posture much like that mentioned as typical of the "rush" display, but somewhat less extreme form. The bird being chased usually flees on foot and if caught may be severely pecked on the back and nape. Most often, such chases involve two males, but sometimes females chase females, and less frequently males will chase females. One case of a mated female chasing away an unpaired male has also been noted (Genelly, 1955). Overt fighting is virtually limited to males and is essentially like that of other quail, with the two birds facing one another, making pecking attacks and short vertical leaps during which they attempt to peck the top of the opponent's head. Between attacks, a series of *squill* calls and associated rapid head-throws that maximally expose the black throat are frequent and no doubt serve as major visual and acoustical threat signals.

Genelly (1955) noted a continued increase in fighting incidence from January until May, with this rise largely reflecting fighting concerned with the defense of the mate. Defense of territory occurred only from March through June, and consisted of fights among unmated males that had established crowing territories and subsequently repulsed other such males.

Starting in July, fighting associated with the defense of the brood occurred, but by October all of the fighting, which gradually diminished until January, was concerned with peck order establishment in the fall and winter coveys. Genelly could find no evidence that California quail actively defend a nesting site, thus the term "nesting territory" is not appropriately applied to the species.

As the mated pairs gradually break away from the covey and locate nesting sites, unpaired males attempt to establish crowing territories in the vicinity of such mated pairs. Genelly first heard *cow* calls uttered by these males in March, and the calling persisted until mid-June. This period corresponds roughly to the period of testis growth plotted by him. The greatest concentration of crowing males was located where nesting pairs were also located. Genelly found only one instance of a mated male uttering a *cow* call and heard a captive female produce it on at least two occasions, so the clear function of the call is that of advertising the location of a sexually active, unmated male. Since laying females that lose their mates through death rapidly attain new mates, the biological advantage of crowing is readily important. However, the localization of crowing males in the vicinity of nesting females may tend to increase the predation rate on such nesting birds.

The gonadal cycle of the female lags about two weeks behind that of males during spring (Genelly, 1955; Anthony, 1970), with adult females either developing slightly in advance of young ones (Genelly) or at approximately the same time (Anthony). Egg laying during Genelly's study in California started the second week of April, with a peak activity the third week in May, while in eastern Washington the peak of laying activity was about a month later, according to Anthony. The rate of egg laying is about 5 per week, at least in captive birds (Genelly, 1955), and the eggs are apparently usually dropped about midmorning. The average clutch size has been reported as 10.97 eggs by Glading (1938b), 13.7 eggs by Lewin (1963), 13.7 (in New Zealand) by Williams (1967), and 14.2 eggs by Grinnell, Bryant, and Storer (1918). Thus, an average figure of 14 eggs in a complete clutch would seem to be a reasonable judgment, which might thus require a total of about twenty days to lay; this plus an additional twenty-two-day incubation period would total forty-two days from the laying of the first egg to the day of hatching (Lewin, 1963). My incubation records indicate that twenty-two or, more commonly, twenty-three days may actually be required for incubation under artificial conditions.

Although renesting is a regular aspect of California quail behavior, the question of the frequency of second broods is not yet fully resolved. Definite instances of second broods have been recorded; McLean (1930) found one such case in a wild bird. Francis (1965) also reported two cases of confined quail in which the male took over the care of the young after about two weeks, when the female remated and began a new clutch, which was subsequently hatched and raised. McMillan (1964) noted that early nests and broods of quail were being cared for by males, while females were presumably freed to raise additional broods. Finally, Anthony (1970) noted that during June and July a larger number of broods were tended by lone males than during August and September, suggesting either that there was high early female mortality or that females left the early broods in the care of males and went on to produce second clutches, the latter of which he believed to be the case. Incubation by males is probably not a regular feature of California quail behavior as long as the female is present; they do not exhibit highly vascularized brood patches as do females (Genelly, 1955). The visual stimulus of an abandoned clutch of eggs may bring about hormonal changes in males that initiate brooding behavior and defeathering adequate to form a simple brood patch (Jones, 1969b).

Broodless males, such as those who have lost their mates, have great interest in young chicks and, if admitted by the parents, make excellent foster parents (Emlen, 1939). However, although crowing males exhibit extreme interest in young broods, they are not allowed to tend them as long as they persist in their crowing behavior, according to Emlen. Parents and chicks gradually merge with unsuccessful adults and eventually with unmated males and with other well-grown broods, forming moderately large aggregations of birds.

Although the percentage of unsuccessful nesting attempts is high in California quail, the combination of persistent renesting, large clutch sizes, and occasional double-brooding usually assures a high ratio of young birds in fall coveys. Nesting losses have been estimated by Sumner (1935) to be about 60 percent, and other studies such as those of Glading (1938b) have revealed losses as high as about 80 percent. In New Zealand, Williams (1967) reported a fairly high nesting success of 62.6 percent, if only nests with completed clutches were considered rather than all indications of nesting attempts being considered. His figures also indicate a fairly high incidence of egg fertility (93.8 percent) and hatchability of fertile eggs (89.8 percent). Anthony's studies indicate a surprisingly high survival rate of chicks, with an estimated 25.8 percent mortality during the first fifteen weeks of study. Edminster's review of other studies (1954) suggests that a chick loss of about 45 to 50 percent may be normal. Over an eight-year period, the yearly fall age ratio of a quail population studied by Raitt and Genelly (1964) varied from 0.56 to 2.22 immatures per adult, or a yearly average of from about one to five young reared per adult female, allowing for a somewhat unbalanced sex ratio in adults. Perhaps an over-all average fall age ratio would be about 1.46 young per adult (Emlen, 1940), or about three young raised per female.

Vocal Signals

A complete analysis of the vocal repertoire of the California quail has recently been provided by Williams (1969), whose terminology will in general be followed here.

Social integration calls include the contact call or *ut, ut* notes and the separation ("assembly") *cu-ca-cow* call. The *ut, ut* notes serve to keep individuals of a group in contact and are given frequently as the birds move about while foraging. When birds are separated visually, they may utter the call in a louder version, but it soon leads to the *cu-ca-cow* call. This loud, somewhat melodious call (sometimes written as *chi-ca-go*) is produced almost identically by both sexes, although there is a certain degree of individual variation in the call. Thus, males can definitely recognize the call of their own mates and will preferentially respond to them. Besides serving as a general separation call the *cu-ca-cow* plays an important role in reproduction, by serving to keep the pair together. In spring the call increases in frequency even in birds that are not separated, when unpaired birds of both sexes begin to use it. However, paired females do not use it unless separated from their mates, and unpaired males soon change from this call to the *cow* crowing call described earlier. This call is much like the last syllable of the separation call, but is uttered from a conspicuous, usually elevated, position. The call is repeated fairly often, averaging from about three to eight per minute. Williams established that the rate of *cow* calling was under testosterone control and was associated with relative aggressiveness. Thus the functional and hormonal origin of the call and the associated establishment of crowing territories is analogous to the territorial behavior of unmated male songbirds.

The *squill* call (called the "sneeze" by Williams) was so named by Sumner (1935), who described it as a high-pitched staccato whistle, used in a situation of defiance to other males. The call is limited virtually entirely to males and occurs only during the breeding season. Somewhat in contrast to the related *meah* call of the Gambel quail, its utterance does not indicate a mutual "stand-off," but rather it is associated with extreme threat and attempted social dominance. The neck-stretching caused by the head-throw raises the pitch of the vocalization to a near whistle, no doubt because of the increased tension on the tympanic membranes. A second aggressive call of the male is the *wip, wip* call which often precedes attacks on other males and may alternate with the *squill* call. It may also be uttered toward strange females, but I have never seen a male perform a *squill* call toward a female. Likewise, the *wip, wip* call has not been reported for females, which utter only *ut, ut* or *cu-ca-cow* calls in this situation.

When feeding, California quail utter soft and repeated *tu, tu* notes, which stimulate pecking by other birds. During the sexual tidbitting display of males to females this same call is uttered.

The calls associated with predator avoidance are several, of which the alarm *pit, pit* notes are perhaps most common. With almost any disturbance these metallic-sounding calls are uttered, especially before the birds begin to flee. When actually fleeing on foot they are more likely to utter a series of *chwip, chwip* sounds that are perhaps a variant of the earlier call. The avian predator alarm call is a low, throaty *kurr, kurr, kurr,* which may stimulate freezing or fleeing behavior by other birds. Following such disturbance a soft *put, put* series of notes may be produced, which may prolong the freezing behavior. When held in the hand, adults of both sexes often utter a loud, downslurred *pseu, pseu* note much like those of other New World quails.

Williams reported that prior to copulation or during it females sometimes uttered soft peeping calls, and males usually produced *ut, ut* notes that changed to *wip, wip* sounds during treading. When building her nest, the female uttered a low, repetitive *pa, pa, pa* series of notes, while the male uttered rather different sounds as he handled nesting material.

No special calls other than contact *ut, ut* calls were associated with incubation, and during brooding of young chicks the parents both uttered low *mo, mo, mo* notes when the chicks became scattered. Chicks that are lost utter a loud distress whistle, to which the adults respond with the *cu-ca-cow* call, especially from the male. Adults also uttered the food call when attracting young to a source of food.

In total, Williams found fourteen adult call types in the California quail. Of these, eleven were typical of both sexes, and three characteristic of the male only. Two of the fourteen were associated with social contact, five with alarm responses, six were believed to have reproductive significance (including two agonistic calls), and one was associated with parental behavior. Most of the California quail's calls have their counterparts in the bobwhite. However, Williams related the absence of a call functioning to space winter coveys (as the *koi-lee* is reported to do for the bobwhite) to the fact that winter coveys of the California quail are generally larger than in bobwhites and sometimes tend to come together into very large wintering flocks.

EVOLUTIONARY RELATIONSHIPS

The probable evolutionary history of the California quail has been discussed in the earlier account of the Gambel quail.

26

Bobwhite

Colinus virginianus (Linnaeus) 1758

OTHER VERNACULAR NAMES

AMERICAN colin, Codorniz Común, Cuiche Común, partridge, quail.

RANGE

Virtually all of the eastern United States north to southern Maine, New York, southern Ontario, central Wisconsin, and central Minnesota, west to southeastern Wyoming, eastern Colorado, eastern New Mexico, and eastern Mexico south to Chiapas and adjacent Guatemala, but excluding the lowlands of Yucatán. Also existing as isolated populations in Sonora (largely extirpated) and as introduced populations in the Columbia and Snake river basins of Washington, Oregon, Idaho, and northwestern Wyoming (Bighorn and Shoshone river valleys). Currently being reintroduced into southern Arizona.

SUBSPECIES (*ex A.O.U. Check-list* and Aldrich, 1946; Mexican Races from *Check-list of Birds of Mexico*)

√ *C. v. virginianus:* Eastern bobwhite. Resident of the southern Atlantic seaboard north to Virginia southwest to north central Georgia, southeastern Alabama, and northern Florida.

C. v. marilandicus (Linnaeus): New England bobwhite. Resident of New England north to southwestern Maine southwest to east central New York, Pennsylvania, and central Virginia and south to southern Maryland and Delaware (not in *A.O.U. Check-list;* part of *C. v. virginianus*).

C. v. mexicanus (Linnaeus): Interior bobwhite. Resident of much of eastern United States east of the Great Plains excepting the Atlantic Coast (not in *A.O.U. Check-list;* part of *C. v. virginianus*).

C. v. floridanus (Coues): Florida bobwhite. Resident of most of peninsular Florida.

C. v. texanus (Lawrence): Texas bobwhite. Resident of most of southern Texas adjacent to New Mexico and northern Mexico including parts of Coahuila, Nuevo León, and Tamaulipas.

C. v. taylori Lincoln: Plains bobwhite. Resident of the Great Plains from South Dakota southward to northern Texas and eastward to western Missouri and northwestern Arkansas. Introduced populations in Washington, Oregon, and Idaho in the Columbia and Snake river basins.

C. v. ridgwayi Brewster: Masked bobwhite. Resident in central interior Sonora and formerly north to southern Arizona. The fate of recent Arizona restocks (near Arivaca and in Altar Valley) is still uncertain.

C. v. cubanensis (Gray): Cuban bobwhite. Resident in Cuba and the Isle of Pines.

C. v. maculatus Nelson: Mottled or spotted-bellied bobwhite. Resident from central Tamaulipas south to northern Veracruz and west to southeastern San Luis Potosí.

C. v. aridus Aldrich: Jaumave bobwhite. Resident from the northern part of southeastern San Luis Potosí to central and central western Tamaulipas.

C. v. graysoni (Lawrence): Grayson bobwhite. Resident from southeastern Nayarit and southern Jalisco on the Mexican tableland south to the Valley of Mexico, Morelos, southern Hidalgo, and central southern San Luis Potosí.

C. v. nigripectus Nelson: Puebla bobwhite. Resident in the plains of Puebla, Morelos, and Mexico.

C. v. pectoralis (Gould): Black-breasted bobwhite. Resident in central Veracruz at elevations of from five hundred to five thousand feet along the eastern base of the Cordillera.

C. v. godmani Nelson: Godman bobwhite. Resident in the lowlands of Veracruz from sea level to fifteen hundred feet and intergrading with *minor* in Tabasco.

C. v. minor Nelson: Least bobwhite. Resident on grassy plains of northeastern Chiapas and adjacent Tabasco.

C. v. insignis Nelson: Guatemalan bobwhite. Resident in the valley of the Río Chiapas (Río Grijalva) in southern Chiapas and adjacent Guatemala.

C. v. coyolcos (Müller): Coyolcos bobwhite. Resident along the Pacific coast of Oaxaca and Chiapas in the vicinity of the Gulf of Tehuantepec.

C. v. salvini Nelson: Salvin bobwhite. Known only from the coastal plains of southern Chiapas near the Guatemalan border.

C. v. thayeri Bangs and Peters: Thayer bobwhite. Resident in northeastern Oaxaca.

C. v. atriceps (Ogilvie-Grant): Black-headed bobwhite. Resident from the interior of western Oaxaca (Putla) northward along the coast to central Guerrero (Acapulco) and probably south to the range of *harrisoni.*

C. v. nelsoni Brodkorb: Nelson bobwhite. Known only from extreme southern Chiapas; of doubtful validity (Edwards and Lea, 1955).

C. v. harrisoni Orr and Webster. Recently described (1968) from southwestern Oaxaca near the coastal plain.

MEASUREMENTS

Folded wing (United States forms): Adults, both sexes, 98-119 mm (sexual differences negligible).

Tail (United States forms): Adults, both sexes, 49-70 mm (males average 3 mm longer than females).

IDENTIFICATION

Length, 9.5–10.6 inches. The sexes are very different in appearance, and males vary greatly in coloration in different parts of the species' range. Males of most races, however, have a white eye-stripe that extends from the bill through the eye back to the base of the neck, with brown to brownish black coloration above. The ear region is blackish to hazel brown in males, and this feathering extends backward below the white eye-stripe and expands under the throat to form a blackish chest collar under the white chin and throat of most races. However, in some populations the chin and throat are also black, and the lower chest may be either blackish or brownish. In the northern populations the breast and abdomen are irregularly barred with black and white in males, but in southern Mexico all underparts are generally darker and lack white markings. Females of all races have buffy chins, upper throats, and eye-stripes, and buffy tones likewise replace the white underpart coloration of males. Females also lack black collars and in general are more heavily marked with brown and buff barring or mottling both above and below.

FIELD MARKS

Except in some parts of Mexico, the presence of a white throat and a white eye-stripe that contrasts with an otherwise brownish to blackish head will serve to identify male bobwhites. Likewise, no distinct crest is present in this species. Bobwhites most closely resemble the black-throated bobwhites of the Yucatán peninsula but are geographically isolated from them. Gray partridges might be confused with bobwhites, but the gray partridge has no white or pale buff on the head and also has a uniformly grayish chest. The whistled *bobwhite* location call of males in spring is distinctive (but also occurs in the next species), and similar whistled notes serve as separation calls in reassembling scattered coveys.

AGE AND SEX CRITERIA

Females have buffy chins and upper throats, as compared with the white (black in *ridgwayi* and some Mexican races) chins and upper throats of males. The whiter chins of males appear to some extent even in the juvenal plumage. The beak coloration (pale yellow present at the base of the lower mandible in females; males uniformly black) us useful in determining sex of birds as early as 6 to 8 weeks old (Loveless, 1958). Sex of birds at least eight weeks old can be determined on the basis of the central portion of the upper middle wing coverts (Thomas, 1969). Males have fine, black, sharply pointed and well differentiated markings here, whereas females have wider, dull gray bands that do not contrast sharply with the rest of the feathers.

Immatures can often be identified by the fact that their outer two primaries are more pointed than the others (Stoddard, 1931), and the upper greater coverts of the first seven primaries have buffy tips (Leopold, 1939). A small percentage of birds may still be of questionable age by these two criteria, in which case first-year birds may be identified by using the seventh upper primary covert, which is usually brownish with buffy tipping and is somewhat ragged. In adults this feather is darker, sleeker, and has more whitish downy tipping at the feather base (Haugen, 1957).

Juveniles have whitish mottling on the tail feathers and the primaries also have mottled buffy edgings. Pale shaft-streaks are also evident on the upperparts, producing a distinctive light over-all coloration.

Downy young (illustrated in color plate 110) of bobwhites may be distinguished from the *Callipepla* group by their lack of a crest and distinctive spinal stripe and from *Oreortyx* young by their more buffy faces and underparts, as well as their lack of clear black coloration dorsally. The middorsal stripe of bobwhites is russet to chestnut and only slightly darker laterally than in the middle, and the pale stripe immediately below is tinged with brown. A narrow, discontinuous dark stripe extends from the back of the eye to beyond the ear region, where it merges with the darker "shoulder" region. See the black-throated bobwhite account for distinction from that species.

DISTRIBUTION AND HABITAT

The total distributional range of *Colinus virginianus* is a remarkably broad one, extending from the southern part of Maine on the east coast in a nearly unbroken series of populations to the Texas-Mexico border, and southward along the eastern foothills of the Sierra Madre Oriental almost to the Río Usumacinta, and to the Chiapas-Guatemala border in the highlands and Pacific slope. The northern limits of the species' range are extreme southern Maine (Aldrich, 1946; Palmer, 1949), Massachusetts (Ripley, 1957), southern New York (Brown, 1956), southern Ontario (Clarke, 1954), the southern half of Michigan's Lower Peninsula (Janson, 1969), southern Wisconsin (Gromme, 1963), and southern Minnesota, where it is now largely limited to the extreme southeastern part of the state in the Mississippi valley (Longley, 1951).

The western limits of the species' native range are in wooded or brushy river valleys from South Dakota southward along the western limits of mixed-grasses prairies to western Oklahoma and Texas. In Nebraska the bird occurs along wooded river valleys (Platte, Republican, Frenchman) all the way to the Wyoming and Colorado borders (Mohler, 1944; Aldrich, 1946). In eastern Wyoming it is probably native only to the North Platte valley but has recently been well established near the Shoshone and Bighorn rivers in north central Wyoming (Our feathered friends, Wyoming Game and Fish Commission). In eastern Colorado the bobwhite is a local resident all the way west to the edge of the foothills (Bailey and Niedrach, 1967), and in extreme eastern New Mexico the species is largely restricted to the plum thickets or similar low shrubby growth (Ligon, 1961).

In the Oklahoma panhandle the bobwhite is limited largely to river bottom habitats, where tree thickets grow adjacent to pasture lands and relatively dense ground-level cover exists, but it is virtually lacking from the short-grass and sand sage (*Artemisia*) habitats utilized by the scaled quail (Schemnitz, 1964). In western Texas the range extends to about the

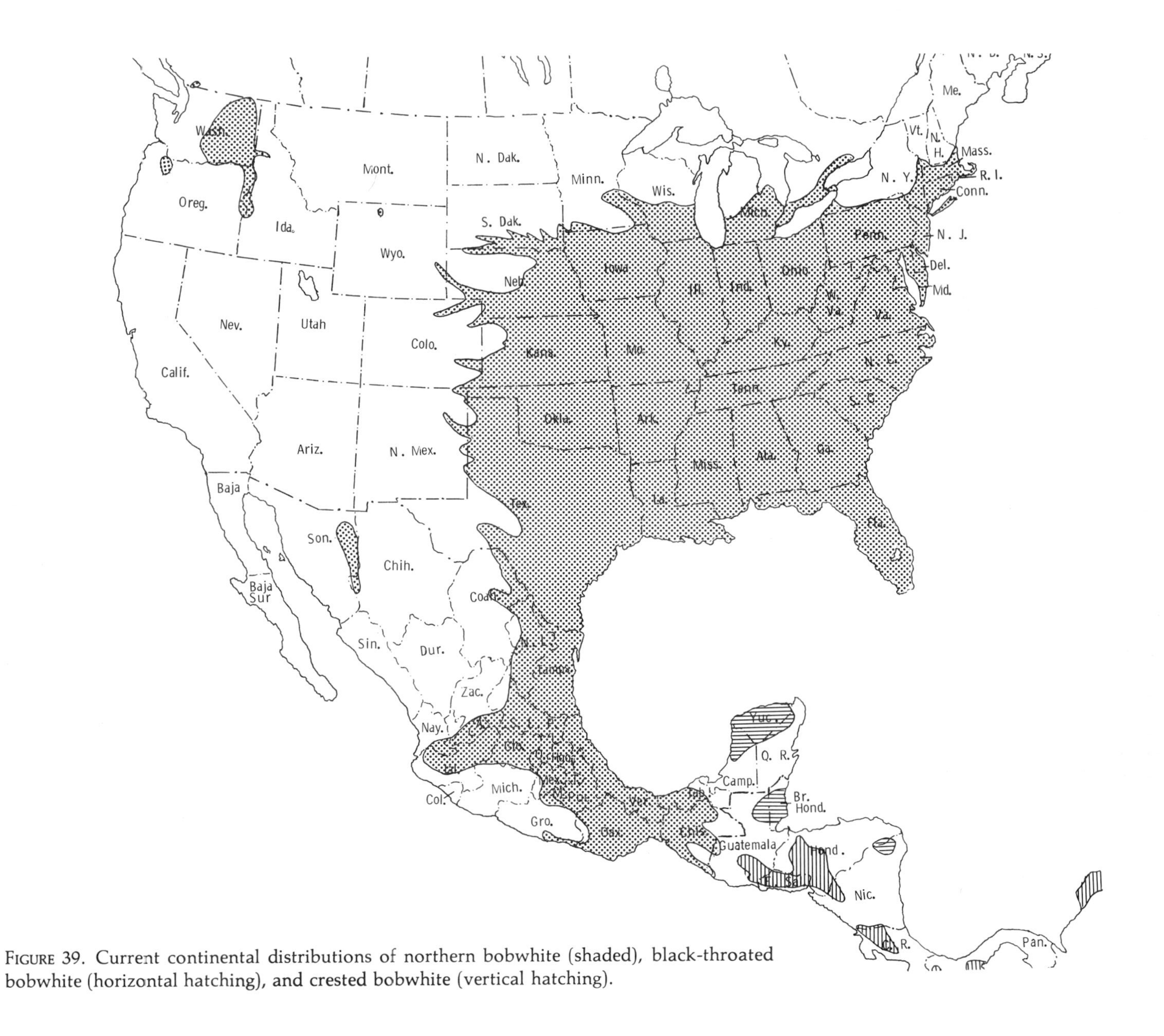

FIGURE 39. Current continental distributions of northern bobwhite (shaded), black-throated bobwhite (horizontal hatching), and crested bobwhite (vertical hatching).

102d meridian; in western and southern Texas the more arid-adapted Texas bobwhite replaces the plains bobwhite, and the birds exist in fair populations wherever excessive grazing does not occur (Principal game birds and mammals of Texas, 1945).

Except for the extirpated Arizona masked bobwhite population (Ligon, 1952), all the more western populations of bobwhites are the result of introductions. In 1970, an attempt to reintroduce the masked bobwhite into southern Arizona was begun by releasing 356 hand-reared offspring of wild birds that had been captured in Sonora during 1968. The success of this effort is still unknown, but well-established populations of bobwhites do occur in Washington, Oregon, and Idaho. In Washington the bobwhite is widely established in the Columbia River basin, and also occurs on a few islands (such as Whidbey) of Puget and Washington sounds, where it was initially introduced in 1871 (Jewett et al., 1953). Birds on the adjoining mainland may barely reach the British Columbia border in the vicinity of Huntingdon. The interior range is more restricted now than formerly, and the bird is presently best surviving in irrigated areas and river valleys such as the Yakima valley (Larrison and Sonnenburg, 1968). In Idaho the bobwhite was first introduced in the Boise valley in 1875, and presently it is found on the lower Boise, Payette, and Weiser river valleys (Upland game birds of Idaho, 1951). In Oregon, where the bobwhite was first released in 1879, the species is best established in the Willamette Valley, as well as near the Columbia River in Morrow and Umatilla counties, and in the Snake River drainage of Malheur County (Masson and Mace, 1962).

The Mexican distribution of the bobwhite has been plotted by Leopold (1959), whose map has been the basis of my own indication of the species' range except the southern parts of Mexico. Further, the present known range of the masked bobwhite in Sonora is much more restricted than is shown by Leopold; it is now believed to be restricted to three small areas there, and the total population may number between four hundred and one thousand birds.

The bobwhite's range in Guerrero has been questionable; Friedmann, Griscom, and Moore (1950) indicate that *atriceps* probably ranges into that state but do not list *nigripectus* for it. Leopold shows only one specimen record for the state, near the Oaxaca boundary, which is presumably referable to *atriceps*. I heard and saw a white-throated male bobwhite calling on the outskirts of Iguala on June 1, 1970, which most probably represented *nigripectus*. From the vicinity of Acapulco southward along the coast toward the Oaxaca border the species is fairly common in brushy habitats. I saw more than thirty in the area near Acapulco and inland as far as Xaltianguis. I also examined several live males that had been captured near Acapulco, and these all had the uniformly dark head color (no superciliary stripe) and a

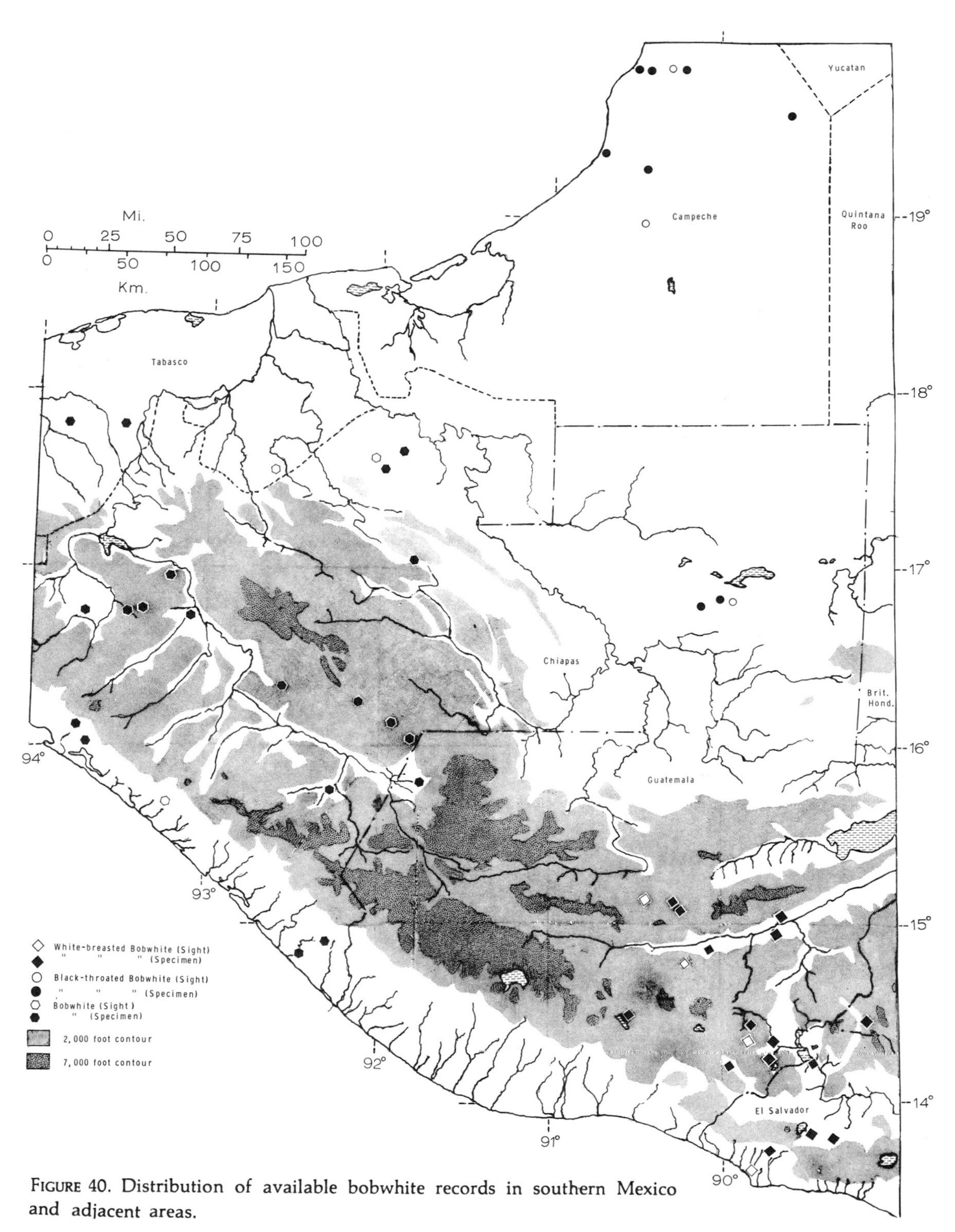

FIGURE 40. Distribution of available bobwhite records in southern Mexico and adjacent areas.

solid black chest, as is typical of *atriceps*. Of two males collected near Xaltianguis, about fifty kilometers inland, one had a slightly light chestnut breast and one had a streaked black and chestnut breast rather than a uniformly black chest, but their heads had no more white present than was typical of the coastal birds. They probably more closely approach the recently described form *harrisoni* (Orr and Webster, 1968), from southwestern Oaxaca. However, I believe that the reduced degree of melanism and resultant plumage similarities shown in the birds from these two areas is of independent origin and simply a reflection of local adaptation to more arid climates.

Although I did not see bobwhites in the vicinity of Copala, Guerrero, I was informed that they occur near there by local residents. Ten miles west of the Guerrero-Oaxaca line near Caljinicuilapa I saw a pair of bobwhites, the male of which appeared to be typical *atriceps*. Another pair was seen along the Río Verde (thirty-one miles southeast of Pinotepa Nacional), and between Pinotepa Nacional and Putla (the type locality of *atriceps*) I saw a total of twenty birds during a single trip. I would judge that Putla represents the interior limit of this subspecies' range, since high and wet country farther to the north is clearly unsuited to bobwhites. Presumably, intergrade populations between *atriceps* and *harrisoni* occur between the Río Verde and San Gabriel Mixtepec, but this road was impassable during the summer of 1970 and thus the area in question could not be visited. Binford (1968) believed that the species occurs along the entire length of the Pacific region of Oaxaca, occurring in savanna and arid tropical scrub habitats, as well as in altered habitats within the general range of tropical deciduous and tropical evergreen forest zones. In the arid interior uplands a white-throated form, *thayeri*, also occurs, and Blake (1950) collected it on the Atlantic drainage near Tutla. In the vicinity of the Gulf of Tehuantepec the coastal population *coyolcos* occurs at least as far west as the town of Tehuantepec. In the Isthmus region the type locality (Chivela) of *thayeri* suggests a population connection with lowland Veracruz bobwhites, but at the eastern edge of the Isthmus and just across the Oaxaca-Chiapas border, specimens referable to *insignis* have been collected on grasslands near Monserrate (Edwards and Lea, 1955). Thus, near the Oaxaca-Chiapas border three rather distinctly different populations probably intergrade, the white-throated race *thayeri*, and the black-throated races *coyolcos* and *insignis*. There is probably a good deal of individual variation in throat and body coloration among birds from this area, judging from remarks made by Ridgway and Friedmann (1946), and many specimens may not be identifiable as to subspecies.

Bobwhites of the race *insignis* are extremely common in the central plateau of Chiapas; I observed them in numerous locations from the outskirts of Cintalapa to a point a few miles west of the Lagunas de Monte Bello, close to the Guatemala border, and I was told by local hunters that they are abundant in the upper reaches of the Río Chiapa (Río Grijalva) all the way to the Guatemala border. At least four of nearly fifty males which I observed closely had no indication of a white superciliary stripe, thus the trait cannot be regarded as a criterion of this subspecies. The interior limits of *insignis* and the possible intergradation of this race with the white-throated race *minor* remain uncertain. Of five male specimens in the University of California collection that have been collected at El Real, Chiapas, all have black throats, but one (MVZ #98109) has "considerable white in the malar regions and laterally on the throat."* Thus it is possible that there has been limited population contact between these highland populations and the lowland populations of *godmani* and *minor* that occur along the wet Caribbean-facing slopes of Chiapas and Tabasco.

Berrett (1963) reported that bobwhites were commonly observed in open savannas of western Tabasco; specimens he examined from that area were closer to *minor* than to *godmani*. He noted that little suitable habitat for bobwhites occurs between that area and eastern Tabasco in the vicinity of Macuspana. East of Macuspana, just across the Chiapas border, fairly extensive savanna occurs between the Río Usumacinta and the wet forests behind Palenque. This area, which is the type locality of *minor*, still supports bobwhites. I saw several pairs in that vicinity and was told by local residents that the birds are hunted to some extent. It is difficult to be certain whether these savannas are partially of natural origin, but when the forests are removed through burning an excellent growth of tall grasses that are highly suitable for grazing purposes can be attained. Bobwhite habitat is, if anything, improving rather than deteriorating in this area as the incidence of well-managed cattle ranches increases.

POPULATION DENSITY

It has been generally agreed that Leopold (1933) was correct in assigning a maximum (fall) quail density of one bird per acre, which he believed represented a saturation point of the species rather than a carrying capacity of the land. He believed that the area of the species' probable optimum range, which centered on the states of Missouri, Illinois, Indiana, and parts of Iowa, was most likely to support populations that would reach but not exceed the saturation point, and he further noted that populations in the more southern states of Mississippi and Georgia were also known to attain this

*N. K. Johnson, 1969: personal communication.

population density. However, on the northern and western parts of the bobwhite's range the populations tended to fluctuate and along the western border of the species' range its density at times exceeded the saturation point in the judgment of Leopold. He noted one Texas estimate of more than two bobwhites per acre over several sections of land in Kenedy County during 1930. In Texas the highest average breeding densities are attained in sandy mesquite semiprairies, pine-oak woodland with interspersed small farms, and transitional coastal prairie uplands, particularly the semiprairies, where early fall densities are generally one per four or five acres but sometimes up to a bird per acre (Principal game birds and mammals of Texas, 1945).

Edminster (1954) suggested that over the best quail range, fall densities may reach from two to ten acres per quail and from ten to fifty or more acres per bird in marginal range. Spring population densities are approximately half the fall figures, or up to a pair per four acres.

In a Kansas study area of about 640 acres, Robinson (1957) estimated that during 1952 a breeding population of 102 birds (with thirty-six mated pairs) was present, while in 1953 the breeding population was 91 birds (with thirty-two mated pairs). Thus, nesting densities of one nest per 20 acres might be expected from such late spring densities. He estimated the maximum carrying capacities of the land for bobwhites to be fifty-three or fifty-four coveys per section during late autumn, since at least 12 acres of habitat are needed to support a single covey. Because his fall coveys consistently averaged 11 to 13 birds, this would agree with other estimates of about one bird per acre as a maximum fall density. It should be noted, however, that he regarded this maximum density to be determined by the carrying capacity of the land rather than to represent a saturation point determined by the species. A density in excess of one bird per acre has recently been reported by Kellogg, Doster, and Williamson (1970).

HABITAT REQUIREMENTS

Edminster (1954) classified the cover types used by bobwhites into four general groups: grasslands, croplands, brushy habitats, and woodlands. He regarded grasslands to be of value primarily during the spring and summer, when they provide nesting cover, some feeding cover, and limited roosting cover. Croplands receive major use during summer and fall, when they provide feeding, loafing, dusting, and limited roosting sites. Brushy areas and woodlands are used through the year for escape and roosting cover but are vital during fall and winter for feeding. Edminster believed that from 30 to 40 percent of the land area should be in grassland,

40 to 60 percent in crop fields, 5 to 20 percent in brushy cover, and 5 to 40 percent in woodland cover for ideal habitat, with a maximum of habitat interspersion and edge margins between habitat types.

Casey (1965) reviewed previous analyses of bobwhite habitat requirements and concluded that three major vegetative types must be present, including grassy nesting cover, cultivated crops or a similar source of food, and brushy cover. He believed that woodlands are not necessary if a brushy cover equivalent to a woodland understory is present. He further believed that a vital habitat factor is the presence of a brushy or woody covey "headquarters," using the earlier concept proposed by Robinson (1957). Such a headquarters must have protective vegetation to provide loafing cover during midday and be separated by about 140 yards from any other covey headquarters. Robinson has found that among ten such headquarters that were in continuous woody vegetation the mean distance between adjacent headquarters was 138 yards. He suggested that such headquarters should consist of areas at least 15 yards square (225 square yards), although some reports indicate that dense woody clumps as small as six feet in diameter might serve, too.

Roosting cover requirements for bobwhites vary somewhat between summer and winter (Rosene, 1969), with the typical roosting behavior serving in winter to maintain body heat through the use of a disk-like formation of birds oriented with their tails together and bodies touching on both sides. Quail use the same circular formation in summer, too, but then the importance of the formation for heat-retention is reduced. The ideal size of such a roosting disk is ten to fifteen birds, and thus the behavior largely regulates the size of winter coveys, a situation in marked contrast to the southwestern desert quail species. Although coveys much larger than fifteen birds will form two such roosting disks, coveys that become smaller will join with nearby groups to maintain this minimum roosting group size. Rosene noted that in the southeast, good winter roost sites are usually on gentle slopes with good drainage, with herbaceous vegetation about two feet high, with bare ground below and exposed sky above. Similarly, in southern Illinois, the sites selected for roosting were usually on medium to low elevations with good drainage, often with south or southwesterly (rarely east or north) exposures that remained warm late in the afternoon, and on bare ground or ground covered only with duff (Klimstra and Ziccardi, 1963). Associated vegetation was typically herbaceous, averaging fifty-nine centimeters high, with relatively little light obstruction. Wheat stubble cover resulting from combining with associated weedy herbs provided ideal roosting cover, and limited burning or grazing may also improve grassland cover for roosting purposes.

Nesting cover requirements are essentially open herbaceous cover with nearly bare ground. The vegetation is usually under twenty inches high, and the stems are sufficiently far apart for the birds to walk through easily. Dead herbaceous material is needed to make the nest lining; thus areas that were not burned the prior spring are preferred over burned areas. Nests are usually within fifty feet of cover edges or other bare ground situations (Rosene, 1969).

To a much greater extent than is the case with the desert-living quails, water in the form of dew or surface water is needed by bobwhites. In the more arid parts of the species' range, the bobwhite becomes increasingly dependent on irrigated areas, river valleys, or other relatively moist habitats. Finally, like all quail, suitable dusting sites are needed in the form of dry and rather powdery soil. Roadsides, field edges, or burned areas all provide such dusting sites, which the birds may visit daily if weather permits.

FOOD AND FORAGING BEHAVIOR

Literally dozens of papers have been written on the food consumption of bobwhites, and it would be impossible to summarize all of them in the available space. Rosene (1969) has provided a recent summary, and the following discussion is based largely on his review.

The animal portion of the bobwhite's diet varies from about 30 percent in summer to only about 5 percent in winter, with the availability of insects largely determining the incidence of foods from this source. However, in southern Florida, where insects are available the year around, the cycle of insect use is similar, indicating a preferential use of insects according to protein needs, which are highest during the period of reproduction.

Based on a study of 1,400 quail crops obtained in Alabama, Rosene concluded that eight of the fourteen most important plant food items were seeds of legume species, and seeds of all types made up 93 percent of the fall diet. Over 3,000 samples obtained from four different soil-type areas of Alabama indicated some regional differences in food consumption. On the sandy coastal plains soils acorns almost equalled legumes in importance during November, but through the winter the use of legume seeds increased to as much as 62 percent by February. In the dark clay "black belt" acorns were not important, and legume seeds contributed over half of the November through February foods. In the red soils of the Piedmont and the red limestone valley soils of northern Alabama legume seeds also provided more than half of the food by volume.

To the west and north, the importance of cultivated grains and weedy herbaceous plants becomes more evident. In Texas, important winter foods in the six different regions varies somewhat, but in four of these regions doveweeds (*Croton* spp.) are most important, and they are among the top five food sources in the other two regions. Danglepod (*Sesbania*) and panic grass (*Panicum*) were the primary food sources in these two regions but had reduced importance elsewhere (Principal game birds and mammals of Texas, 1945).

Winter foods of major importance in Oklahoma include weedy herbs such as ragweed (*Ambrosia*), sunflower (*Helianthus*), and trailing wild bean (*Strophostyles*), as well as acorns and cultivated plants such as sorghums and lespedezas, judging from various studies summarized by Rosene. Robinson's study of Kansas bobwhites (1957) indicated that during a nine-month period three plants, sorghum, wild beans, and foxtail millet (*Setaria*), were most important, and all of these were eaten during most of the nine months.

In Missouri, fall and winter foods vary in different regions, but on a state-wide basis the five most important seed-producing plants are probably Korean lespedeza, corn, ragweed, sorghum, and oats (Korschgen, 1948).

In the northern parts of the bobwhite's range, especially the "corn belt," the availability of corn or other grain is clearly of some importance for winter survival. In Nebraska corn is perhaps the most important winter food (Damon, 1949), and in Indiana the four most important fall foods were corn, sassafras, Korean lespedeza, and ragweed (Reeves, cited by Rosene, 1969). Winter foods in southern Illinois include, in diminishing importance by volume, corn, soybeans, Korean and common lespedeza, acorns and wheat (Larimer, 1960).

Bobwhites typically have two foraging sessions a day, one in early morning and one in late afternoon which lasts until dark. Little if any feeding is done when the vegetation is wet following rain or heavy dew, and the birds move only as far from their roosting cover as is needed to obtain adequate food. Birds of a covey feed together without aggression, and males may attract their mates to a choice morsel of food by using the tidbitting display. Grit may be picked up at the time of foraging, or searched out separately along roadways or cuts.

MOBILITY AND MOVEMENTS

Bobwhites are among the most sedentary of quails, and virtually no major seasonal movements are normally performed. Some early records of "migrations" were no doubt the result of dispersals following unusually high fall populations (Rosene, 1969). Perhaps the nearest approach to a true migration may be seen in the Smoky Mountains, where, at elevations of from 3,500 to 6,500 feet, bobwhites occur on "grass balds" during the summer but are rare or absent there from September through April, when they move to lower grounds (Stupka, 1963).

During the winter covey period, each covey occupies a range which is large enough to fulfill its roosting, foraging, and escape-cover requirements but which rarely exceeds 50 acres. Rosene (1969) estimated the covey ranges of more than one thousand coveys in Alabama and South Carolina and found averages in four areas that ranged from 8.2 to 17.9 acres. Farther west and north the winter covey ranges may tend to be somewhat larger; Schemnitz (1961) summarized studies from Missouri and Texas that indicated an average winter covey range of 24 acres, and one from Oklahoma reported an average covey range of almost 50 acres. Robinson (1957) believed that a minimum of 12 acres was required to support a covey of bobwhites during the critical season in Kansas.

With the coming of spring, coveys gradually move from their winter range into the nesting range. In some areas, particularly in the south, these movements may not be very great. In one Kentucky study (Wunz, cited in Rosene, 1969), six of nine coveys moved less than one-quarter of a mile between late winter and early spring, and none moved more than three-quarters of a mile. Of thirty-four birds, twenty-four moved less than one-quarter of a mile. Similarly, in Florida all but one of twenty birds moved less than one-quarter mile between April 1 and mid-June (Loveless, 1958), and in Missouri most quail move less than one-half mile during the spring period (Murphy and Baskett, 1952).

In one Wisconsin study (Kabat and Thompson, 1963), movements of marked quail observed between April 8 and May 26 averaged 0.6 miles from the winter range, while between May 27 and June 23 the average distance for marked birds was 1.3 miles from the winter range. This would indicate that a considerable number of birds, perhaps unmated males, continue to move about for some time after the breakup of coveys. Robinson (1957) noted that movements of males during the breeding season were almost twice as far as during the nonbreeding season, with females' movements averaging only slightly less than those of males and the difference between yearling and adult birds being insignificant.

Summer movements by mated pairs and pairs with broods are relatively negligible. Studies of summer mobility in Missouri (Murphy and Baskett, 1952) and in Florida (Loveless, 1958) indicate that nearly 90 percent of the birds moved less than half a mile. In both instances, records of longer movements were believed to have been the result of movements of unmated males.

By fall, with the growth of the young completed, and the integration of the broods into coveys, considerable social reorganization occurs. Unmated males and unsuccessful pairs probably attach themselves to pairs with well-grown young, and members of individual broods may break up and become affiliated with different fall coveys. This period of instability has been called the "fall shuffle." Agee (1957) investigated this phenomenon in Missouri and, surprisingly, found that fall movements (0.14 miles) averaged less than summer movements (0.39 miles) and were only somewhat greater than average winter movements (0.08 miles). He found that whistling males tended to join coveys near their summer ranges, with eleven of nineteen males apparently joined to the first family group they encountered. Of seven family groups, five had eventual winter ranges that overlapped their summer brood ranges, and a maximum movement of 200 yards was noted. Four fall coveys were developed from two families each plus unmated males and apparently unsuccessful pairs, while one covey was comprised of young from only one family. No quail in or with a brood moved more than 710 yards during the fall or winter, and most moved less than 400 yards.

In contrast to these findings, the studies of Duck (1943) indicate that in some areas fall movements may be considerable. In twelve or thirteen counties of northwestern Oklahoma, there is a distinct shift from summer ranges in sagebrush uplands and mixed grasslands to winter ranges in canyon bottoms and dunelands. Eleven quail that were banded during August and September and were recovered in December had moved an average distance of 9.7 miles and one was found 26 miles from the banding point, which is the maximum known case of a seasonal movement of bobwhites that I have encountered in the literature.

Yearly movements between successive winters provide a general index to bobwhite mobility traits; Kabat and Thompson (1963) noted that the average distance moved by both sexes between successive winters in Wisconsin was only 0.78 miles, with males moving significantly farther than females. In no case was a movement of more than four miles recorded among more than one hundred birds for which such records were obtained.

In summary, it would seem that in general bobwhites are not highly mobile, even during the fall period. Indeed, such mobility and potential range extension as does occur may be related more directly to late spring and summer movements by young birds, particularly males.

SOCIAL AND REPRODUCTIVE BEHAVIOR

During the winter the social unit is the covey, which as mentioned earlier tends to average from about 10 to 15 birds, largely because of the need for efficient temperature maintenance during roosting. Kabat and Thompson (1963) noted that coveys drop in average size from about 17 birds in November to 7.5 birds by late March, representing a 56 percent winter loss. Other studies indicate covey sizes of from about 12 to 15 birds as typical, suggesting that covey size is a reflection of behavior rather than a possible index of population density. There appears to be no definite indication of specific age or sex structure in these winter coveys; males or females may predominate, and the size of the covey bears no apparent relationship to its age composition.

With spring, however, social structuring of the covey begins to develop. Rosene (1969) considered the breeding period to begin with the first *bob-white* whistling, which may be as early as January in the South and early March in the northern states. However, if the bobwhite is like the western quails, much pair formation will have occurred before whistling is well under way. Further, it is most unlikely that pair formation under natural conditions is normally characterized by the male's performance of the elaborate display described by Stoddard (1931). However, his description is worth quoting, since it is the typical posture elicited when a male in breeding condition is initially exposed to either a strange female or male: "The display is a frontal one. The head is lowered and frequently turned sideways to show the snowy-white head markings to the best advantage, the wings are extended until the primary tips touch the ground, while the elbows are elevated over the back and thrown forward, forming a vertical feathered wall. The bird, otherwise puffed out to the utmost in addition to the spread, forward-thrust wings and lowered, side-turned head, now walks or advances in short rushes toward the hen, and follows her at good speed in full display in case she turns and runs." I have never seen the head-turning described by Stoddard, but otherwise his description agrees with my own observations. The similar if not identical responses of males to other males clearly indicates the aggressive nature of this display and its probable function in initial establishment of social dominance. Males in the same cage will not hold this posture long, but rather engage in actual fighting if they are roughly equal in social rank, but when prevented from fighting by cage walls will often perform the display whenever they are allowed to see one another.

Stokes (1967) has studied this "frontal" display and concluded that its function is aggressive rather than sexual, serving to establish social dominance. Only when a female fails to respond in kind does a male accept her as a female. Strictly sexual displays of the male bobwhite include lateral display, bowing, and tidbitting. During lateral display the male walks slowly about the female, with tail fanned and its upper surface tilted toward her. The flank feathers are held loosely and drooped toward the ground and the head is somewhat lowered, but the wings are not distinctly drooped. Lateral display is silent and is usually brief. Bowing is closely associated with lateral display and consists of incomplete pecking movements, while the body is held horizontally and the bird walks around the female. During the breeding season the food call of the male is used in conjunction with pecking movements, which collectively serve as a tidbitting display and attract females, especially the male's mate. Tidbitting probably serves as a major means of pair bond maintenance, since it extends well beyond the period of actual pair formation. Female displays include wing-quivering movements and an inconspicuous lateral presentation display. Copulation is not preceded by any specific precopulatory behavior but is often preceded by female presentation behavior and is initiated by crouching on the part of the female. The female calls during copulation, but no definite postcopulatory display is present (Stokes, 1967).

Nest-building, performed by both sexes, is initiated by the digging of a scrape a few inches deep and four or five inches in diameter (Rosene, 1969). This scrape is then filled with dead leafy materials, so that the bottom of the nest is nearly level with the adjacent soil. Grasses or other herbaceous plants are arched over the top of the completed nest, effectively concealing it. The first egg is usually deposited one or two days later, and the egg-laying rate is approximately 1 per day with about eighteen to twenty days needed to complete a clutch of about 14 eggs. The average clutch size has been variously reported as 14.4 (Stoddard, 1931), 12.5 (Schemnitz, 1964), and 13.2 (Klimstra and Scott, 1957). There may be yearly variations in this, and in addition late clutches tend to have fewer eggs than do early-season clutches (Stoddard, 1931). Hatching typically occurs on the twenty-third day after incubation is initiated.

Robinson's study (1957) indicated that in Kansas during 1952 some nesting attempts were begun in early April or mid-April, while male calling did not become common until late May and early June, so that there was a lag of about a month between the peaks of nesting activity and calling. Peak calling occurred in mid-June, which was near the period (late June) Robinson estimated to be the time of maximum hatching. Fatora, Provost, and Jenkins (1967) also noted that male calling reaches a peak about a week before hatching. Robinson thought that "in addition to unmated males, mated males whistle in the breeding season, especially at the time of emergence of the young." However, Stoddard (1931) concluded that the whistle is "largely" that of unmated males, while Rosene (1969) thought that mated males "may or may not" whistle while the female is on the nest. Perhaps the best answer to this question comes from Robeson (1963), who compared the whistling behavior of a definitely unmated male and an apparently mated male. He found that the unmated male usually uttered six or more calls per minute and called from eight to ten minutes, with the last note of the *ah-bob-white* call being loud and piercing. The bird almost always responded to a whistled covey call and was highly mobile, moving up to one-quarter mile in three hours. By contrast, the apparently mated bird called four or less times a minute, for durations of two minutes or less, and the last note of the call was soft and subdued. It was not observed to respond to the covey call, and was wholly sedentary.

From these and other reports, it would seem that nearly all the calling by male bobwhites is attributable to unmated birds that are announcing the locations of their whistling territories. These birds tend to establish such territories as close as possible to those of mated pairs, thus accounting for the positive relationship between the locations of calling males and nesting sites (Klimstra, 1950a). Such males with established whistling territories forceably expel other males from the immediate area and these nonterritorial birds, presumably most often yearlings, are no doubt responsible for the considerable summer movements recorded among males. In all likelihood, males which fertilized their mates early in the breeding season will have been past the peak of their fertility by the latter part of the female's incubation period. Should her nest be destroyed at that time, the availability of "surplus" whistling males still in maximum breeding condition makes a rapid remating and initiation of a fertile second clutch highly likely. Such a possibility would seem to provide the adaptive function of unmated males' whistling and more than counterbalance the potentially dangerous effect that their conspicuous presence near active nests might provide. The rapid decline in whistling at or shortly before the time of hatching probably is an indication that these birds are passing out of their reproductive condition. The gonadal cycle may be somewhat independent of the molt cycle as to hormonal control (Watson, 1962c), but it is probable that mated males would be first to go out of reproductive condition. At least in the case of males that have been participating in incubation (which may be about 25 percent of the nests judging from Stoddard's data), prolactin levels are undoubtedly high (Jones, 1969a). The birds' abilities for further gamete production are as a result probably quite limited, since high prolactin levels have been found to interfere with sperm production in such birds as phalaropes and white-crowned sparrows.

It is typical for females to renest at least once if their first attempt is unsuccessful, and perhaps as many as two or even three renesting attempts may be made. However, not only are renests somewhat smaller in average clutch size, but also the likelihood of successful hatching declines during summer (Rosene, 1969). There is so far no indication that bobwhites ever normally have second broods under natural conditions, but in a captive situation three different pairs were observed to produce a second brood by the male's undertaking brooding responsibilities when the young were about two weeks old, and the female's then starting a second clutch (Stanford, 1953). It is possible that such behavior also occurs in wild populations where there is an unusually long breeding season, such as in Mexico, but this situation would be unlikely over most of the United States bobwhite range.

Although nesting losses may on the average be as high as 60 or 70 percent, persistent renesting attempts by females is likely to result in at least half of the adult females in a population bringing off a brood. Hatchability of eggs is usually high, and in Wisconsin and Iowa the initial brood size may be between 13 and 16 chicks (Klimstra, 1950b; Kabat and Thompson, 1963). Most chick mortality probably occurs during the first two weeks, and by late October and November the average brood size may be reduced to about 8.5. By that time the broods have been joined by unmated males and unsuccessful pairs, and the resulting fall coveys will have grown to about 12 to 17 birds. Fall age ratios in hunter-kill samples may range from as high as 85 percent juveniles (6.6 young per adult) to as low as 72 percent juveniles (2.4 young per adult), judging from a survey by Kabat and Thompson (1963). In general, about 80 percent of the fall population can be expected to consist of juvenile birds, which figure thus also roughly corresponds to the average annual mortality rate of the species. The resultant life expectancy for a bobwhite is less than a year; therefore relatively few birds survive to breed more than once.

Vocal Signals

The recent paper by Stokes (1967) provides a complete summary of the vocalizations of the bobwhite, which are perhaps the most diverse and complex of those of any United States species of quail.

The bobwhite call, already mentioned, is limited almost exclusively to males during the breeding season, particularly unmated ones.

Group movement calls used by both sexes are a series of increasingly louder *hoy*, *hoy-poo*, and *koi-lee* or *hoyee* notes that have been called the separation call (Stokes, 1967), scatter call, and covey call (Stoddard, 1931). Stokes has established that it not only functions to reunite separated pairs but also probably serves to space coveys, to attract unmated males to unmated females, and to repel intruders. Softer contact notes, *took* and *pitoo*, are used when the birds are feeding together. However, the typical food-finding call is a soft *tu-tu-tu-tu* series of notes uttered with the bill pointed toward the source of food. This is used both by the male during the tidbitting display and by parents directing young to food.

When frightened by ground predators, a soft, musical *tirree* is initially uttered, but this usually quickly changes to an *ick-ick-ick* or *toil-ick-ick* as the birds become more alarmed. These latter notes are similar and no doubt correspond to the repeated *pit* or *chip* notes of *Callipepla* species. As the source of danger disappears, a soft *tee-wa* note may be uttered. The avian alarm note is a throaty *errrk*, and a loud, down-slurred distress *c-i-e-w* is produced when the birds are held in the hand. A somewhat similar but softer *psieu* note is uttered by adults during distraction display, which may be followed by repeated, staccato *tip* notes. Females may utter a "take-cover" call when a brood is disturbed, causing them to hide and freeze.

Agonistic calls of the bobwhite are greater in number than those of *Callipepla;* Stokes has recognized four different calls functioning in this situation. These are the "caterwaul," *squee, hoy,* and *hoy-poo.* Of these, only the caterwaul and *squee* are limited to the agonistic situation, while the *hoy* and *hoy-poo* have group and pair contact functions as well. Both caterwauling and the *squee* may be performed by both sexes but are more frequent in males. The *squee* note, a long series of whining or muttering-like sounds, is indicative of a thwarted attack or a balance between attack and escape tendencies. The caterwaul, however, is a loud, raucous call sounding like *h-a-o p-o-o w-e-i-h'* that is clearly indicative of a dominant status and a strong attack tendency and is often associated with frontal display. Rarely do males utter this call toward strange females, but it is typically elicited when a strange pair is visible, and less often when a single rival male is seen. Its nearest functional equivalent in *Callipepla* is the head-throw of the scaled quail or the *squill* of the California quail, although the associated postures and sounds are quite different from either of these.

Stokes has mentioned several additional calls typical of parent-young interactions, including a "broody call" of the parents, two different alarm notes, as well as the "take cover," distraction, or "decoy ruse" call, and the food-finding calls already mentioned. Chicks have at least two calls, a "contentment" note and a distress or separation call.

EVOLUTIONARY RELATIONSHIPS

There can be little doubt that the nearest living relatives of *Colinus* are the species of *Callipepla (sensu lato).* Holman (1961) has indicated that on the basis of skeletal structure these species might be considered congeneric, and I (1970) have indicated that the same conclusion might be made on the

↑ Hybrid Barred x Scaled Quail

↓ Hybrid Mountain x California Quail

↑ Hybrid California x Scaled Quail

↓ Hybrid Scaled x Gambel Quail

↑ Hybrid Bobwhite x Gambel Quail

↓ Hybrid Bobwhite x California Quail

↑ Hybrid Bobwhite x Scaled Quail

↓ Elegant Quail, Male and Female

↑ Gambel Quail, Male and Female

↓ Gambel Quail, Male

↑ California Quail, Covey

↓ California Quail, Covey

↑ Bobwhite, Covey

↓ Masked Bobwhite

New Mexico Dept. of Game & Fish

↑ Godman Bobwhite, Male

↓ Guatemalan Bobwhite, Male

Black-throated Bobwhite, Male

↑ Black-throated Bobwhite, Male

↓ Black-throated Bobwhite, Male and Female

Spotted Wood Quail Habitat, Chiapas

U. S. National Zoological Park

↑ Spotted Wood Quail, Male

↓ Spotted Wood Quail, Female

↑ Singing Quail Habitat, Chiapas

↓ Mearns Harlequin Quail, Female

↑ Mearns Harlequin Quail, Male

↓ Ocellated Quail, Male

Colo. Game, Fish & Parks Dept.

↑ Gray Partridge, Male

↓ Gray Partridge, Male

↑ Chukar Partridge

↓ Chukar Partridge

basis of hybridization evidence. Were it not for the taxonomic problems at the species level existing within the bobwhites, this would probably be the best treatment, but considering that three fairly distinct populations of bobwhites exist and at least for the present are best regarded as full species, the application of the generic name *Colinus* to this population complex seems the most practical method of emphasizing their close relationships to one another without too seriously obscuring the relationships of the bobwhite group to the more typically crested quails of the arid Southwest.

Among the *Colinus* × *Callipepla* hybrids so far produced (involving Gambel, California, and scaled quails), only those with one California quail parent have exhibited any fertility beyond the F_1 generation so that second generation (F_2) hybrids have been hatched and have survived to maturity. It seems reasonable to believe that the ancestral *Colinus* type diverged from an ancestral *Callipepla* well before any splitting of the latter's gene pools into populations representative of any of the living species. The southernmost point of current common contact between the genera is southern Mexico, and this area would seem to be a possible region of origin for the genus *Colinus*. Possibly the Isthmus of Tehuantepec served as an initial extrinsic isolating factor that split the early *Colinus* population into northern (pre-*virginianus*) and southern (pre-*cristatus, nigrogularis*) segments, or perhaps the mountainous highlands of northwestern Guatemala provided such a barrier, but at least at present the latter group of mountains seems to be the primary barrier between the *insignis* population of *virginianus* and the *incanus* population of *cristatus* (see figure 41). Curiously, no such major barrier separates the coastal populations of *virginianus salvini* and *cristatus hypoleucus*, which are presently separated only by about three hundred kilometers of Guatemala coastal plain between Chiapas and El Salvador.

Assuming that *Colinus* originated in the area of what is now interior Chiapas, the pre-*virginianus* stock probably followed river systems northward to the coastal plain of the Caribbean, where it then moved northward along the Gulf Coast ultimately reaching what is now the eastern half of the United States, where its northward expansion was ultimately limited by cold winters and its western limits set by the arid climates and resulting absence of woody vegetation. The birds also dispersed from the Chiapas highlands to the Pacific coast of Mexico, and northwestward along that coastline in savanna or similar habitats until blocked from further expansion by the arid coastal desert of Sonora, with the interior Sonoran masked bobwhite population representing the point of maximal northwestern expansion. This population was evidently subsequently isolated from the other black-throated and coastal-dwelling populations by extinction of populations between Sinaloa and Guerrero. The Valley of Mexico and adjoining temperate uplands were likewise colonized, probably through movement upward along river systems draining into the adjacent Gulf coastal plains. There birds exhibit the white-throated and fairly light-bodied characteristics of the Atlantic coastal populations, rather than the black-headed and generally dark-bodied condition typical of Pacific coastal birds and those of the Chiapas highlands.

South of the Guatemalan highlands, the ancestral *Colinus* stock probably followed coastal plains and arid highlands southward and eastward, perhaps initially giving rise to a Caribbean coastal population that subsequently developed into *nigrogularis*, as well as a series of more southerly populations that ultimately crossed the Panama Isthmus and spread out over a considerable portion of northern South America. For reasons not presently clear, these populations acquired (or more probably retained) a more distinctly crested condition in males than did those occurring farther north, but this is of minor taxonomic importance. Local adaptations also modified the degree of body darkness, especially the amounts of brown and yellow feather pigments. Maximal loss of pigmentation occurred in the arid Guatemala highlands and adjacent El Salvador, while many of the more southerly populations acquired a fairly dark coloration.

27

Black-throated Bobwhite

Colinus nigrogularis (Gould) 1843

OTHER VERNACULAR NAMES

BLACK-THROATED quail, Codorniz Garganta Negra, Cuiche Yucateco, Yucatán bobwhite.

RANGE

The Yucatán peninsula, the Lake Petén district of Guatemala, and coastal portions of British Honduras and Honduras, to extreme northeastern Nicaragua.

MEXICAN SUBSPECIES

C. n. caboti Van Tyne and Trautman. Resident of eastern Campeche and Yucatán except for the Progreso region.

C. n. persiccus Van Tyne and Trautman. Resident of the Progreso region of the Yucatan peninsula.

MEASUREMENTS

Folded wing: Adults, both sexes, 95–104 mm (sexual differences negligible).

Tail: Adults, both sexes, 50–59 mm (sexual differences negligible).

IDENTIFICATION

Length, 7.5–8.5 inches. The sexes differ considerably in appearance. This species somewhat resembles the bobwhites of adjacent Mexico in that both sexes are uncrested, have brownish crowns, and a white (buffy in females) eye-stripe extending from the forehead down the sides of the neck. Males of this species in addition have a mottled brown crown, black in front of the eyes, and a black throat, which is separated from the black eye-stripe by a broad whitish stripe that passes through the ear region and down the neck, where it separates the black throat from the rest of the underparts. The underparts of males are contrastingly colored with a "scaly" black and white pattern that includes most of the lower surfaces. The upperparts are complexly patterned with buff, brown, and black, as in the bobwhite. Females generally resemble female bobwhites, but the nape feathers are edged with gray, the ear feathers are streaked with buff, and the breast feathers are white with dark shaft-streaks and cross-bars that organize the feather pattern into two rows of terminal white spots.

FIELD MARKS

This species closely resembles the bobwhite, and any bobwhite-like bird on the Yucatán peninsula or at least east of the Usumacinta river can safely be identified as this species. Males utter a *bobwhite* location call that cannot be readily distinguished from the bobwhite's corresponding call.

AGE AND SEX CRITERIA

Females have buffy chins and upper throats rather than black chins and throats as do adult males.

Immatures have the upper greater coverts of the primaries buffy-tipped. They can probably also be identified using the criteria mentioned under the bobwhite account.

Juveniles have narrow white shaft-streaks present in the scapular, interscapular, and upper back feathers (Ridgway and Friedmann, 1946).

Downy young (illustrated in color plate 110) of this species are extremely similar to downy bobwhites but appear to have a somewhat less fuscous and more reddish to chestnut tone dorsally, which is evident both in the mid-dorsal stripe and in the facial areas. The postocular stripe is also more conspicuous and more nearly continuous in this species than in the bobwhite. In the closely related crested bobwhite of South America a trend toward a dark blotch behind the ear in the pale superciliary area also becomes apparent but is much less visible in the black-throated bobwhite and can scarcely be detected in the bobwhite.

DISTRIBUTION AND HABITAT

The Mexican distribution of this bobwhite has been depicted very well by Leopold (1959). I can find no locality records outside the range indicated by him, with the exception of 1965 sight records near the ruins of Coba and Tulum, Quintana Roo (Lee, 1966). These localities are substantially east of any records shown by Leopold or the easternmost locality listed by Paynter (1955) and may reflect a recent range extension related to the forest clearing that is rapidly occurring in Quintana Roo at the present time. The southernmost sight record listed by Paynter is one for the Pixoyal vicinity, about twenty-five miles north of Escarcega, Campeche. The southernmost specimen record I know of is for San Dimas, Campeche (Storer, 1961). I have not seen the bird south of the city of Campeche, but the hilly, calcareous land with dense tropical forest scrub that is being increasingly burned for agricultural purposes should support moderate populations of the species. At approximately the vicinity of Escarcega, this hilly and shallow-soil land gives way to flat, deep-soiled low country that drains into the Laguna de Terminos. From this point southwestward to the Río Usumacinta, Tabasco, I found no evidence of any bobwhites during four trips. The land consists largely of dry or wet lowland evergreen forest. However, an area of savanna-like vegetation occurs from about six miles east of the Río Usumacinta for about eight miles along highway 186, where it merges with lowland evergreen forest. This grazing land would seemingly support bobwhites, but it is associated with a very high rainfall and extremely poor drainage; thus the periodic flooding during the summer would probably effectively prevent utilization by bobwhites, populations of which do occur (*Colinus virginianus minor*) only a few miles west of the Río Usumacinta on hilly ground. Therefore there appears to be a barrier of at least sixty miles of unsuitable habitat between the ranges of the black-throated bobwhite and the least bobwhite, and there has been no recent past contact between them or likelihood of future natural contacts. The accompanying range map illustrates the relationship of the ranges of the black-throated bobwhite with those of *virginianus* and also the Guatemalan bobwhite (*cristatus incanus*).

The two Mexican subspecies of the black-throated bobwhite occupy slightly different climatic and habitat situations, although they no doubt intergrade. Along the northern, relatively arid coast, the subspecies *persiccus* occurs in coastal scrub forest (Paynter, 1955), where the total annual rainfall may be as little as nineteen inches near Progreso (Edwards, 1968). From about twenty kilometers inland toward Mérida (where the annual rainfall is about thirty-six inches) the subspecies *caboti* occurs in the open deciduous forest, small cultivated areas (milpas), and particularly in the extensive henequen or sisal (*Agave fourcroydes*) plantings (Paynter, 1955). Indeed, the map of henequen-producing areas shown by White and Foscue (1939) corresponds fairly well to the probable areas of maximum abundance of this subspecies. Henequen is one of the few economically important plants that grows well on these arid, shallow, and porous limestone soils, and the associated weed cover provides an excellent seed food source, while the spiny leaves and ease of terrestrial movement under the plants provides perfect protective cover for the birds.

In spite of the small home ranges these birds appear to have, they are able rapidly to colonize new forest openings that are associated with the building of any new roads into the heavy forest cover. Indeed, it is probable that the birds follow newly-constructed roads and soon colonize the milpas that invariably develop within a short time. Leopold (1959) noted that he found the species in all the open, weedy fields he saw in southeastern Yucatán, no matter how remote they appeared to be from other clearings.

In the Petén district of Guatemala the species does not presently occur in the vicinity of Tikal but is present at Laguna Petén Itza and is common in the savanna country south of that lake (Smithe, 1966). These savanna-like clearings in the evergreen tropical forest are probably artificially maintained by agricultural activity, rather than being natural results of climatic or soil conditions (Lundell, 1937). Apparently the bird is fairly common on these savannas near La Libertad and San Francisco (Van Tyne, 1935; Saunders, Halloway, and Handley, 1950).

In British Honduras the species is locally common, occurring on lowland pine ridges, especially in mixtures of oak, pine, and palmetto overgrown with tall grasses (Russell, 1964). In eastern Honduras it occurs fairly commonly on lowland pine savannas of the Mosquitia area and also occurs in the corresponding portion of Nicaragua (Monroe, 1968).

POPULATION DENSITY

Few objective figures on population density are available for this bird. Paynter (1955) noted that the bird occurs in "unbelievable numbers" in the *henequen* fields near Mérida. Klaas (1968) indicated that in the coastal scrub near Sisal he observed a pair of quail about every hundred paces during a two-mile walk in August, when the birds were breeding. I have tried to estimate density based on singing males heard during June near Mérida, and in one case, where a henequen field was bounded on two sides by low tropical forest, I observed one definite pair and heard three singing males in an area of 2.6 acres, or about two birds per acre. Just beyond the limits of the paced area at least three more males were singing. The average estimated distance between the six singing males was about 140 yards. In such areas where henequen fields are bounded on one or more sides by scrub or forest vegetation, I would estimate the population to commonly exceed one bird per acre.

HABITAT REQUIREMENTS

The basic bobwhite requirements of a weedy seed supply for food, brushy or woody escape cover, and fairly open grassy or herbaceous cover for nesting and foraging is well met in the northern Yucatán peninsula. If any factor might be limiting, it would be a supply of water or succulent food during the relatively long dry season, which extends roughly from November to May. Virtually no surface water is to be found even during the wet season, but the availability of herbaceous leafage at that time makes a supply of free water unnecessary. However, during the latter part of the dry season, by April or May and sometimes into early June, the herbaceous vegetation has long since dried up and the birds may suffer considerably from lack of water. During such times they may be easily caught by setting out walk-in traps using water for bait.

FOOD AND FORAGING BEHAVIOR

No doubt a variety of weedy herbs provides the seed supply of bobwhites, although not many specimens have been examined for their food contents. Leopold (1959) noted that assorted weed seeds were present in the crops he examined, particularly the seeds of a species of tick trefoil (*Desmodium*) which is commonly found in newly cleared fields.

If my experiences near Mérida are typical, it would seem that the birds leave the cover of the forest scrub before daylight and move out into the henequen fields to feed, where they remain until about 8:00 A.M. By that time the sun is already becoming uncomfortably hot, and the birds gradually walk back to the shady cover of the forest, where they remain through the hottest part of the day. Again at about 5:00 P.M., they will return to the fields for a second feeding session, which may last until dark. They probably return to the cover of the forest to roost at night, although this is somewhat uncertain.

MOBILITY AND MOVEMENTS

Probably the black-throated bobwhite differs very little from the United States races of bobwhites as to its normal mobility. The home ranges of the birds I observed appeared to be very small (during the wet season at least), with only short daily movements between forest cover and open fields. Flights are rarely very long, since normally the birds do not have to move far to get into heavier cover when flushed. Like the United States bobwhites, the birds hold well for pointing dogs (Leopold, 1959), but except in the youngest of henequen fields they occur in habitats that are virtually impossible to penetrate on foot.

SOCIAL AND REPRODUCTIVE BEHAVIOR

During the nonbreeding (dry) season, the birds are found in coveys of about the same size as the United States bobwhite. Stone (1890) indicated that groups of twelve to twenty birds were typical in his observations near Izamal. Probably these coveys are maintained until the start of the rainy season, which varies somewhat in different years but usually begins between April and June. To what extent the reproductive cycle of the birds may be tied to increasing photoperiods or to increased availability of green foods as a proximal stimulus for gonadal development is somewhat uncertain. However, it is interesting that within forty days after I put a captive but wild-caught pair of birds under a long-day artificially lighted situation, the female laid an egg. The small size of the egg suggested that it was perhaps the first she had ever laid, and she had been trapped during the previous dry season and held in captivity several months without laying. One other female also placed on a long-day photoperiod laid her first egg in thirty-five days, but a third female which had been pecked severely did not lay any eggs.

The breeding season begins with the breakup of coveys and the establishment of singing posts by unmated males. Chapman (1896) was apparently the first to notice that the male's whistled *bob-white* call is virtually the same as that of the United States form. The call intervals of individual males probably varies somewhat, but the average of twenty-two such intervals that I timed was seventeen seconds, with ranges of thirteen to twenty-five seconds. Much more commonly than in the United States races the male omits the first note of the *ah-bob-white* call, but sometimes a single male will utter both call-types.

The period of breeding is greatly prolonged and is presumably dependent on the relative length of the rainy season. Paynter (1955) found evidence of breeding in *caboti* as early as April 30 and reported a female with eggs taken in late June. The specimens of *persiccus* that Klaas (1968) collected in August were in full breeding condition, and juvenile birds have also been collected as late as early December (Paynter, 1955). It would seem clear that nesting must occur over a four- or five-month period, allowing plenty of time for renesting and possible double-brooding. One female that I inadvertently collected (by hitting it with a car) near Uxmal on June 11, 1969, had an egg in her oviduct, and I observed a brood of young only a few days old near Mérida on July 19, 1970. I was also informed of the finding in mid-July of a nest near Motul with twenty-two eggs in it, and several natives indicated that nests were to be found in the henequen fields during that month.

The nests are apparently often situated under henequen plants in the Mérida area, but I have not seen any personally and have no information on normal clutch sizes. The eggs I have seen that were laid by captive birds have ranged from pale buff color to white and have been slightly smaller than those of United States bobwhites. The average measurements of fourteen eggs are 22.7 mm (with a range of 21.5–23.5) by 30.5 mm (with a range of 30–32) mm. These eggs were among sixteen laid by a single wild-caught female between July 7 and September 1. Two persons in Mérida told me that very small dark specks may sometimes be present on the eggs, and I have also seen such marking on some of the eggs obtained from our captive birds. The incubation period has not been reported previously, but eggs in our laboratory have hatched simultaneously with those of other *Colinus* forms, at twenty-three to twenty-four days.

Vocal Signals

A recent study (Cink, 1971) on the vocalizations of this species indicates that it has many features in common with *virginianus* and somewhat fewer in common with *cristatus*. Contact calls of the black-throated bobwhite consist of *hoy* and *hoy-poo* notes that in both cases are of shorter duration than in *virginianus*. Likewise, the *bob-white* call of the black-throated bobwhite has an average duration that is only about half as long as in *virginianus* (0.9 seconds compared to 1.9 seconds). The prolonged and multiple-syllable caterwaul call of the bobwhite is represented in the black-throated bobwhite by a single *churrr* element that is similar to that found in *cristatus*.

Calls of recently hatched chicks exhibit comparable similarities and differences to those of adults. Cink found that the "lost" calls of isolated chicks are higher pitched than those of the bobwhite, probably a reflection of the smaller size of the chicks. In both of these species the calls are single-note whistles that are produced in long series. In contrast, the lost calls of *cristatus* consist of a "rolling" tune of higher and lower frequency notes that are distinctly more melodious. Hybrid chicks exhibited intermediate call types.

Cink concluded that although the calls of *nigrogularis* show more features in common with *virginianus* than they do with *cristatus*, the three forms are sufficiently different in vocalizations so as to best be considered allospecies in the absence of suitable tests for possible reproductive isolation. Limited studies on the effects of playbacks of female separation calls indicated virtually no response on the part of unmated males to female calls of the other species.

EVOLUTIONARY RELATIONSHIPS

The geographic distributions of the three major bobwhite types (*virginianus*, *nigrogularis*, and *cristatus*) in southeastern Mexico and Guatemala present an interesting problem of evolution and geographic isolating factors. As mentioned, the black-throated bobwhite is effectively isolated from *virginianus* by an extensive area of wet, tropical lowlands that has doubtless been in existence for a very long period. It is difficult to imagine that the Yucatán population of bobwhite originated by a separation from a common ancestral population in the lowland Campeche Gulf area, and I thus regard the fairly close geographic proximities of these two populations as fortuitous.

Considering the current range of the black-throated bobwhite as a whole, it must generally be accepted that it centers on the Caribbean, extending all the way southward to approximately 15° north latitude. There, it is separated from the interior bobwhite populations of Guatemala and Honduras by climatic and topographic barriers. In Guatemala (see range maps), the known specimen localities of the white-breasted bobwhite (*C. cristatus incanus*) extend easterly almost to Lake Izabal, in the rain-shadow valley of the Río Montagua (Saunders, Halloway, and Handley, 1950), where it occurs in arid tropical scrub vegetation. The more westerly race *hypoleucus* is more typical of grasslands and cultivated areas of Guatemala and El Salvador (Saunders, Halloway, and Handley, 1950; Dickey and van Rossem, 1938). In Honduras the race *sclateri* is typical of the interior highlands, but *leylandi* is found below nine hundred meters in arid Caribbean slope valleys all the way to the coast (Monroe, 1968). Thus there are few topographic barriers between the current ranges of *cristatus* and *nigrogularis*, and their ecological distributions in eastern Guatemala, Honduras, and British Honduras are arid tropical scrub valleys and lowland savannas, especially pine-dominated ones.

It seems probable to me that the black-throated bobwhite evolved in the general area of what is now British Honduras after being isolated from an interior, arid-adapted bobwhite-like ancestor, and subsequently adapted to soil-determined savanna-like openings in this area. Natural disruptions of the tropical forests such as forest fires of the Petén area and Yucatán peninsula probably gradually allowed the population to move northward into the peninsula, as well as moving southeasterly in natural savanna-like areas to what is now eastern Nicaragua. Regrowth of the tropical forest isolated and probably eliminated many of these local populations along the entire range, thus bringing about the quite disjunctive distribution pattern that now exists and allowing a certain degree of subspeciation to develop.

The apparently completely allopatric distributions exhibited by the three major bobwhite types pose a problem in taxonomy. Perhaps they should be regarded as allospecies (Amadon, 1966), to emphasize the obviously very close relationships existing among them. Certainly I agree with Monroe (1968) that *leucopogon* cannot be considered a valid species and with Mayr and Short (1970), who regard the entire *Colinus* group as comprising a superspecies complex. Holman (1961) remarked that the recent species of *Colinus* exhibit fewer interspecific skeletal differences than do those of *Odontophorus* or "*Lophortyx*," with *nigrogularis* and "*leucopogon*" each having only 4 (out of 109 total characters examined) unique characters, while *virginianus* had 2 unique characters. Mayr and Short (1970) concluded that *nigrogularis* should probably be considered conspecific with *virginianus*, and the greater similarities in vocalizations that occur between these two forms than exist between *nigrogularis* and *cristatus* would favor that viewpoint (Cink, 1971).

28

Spotted Wood Quail

Odontophorus guttatus (Gould) 1838

OTHER VERNACULAR NAMES

BOLONCHACO, spotted partridge, thick-billed wood quail.

RANGE

Forested parts of the subtropical zone of southeastern Mexico south through Guatemala, British Honduras, Honduras, Nicaragua, and Costa Rica to extreme western Panama.

MEXICAN SUBSPECIES

O. g. guttatus: Spotted Wood Quail. Forested parts of Oaxaca, Chiapas, Veracruz, Tabasco, and Campeche in Mexico southward to Panama.

O. g. matudae Brodkorb: Matuda spotted wood quail. Known only from one locality in Chiapas. Of doubtful validity (Edwards and Lea, 1955; Blake, 1958).

MEASUREMENTS

Folded wing: Adults, both sexes, 134–54 mm (males average 144 mm, females, 140 mm).

Tail: Adults, both sexes, 61–76 mm (males average 72 mm, females, 67 mm).

IDENTIFICATION

Adults, 10–11.5 inches long. The sexes are quite similar in appearance. This is a tropical rain forest species that is rarely seen. Both sexes have a large, bushy crest that is bright orange in males and black in females. They also have black throats conspicuously marked with white shaft-streaks that produce a distinctively streaked appearance unlike any other *Odontophorus* species. Otherwise the upperparts are generally dark brown, with irregular black markings, especially on the wings. Both sexes have reddish brown and olive brown phases, which colors are extensive on the underparts, interrupted only by small rounded or teardrop-shaped lighter spots that are narrowly edged with dusky or black.

FIELD MARKS

More often heard than seen, this forest-dwelling species has a loud call of six notes, *wheet-o-wet-to-wheo-who*, repeated steadily with the last syllable sometimes changed to *to-whao*, and which may actually be a duet (Wetmore, 1965). It also utters mournful whistles when a flock is scattered and has repetitive *gahble-gahble* or *ga-gobble* calls (Slud, 1964). The distinctively streaked black throat of both sexes, and the bushy black or bright orange crest would serve to separate this species visually from all others in the region. Several additional species of *Odontophorus* occur farther south in Central America, but these all lack the streaked black and white throat pattern of the spotted wood quail.

AGE AND SEX CRITERIA

Females have dark brown, rather than bright orange, crests, or the orange buff coloration is limited to a shaft-streak. An erythristic phase is common, however, in which the crest is entirely fuscous and the female is darker overall.

Immatures evidently have the usual condition of two relatively pointed outer primaries (difficult to ascertain), but immature birds of both sexes also have rusty brown crowns that are tipped or vermiculated with brown or fuscous, and they are somewhat more rufous-colored ventrally than are adults.

Juveniles have black-edged breast feathers with buffy shaft-streaks that widen into bars. Reddish brown crest feathers begin to appear before the head has lost all of its down on the face and throat.

Downy young (illustrated in color plate 110) of this species can be distinguished from *Dendrortyx* downies by their olive gray rather than bright Naples yellow underparts (especially abdomen and throat). Compared to the very similar singing quail they appear to be darker and slightly more yellowish rather than buffy white below and have dark chin markings that are lacking in the singing quail. According to Wetmore (1965), the description given by Ridgway and Friedmann (1946, p. 371) for the natal plumage of *O. erythrops melanotis* actually refers to this species.

DISTRIBUTION AND HABITAT

The distribution of the spotted wood quail in Mexico has been plotted by Leopold (1959), who indicates that its distribution is more or less coextensive with that of tropical rain forest in addition to its occurring in portions of tropical evergreen forest. The species' northern limits are at about Potrero, Veracruz (Brodkorb, 1939). It is reported to be fairly common at elevations of three hundred to thirteen hundred meters in both tropical rain forest and cloud forests of the Sierra de Tuxtla, Veracruz (Andrle, 1967). In Tabasco there are no definite records of the species' occurrence (Berrett, 1963), but Ridgway and Friedmann (1946) listed it as occurring in the state. It has been recorded for a few localities in Quintana Roo and also was reported as being fairly common near Aquada Seca, Campeche, by Paynter (1955), who indicated that its habitat consists of dense rain forest with an open understory.

In Oaxaca the species was reported by Binford (1968) as an uncommon permanent resident of the Atlantic region in tropical evergreen forest from the Isthmus of Tehuantepec northwest to Taxcalcingo, at elevations of from 250 to at least 1,500 feet. No specimens are known from east of the Isthmus, but it is probably present in suitable habitats there as well. In Chiapas it occurs in humid forests in low and middle altitudes of the north, northeast, and northwestern portions of the state (Alvarez del Toro, 1964). Near Soconusco it was collected by Brodkorb (1939) at 750 meters elevation and was regarded as representing a new subspecies (*matudae*). Paynter (1957) reported it from the Selva Lacandona of eastern Chiapas, and it was listed by Goodnight and Goodnight (1956) as occurring in tropical rain forest at Palenque.

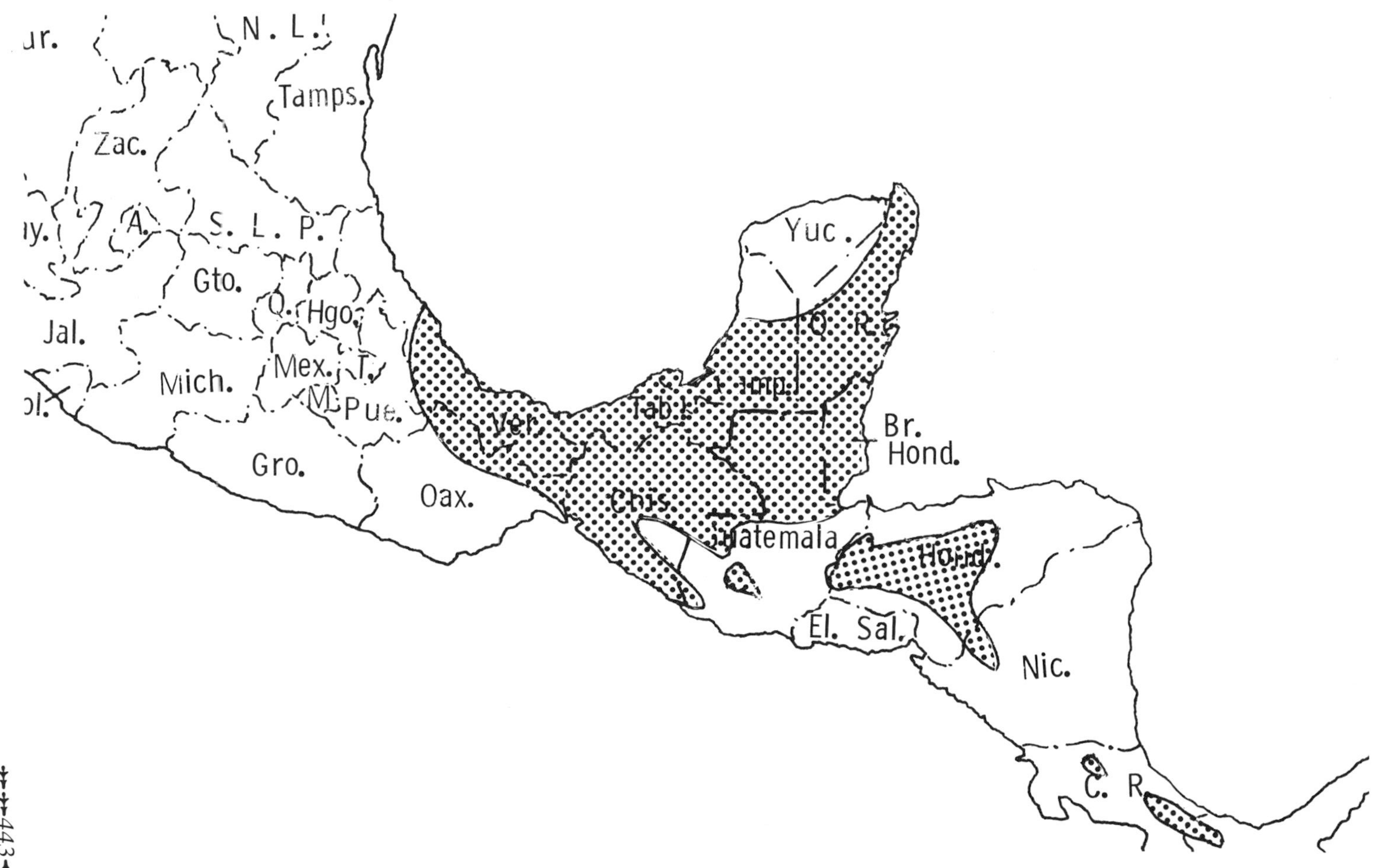

FIGURE 41. Current distribution of the spotted wood quail.

In Guatemala the spotted wood quail is characteristic of heavy rain forest, being found at lower elevations of from about 1,600 to 6,000 feet (Saunders, Halloway, and Handley, 1950). In British Honduras it occurs both in tall rain forest and high second growth (Russell, 1964). Griscom (1932) noted that although the bird occurs at fairly low elevations near the northern part of its range, it is primarily characteristic of cloud forest from Nicaragua southward, being replaced by *O. melanotis* at lower elevations. In Honduras it is most common above 600 meters and occurs at up to 2,000 meters, mainly in montane rain forest or cloud forest, but is less frequently found in lowland rain forest (Monroe, 1968). In Costa Rica it occurs from the middle of the subtropical belt upwards to timberline (Slud, 1964). Finally, in western Panama it is apparently confined to the Pacific slope and is mainly found between 1,250 and 2,100 meters in subtropical forest areas (Wetmore, 1965). It evidently occurs just below the Panama cloud forests, which on Cerro Pando lie between elevations of 2,100 and 2,290 meters along ridge-tops, and below which a montane rain forest extends down to about 1,800 meters (Myers, 1969). The rufous phase is apparently more common at the southern end of the species' range than it is in the north (Brodkorb, 1939).

POPULATION DENSITY

Not surprisingly, virtually nothing is known of population densities of this elusive, deep-forest species. It was recorded as being present in a census of lowland rain forest in Veracruz (*Audubon Field Notes*, 7:352–353, 1953), but evidently in numbers too few to estimate. It was also listed for an area of cloud forest near San Cristobál de las Casas (*Audubon Field Notes*, 8:374, 1954), but the species is not known to occupy these high elevations in southern Chiapas; thus, the record probably refers to singing quail or the buffy-crowned tree quail, which are both present in that region.

HABITAT REQUIREMENTS

Leopold (1959) has admirably summarized the habitat needs of this species: "The future status of the wood quail will depend entirely upon whether some of the dense rain forest is preserved. The bird disappears when the canopy is broken and a brushy understory springs up." There can be no doubt that this is the case. I was told that the bird was quite common on the lower part of the road from Ocozocoautla to the Presa Nezahualcoytl reservoir when this road had just been cut through the dense evergreen forest. However, it is now quite rare near the road where small milpas have broken the forest cover, being limited to a very few protected forest pockets occurring on slopes too steep to cultivate. In the southern part of its range its habitat needs may be somewhat different, since Slud (1964) reports it as being typical of forest borders and secondary woodland as well as scrub and thickets, rather than tall forest.

FOOD AND FORAGING BEHAVIOR

Leopold (1959) reported finding crop contents of small bulbs and soft rootlets, as well as the larvae, pupae, and adults of insects, mostly of dipterans and coleopterans, as well as miscellaneous seeds and the white meat of a large nut or seed. Most of these foods are evidently obtained by scratching in the soft forest floor; the birds' strong toes are well adapted for such scratching. A pair of adults that I had in captivity for a few weeks preferred fruits, such as grapes, and softened grains to hard grains and dried beans.

Wetmore (1965) reported that the scratchings of this species are roughly circular and thirty centimeters or more across, with the leaves cleared away to expose the bare ground. Such depressions may be spread over an area several meters across where the forest floor is level. Referring to a related species, Skutch (1947) noticed that the birds scratched with long, deliberate strokes, using either one foot or the other, but only one at a time, then picked up material exposed by their actions. Birds fed side by side in apparent harmony, sharing the food they found and constantly uttering soft and low liquid sounds. Remnants of banana pulp were apparently favored foods.

MOBILITY AND MOVEMENTS

There is little reason to believe that these birds have extensive movements or are highly mobile, although Slud (1964) suggested that some seasonal vertical migration might occur in Costa Rica. The birds generally run rather than fly when they are frightened or may hide if they are not detected. Following such disturbance and a scattering of the group, they reunite with one another by uttering mournful-sounding whistles (Wetmore, 1965). Leopold (1959) noted that he could make the birds fly only by firing a gun, but apparently when they are approached by a dog they regularly fly up into a tree, from which they sit and watch the dogs below. Lowery and Dalquest (1951) mentioned the use of dogs in capturing these birds in Veracruz, and I was told the same method was sometimes used in Chiapas. Leopold (1959) noted that when the birds did fly they usually would not cover more than one hundred yards.

SOCIAL AND REPRODUCTIVE BEHAVIOR

During the nonbreeding season the birds move in fairly small coveys; Leopold (1959) indicated that they consisted of from five to ten birds in his experience. Alvarez del Toro (1952) reported somewhat larger coveys of from six to twenty birds.

On the Yucatán peninsula the bird is reported to breed in May and June (Paynter, 1955), corresponding to the beginning of the wet season. However, no nests have been found in the wild to my knowledge, either in Mexico or elsewhere in the bird's range.

So little is known of the reproductive biology of this species that a summary of what is known of the breeding biology of the other species of the genus might be presented, on the assumption that it is similar to the spotted wood quail.

Recent observations on a captive breeding of the spot-winged wood quail (*O. capuiera*) by Flieg (1970) indicate that in that species a domed nest is built, some forty to fifty centimeters across. Three birds (two males and a female) cooperated in gathering the material and throwing it backwards toward the nesting site. A total period of about three days was required to build the nest. Five eggs were then laid at daily intervals. When these were removed a second clutch of three eggs was laid about two weeks later. The incubation period was determined to be from twenty-six to twenty-seven days, and it required three and one-half months for the young to attain full size and an appearance very similar to that of their parents.

An excellent study of the general and nesting behavior of the marbled wood quail (*O. gujanensis*) has been provided by Skutch (1947), based on observations in Costa Rica. In addition to the mutual harmony the birds showed while foraging, he also observed reciprocal preening (allopreening) behavior among a group of six or seven birds, a behavioral trait not otherwise reported for the New World quails.

During ten years, Skutch found three nests, in the months of January, April, and June. All of them consisted of well-enclosed chambers, (in one case five inches high, five inches wide, and ten inches long), roofed over with dead leaves, twigs, grasses, etc., which had round entryways about four inches in diameter in the side. One was in a depression at the base of a mound produced by the roots of an uprooted tree, another was at the foot of a gentle slope near a road but in second-growth woodland, and the third was at the base of a fig tree, between the ridges of its roots. In this last nest there was a tubular cavity about nine inches long sloping downward to the base of the nest. Two of the nests contained four eggs each, and the other set was destroyed before the clutch was completed.

One of the nests was studied intensively, and only a single marked bird, probably the female, incubated. Except for a single feeding period of from nearly two hours to somewhat more than three hours, she remained on the nest continuously during the daylight hours. Each morning the presumed male would arrive and call his mate from the nest, but he would stop short of it. A third bird was with the male toward the end of the incubation period. Hatching occurred between twenty-four and twenty-eight days after incubation had begun, and the female led the young away from the nest when they were less than twenty-two hours old. On the morning of departure the male arrived and called repeatedly, and the female and young then emerged from the nest. While the female led away three of the chicks, the male remained behind to look after a laggard. Skutch noted several occasions when young chicks were seen with five or six grown adults, one of which typically would perform a distraction display as the chicks and other adults disappeared in the brush. Skutch thought that it was perhaps the male which took the responsibility for such distraction behavior.

Vocal Signals

The distinctive call of the *bolonchaco, cobán chaco, bulub'tok,* or *totoloschóco* is indicated by these various local names, all of which represent interpretations of its typical call. Leopold (1959) said the call consists of six notes, repeated frequently, and is loud and strong. Wetmore (1965) reported that the usual series of phrases sounds like *wheet-o-wet-to-wheo-who,* with the last syllables sometimes changed to *to-whao.* This call is heard primarily at sunrise and often again near dark, so it probably functions in the same manner as the tree quail's "song," serving to notify other birds of the position of a covey or perhaps to gather them together for roosting at night.

Tape recordings made by L. I. Davis in Chiapas and filed in the Laboratory of Ornithology's Library of Natural Sounds indicate that at least two song types are present. One type (recorded in May, 1957) are of uniformly spaced *to-wet'* notes, uttered at 0.75-second intervals, with unbroken series of up to 31 such notes represented in the recordings by Davis. The second song type (recorded in April, 1958) was preceded by ten plaintive *wee-oh'* notes uttered at 1.8-second intervals, which led directly into a prolonged "song" consisting of repetitive and distinctively cadenced phrases, each lasting about 1.5 seconds and the individual phrases sounding like *whet'-o-wet, whe'-oo* (or *bo'-lon, cha'-co*). It seems probable that one bird sang the preliminary sequence of notes and a second bird sang all or part of the complex note phrases. The similarities of this song type to that typical of the singing quail are clearly apparent.

I was not fortunate enough to hear this call under natural conditions, but I talked with a man who had a pair of wood quails in captivity. According to him the birds sang both at dawn and at dusk, during the daytime on cloudy days, and often just before a rainstorm. Both sexes of the pair sang simultaneously but had recognizably different voices, and the song usually lasted about two minutes. For a time, when the male was sick, the female would not sing, suggesting that the antiphonal calls may play an important role in pair bond maintenance as well as presumably serving as a pair contact call. Wetmore (1965) also noted that the call probably is uttered as a duet, as occurs in various other wood quails. Chapman (1929) thus described how a presumed pair of Colombian marbled wood quail (*O. gujanensis marmoratus*) faced each other and sang a song in unison, with one bird singing *corcoro* and the other ending *vado* so perfectly that the entire *corcorovado* sequence could almost have been coming from a single bird. Wetmore (1965) confirmed this and noted that when he collected the female of a pair the male continued to sing the first part of the song alone, until it apparently obtained a new mate some time later, and the complete song was once again heard.

When disturbed, the birds utter "mournful whistles" (Wetmore, 1965). These disturbance notes are very much like those of tree quails, being rapidly repeated whistling notes of varying pitches and amplitudes, occasionally interspersed with more rattling sounds. The birds usually raise their crests when uttering such calls, exposing the orange red feathers of the male or the more fuscous feathers of the female. When held in the hand, both sexes often utter strong and rapidly repeated piercing whistles of the typically down-slurred type characteristic of New World quails.

EVOLUTIONARY RELATIONSHIPS

Holman (1961) regarded the genus *Odontophorus* as the most primitive of the group of genera that also included *Dactylortyx*, *Cyrtonyx*, and *Rhynchortyx*. All of these species are terrestrial forms that are typical of forests or woodlands and probably obtain much of their food (not definite for *Rhynchortyx*) by scratching for soft vegetative materials such as rootlets, bulbets, and the like. Holman believed that the pelvic skeletal condition of *Odontophorus* exhibited strong affinities to that of *Dendrortyx*, and both are presumably more primitive than the other three genera of the group.

It would seem that the pattern of evolution of the *Odontophorus* group was from a tree quail–like ancestral type that became more highly terrestrial and developed structural modifications that improved its foraging efficiency on the floor of wet tropical forests. This niche exploitation evidently opened the way to considerable range spread and speciation through the tropical forest of the New World, and the group must be regarded as the most successful of the New World quail genera on the basis of over-all range and number of extant species. Most of the species, however, exhibit allopatric distribution patterns, since presumably niche opportunities for foraging in this way are limited. Where more than one species does occur in a common area, there are apparently altitudinal differences that reduce interspecies contacts. Thus, for example, from Nicaragua southward to Costa Rica the spotted wood quail occurs with the dark-backed wood quail (*O. melanotis*), but there the latter occupies the tropical zone while the spotted wood quail occurs in cloud forest (Griscom, 1932). In Costa Rica and western Panama the black-breasted wood quail (*O. leucolaemus*) and the marbled wood quail (*O. g. castigatus*) also occur. Here the marbled wood quail occupies the tropical zone forests and the black-breasted wood quail occurs at intermediate elevations (in Panama) of from 1,350 to 1,600 meters (Wetmore, 1965). Slud (1964) gives the distribution of *leucolaemus* in Costa Rica as including upper subtropical and lower montane zones. Thus in the Dota region there is perhaps some contact with the spotted wood quail, which ranges from the subtropical zone to timberline.

The center of distribution of the genus *Odontophorus* would seem to be extreme northwestern South America, with five species occurring in both Colombia and Panama. To the north of this region, four species occur in Costa Rica, two in Nicaragua and Honduras, and only *guttatus* occurs in British Honduras, Guatemala, and Mexico. To the south and east, four species occur in Bolivia and Ecuador (assuming that *melanonotus* and *speciosus* are conspecific), three occur in Brazil and Venezuela, and one in Surinam, Guyana, French Guiana, Argentina, and Paraguay.

The two species with the largest ranges, the marbled (*gujanensis*) and spot-winged (*capueira*) wood quails, have allopatric lowland distributions and seem to represent a superspecies. Another large group of apparently allopatric populations is the highly variable *erythrops* group, which extends from Honduras southward via Colombia (including *hyperythrus*) to western Ecuador, where it is represented by *melanonotus*, and continuing on into Peru and Bolivia as *speciosus*. Some or all of these should probably be considered conspecific (Meyer de Schauensee, 1966).

In Panama and northern South America a group of generally white-throated (except for *atrifrons*) species occur in the subtropical zone, including the black-breasted (*leucolaemus*), Tacarcuna (*dialeucus*), gorgeted (*strophium*), Venezuelan (*columbianus*), and black-fronted (*atrifrons*) wood quails. Some of these are of questionable specific rank and probably should be merged, as Hellmayr and Conover (1942) suggested for *columbianus* and *strophium*. Certainly *strophium* and *dialeucus* are also close relatives and probably represent geographic replacement types. The relationships of the melanistic forms *atrifrons* and *leucolaemus* to these populations and to one another are less clear at present.

The two remaining species are gray-throated, with chestnut underparts variably spotted with white. These are the apparently closely related stripe-faced wood quail (*balliviani*) of Peru and Bolivia and the more tropical starred wood quail (*stellatus*), which ranges from Bolivia to eastern Ecuador.

The spotted wood quail is in my view probably not so closely related to these two latter species with similarly spotted underparts as it is to the white-throated species group, particularly *dialeucus*. The relative geographic relationships between *guttatus* and *dialeucus* would also support a possible common origin, with the lowlands of the Panama isthmus providing a possible barrier. Only in the northern part of the spotted wood quail's present range is it adapted to lowland tropical rain forest, and there its niche in higher and also in somewhat drier habitats is taken over by *Dactylortyx*.

29

Singing Quail

Dactylortyx thoracicus (Gambel) 1848

OTHER VERNACULAR NAMES

CHIVISCOYO, Cinco Real, Codorniz, Chifladora, long-clawed quail, long-toed partridge, long-toed quail.

RANGE

Mountainous areas from Mexico to El Salvador and Honduras.

MEXICAN SUBSPECIES (Based on Revision by Warner and Harrell)

D. t. thoracicus: Veracruz singing quail. Resident in northeastern Puebla and central Veracruz.

D. t. devius Nelson: Jaliscan singing quail. Known only from heavy oak forests of Jalisco.

D. t. sharpei Nelson: Yucatán singing quail. Resident in tropical forest lowlands of Cameche, Yucatán, and Quintana Roo.

D. t. chiapensis Nelson: Chiapan singing quail. Resident in mountain forests of central Chiapas.

The following Mexican races have recently been described but are not yet verified:

D. t. pettingilli Warner and Harrell. Resident in forests of the Sierra Madre Oriental of southwestern Tamaulipas and southeastern San Luis Potosí.

D. t. melodus Warner and Harrell. Resident in mountain forests of central Guerrero.

D. t. ginetensis Warner and Harrell. Resident in mountains near the Chiapas-Oaxaca border.

D. t. moorei Warner and Harrell. Known only from mountain forests of Cerro Brujo and Distrito Comitán, Chiapas.

D. t. dolichonyx Warner and Harrell. Resident in the forests of the Sierra Madre de Chiapas, Chiapas.

D. t. edwardsi Warner and Harrell. Known only from the cloud forests of Chiapas near the Oaxaca border.

D. t. paynteri Warner and Harrell. Resident in the lowland rain forest of south central Quintana Roo.

MEASUREMENTS (Mexican Races Only)

Folded wing: Adults, both sexes, 113–37 mm (males average 4 mm longer than females).

Tail: Adults, both sexes, 45–56 mm (males average 2 mm longer than females).

IDENTIFICATION

Adults, 8–9 inches long. Sexes quite similar in appearance. This species has relatively large feet and unusually long claws, which are used in scratching and digging for food. Both sexes have short, bushy crests and differ mainly in head color. Males have a mostly cinnamon head except for a brown crown, while in females the cinnamon areas are replaced with a grayish white. The upperparts of both sexes are finely mottled with browns and grays, and the upper wing surfaces are much more heavily marked with black and whitish coloration. The underparts and flanks are mostly brownish gray, fading to white or buffy on the abdomen, and the chest and flanks are rather broadly striped with whitish shaft-streaks.

FIELD MARKS

This forest-dwelling species is rarely seen but often heard and can be most easily identified by its thrush-like call, which usually consists of a series of single whistled notes that gradually increase in loudness and rate of repetition, followed by three to twelve rapid phrases varying in pitch and sounding like a repeated *pitch-wheeler* (Sutton, 1951) or *che-va-lieu-a* (Gaumer in Boucard, 1883). It also utters a low-pitched twittering call during foraging and has a loud, sharp alarm whistle (Lefebvre and Lefebvre, 1958). If seen, the bright tawny-cinnamon color of the male's head and the pale gray color of the female in these same areas would serve to identify it.

AGE AND SEX CRITERIA

Females have gray on the chin, throat, cheeks, and above the eyes, whereas adult males are tawny orange in these areas.

Immatures have the two outer primaries relatively pointed, and in addition the basal half of the mandible of first-fall birds is paler than that of older birds (Dickey and van Rossem, 1938). The upper greater primary coverts are reportedly spotted near the tip in young birds, but Warner and Harrell (1957) indicated that this criterion and the outer primary shape are not diagnostic for immature birds. They reported, however, that the base of the bill in young birds is light-colored.

Juveniles (at least of *devius* and *chiapensis*) evidently resemble adult males, but the tawny orange of the head and throat is replaced by cinnamon buff, and the cheeks are somewhat mottled with blackish. The crown and occiput of both sexes is barred or blotched with blackish, more conspicuously so in males than in females (Ridgway and Friedmann, 1946). Warner and Harrell (1957) reported that the sexes are not certainly separable but that juvenile birds have black spotting on the breast, sides, and flanks, while the head and neck are colored as in adult females.

Downy young (illustrated in color plate 110) of the singing quail are most like those of *Odontophorus* and *Dendrortyx*, in that all three species have dark chestnut crown and back patterns and strong preocular and postocular stripes. Downy singing quail and spotted wood quail have faint buffy lines along both sides of the rump but otherwise lack dorsal patterning. Perhaps the downy singing quail can be distinguished from the spotted wood quail by its somewhat more rufous dorsal tones and its brighter and clearer superciliary stripe, but too few specimens are available to be certain of this.

DISTRIBUTION AND HABITAT

The Mexican range of the singing quail has been plotted by Leopold (1959), and the total range of the species was illustrated by Warner and

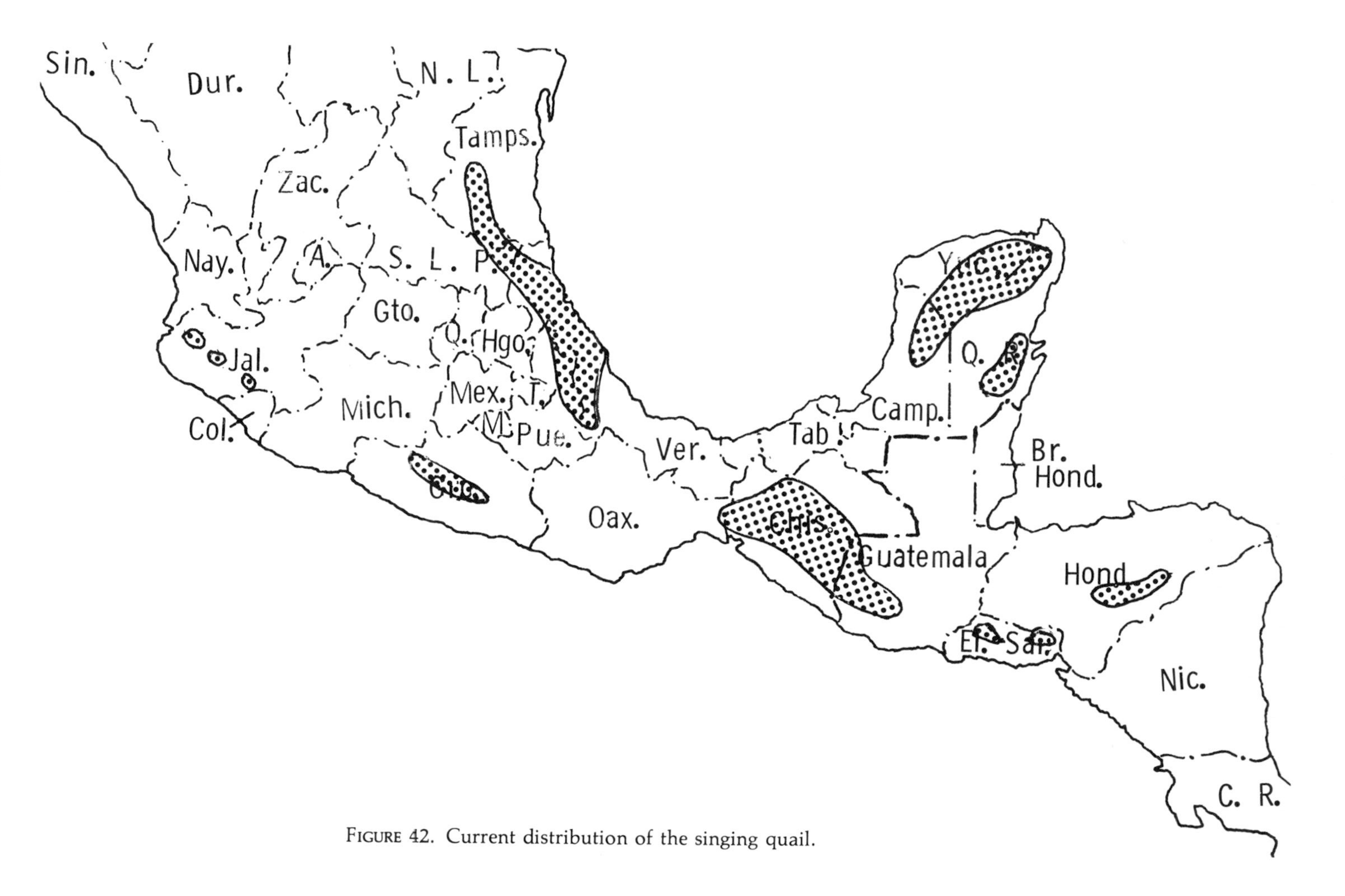

FIGURE 42. Current distribution of the singing quail.

Harrell (1957). Both maps indicate a discontinuous range which is largely but not entirely correlated with the distribution of cloud forest in Mexico and northern Middle America, as mapped by Martin (1955). The singing quail reaches its northern limit of range at the northern latitudinal limits of cloud forest, at 23° north, in southwestern Tamaulipas. At that locality cloud forest is developed at elevations between three thousand and five thousand (nine hundred and fifteen hundred meters), occurring above tropical deciduous forest and below oak-pine forest (Martin, 1955). The singing quail is most numerous in the cloud forest zone but does occur during the breeding season as low as one thousand feet and at least as high as seven thousand feet (Martin, 1955). Warner and Harrell (1957) indicated an altitudinal range of from three hundred to about six thousand or seven thousand feet in this area, occurring in semideciduous tropical forests, oak–sweet gum and beech forests, and pine-oak forests, plus one record from oak-madroño forest. In San Luis Potosí the altitudinal distribution and habitat occurrence is similar; the species has been recorded by several investigators in the vicinity of Xilitla in cloud forest. This locality is about 150 kilometers south of the Tamaulipas limit of cloud forest, and locality records between the two points suggest that the bird has an uninterrupted distribution in the subhumid forests between them.

Another major area of cloud forest in eastern Mexico is associated with the Sierra Madre Oriental in eastern Hidalgo, northeastern Puebla, and central Veracruz south to near the Oaxaca border. In this area the singing quail occurs in Puebla and Veracruz at altitudes similar to those mentioned, including evergreen tropical and probably semideciduous tropical forests of adjacent lowlands (Warner and Harrell, 1957).

In the Sierra Madre Occidental the singing quail has a much more restricted range, perhaps because of the generally drier conditions prevailing there. A single specimen has been taken in Jalisco, in heavy oak forest above 3,850 feet (Warner and Harrell, 1957), and another was obtained at La Cumbre, near Autlán (Schaldach, 1963). Schaldach has also reported hearing the birds in tropical deciduous forest habitat of Colima and considered them peculiar to that habitat.

In Guerrero, the singing quail is known only from pine-oak-fir forest of from six thousand to nine thousand feet elevation, in the vicinity of Omilteme (Warner and Harrell, 1957).

In Oaxaca, the species is a fairly common resident of the Pacific region in the Sierra Madre de Chiapas, occurring in humid gallery forest, tropical semideciduous forest, and the extreme lower edge of cloud forest, at elevations of from 800 to 4,900 feet (Binford, 1968).

A large number of locality records exist for Chiapas, and Warner and

Harrell recognize five different subspecies from that state. The most widely ranging form, *dolichonyx*, occurs widely in the Sierra Madre de Chiapas between four thousand and nine thousand feet elevation. Near the Guatemala border, this form is replaced by *calophonus* on the Volcan de Tacaná, at three thousand meters (ten thousand feet) elevation. In the interior highlands of Chiapas, the race *chiapensis* occurs in humid and semideciduous forests and also in pine forests (Alvarez del Toro, 1964). Finally, the race *moorei* is known from two mountain forest localities of central Chiapas, and *edwardsi* from a single cloud forest locality near Monserrate (Warner and Harrell, 1957).

On the Yucatán peninsula, the race *sharpei* occurs in tropical evergreen climax forest in Campeche lowlands, in tropical deciduous forests of Yucatán, and in mixed tropical evergreen and dedicuous forests of northern Quintana Roo (Warner and Harrell, 1957). In southern Quintana Roo, in evergreen rain forest, a surprisingly light-colored race *paynteri* apparently occurs, although the variation in coloration found in this area is still not well understood (Paynter, 1955).

To the south of Mexico, the singing quail is primarily associated with cloud forests. In Guatemala it is found in the cloud forests of the Pacific Cordillera, primarily at elevations of from 7,000 to 8,500 feet (Saunders, Halloway, and Handley, 1950). Warner and Harrell (1957) refer this population to *calophonus*, and indicate that it ranges up to 10,000 feet on the Volcan Tacaná. In Honduras the bird is uncommon and is confined to cloud forests above 1,300 meters (Monroe, 1968). In El Salvador the bird is typically associated with oak forests of between 2,500 and 4,000 feet elevation in the arid upper tropical zone, but they also utilize coffee groves of this same climatic zone (Dickey and van Rossem, 1938).

POPULATION DENSITY

Warner and Harrell (1957) estimated that during each of two years there were approximately 3.5 pairs per one hundred acres in a climax oak–sweet gum forest of southern Tamaulipas. LeFebvre and LeFebvre (1958) noted four and five pairs present in a twenty-acre plot of partially lumbered cloud forest and reported that elsewhere in cutover sections of forest there was at least 1 pair per twenty acres. Some more recent studies of the same area have been made, which suggest somewhat higher densities. During two years, a thirty-acre area of oak–sweet gum cloud forest was censused, and similar figures of 4 and 4.5 males (13 to 15 presumed pairs per one hundred acres) were judged to be present (*Audubon Field Notes*, 20:648–649, 1966; 19:599–600, 1965). The species also appeared in a census of pine-oak forest near San Cristobál, Chiapas, where a single territorial male was judged present on a fifteen-acre study area (*Audubon Field Notes*, 13:478, 1959). These figures would suggest that a population density of up to 1 bird per three or four acres might occur in favorable habitats, and certainly at least 1 per ten acres.

HABITAT REQUIREMENTS

Leopold (1959) has summarized the habitat needs of this species very well as follows: "The best habitat for singing quail is cool, moist forest that has been neither grazed nor burned. This perhaps is because of their preference for feeding in deep, rich litter, which would be compacted or destroyed by either grazing or burning." In such habitats, both food and protective cover are fully provided, and any moisture that may be needed can readily be obtained from succulent vegetation or insects.

The observations of LeFebvre and LeFebvre (1958) indicate that the species can tolerate at least some disturbance of climax forest, and with partial lumbering they were evidently at least maintaining their past population densities. However, they noticed that quail did not occur where the terrain was rocky and uneven or where little or no soil or leaf litter had accumulated among the moss-covered rocks. Their observations, and earlier ones by Dickey and van Rossem (1938), suggest that establishment of forest edges produced by lumbering or forest alteration by coffee-planting activities may to a limited extent benefit the species.

FOOD AND FORAGING BEHAVIOR

As might be expected, the foods of singing quail contain a variety of fleshy vegetable materials available in the litter and soil. Leopold (1959) noted onion-like bulbs, some small seeds, and the larvae and pupae of insects. Coffee berries may be preferred foods where they are available (Dickey and van Rossem, 1938). Arthropods, such as beetles, centipedes, crickets, and grubs, may also be taken in some numbers (Warner and Harrell, 1957). In disturbed areas the weedy pokeweed (*Phytolacca*) may be utilized fairly heavily according to LeFebvre and LeFebvre (1958), who also noted a variety of insect and spider materials in two specimens examined.

The mode of feeding is evidently much like that of *Odontophorus*. The bird reportedly leans to one side, lifts the opposite leg, and extends the foot far forward, even beyond the bill. With a single strong backward thrust it then tears and scatters the litter, after which it picks up edible materials (Warner and Harrell, 1957). A manner of foraging more like that of the domestic hen (*Gallus gallus*) was described by LeFebvre and LeFebvre (1958), who noted that the birds sometimes made four or five strokes with first one foot and then the other before they began to pick up materials. Often these workers noted that a paired male would apparently call the female to him after finding a large insect or other morsel, suggesting that tidbitting behavior is as well developed in the singing quail as in the United States species of terrestrial quails.

MOBILITY AND MOVEMENTS

The strong legs of the singing quail serve not only for foraging but also as their principal means of escape. They may squat when frightened, but when pressed closely they quickly run for cover, often in a zigzag manner (Warner and Harrell, 1957). If they are flushed, they take off suddenly and are quite agile in the air, banking around rocks and trees. Warner and Harrell reported that the flight distance varied with proximity to cover but was usually fifty to seventy-five yards and often less. In a few cases they have been observed to fly into trees and remain there at heights of from four to twelve feet (LeFebvre and LeFebvre, 1958). It is doubtful, however, that they normally roost in trees.

SOCIAL AND REPRODUCTIVE BEHAVIOR

During the period between the appearance of the first broods and the start of the next breeding season, singing quail are typically found in small flocks of from four or five to as many as a dozen birds (Warner and Harrell, 1957). Such flocks are presumably basically family groups or multiples of families that have merged. In Mexico the breeding season lasts at least from March to August (Leopold, 1959), possibly even from February through October (Warner and Harrell, 1957). With the onset of the breeding season the coveys break up, and song begins in northern Mexico in March. This reaches its peak in April and May and gradually diminishes, but records for singing have been obtained as late as December 26 (Warner and Harrell, 1957). In the Yucatán peninsula the breeding season is clearly associated with the wet season, which typically begins in May. The few available records from that region indicate that nesting extends from early May to early August (Paynter, 1955).

No nests have yet been described, although Paynter (1955) reported that a female collected in early May was incubating five eggs. The number of young in the broods is small, usually only from two to four (Warner and Harrell, 1957). Both parents protect and care for the young when they leave the nest, and at least the female will perform injury-feigning when defending the brood (LeFebvre and LeFebvre, 1958).

Vocal Signals.

The "song" of the singing quail is justly famous and certainly one of the most impressive of all quail vocalizations. The first description of it (Boucard, 1883) is hard to improve upon: "Sings at nightfall, a low whistle, repeated three times with increasing loudness, followed by *'che-vă-lieu-a'* repeated three to six times in rapid succession. The tone is musical, half sad, half persuasive, beginning somewhat cheerful, and ending more coaxingly." A more recent description has been provided by Warner and Harrell (1957): "The first part of the song is a series of about four loud, penetrating whistles, which increase in frequency and pitch and seem to be an announcement of the start of a song; these whistles often are repeated by other birds. Sometimes an imitation of these notes will initiate singing in nearby birds. The last of these notes is followed immediately by the second part of the song, a series of three to six rapid phrases, each made up of notes of differing pitch, the middle ones higher and more definitely accented. . . . These phrases are followed by a low twittering which is often not audible or may be absent. There is no evidence that the female sings." However, an absence of female participation in the song would be surprising, considering the situation found in the tree quail group as well as the wood quails.

Tape recordings by L. I. Davis made in various parts of Mexico and filed in the Laboratory of Ornithology's Library of Natural Sounds indicate that considerable variation occurs in the vocalizations. The preliminary notes may be repeated only a few times or uttered as many as twelve times before the complex phrasing begins. The preliminary notes begin in a slow, measured cadence, spaced about two seconds apart, but soon accelerate to nearly two per second just before the "song" proper begins. This typically consists of up to as many as twelve melodic *che'va-lieu'-a* phrases, each phrase lasting about one second and almost merging with the next. There seem to be two major song types, one more pulsed, less melodious, and sounding like a repeated *pa-che'-va* (or *pitch-wheeler*), indicating that the more typical phrasing may actually be a duet. In one sequence (February 17, 1961), it sounds as if the bird singing the preliminary notes also sings the *lieu'-a* portion of the song, while a second bird apparently sings the *che'-va* portion. Sonagraphically, these two portions appear very similar, each consisting of two major rising and falling notes that fluctuate in fundamental frequency between about 1,500 and 2,800 Hz, with almost no harmonic development.

The observations of H. E. Anthony (quoted by Griscom, 1932) suggest that antiphonal duetting is a common feature of the song of this species, although he was unable to determine if the birds concerned were of the same sex. "At Finca Perla, two of these birds were kept in separate cages on different sides of the house. I was told that it was for the sake of their 'song,' which they would sing in the early morning, but that there would be no song, if the birds could see each other. I had often listened to the calls of this quail, from the edge of the forest, and wondered at its mellow richness, reminiscent of the flute-like call of the plumed partridge of the Sierra Nevadas. I was not, however, expecting a duet, such as I witnessed several times at this place. In the early morning, when the calls were most frequent on the mountain sides, one of the cage birds would utter its invitation, 'cua-cua-cua' at intervals of two or three seconds. Soon the other bird responded, 'cua-kaka-wak-cua-kaka-wak,' while the other joined in perfectly in time and tone, 'cua-cua-cua,' both continuing for some twenty or thirty seconds and stopping in exact unison. It seems hard to believe that two birds were calling, so perfect was the time and tone." This description certainly suggests that the "invitation" notes are not sung by the same bird that sings the *cua-kaka-wak* or *che-va-lieu-a* sequence. If Wetmore's (1965) observations on *Odontophorus* provide any clues, one would guess that the male is responsible for the first section and the female for the latter part of the call. LeFebvre and LeFebvre (1958) indicated that a "twitter call," uttered during foraging or when disturbed by humans, might also be a duet. Warner and Harrell thought that this twittering note served as a location call, and they also once reported a call from a flying bird.

When the birds are alarmed, especially when adults are tending young, a sharp whistle may be uttered. At least the male is known to produce this call, and both sexes also utter clucking notes when their young are threatened (LeFebvre and LeFebvre, 1958). A hand-held distress call has not yet been described, and since I have not handled any live birds I was unable to learn whether such a call is typical.

EVOLUTIONARY RELATIONSHIPS

The observations of Holman (1961) suggest that the nearest living relative of *Dactylortyx* is *Cyrtonyx*, while it is less closely related to *Rhynchortyx* and most distantly related to *Odontophorus*. I would instead suggest that *Dactylortyx* evolved from an *Odontophorus*-like ancestor through becoming even more highly specialized for scratching and digging through the increased modification of its legs, toes, and claws and presumed associated muscular modifications. By developing an ecological tolerance for somewhat drier habitats than those typical of *Odontophorus*, it has extended its range considerably farther north than have most of the wood quails, and in Mexico it occupies a much broader altitudinal range than does the spotted wood quail.

30

Harlequin Quail

Cyrtonyx montezumae (Vigors) 1830

OTHER VERNACULAR NAMES

BLACK quail, Codorniz Encinera, Codorniz Pinta, crazy quail, fool quail, massena quail, Mearns quail, Montezuma quail, painted quail, squat quail.

RANGE

Southwestern United States south to Oaxaca, Mexico. The doubtfully distinct species *C. ocellatus* (Gould) extends from southern Oaxaca to Nicaragua.

SUBSPECIES (*ex Check-list of the Birds of Mexico*)

C. m. montezumae: Massena harlequin quail. Resident from Michoacán, Oaxaca, Distrito Federal, Hidalgo, and Puebla to Nuevo León and west central Tamaulipas.

C. m. mearnsi Nelson: Mearns harlequin quail. Resident in western central Texas, central New Mexico, and central Arizona south to northwestern Mexico, including northern Coahuila (*Condor* 57:162).

C. m. merriami Nelson: Merriam harlequin quail. Known only from the eastern slope of Mount Orizaba, Veracruz.

C. m. sallei Verreaux: Sallé harlequin quail. Resident from Michoacán south through Guerrero to east central Oaxaca.

C. m. rowleyi Phillips. Recently described from Guerrero (1966); not yet verified.

MEASUREMENTS

Folded wing: Adults, both sexes, 110–31 mm (males average more than 120 mm, females under 120 mm).

Tail: Adults, both sexes, 47–63 mm (no consistent difference in the sexes).

IDENTIFICATION

Adults, 8–9.5 inches long. The sexes are very different in appearance. Males have a beautiful facial pattern of black or bluish black and white and a soft, tan crest that projects backward and downward over the nape. The upperparts of males are grayish to olive brown, extensively spotted and marked with black, white, and buffy markings. The sides and flanks are dark grayish, with numerous rounded spots of white, cinnamon, or rufous brown, depending on the population. The breast is unmarked brown, grading gradually to black on the abdomen and undertail coverts. Females are generally cinnamon-colored, with blackish markings extensive on the back. The female has a small, buffy crest that is less conspicuous than the male's and a mottled brown and buffy face with a whitish chin and throat. The upper surfaces of the back and wings are extensively mottled, and the underparts are mostly buffy with black flecks or streaks in the abdominal region.

Ocellated quail may be distinguished from the harlequin quail of southern Mexico as follows: southern populations of harlequin quail have the white lateral spotting reduced to their anterior portions, while the posterior flank spots are dark chestnut. Male ocellated quail also have their midbreast and upper abdominal areas a much lighter, generally buffy or slightly tawny color and instead of gray flanks with chestnut spotting have chestnut flanks with black and gray cross-markings. Females have somewhat paler and more pinkish breasts than do female harlequin quail of southern Mexico and differ from the more northerly populations by having buffy rather than white shaft-streaks on the upperparts.

FIELD MARKS

Males are unmistakable if their distinctively patterned face can be seen or if their extensively spotted flank pattern is visible. Females are more uniformly cinnamon-colored below than are other species of quails. Unlike the scaled quail of the same region (which occurs in more open habitats), harlequin quail will rarely run and instead tend to crouch and hide. They are rarely found far from pine-oak woodlands throughout their entire range. The distinctive call consists of a series of uniformly paced whistling notes, slowly descending in scale. It has not yet been determined whether the ocellated quail has an exactly comparable call, but current evidence would suggest that it does not.

AGE AND SEX CRITERIA

Females lack the black and white ornamental patterning of the face and throat of adult males, having instead a white or buffy chin and throat.

Immatures may be recognized by the upper greater primary coverts (Petrides, 1942; Leopold and McCabe, 1957), which are edged with buffy or barred near the base with buff, whereas in adults they are spotted with whitish (males) or barred with wide white markings. Also in immatures the outer two coverts are pointed rather than rounded. The condition of the outer primaries does not appear to be very useful in determing age.

Juveniles initially resemble adult females, and young females continue to do so but may be recognized by the transverse barring on the head rather than longitudinal striping as in adults. Juvenile harlequin quail males soon acquire dark underparts and flanks, but whereas adult males have a double row of white spots on a dark background in young males these feathers are pale, with a double row of dark. The head remains juvenile-like for some time (Swarth, 1909).

Downy young (illustrated in color plate 110) of the harlequin quail (and presumably also the ocellated quail, which is undescribed) may readily be recognized by the patch of ochraceous buff on the rear of the wings, and the relatively unpatterned back, which varies from argus brown in the middorsal area to a cinnamon buff which forms two incomplete stripes just below the darker middorsal area. The crown is a light chestnut and the rest of the head is pale cinnamon buff, with a narrow dark line extending from the back of the eye to the posterior tip of the crown.

DISTRIBUTION AND HABITAT

The United States distribution of the harlequin quail is limited to parts of Texas, New Mexico, and Arizona. In Texas it is currently a rare, local resident, mostly west of the Pecos River, although possibly a few still persist on the Edwards Plateau (Peterson, 1960). The bird may formerly have occurred in all the counties west of the Pecos River except El Paso County and eastward through Crockett and Val Verde counties as far as about the ninety-ninth parallel, but by the mid-1940s as a result of overgrazing the bird was found in normal numbers only in the Davis Mountains and parts of the Big Bend (Principal game birds and mammals of Texas, 1945).

In New Mexico the species was once fairly common in the southwestern part of the state, especially near the headwaters of the Gila, San Francisco, and Mimbres rivers (Bailey, 1928). Now its range is greatly restricted, and it occurs only where rank grasses still grow, particularly near the summits of mountains in the Capitan, Sacramento, San Mateo, Black, and Mogollon ranges, and in extreme southwestern New Mexico near the Arizona and Mexico borders (Ligon, 1961). Ligon indicates a surprising altitudinal range for the species, with the bird occurring at as high as twelve thousand feet in summer and at from five thousand to eight thousand feet during the winter months.

Arizona's population of harlequin quail is found from the Mogollon rim area south to the Mexico border, occurring most commonly in the oak-grassland zone and, to a limited extent, in the pine forests as well (Bishop, 1964). Most of the recent sight records Bishop lists are for Cochise, Santa Cruz, and eastern Pima counties.

The Mexican range of the harlequin quail has been mapped by Leopold (1959), who concluded that it occurs in essentially all the pine-oak upland vegetation from Sonora, Chihuahua, and Coahuila south to near the Isthmus of Tehuantepec in Oaxaca. Binford (1968) reported that in Oaxaca the bird occurs at elevations of from thirty-five hundred to ten thousand feet in air and upland oak scrub vegetation. The southeasternmost locality in the bird's range is La Cienguilla, Oaxaca. South of the Isthmus in comparable vegetation the ocellated quail occurs in Chiapas. Binford indicated that the northwesternmost locality records for the ocellated quail are near Tapanatepec and north of Santa Effiginia. Thus the two populations are isolated by the tropical lowlands of the Isthmus of Tehuantepec and represent allopatric replacement forms occupying the same habitats and foraging niches. The somewhat intermediate male plumage traits of the Sallé harlequin quail, occurring mostly in Guerrero and Oaxaca, further brings into question the validity of considering the ocellated quail as a distinct species. Thus, it shows the reduction of melanism on the underparts which is so

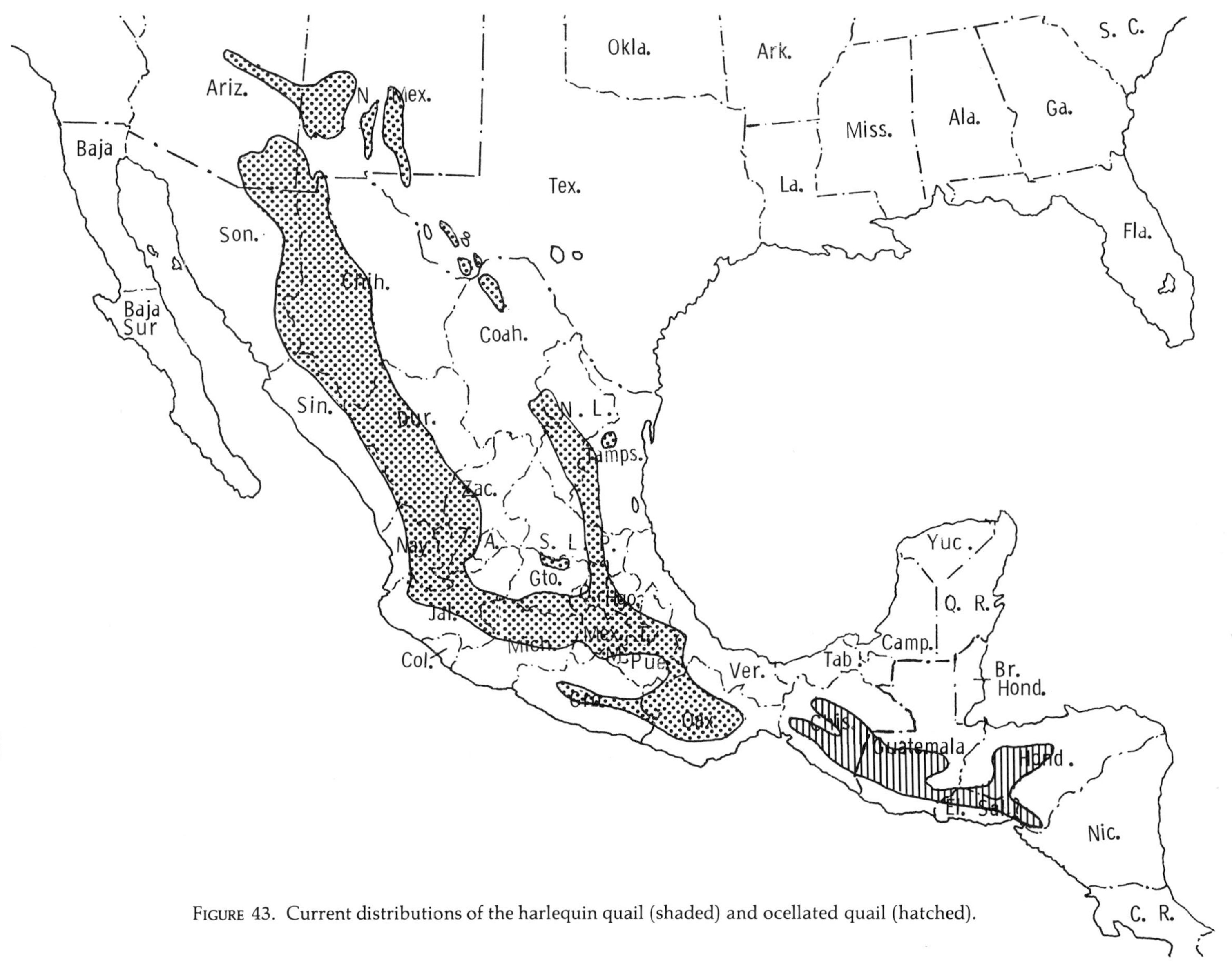

FIGURE 43. Current distributions of the harlequin quail (shaded) and ocellated quail (hatched).

strongly evident in the ocellated quail, as well as the replacement of white spots on the flanks with brown markings.

In Guatemala the ocellated quail occurs in drier parts of the central highlands, on grassy slopes, in fields, and in open pine woodlands mainly between 5,000 and 10,000 feet (Saunders, Halloway, and Handley, 1950). In Honduras it is found in highland (above 750 meters) pine areas, mainly in open or semiopen situations where there is a heavy undergrowth or long grass (Monroe, 1968). In El Salvador it has been reported at elevations of between 6,400 and 7,000 feet in open pine forests, especially where dense bracken, blueberry vines, and similar undergrowth is well developed (Dickey and van Rossem, 1938). The limit of the species' range is reached in southern Honduras and northern Nicaragua.

POPULATION DENSITY

Two estimates of population density were provided by Leopold and McCabe (1957). One was an estimate of twenty-six birds per section (27 acres per bird), based on a count of at least forty-five birds on 1,120 acres made by Wallmo (1951) in Arizona. In northern Chihuahua, Leopold and McCabe estimated that at least twenty-eight to thirty adults per section occurred in fairly well populated range, or 21 to 24 acres per bird. Bishop (1964) reported that one study area in Arizona consisting of about 120,000 square yards (24.8 acres) had five pairs at nesting time, or 5 acres per pair. Another study area of about 33 acres had nine pairs in mid-July, or 3.7 acres per pair. Thus, in favored habitats considerable population densities can occur. Fall population densities were estimated by Bishop in two areas. One area of 130 acres had a minimum of forty-five birds, while another of 160 acres had sixty-two birds; thus fall densities may sometimes reach about three acres per bird. Bishop estimated that over a large area the oak-juniper habitat might have averaged about forty birds per square mile in early December of 1963.

Harlequin quail have been reported on various Mexican breeding bird censuses of the Audubon Society, but on none of these has the population been particularly high. Thus, on both a cactus-acacia grassland and a piñon pine–oak woodland area of Durango, the estimated breeding population was 1 male per thirty acres (*Audubon Field Notes*, 18:560-561, 1964), while on a pine-oak-mesquite grassland ecotone area of fifteen acres the population was also estimated at 0.5 males (*Audubon Field Notes*, 11:449-450, 1957). Such low breeding densities probably reflect habitat disturbance, particularly grazing effects.

HABITAT REQUIREMENTS

Leopold and McCabe (1957) concluded that the harlequin quail is an "indicator species" of the pine-oak vegetative zone in Mexico but emphasize that it is neither the pines nor the oaks by themselves that comprise ideal quail habitat. Rather, the understory characteristics represent the critical factor, particularly the presence of bulb-bearing forbs and sedges. These plants can tolerate some periodic burning or limited logging but are severely affected by grazing. Grazing also probably reduces cover for escape and nesting, but the critical factor is the loss of plants upon which the harlequin quail depends for both food and moisture.

Bishop (1964) agreed that the harlequin quail could probably get enough moisture from its succulent foods to survive without other free water and noted that in many areas of southern Arizona such water is lacking except during the summer rainy season. He did, however, observe at least one bird drinking from a puddle after a thundershower, and noted that the possible dependent relationship of reproduction to available water in the free state as well as in succulent foods is still not known.

FOOD AND FORAGING BEHAVIOR

Only a few studies of the foods of the harlequin quail have so far been made. Martin, Zim, and Nelson (1951) noted that in a sample of birds collected primarily in winter from Texas, Arizona, and New Mexico, chufa or nut grass (*Cyperus*) tubers were most important, followed by oaks (acorns), bulbs of wood sorrel (*Oxalis*), and brodiaea (*Brodiaea*) and sunflower (*Helianthus*) seeds. About 70 percent of the winter food samples were of plant origin, with various insects and other arthropods comprising the animal food.

Leopold and McCabe (1957) provided a complete summary of food items found in harlequin quail, based on their own observations and previous studies. They estimated that about 40 percent of the summer foods eaten were of vegetable origin. Although acorns were listed in seven different studies, the major food item would appear to be bulbs, from various lily species (*Echeandia*, *Brodiaea*) and especially from the nut grass *Cyperus esculentus*. Other succulent foods that are dug up are the bulbs of wood sorrel and the tubers of buttercups (*Ranunculus*). Seeds of legumes, grasses, piñon pine (*Pinus edulus*), and forbs are used, as well as the fruits of juniper (*Juniperus*), ground cherry (*Physalis*), sumac (*Rhus*), caltrop (*Kallstroemia*), and various ericad shrubs (*Arbutus*, *Kalmia*). During the summer rainy season a variety of insect life is also taken, especially beetles and the larval stages of moths and butterflies.

A monthly analysis of harlequin quail food consumption in Arizona has been made by Brown (1969a), who noted that, by weight, plant material comprised from 90 to more than 99 percent of the monthly samples, with animal materials being of significance only from June through September, when beetles in particular were consumed. The two primary vegetable food sources were wood sorrel bulbs, which occurred in large amounts from June through January, and nut grass (*Cyperus esculentus*) bulbs, which were equally important from January through April. In April and May seeds (*Paspalum, Lotus*) and buds (*Gilia*) were taken in limited amounts, and during July and August the tubers of morning glories (*Ipomoea*), seeds of *Glactia*, and fruits of manzanitas (*Arctostaphylos*) also appeared in the diet.

A similar seasonal food analysis has been provided by Bishop and Hungerford (1965), based on the study of 221 crop contents. Throughout the year the major foods were acorns, bulbs of wood sorrel, seeds, sedge tubers, and insects. During the winter months of January through March, wood sorrel bulbs were the primary food, with other plant materials such as acorns, seeds, and tubers of secondary importance. In April, May, and June an increasing amount of nut grass or sedge tubers were taken, as well as green acorns, and the importance of wood sorrel began to decline. From July through September insects and green acorns made up the bulk of the foods, with *Oxalis* and *Cyperus* of minimal significance. However, from October through December these two food sources, as well as acorns, again became the predominant sources of food intake.

In summary, it would appear that for all except the summer months, the availability of *Oxalis* and *Cyperus* underground parts is crucial to the survival of the harlequin quail, with acorns and other seeds or fruits of secondary importance.

The typical foraging behavior of these quail has been described by various writers. Leopold and McCabe (1957) noted that the birds typically dig a hole about two inches long, an inch across, and two or three inches deep, while extracting bulbs. They do not eat the dried hulls, and leave them near these diggings. When eating acorns, the birds also open the hull and remove the meaty center.

Bishop (1964) also noted that when *Oxalis* bulbs are dug up the birds make cone-shaped holes, with one side of the cone dug away and the bulb hulls left in the hole. When searching for foods nearer the surface the birds made fan-shaped depressions about one-eighth inch deep in duff and litter under bushes and trees, which sometimes covered several square yards in area. He noted that the birds often scratched with one foot and then the other, with frequent pauses to examine the scratched area for foods. Often the members of a covey fed so closely together that they touched one another, apparently without hostility, with up to eight feeding in a circle only fourteen inches in diameter. He observed that birds apparently fed throughout the day, and only those that were collected after 3:00 P.M. had full crops.

MOBILITY AND MOVEMENTS

Nearly all observations of harlequin quail indicate that they are not highly mobile. In spite of their strong legs they do not run when disturbed but rather tend to squat and "freeze." When flushed, they usually fly only fifty to one hundred yards (Leopold, 1959). Bishop (1964) noted that the birds were usually less than twenty feet away when they were flushed and flew no more than one hundred yards, after which they would run rather than fly again. At least on the winter range, coveys apparently return day after day to the same foraging place, and the covey home range may be no more than two hundred yards in radius (Leopold and McCabe, 1957). It is not uncommon to find a covey using the same fifteen yards of a canyon area on consecutive days or at greater intervals (Miller, 1943).

In New Mexico as well as elsewhere a definite altitudinal movement between summer and winter has been noted (Ligon, 1961; Leopold and McCabe, 1957), however, these appear to be relatively short movements, probably not exceeding a few miles. Bishop's (1964) study did not indicate such a seasonal migration; areas which contained birds prior to the nesting season had all supported coveys during the previous hunting season. As the nesting season approached the birds moved decreasing amounts, and he found no evidence that either member of a pair moved more than 150 yards from a nest site. Shortly after hatching the brood range was even less than this; as the chicks grew it gradually increased but even then did not exceed an area of more than 200 yards' radius.

SOCIAL AND REPRODUCTIVE BEHAVIOR

While in coveys during the nonbreeding season, the birds form small flocks that probably represent family groups. Leopold and McCabe noted that the average covey size of sixty-two coveys was only 7.6 birds, and rarely have groups of more than 25 ever been reported. These coveys spend the day following a usual activity pattern of morning and evening foraging, with the intervening hours spent resting, dusting, and preening, with some digging for food. During rainy weather they may remain huddled together, and at night they roost on the ground, often facing outward in a semicircle around a rock or a grass clump (Bishop, 1964).

Pairing evidently occurs well before the nesting season actually is underway. Records summarized by Leopold and McCabe (1957), and observations by Bishop (1964), indicate that most pairing in Arizona may occur during March through May, beginning as early as February. In spite of this early pairing, gonadal development does not usually begin until June, with the earliest Arizona records for broods occurring about mid-June, and eggs being found as late as September 20 (Wallmo, 1954). Bishop (1964) concluded that during his study few females began laying before June 28, and most laying probably occurred during July, or about four months after pairing was initiated. It is believed that nesting in this species is adaptively timed so that broods appear soon after the summer rains have provided new green plant growth and an abundance of insects, although the physiological mechanism of such timing is still obscure (Leopold and McCabe, 1957).

Although lone, presumably unpaired, males began to appear as early as mid-May, Bishop did not hear any male calling until mid-June. Most male calling occurred from late July to mid-August, or during the peak period of incubation. Bishop believed that the majority of calling males were mated ones, but Leopold and McCabe said calling during the breeding season is largely and perhaps entirely by lone males. Bishop indicated that he often heard males calling from fifty to one hundred yards away from nest sites, but attraction to nesting sites is typical of unpaired male quail and need not indicate that the calling bird is the mate of the nesting female. A peak of male calling during incubation on the part of unmated males is also characteristic of several of the United States species (see California quail account), and the incidence of male calling is probably correlated with the gonad cycle.

The participation of the male in nest-building, incubation, and nest defense is still slightly uncertain. One study of captive birds indicated that the male may help to construct the nest, which would be in agreement with observations on *Odontophorus*, which also builds a domed nest. Prior to building the nest a scrape is made, which may be one to three inches deep (Bishop, 1964). The cavity may be five or six inches wide and is lined with vegetative material such as grass or oak leaves and often some down (Wallmo, 1954). The sides of the cavity usually consist of grass stems which may appear to be woven together, and which are roofed over the top of the scrape to form a chamber four or five inches high. The side entrance to the nest is often well hidden by a mat of grass stems that hang down over the entrance. Bishop reported that this mat acts like a hinged door, so that it falls back into place whenever the female enters or leaves the nest.

The average clutch size is reported by Leopold and McCabe to be 11.1, with an observed range of 6 to 14 eggs (Leopold and McCabe, 1957). The egg-laying rate of wild females is as yet unrecorded, but three captive females in the collection of F. S. Strange laid 87 eggs during a sixty-one day period, averaging about three days per egg. During 1967 and 1968, egg laying by his birds consisted of the following monthly totals: 7 in May, 45 in June, 42 in July, 20 in August, and 6 in September.

Bishop never observed males on or very near the nest, but Willard (in Bent, 1932) reported seeing males sitting on eggs in about half of the nests he examined. Males have also been reported sitting close beside incubating hens and without question remain with the female to help guard and rear the young.

The incubation period is probably twenty-five to twenty-six days, which is in general agreement with *Odontophorus* but longer than the incubation periods of other United States quails (Leopold and McCabe, 1957).

Both parents actively participate in brood care; Leopold and McCabe (1957) reported two instances of injury-feigning on the part of the male. The decumbent crest of the male is spread laterally during such disturbances. In eight of ten observed cases, Bishop (1964) noted that pairs with broods under a month old acted in the same fashion, with the female being first to expose herself and attempt to lead intruders away from the brood by feigning a broken wing. If necessary, the male may also appear and behave similarly, after first sending the chicks into hiding by uttering a series of moaning cries. In two instances the male was evidently the first to expose itself and perform distraction displays.

When newly hatched, the birds are fed insects, seeds, and bulbs by the parents, but by the time they are two weeks old they begin to forage for themselves (Bishop, 1964). Probably little brood mixing occurs, since the average reported brood sizes of 6.6 to 8.4 young is not much below the average clutch size (Leopold and McCabe, 1957). However, some broods containing two age-classes have been seen (Wallmo, 1954).

Probably little merging of family units occurs during the fall. Brown (1969b) noted that before the hunting season 70 coveys containing 451 birds occurred on 2.95 square miles, indicating an average covey size of 6.4 birds. These 23.7 coveys per section were thought to be the result of a breeding population of about twenty-four breeding pairs per section. Hunting seasons in Arizona during the years 1965 through 1969 have provided age and sex ratio population data not previously available for the species. Of 4,095 birds shot during these years, 71.5 percent have been young and 56.4 percent have been males (Brown, 1970). This age ratio would represent a juvenile-to-adult ratio of 2.5:1, or more than 5 young raised per adult female on the average, assuming that young birds are not more vulnerable to shooting than are adults. Comparisons of age ratios based on wing samples with those based on average covey sizes of well-grown broods are in close agreement, suggesting that coveys do consist of family units and probably little differential age vulnerability to shooting exists, judging from data presented by Brown (1969a).

Vocal Signals

The vocalizations of the harlequin quail are neither so loud nor so varied as those of *Odontophorus* and *Dactylortyx*, but this is not surprising in view of the relatively more open habitat which the harlequin quail uses and its probable greater reliance on visual signals. Certainly, more plumage dimorphism exists in this species than in any other of the species of the other genera in this subgroup.

Leopold and McCabe described the separation or assembly call of the harlequin quail as a low quavering whistle, with the separate notes slowly descending in pitch. Fuertes (1903) described it as owl-like, and Bishop (1964) reported that it is higher in pitch but lower in volume than the calls associated with the breeding season. The call is uttered by chicks as well as adults of both sexes, although Bishop (1964) indicated that in contrast to Leopold and McCabe he had never heard males produce the call.

Recordings of the separation call made by L. Irby Davis in Jalisco and filed in the Laboratory of Ornithology's Library of Natural Sounds indicate that this call consists of from six to nine uniformly spaced notes, with each lasting about 0.3 seconds, and the entire series lasting about 2.5 seconds, during which time the fundamental frequency gradually drops from about 4,000 Hz to 3,500 Hz. Eight such call sequences occurred during a 67-second recording period, or about one every 8 seconds.

The second major call is produced by males during the breeding season and is probably an indication of the location of unmated males. Leopold and McCabe (1957) said that it is a high-pitched *buzz* sound that ascends in pitch rapidly to an inaudible level. In contrast, Bishop described it as a descending whistle combined with a buzzing sound, which can be heard up to 200 yards away under favorable conditions. According to him, a similar call is produced by females, a series of nine high-pitched, low-volume notes of descending pitch, audible up to 150 yards away and resembling the call of the canyon wren (*Catherpes mexicanus*). Levy, Levy, and Bishop (1966) found that males began to respond to recorded playbacks of this call in June and their period of strongest response was about the beginning of August, or during the period of maximum nesting activity. In contrast to Gambel quail, male harlequin quail would respond throughout the day to such playbacks. Further, although the Gambel quail that were attracted were clearly unmated males, these authors apparently believed that mated male harlequin quail could also be attracted by such calls.

In addition to these two call-types, a few other vocalizations have been noted. Conversational or contact notes have been mentioned by a few workers as occurring when birds were in a covey or foraging, and sometimes a squealing call is produced when they are flushed (Leopold and McCabe, 1957). Bishop (1964) mentioned that he frequently heard a moaning-crying sound produced by adults when their young were in danger, and he heard the same distress call when he picked up crippled or captive birds.

I have had little experience with the harlequin quail and thus cannot evaluate their vocal similarities to other species. However, while in Chiapas I inquired of several people as to the calls of the ocellated quail. In the vicinity of San Cristóbal and southward toward the Guatemala border, where at least until recently the species was fairly common in pine and pine-oak forests, the local vernacular name for the bird is *colonchango*, which I was told referred to the call of the male. A woman in Comitán who had frequently kept the species in captivity told me further that the male has a beautiful whistled song, which sounded to her like *pico-de-oro*. A man who had obtained a male as a young bird some six months previously told me that it had just begun to sing about two weeks previously and had two different calls. One was the *col-on-chang'-o* song, which no doubt corresponds to the *pico-de-oro* vocalization, and the other was a vibrating and whistled *preeet*. This latter call is probably equivalent to the buzzing call of the harlequin quail or perhaps to the separation call. While handling the bird I was unable to stimulate it to utter any distress calls. Because of its song, the ocellated quail is far more highly valued as a cage bird in that part of Chiapas than is the local bobwhite, which is much more readily available and thus more frequently seen as a cage bird.

EVOLUTIONARY RELATIONSHIPS

Most of the anatomical specializations that are exhibited by the harlequin quail are related to its digging behavior associated with foraging. Miller (1943) has mentioned its arched back, strong legs, long claws, and dorsally narrowed pelvis, which are all associated with the strong leg muscles related to its digging abilities. The posterior iliac crest of *Cyrtonyx* is the most highly developed of the entire group and exceeds that of *Dactylortyx* (Holman, 1961). Further, in this species the dorsal surface of the postacetabular ilium is narrow anteriorly, and it gradually narrows posteriorly to form a highly elongated, narrow dorsal roof of the posterior process. *Dactylortyx* and *Ryhnchortyx* are like most other New World quail genera in having a moderately broadened anterior face of the postacetabular ilium that narrows abruptly posteriorly, but in these the posterior process of the ilium forms a moderately long, narrow dorsal roof, rather than a short and broad roof (Holman, 1961). *Odontophorus* is variable with regard to this character, suggesting that an evolutionary trend may be traced from *Odontophorus* through *Dactylortyx* and *Rhynchortyx* to *Cyrtonyx*. The angle of the ischium relative to the iliac crest is also greater in *Cyrtonyx* than in the other genera (Holman, 1961), which is probably also related to muscular digging adaptations.

From these considerations, as well as distributional patterns, ecological and behavioral considerations, and plumage comparisons, I would judge that *Cyrtonyx* evolved from a *Dactylortyx*- or *Odontophorus*-like ancestral type in a forested or woodland environment and gradually became increasingly efficient at surviving in more xeric habitats than were its ancestors. It is the only species of the *Odontophorus* subgroup that has become fully emancipated from a fairly dense forest habitat and thus has extended its range much farther to the north in arid climates than have any of these.

31

Gray Partridge

Perdix perdix (Linnaeus) 1758

OTHER VERNACULAR NAMES

BOHEMIAN partridge, English partridge, European partridge, Hungarian partridge, Hun, Hunkie.

RANGE

Native to Europe and Asia but introduced into North America and now widely established in southern Canada and the northern United States (see distribution map). The North American population was probably derived from stock representing several different geographic races.

MEASUREMENTS

Folded wing: Adult males, 144–57 mm; adult females, 146–54 mm (males average 152 mm; females, 150 mm).

Tail: Adult males, 78–84 mm; adult females, 76–80 mm (males average 80 mm, females, 78 mm).

IDENTIFICATION

Adults, 12–13 inches long. Sexes similar in appearance. The head color of adults is tawny cinnamon except for an uncrested buffy brown crown and ear-patch. The breast and upper abdomen is a finely vermiculated gray which is interrupted by a chestnut brown horseshoe marking in males (smaller or absent in females), and the gray flanks are similarly interrupted by vertical chestnut barring. The upperparts are grayish to brownish, with darker mottling in the wing region and with conspicuous white shaft-streaks on the scapulars. The upper tail coverts and two central pairs of tail feathers are heavily vermiculated and barred, while the other tail feathers are rusty brown.

FIELD MARKS

In flight, the rusty tail feathers are spread and are usually conspicuous; otherwise the impression is one of a grayish brown bird without bright markings. Chukar partridge also exhibit rusty outer tail feathers in flight but in addition have conspicuous white throats. The bobwhite occurs in some of the same regions as the gray partridge but is smaller and shows a grayish tail when flushed. In spring a raspy *tur-ip* call may be heard (Godfrey, 1966), which has also been described as a "rusty-gate" or *keee-uck!* call (McCabe and Hawkins, 1946).

AGE AND SEX CRITERIA

Females may sometimes but not always be identified by the scapulars and median wing coverts, which have a wide buff stripe along the shaft and two to four buff crossbars; in males the feathers are darker and have only a narrow buff stripe along the shaft (McCabe and Hawkins, 1946). Furthermore, the scapulars of males are yellowish brown with very fine wavy black lines running across each feather and with a chestnut patch near the outside edge. Females have scapulars that are blackish at the base with about two light yellow crossbars, and only the outer parts of the feather are vermiculated (Lodge, quoted in Bannerman, 1963).

Immatures have the usual condition of pointed outer primaries and, at least for a time, have yellow rather than blue gray feet (Edminster, 1954). In immatures the outer two primary coverts from the juvenal plumage are also retained; the ninth covert is typically pointed rather than rounded and, although it is like that of adults in being brown with white barring, is only rarely edged with white at the tip (Petrides, 1942).

Juveniles have yellow feet and tail feathers that are much like the adult's, but the rectrices are tipped with buff and have subterminal dark bars and spots, while the central feathers are speckled and barred with dusky (Ridgway and Friedmann, 1946). White shaft-streaks are conspicuous on the breast, neck, and interscapular regions (McCabe and Hawkins, 1946).

Downy young (illustrated in color plate 61) of this species are highly distinctive; the head is buffy yellow, with a slightly darker and more rufous crown, while scattered over the sides and top of the head are a large number of dark brown spots which tend to be arranged into anterior-posterior stripes. The largest of these black markings is on the nape, and another large stripe extends from below the eye back toward the "shoulder" region and forward almost to the beak. The throat and underparts are a pale yellow, and patches of rufous occur at the rear edges of the wings and in the rump region, but the dorsal part of the body is only faintly patterned with fuscous and buff streaks.

DISTRIBUTION AND HABITAT

The present distribution of this introduced species is a highly disjunctive one, a reflection in part of the patterns of introduction. However, four fairly discrete populations can presently be recognized. The earliest established populations were those of the Pacific northwest, with birds being first released before 1900 in California and Washington. In the early decades of the 1900s there were additional and successful releases in Washington and successful introductions in Oregon, Idaho, and Montana (Yocom, 1943). The species was also introduced during 1911 in Utah (Porter, 1955) and during 1923 in Nevada (Gullion and Christensen, 1957). This population currently is largely restricted to the high, relatively arid intermountain region between the Cascade and Sierra ranges and the Rocky Mountains between forty degrees and fifty degrees north latitude. Low to moderate populations also occur in western Washington on the western slope of the Cascade ranges (Yocom, 1943) north to extreme southern British Columbia (Guiguet, 1955) and south to the Willamette Valley (Masson and Mace, 1962). Except for these most westerly populations, the birds are generally associated with grassland and semidesert vegetational types. In Oregon they are most abundant on bunch grass and sagebrush areas adjacent to wheat and other farmlands (Masson and Mace, 1962), and in eastern Washington they commonly occur in arid areas dominated by bunch grass and sagebrush where farms also occur (Yocom, 1943). In northern Nevada they are limited largely to habitats along stream bottoms and near pastures and hayfields where willows, berry-bearing bushes, and grasses are abundant

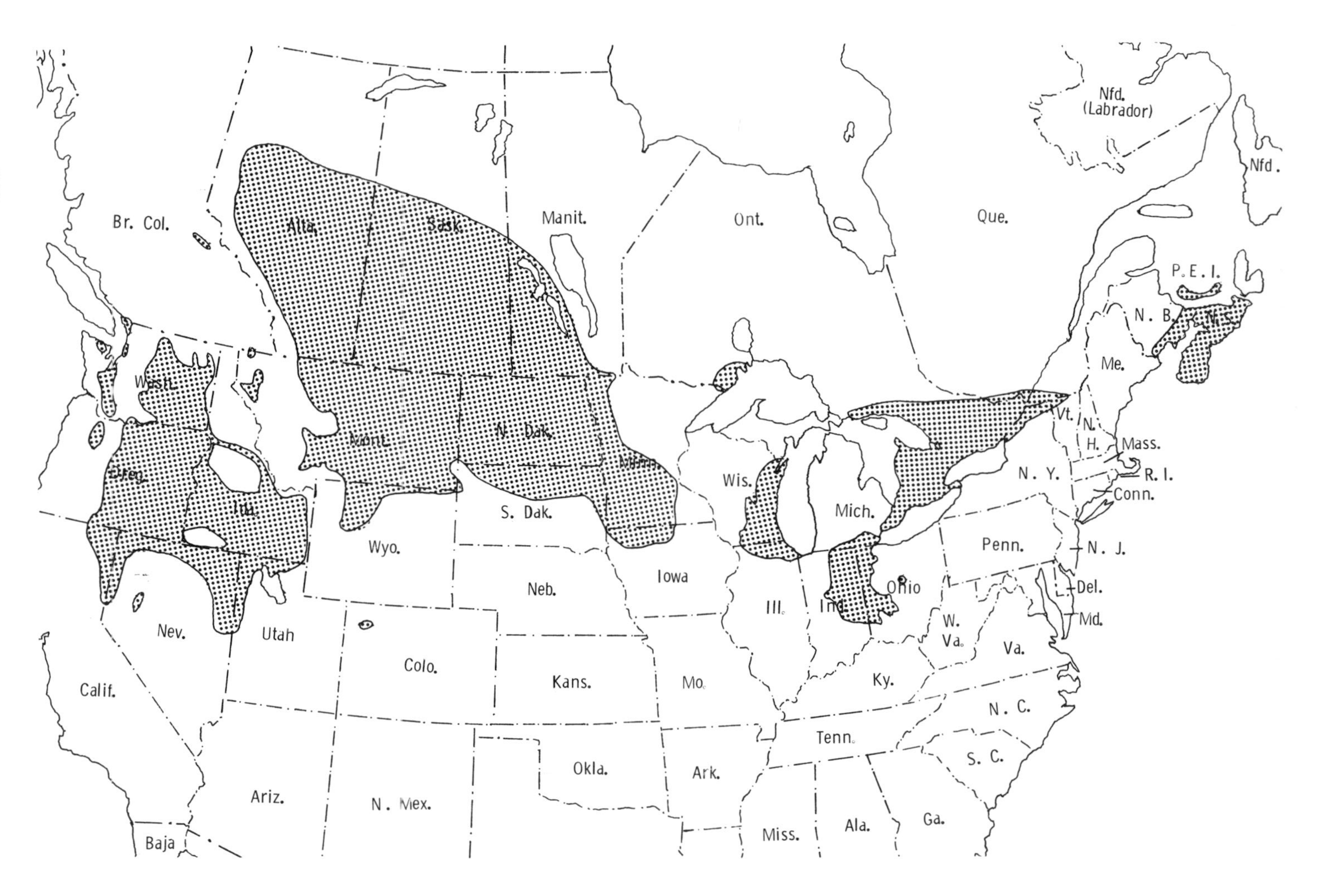

FIGURE 44. Current North American distribution of the gray partridge.

(Gullion and Christensen, 1957). In Utah they are generally found where alfalfa, wild hay, and grain grow near streams, with sagebrush nearby (Porter, 1955). In Idaho they are widely distributed throughout agricultural areas, but broods have been seen as far as fifty miles from agricultural lands in the aspen zone (Upland game birds of Idaho, 1951). This Pacific northwest population has undergone considerable retraction of range; it is now gone from the southern part of interior British Columbia, and it is probably a good deal less common throughout the intermountain region than it once was. The current yearly hunter harvest is probably under two hundred thousand, with most of this kill occurring in Oregon and Idaho.

The second major population segment is the Great Plains population, which extends from the "prairie provinces" of Alberta, Saskatchewan, and Manitoba (Rowan, 1952), southward across eastern Montana, northwestern Wyoming, the Dakotas, western Minnesota, and northwestern Iowa. This population has its origin in limited but highly successful releases that began in Alberta in 1908, supplemented by releases in Montana, North Dakota, and Manitoba during the next few decades. The Alberta releases were so successful (the first season being held in 1913) that Saskatchewan was colonized by birds of this source, and a season there was established in 1927, followed by one in Manitoba in 1931. Both Montana and North Dakota also benefited from the Alberta releases, and open seasons were established in 1929 and 1934, respectively (Johnson, 1964). A limited season was initiated by South Dakota in 1937, and in 1939 by Minnesota. Iowa first introduced the bird in 1910, but it has never extended its range beyond the north central part of the state (Green and Hendrickson, 1938). Although Nebraska began to release partridges as early as 1907, the bird has never become established and at present is only rarely encountered in the state. Throughout this extensive area, by far the largest contiguous portion of the gray partridge's range in North America, the bird is associated with small grain cultivation (wheat, oats, and barley) on high quality soils, moderate spring precipitation, severe winters, and adequate amounts of available nesting cover in the form of native grasslands or hayfield pasturelands. The average yearly hunter kill is in excess of four hundred thousand birds, with the largest current harvests in Alberta and Saskatchewan and somewhat smaller harvests in Montana and North Dakota. The population and hunter kill in Minnesota has declined considerably since the late 1950s, and the same is probably true of South Dakota.

The third population segment is the Great Lakes area, including eastern Wisconsin, southern Michigan, eastern Indiana, western Ohio, southern Ontario, and northern New York. This population was extensively studied by Yeatter (1935) in Michigan, where the birds were first released in 1911.

Releases at about the same time in Wisconsin, Indiana, and Ohio were also relatively successful. In spite of considerable efforts by the Michigan Department of Conservation in releasing birds between 1930 and 1940, nearly all these releases were failures, with the best successes occurring on light-textured soils along the southern border of the state. In contrast, Wisconsin's introductions were much more successful. After first being introduced in 1908, the birds gradually extended their range northward at a rate of about 4 miles per year, until they had moved 102 miles north in twenty-nine years. Between 1944 and 1954 the birds further extended their northern range at a rate of about 8 miles a year and also moved southwesterly at a rate of about 1 mile a year (Resadny, 1965). Apparently the Wisconsin population is now fairly stabilized, with limitations of soil and land use restricting further range extension. The birds are most abundant on red clay soils, particularly on flat lands that have few woodlands and are about 65 percent cultivated. They thrive where about half the land is planted to hay and small grains and do no better on large farm acreages than on smaller farming units (Resadny, 1965).

In Ohio a fairly extensive release program was carried out between 1909 and 1940, and by the late 1920s the birds were well established in lacustrine limestone and glacial limestone soils of western Ohio. The population probably peaked in the mid-1930s, and by 1965 had apparently all but disappeared from the state. However, surveys in the late 1960s indicate that the birds have been seen in twenty counties and may not be quite so rare as had been thought (Bachant, 1969). In Indiana the status of the gray partridge is still moderately favorable, with the birds being hunted to some extent over much of the northeastern part of the state. Wright (1966) has recently reviewed the status of this species in Indiana. The Illinois population is a southern extension of the large Wisconsin population and is limited to the northeastern corner of the state where moderate numbers are harvested each year. Early attempted introductions in New York were failures, but nearly thirty thousand birds were released between 1927 and 1932. Of these releases, only those birds in the St. Lawrence Valley prospered to the point that a limited season was possible in 1952 (Brown, 1954). The New York population is limited largely to areas with soils of limestone origin, and the best densities occur in areas of 30 to 45 percent croplands, with large areas of pasture and hay present. Major factors favoring the species there include dry weather during the hatching and brooding seasons, large areas planted to grain crops, ample nesting and brooding cover, and the presence of few pheasants and fairly light hunting (Brown, 1954). Little information is available as to the density and geographic range of the southern Ontario and adjacent Quebec population, but it is of interest that Yocom (1943) indicates no eastern Canada population, whereas Aldrich and Duvall (1955) report an extensive one extending all the way to the mouth of the St. Lawrence River. I have accepted Godfrey's estimation (1966) of the eastern Canada distribution, which indicates that most of the area south of forty-nine degrees north latitude is occupied range. No information is available as to the size of the Canadian yearly hunter kill, but probably about fifty thousand birds are currently harvested annually in the Lake States, with the majority of these taken in Wisconsin.

There are also established gray partridge populations on Prince Edward Island, southern New Brunswick, and Nova Scotia, but they are probably fairly small and no details as to habitats utilized are available to me. These populations apparently date from introductions made in the late 1920s.

POPULATION DENSITY

Most density figures for United States populations of gray partridge come from the Lake States area. Yeatter (1935) reported spring populations of 4.4, 11 and 13.3 acres per bird on three 160-acre study areas in southern Michigan. During nine years of study on a Faville Grove study area in Wisconsin, fall populations varied from an estimate of 7.5 to 26 acres per bird, averaging 15 acres per bird over the entire period. Since winter losses averaged 40 percent, expected spring densities would be nearly 30 acres per bird. Such breeding densities are far below those reported for England, where estimates of a pair per 8 or 10 acres are not uncommon (McCabe and Hawkins, 1946). The nearest comparable figures I have found are for North Dakota, where estimates of from 3.5 to 5.3 acres per bird during February have been reported on study plots of a game refuge (Hammond, 1941). It would seem probable that densities in the prairie provinces of Canada may exceed these, at least during favorable years. In England, May densities vary from 1.9 to 10.7 acres per pair, with densities of less than 5 acres per pair considered high (Jenkins, 1961).

HABITAT REQUIREMENTS

In spite of numerous attempts to introduce the gray partridge in virtually all parts of North America, no clear agreement on what constitutes ideal partridge habitat is yet available. Correlations with soil types have not proven highly successful, but the birds are typically associated with highly fertile soils associated with natural grassland and avoid both extremely sandy and heavy clay soils. Topographic conditions associated with high populations are usually flat or gently rolling lands, with the birds sometimes occurring at elevations of up to about five thousand feet in the bunch grass hills of Washington (Yocom, 1943). Favored climates are those with fairly short growing seasons and limited precipitation during the incubation and brooding periods. Severe winters are normally no serious limitation as long as snowfall is not so great that it makes grain or other seeds unavailable (Westerskov, 1965).

Perhaps the most important aspects of habitat needs of the gray partridge are those related to vegetation. Combinations of croplands, particularly small grain crops, and herbaceous cover in the form of native grasses, hayfields, or weedy herbaceous growth provides necessary nesting and escape cover. Woody cover is little utilized, and the birds seemingly avoid extensively wooded areas. Brushy areas may be used for winter shelter, and nests may sometimes be located in brushy edges, but the birds are surprisingly independent of such cover sources during most parts of the year.

The preferred nesting cover of gray partridges is clearly native grasslands or hayfields, where an abundance of dead herbaceous plant growth is to be found. Yeatter's study (1935) of 143 nest sites indicated that hayfields and grainfields accounted for more than half of the nest locations. Yocom (1943) noted that about 60 percent of 68 nests were located in hayfields, with alfalfa providing preferred nesting cover. McCabe and Hawkins (1946) also noted that hayfields provided cover for more than half of 427 nests and that alfalfa was the plant species immediately surrounding nearly 50 percent of 403 nest sites located. Most birds selected locations fairly near the edges of hayfields for nesting and were rarely more than one hundred feet from the edge, as had been earlier noted by Yeatter.

Brooding cover is essentially like nesting cover: hayfields, grainfields, or natural grasslands are all utilized. Evidently the young birds do not require a nearby source of water (Yocom, 1943), provided that succulent vegetation and insect foods are available. However, during hot weather they may move to brushy or woody cover for shade during the middle of the day.

During winter the birds may roost in the manner of bobwhites or may plunge into a snowdrift to spend the night. They are also able to tunnel under the snow, at least to a depth of a foot, to obtain food (McCabe and Hawkins, 1946; Westerskov, 1965).

Although free water is probably not essential to partridges, a supply of grit is definitely needed, particularly at times when the diet is composed primarily of grain and seeds (Trippensee, 1948).

FOOD AND FORAGING BEHAVIOR

The food intake of gray partridges comes from three primary sources, cultivated grains, seeds of various weedy herbs, and green leafy materials. Only during summer are insects taken in any appreciable amount, and rarely do they comprise more than 10 percent of the summer diet.

The grain sources utilized vary with locality, but in the Canadian Great Plains population they consist primarily of oats, barley, and wheat, which during the winter represent about 70 percent of the food consumed. Yocom (1943) also reported that these three grains, especially wheat, are major winter food sources in Washington, while in Michigan corn is perhaps the most important grain crop for partridges (Yeatter, 1943). Other cultivated crops, such as buckwheat, soybeans, and peas, may be of secondary or local significance.

The kinds of weed seeds used no doubt vary greatly in different regions but include a wide range of forbs and a few grasses. These are used mainly from late spring until grain crops become available in late summer. Green leafy materials are probably taken as soon as they become available; Yocom (1943) reported their major use during the winter season in the Palouse region of Washington, where moist, mild winters are typical. In the Canadian prairies green foliage is of minor importance in winter but rather is used heavily in spring, when it may represent about 50 percent of the food volume, and is used again in diminishing amounts during the fall (Westerskov, 1966).

MOBILITY AND MOVEMENTS

Under normal conditions relatively short movements are typical of gray partridges. There is no major habitat shift between seasons that requires any great mobility, although flights of from half to three-quarters of a mile have sometimes been noted. Usually, flights are less than a quarter mile in length, and Yocom (1943) noted that during the winter, coveys usually moved less than a quarter mile, rarely as much as half a mile. In Michigan, Yeatter (1935) noted a similar winter mobility that averaged about a fifth of a mile, and 20 percent of the coveys had a cruising radius of no more than one-eighth of a mile. Over the course of a year, Yocom found that a single female had a cruising radius of seven-eighths of a mile.

In spite of their sedentary nature, the gray partridges in Canada exhibited a remarkable rate of range expansion during the early years after their introduction. Leopold (1933) calculated that during the first fourteen years after their introduction in Alberta, a maximum average range extension of twenty-eight miles a year occurred, which is little short of astonishing. Comparable estimates of range extension in Michigan and Wisconsin were only from two to four miles a year during the period shortly after successful introduction.

SOCIAL AND REPRODUCTIVE BEHAVIOR

To a degree surprisingly similar to that of the bobwhites, the basic social unit of the gray partridge is a moderate-sized covey that infrequently exceeds 15 birds, with maximum covey sizes of about 30 birds. Probably the nucleus of each fall covey is a pair and their well-grown young, which usually number about 10 by the time the chicks are two months old (Yocom, 1943). Johnson (1964) has tabulated the average covey sizes of gray partridges by month from midsummer until March as reported from 1938 to 1963 in North Dakota. These figures and those of Hammond (1941) indicate that from the time the broods emerge in July and August, when the covey size is from 12 to 13 birds, there is a monthly decline that averages about a 9 or 10 percent reduction per month, so that by February the average covey size is approximately 7.5 birds. An average covey size of 4.7 birds in March suggests that during that month considerable covey breakup occurs as the birds prepare for nesting.

Pair formation probably begins well before the breakup of coveys, since McCabe and Hawkins (1946) noted that fighting may be seen as early as January, and Yocom (1943) reports the same activity for late January and early February. This fighting behavior is at least in part ritualized into a display during which the birds maintain a distance of about six to eight yards from one another, each alternately chasing and being chased. Once two birds were seen by one observer to run toward one another at full speed, only to stop at the last possible moment and rear up with their beaks and breast almost touching in a nearly vertical stance (Cooke, 1958). The call uttered during such threats, and especially during early morning and evening, is a "rusty-gate" call sounding like *keee-uck!*, with a very metallic tone to the first note and an accent on the second one (McCabe and Hawkins, 1946).

The social displays of the gray partridge have been studied by Jenkins (1961). He noted that coveys remain intact until pairing starts in January or February. Since the aggression that he observed did not appear to be related to defense of a nesting site or the defense of any other specific area, he did not feel that the term territoriality should be used for partridge behavior. Likewise, Blank and Ash (1956) indicated that true territorial behavior is lacking in this species (as well as *Alectoris*), and that the nearest thing to territorial behavior is the stability exhibited in covey structure. Watson found that pairing was achieved by two different methods. Pairing within coveys occurred when a pair of the previous season was re-formed or when a female actively solicited a mate from her own covey, which in no case was found to be her father or one of her brothers. Most of the chasing Jenkins observed was between yearling hens, but sometimes older females would also participate. Since most young males were not chosen by females of their own covey for mates, they left the covey and moved about singly or in groups, displaying to or attacking birds in other coveys. When an unmated cock met a covey it might display before females, which usually resulted in attacks by males within the covey, and sometimes it was able to lure a female away from the covey. Pair formation was apparently a gradual process, and many of the birds pairing for the first time changed their mates several times before a permanent pair bond was established. Often an unmated male would attach itself to a mated pair, remaining fifteen to twenty yards away and frequently displaying or crowing.

Displays mentioned or illustrated by Jenkins included an upright threat posture that resembles an upright alert posture, in which the breast was protruded, exposing the chestnut markings, and the bird stood erect, jerked its tail, and crowed. This posture is virtually identical to that assumed before copulation. Females were not observed to perform this display. Display by the male toward the female apparently emphasized his barred flanks, and the female directed her displays toward this area of the male. She often ran toward the male with her neck stretched and head held low, and directed her bill toward the male's flanks or brown breast markings while making sinuous neck movements. The lateral display of the male consists of a slight tilting of the dorsal surface of the male toward the female, but evidently there is little or no wing-lowering present (see figure 21). Sometimes the female was observed to raise her head and pass it over the flanks and back of the male as she circled him. Eventually she might stand breast-to-breast with him, rubbing her neck along his, pointing her beak upwards, and the two birds might finally rub their beaks together.

According both to Jenkins and to Blank and Ash, copulation is not preceded by elaborate displays and is begun by the female's crouching before the male. The male then approaches her in an erect posture (see figure 21) and grasps her nape, and copulation occurs.

It is not known whether a tidbitting display occurs as a courtship display in the gray partridge, but Jenkins noted that feeding behavior included courtship feeding, suggesting that such a display is present.

Nest-building, according to Yocom (1943), is performed by the female, with the male standing guard. A scrape is dug first, usually about two and a half inches deep and six or eight inches wide. Dead herbaceous vegetation is used to line the scrape, but few if any feathers are used. The first egg is probably laid shortly after the nest is finished, and after the first egg is deposited the clutch is usually covered with leafy materials between visits of the female. The egg-laying rate is probably 1.1 days per egg (McCabe and Hawkins, 1946), and the average clutch size of first nestings is probably between fifteen and seventeen eggs, with somewhat lower figures being reported for England. Lack (1947) concluded that minor annual variations in clutch sizes do occur, that the clutch size is not limited by potential egg production by females, and that hatching success is no less in clutches of twenty than in much smaller ones. He judged that the limits of clutch size in this species are probably those imposed by limits of food available to the young.

The incubation period has been established to be from 24½ to 25 days, and the female is believed to perform all of the incubation. However, in two instances the male has been observed sitting beside the female on the nest, and it is thought that this might occur only at the time of pipping (McCabe and Hawkins, 1946). The rate of nesting failure may be fairly high; three different United States studies have indicated nesting failures of 68 percent, often with mowing of hayfields being a major source of nesting losses. However, partridges are known to attempt to renest regularly, with only a slight average reduction in clutch size.

Following hatching, both parents closely attend the chicks, but, perhaps because of their large number and small size, brood losses are often substantial. Yocom (1943) estimated that almost 50 percent of the brood may be lost during the first two weeks, with chilling apparently being an important mortality factor. Recent extensive studies in England (Blank, Southwood, and Cross, 1967) have clearly indicated that at least there the key mortality factor affecting fall partridge populations is chick mortality. Further, the primary factor associated with variations in chick mortality is the relative degree of insect abundance, whereas unfavorable summer weather was believed to have only a secondary effect on breeding success (Southwood and Cross, 1969). Thus, apparently fall densities in England are related to breeding success in terms of chick survival, whereas spring breeding densities are determined by the habitat, particularly the amount of spring ground cover and the extent to which cultivated fields are broken up by hedge rows or grassy tracts, with a greater degree of habitat interspersion associated with higher breeding densities.

By the hunting season, the juvenile-to-adult ratio may vary from as little as 1.44:1 to as much as 4.35:1, depending on breeding success and chick survival, with a ratio of 3.9:1 perhaps being an average age ratio, judging from data on more than fourteen thousand birds sampled in North Dakota between 1950 and 1963 (Johnson, 1964). This would represent about eight young per pair surviving to the start of the hunting season, which agrees well with the average covey sizes of ten to twelve birds typical for that time of year.

Vocal Signals.

One of the few attempts to summarize the calls of the gray partridge is that of McCabe and Hawkins (1946), who recognized six different calls. One of these is the distress *peep* of chicks. A second "rattle" *peep*, first given by birds when they are about a month old, is transitional between the chick call and the call of adult birds. An excited *kuta-kut-kut-kut* is uttered when the birds are frightened and is accompanied by tail-flicking. Adults of both sexes hiss during the breeding season, especially when the coop of a captive pair is approached or sometimes when birds are being handled. The feeding call is uttered both by older chicks and by adults and sounds like *güp, güp*. When a brooding adult calls toward its young, it utters a low, purring *burruck-burruck*, which when imitated causes the birds to take cover and "freeze." The last of the calls that McCabe and Hawkins recognized was the "rusty gate" crowing call which, judging from Jenkins's observations, is characteristic of unmated males rather than mated ones and is associated with a threatening posture. He also noted that threatening males sometimes uttered a harsh *tit-tik-tik*.

According to Yocom (1943), birds in a covey often utter soft conversational or contact *chrrr* notes when settling down for the night, and when flushed with his mate during the prenesting season, the male nearly always "cackles." Coveys sometimes also utter a series of cackling notes when flushed, or they may remain silent.

EVOLUTIONARY RELATIONSHIPS

Inasmuch as the other probable relatives of *Perdix* that are found in southeastern Asia, Borneo, and Madagascar are not included in the current work, a discussion of the evolutionary relationships of *Perdix* is not appropriate here. It is, however, interesting to compare the similarities of evolutionary adaptation in the behavior and ecology of *Perdix* to those of such New World quail as *Colinus*. Strong similarities of covey behavior, with greatly reduced social aggression during the nonbreeding season, are found in both groups. In addition, in both groups territoriality is poorly or not at all developed during the breeding season, and male hostile behavior is associated primarily with protection of the female from unmated males. Unlike *Colinus* females, female partridges also become aggressive during the spring and may compete actively with other hens for mates, sometimes even stealing them. In both males, and especially young males, are forced to leave their coveys in spring and attempt to seek out mates from other coveys and may make themselves conspicuous by crowing behavior. This behavior probably brings about a certain degree of population mixing and may facilitate range extension. In both groups, strong monogamy is characteristic, probably as a result of a need for both sexes to care for the typically large brood of developing young. In both also, the throat, lower breast, and flank areas are important sources of visual signals in males and are associated with frontal (primarily threat) and lateral (primarily sexual) displays.

32

Chukar Partridge

Alectoris chukar (Gray) 1830
{*Alectoris graeca* (Meisner) in
A.O.U. Check-list}

OTHER VERNACULAR NAMES

CHUKOR, Indian hill partridge, rock partridge (refers to *graeca only*).

RANGE

Native to Eurasia, from France through Greece and Bulgaria (typical *graeca*) southeastward through Asia Minor and southern Asia (typical *chukar*). These two populations should probably be regarded as separate species (Watson, 1962a, b), and all of the introduced United States stock is apparently referable to *A. chukar*. The racial origin of the birds introduced into North America is varied and includes not only Indian stock (probably *A. c. chukar*, as recognized by Sushkin, 1927) but also some Turkish stock (*cypriotes* or *kurdistani*). These Turkish birds probably merged with Indian stock or have disappeared, except in New Mexico and California. The present range of the North American population is from southern interior British Columbia southward through eastern parts of Washington, Oregon, and California to the northern part of Baja California, and east in the Great Basin uplands through Nevada, Idaho, Utah, western Colorado, and

Montana, with small populations of uncertain status in Arizona, New Mexico, western South Dakota, and southern Alberta.

MEASUREMENTS

Folded wing (various races): Adult males, 144–76 mm; adult females, 140–70 mm. Males average 7 mm longer than females of same subspecies.
Tail: 78–105 mm (range of both sexes).

IDENTIFICATION

Adults, 13–15.5 inches long. The sexes are identical in appearance, with white or buffy white cheeks and throat separated from the breast by a black collar or necklace that passes through the eyes. The crown and upperparts are grayish brown to olive, grading to gray on the chest. Otherwise, the underparts and flanks are buffy, with conspicuous black and chestnut vertical barring on the flanks. The outer tail feathers are chestnut brown. The bill, feet, and legs are reddish, and males often have slight spurs on the legs.

Two other closely related species have been locally introduced in some western states and might occasionally be encountered. These include the Barbary partridge (*Alectoris barbara*) and the red-legged partridge (*A. rufa*). All have *chu-kar* calls and red legs, but the Barbary partridge has a reddish brown rather than black collar and a grayish throat and face terminating in a chestnut crown. The red-legged partridge more closely resembles the chukar partridge, but its black neck collar gradually blends into the breast by breaking up into a number of dark streaks, whereas in the chukar partridge the collar is clearly delineated from the grayish breast. Barbary partridges have been unsuccessfully introduced in California (Harper, 1963), and red-legged partridges have been introduced without success in various states including Utah, Texas, and Colorado. They have possibly survived in eastern Washington (Bump and Bohl, 1964).

According to Watson (1962a, b) chukars from Turkey and farther east are specifically different from those occurring from Greece and Bulgaria through western Europe. Birds from the Asia Minor and India populations have been successfully introduced in several states and according to Watson (1962a, b) represent the species studied by Stokes (1961) and identified as *A. graeca*. There is no evidence that wild birds representing *graeca* now occur in North America. Watson states that in addition to a number of minor plumage differences, *A. graeca* differs greatly from *A. chukar* in voice, with males of *graeca* emitting a clear ringing series of whistling notes whereas *chukar* males produce only clucking or cackling sounds.

FIELD MARKS

The striking black and white head pattern of this species can be seen for considerable distances in the arid country which this bird inhabits, as can the contrasting flank markings. In flight the reddish legs and chestnut outer tail feathers are usually visible. The "*chu-kar*" call often provides evidence for the presence of this species.

AGE AND SEX CRITERIA

Females have no apparent plumage differences from males, and measurements must be used. After the third primary (counting from inside) is fully grown (by about 16½ weeks of age) the distance from the tip of the feather to the wrist joint is diagnostic for sex, with males measuring over 136 mm (averaging 139.3 mm) and females measuring under 136 mm (averaging 131.8 mm) when measured properly (Weaver and Haskell, 1968).

Immatures may be recognized by the fact that the length of the upper primary covert for the ninth primary is less than 29 mm long in immatures and is 29 mm or longer in adults (Weaver and Haskell, 1968). Since some chukars molt their ninth primary the first year, determining age by the use of the outer primaries is often difficult, but in general the presence of faded vanes and pointed tips on the outermost or two outer primaries would indicate an immature bird. These feathers may also have a yellowish patch near the tip.

Juveniles may be identified (until about 16 weeks old) by the presence of mottled secondaries, with the innermost ones usually persisting longest (Smith, 1961). Retention of the outermost secondaries of this plumage into the first-winter plumage has been found in one captive bird (Watson, 1963).

Downy young (illustrated in color plate 61) are rather reminiscent of downy scaled quail, but the head lacks a crest or a distinctly recognizable crown patch. Instead, the crown is only slightly darker brown than is the rather grayish face, which has an eye-stripe extending back past the ear region. The underparts are buffy white, and the back pattern is similar to that of the scaled quail and elegant quail.

DISTRIBUTION AND HABITAT

The distribution of this introduced species was recently mapped by Christensen (1970), whose study provided the basis for the range map shown in this book, with minor modifications as seemed to be justified on the basis of recent information. This indicated range is considerably greater than that shown by Aldrich and Duvall (1955) or the range indicated by Edminster (1954). It is probable that continued distributional changes will occur until all of the habitats suitable for this species are eventually occupied. It would seem that much of the arid Great Basin highlands between the Cascade and Sierra ranges and the Rocky Mountains provide the combinations of climate, topography, and vegetation that best suits the chukar partridge, and only very limited success has been achieved in introducing the species to the grassland plains east of the Rocky Mountains.

The history of chukar introductions in the United States has been summarized by a variety of authors, including Cottam, Nelson, and Saylor (1940), Christensen (1954, 1970), and Bohl (1957). All told, at least forty-two states and six provinces have attempted introductions; ten states and one province have had sufficient success to declare legal seasons on the bird. These specific cases may be mentioned individually, to provide an indication of the degree of success that has been attained, as indicated in a summary made by Christensen (1970).

The first state to open a hunting season on chukars was Nevada, which had begun its introductions in 1935 and initiated a season in 1947. From that time through 1967 about 968,000 chukars had been harvested in Nevada. In 1949 Washington declared its first season, eighteen years after first introducing the species. Its total kill of an estimated 1,337,000 birds through 1967 represented the largest harvest of any state. Idaho was the third mainland state (Hawaii had its first season in 1952) to open a season on chukars, starting in 1953, following introductions that had started in 1933. Since then, an estimated 994,000 birds had been harvested through 1967. California followed with an open season in 1954, after an intensive planting program that was started in 1932 and continued through the 1950s in nearly all of the state's counties (Harper, Harry, and Bailey, 1958). An estimated 438,000 birds had been harvested there through 1967. Wyoming's first open season was held in 1955, following introductions that began in 1939. Estimated hunter kills through 1967 were 160,000 birds. Oregon and Utah both opened chukar seasons in 1956, after initially introducing birds in 1951 and 1936, respectively. The total estimated kills through 1967 were 346,000 for Utah and 1,235,000 for Oregon; the latter figure is second only to that of Washington and is based on seven fewer total years of hunting. Colorado and British Columbia had their initial hunting seasons in 1958, in the case of British Columbia only eight years after the initial introduction. Although British Columbia's population is currently limited to the Okanagan and Similkameen valleys and the lower Fraser and Thompson drainages

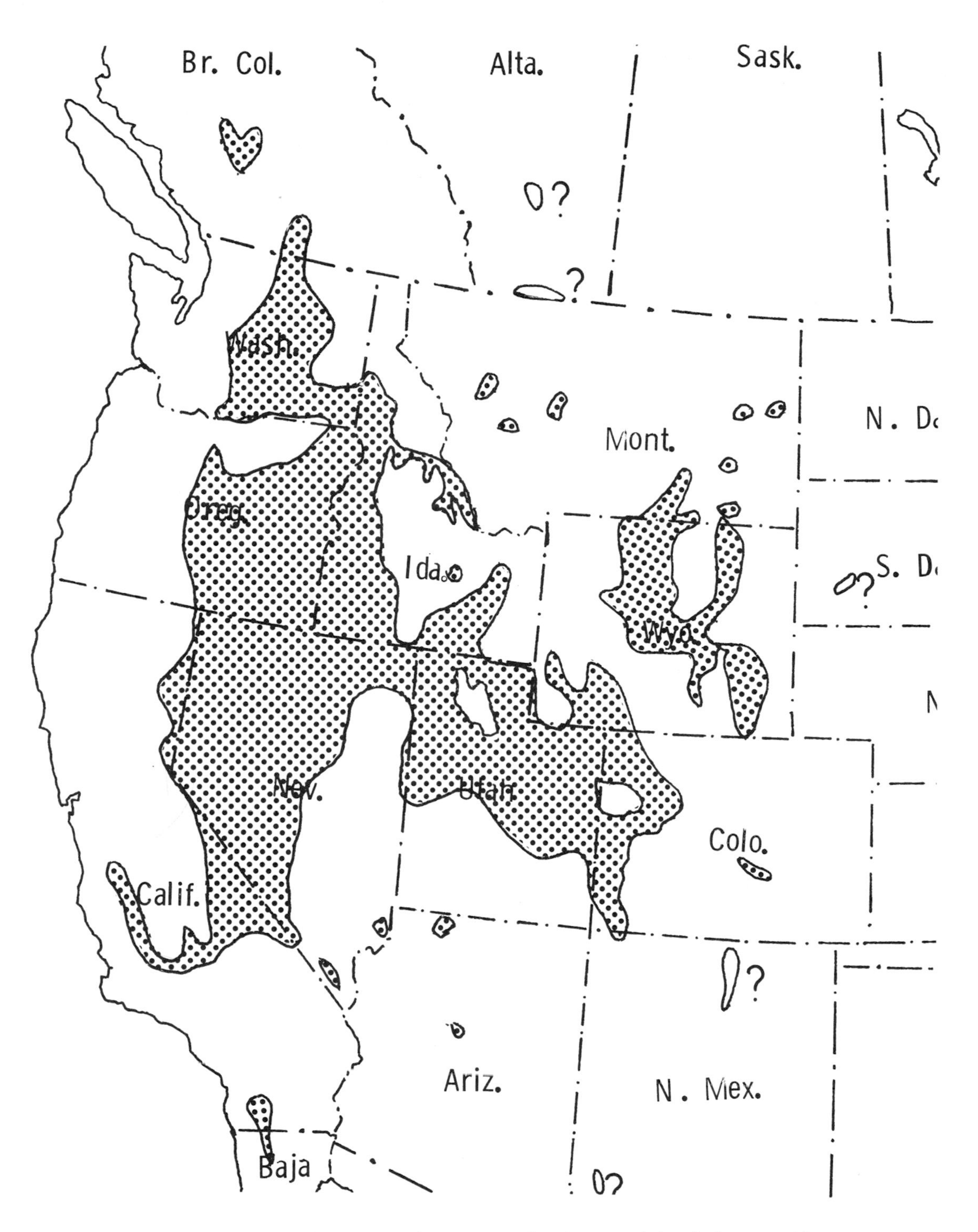

FIGURE 45. Current North American distribution of the chukar partridge.

(Godfrey, 1966), an estimated total of 107,000 birds had been harvested during the ten seasons through 1967. Montana's success with introduced chukar partridges warranted their first open season in 1959, and approximately 20,000 birds had been harvested through 1967.

A very limited degree of success can be indicated for Arizona, which first opened a season on chukars in 1962 and reported an estimated total of 250 birds harvested through 1967. Even more doubtful are South Dakota's efforts, which resulted in a very few birds shot after it opened a season in 1966. Presently the state does not list the chukar as legal game, and its status as a successfully reproducing population there is in doubt. Also in doubt is the bird's status in Alberta's Milk River valley (Godfrey, 1966) and in New Mexico and Texas (Christensen, 1970). There are no recent records of birds surviving in Nebraska in spite of a fairly extensive introduction program. Chukars spread into the Baja area of Mexico from adjacent California and now are well established there (Leopold, 1959). In addition, the Mexican government is rearing the birds in captivity for supplemental releases, and a considerable part of northwestern Mexico might eventually prove suitable for them.

Through virtually all of the chukar partridge's adopted North American range the typical vegetation is a sagebrush (*Artemisia*)–grassland community, although in the southern part of its range in California and Mexico the chukar also occurs in a saltbrush-grassland community type (Christensen, 1970). It ranges in altitude from below sea level in California's Death Valley to as high as twelve thousand feet in the White Mountains. Harper, Harry, and Bailey (1958) noted that in California the bird's distribution generally follows the 5- to 20-inch annual rainfall isohyets, and Christensen (1970) noted that in Nevada habitats the annual precipitation varies from 3.5 inches to about 12 inches. Throughout most of the species' North American range the summers are hot but short, and winters are long and moderately cold. At higher elevations snow may cause the birds to move downward into snow-free areas, but many areas in good chukar range have recorded extreme winter temperatures that are well below zero (Christensen, 1970).

POPULATION DENSITY

Remarkably little information is available on population densities of the chukar, and because of their considerable mobility and tendency to "clump" at natural or artificial watering areas it is difficult to judge populations occurring over broad areas. Moreland (1950) reported that on one study area of 61 square miles a fall population prior to the hunting season was determined to consist of 1,705 birds, which would represent 22.9 acres per bird. He also noted that on one area of 360 acres 37 chukars were flushed, in addition to a variety of other upland game. This suggests that in favorable habitats considerably greater densities might occur, possibly in excess of one bird per 10 acres. Harper, Harry, and Bailey (1958) estimated that on a study area of 60,000 acres a fall population estimate of 6,060 birds was indicated, or approximately 10 acres per bird.

Natural or artificial watering sites for chukar partridges may attract as many as one hundred birds (Harper, Harry, and Bailey, 1958; Alcorn and Richardson, 1951). Assuming that the birds rarely travel more than a mile to water (Harper, Harry, and Bailey), such a water source might be expected to have an effective "range" of about two thousand acres. Thus, visits by one hundred birds might suggest a population density of about twenty acres per bird.

HABITAT REQUIREMENTS

Habitat requirements of the chukar partridge include topographic as well as vegetative characteristics. Foremost among the topographic features that are needed by chukars is the presence of rocky slopes, which the birds use for escape (by running upslope) and roosting cover. Observations in Washington (Moreland, 1950; Galbreath and Moreland, 1953) indicate that optimum range includes from a quarter to half of the area in talus slopes, rock outcrops, cliffs, and bluffs, about half the surface covered by sagebrush and cheatgrass (*Bromus tectorum*), and a small amount of brushy creek bottom habitat as well as the presence of bunch grass (*Agropyron*) and bluegrass (*Poa*). The slopes should exceed a 7 percent grade and should have more than a two hundred-foot elevation range.

In the northern portions of the chukar's range, the amount of snow cover may be a major factor in survival. The birds are known to be able to survive winter temperatures as low as thirty degrees below zero (Moreland, 1950), but several major winter losses have been reported when snow cover more than a few inches in depth has persisted for several weeks (Christensen, 1970).

Nesting cover is little different from that used for foraging purposes and usually consists of sagebrush or a mixture of sagebrush and grassland on mountains several hundred feet above creek bottoms, often on south-facing slopes (Galbreath and Moreland, 1953). The availability of water during the summer months is a significant habitat factor; Harper, Harry, and Bailey (1958) noted that of 317 adult and young chukars seen on two California study areas between April and June, 288 birds (91 percent) were seen within a half mile of water. Further, reproductive success in California appeared to be correlated with normal or above normal late winter and early spring precipitation and associated with improved vegetative growth for food and nesting cover.

Sites for dusting and obtaining grit are no problem in the arid habitats utilized by chukar partridges, and roosting sites are usually abundant. Preferred roosting locations include talus slopes or similar rocky areas, sometimes underneath shrubs or low trees (Bohl, 1957; Christensen, 1970). During winter in Washington, the birds may roost in protected niches and caves on rocky cliff faces (Galbreath and Moreland, 1953). Circle roosting, similar to that of gray partridges and bobwhites, has been noted in various areas.

FOOD AND FORAGING BEHAVIOR

Fairly extensive studies on the foods of the chukar partridge are now available from several states, including Nevada (summarized by Christensen, 1970), Washington (Galbreath and Moreland, 1953) and California (Harper, Harry, and Bailey, 1958). More limited data are available from New Mexico (Bohl, 1957) and Colorado (Sandfort, 1954). However, virtually all of these analyses point to a predominating importance of grasses, especially cheatgrass (*Bromus tectorum*) leaves and seeds, and the seeds of weedy forbs such as Russian thistle (*Salsola*), filaree (*Erodium*), and fiddleneck (*Amsinckia*). In contrast to the western quails, chukars utilize legume seeds little, although the leaves of alfalfa (*Medicago*), clover (*Trifolium*), and sweet clover (*Melilotus*) are highly preferred foods when they are available, and locust (*Robinia*) seeds are sometimes utilized.

On a year-round basis, the seeds of cheatgrass and grass leaves are probably the most important foods, judging from studies in Washington (Galbreath and Moreland, 1953). These are supplemented during the spring by the leaves of various herbs such as dandelion (*Taraxacum*), fringecup (*Lithophragma*), and shepherd's purse (*Capsella*). The crowns and seeds of bunch grass (*Agropyron*), the fruits of serviceberry (*Amelanchier*) and hawthorn (*Crataegus*) are consumed during summer, wheat (*Triticum*) kernels are used during the fall, and various forb and shrub seeds or fruits are eaten during the winter.

Young birds eat the usual array of insect or other animal materials, but adult consumption of animal foods is rarely more than 15 percent by volume. These consist primarily of grasshoppers, crickets, and ants.

Foraging behavior is usually high during midmorning and may extend through the afternoon, with the birds moving widely while searching for food (Christensen, 1970). During hot days, they may feed early in the morning and again in late afternoon, spending the hottest period in shady canyons near a supply of water. Toward evening they again gradually move back into the canyon slopes to spend the night, foraging on the way.

Although the birds are said to be adept at scratching the ground free of litter to expose seeds, they have only a limited capacity to dig through snow. Snow depths as great as eight inches may force the birds out of mountainous areas and into the lower foothills, but the birds can scratch through snow that is only an inch or two deep (Christensen, 1970).

MOBILITY AND MOVEMENTS

Considerable dispersal ability is present in the chukar partridge, and following releases into a new habitat a large number of cases have indicated that the birds may travel extensively before becoming localized. Bohl (1957) listed dispersion records from three release points in New Mexico, which included maximum mobility records of thirty-eight miles in about seven months, twenty-two miles in a year, and thirty-eight to forty miles in a year. Brood movements of ten, eleven, and eighteen air distance miles were also reported from one release site. In California, one banded bird was known to have moved twenty miles in three months, and another banded bird was found thirty-three miles from the point of banding after two years and three months (Harper, Harry, and Bailey, 1958). In Nevada one adult bird was killed twenty-one miles away from the point where it had been caught and banded only ten days previously. All of these records indicate a remarkable ability to move across unfamiliar terrain with surprising speed.

Seasonal movements are known to occur in chukars as well; these often involve altitudinal migrations to lower valley areas during the wintertime, followed by a return to higher elevations in spring (Galbreath and Moreland, 1953; Christensen, 1970). Following the growth of succulent plants after fall rains, the birds may also move into waterless areas that were previously unoccupied during the summer (Christensen, 1970).

Individual daily ranges have not been well studied, but various lines of evidence suggest that the birds may often move about an area as wide as a mile in the course of a day, and Bump (1951) reported that the birds may travel as much as two or three miles to reach waterholes.

SOCIAL AND REPRODUCTIVE BEHAVIOR

From the appearance of broods in late summer until the beginning of pair formation in spring, the social unit of the chukar partridge is the covey. Covey sizes range widely, often from five to forty or more birds, perhaps averaging about twenty. It is possible that, as in the bobwhite, the circular roosting behavior during winter places an upper and lower limit on optimum covey size in this species, but apparently few winter counts of covey sizes have been made.

In late winter the coveys gradually begin to disband as pair formation progresses; Mackie and Buechner (1963) found that in Washington this period of breakup occurred from February through March, with older birds pairing sooner than young ones. Although basically monogamous, pairing of one male with two females may occur at the rate of about 10 percent of total pairings, according to these observers.

Although some earlier authors suggested that after pair formation has occurred the male establishes and defends a breeding territory, recent studies (Mackie and Buechner, 1963; Blank and Ash, 1956) indicate that no true territorial behavior is present, although males will repel other males from the vicinity of their mates. Stokes (1961, 1967) believes that the *chukar* or rally call when uttered by paired birds tends to repel other males; thus it may have some spacing effects. Indeed, Stokes indicated that his limited observations on wild birds suggested that the birds do defend well-defined territories.

As in the New World quail and the gray partridge, pair formation is a subtle process. It may occur only gradually, after some initial shuffling of mates (Stokes, 1961). Several displays and calls are associated with courtship, and these postures will be noted here.

Because the females have plumage identical to that of the males, it is not surprising that initial responses of males to females are aggressive ones. Stokes (1961) has described these postures, and the following description is based on his work. Three postures are usually initially performed by a reproductively active male when first exposed to a female. Head-tilting is the most common aggressive display, during which the bird tilts his head away from the opponent, simultaneously turning sideways so as to expose his barred flanks to the greatest degree. The neck and chin feathers may be raised, and the bird often stands in an erect, stiff posture ("lateral stance"). A more intense form of aggressive display is "circling," in which the dominant bird moves about another while tilting his head, again exhibiting his flank feathers. The most extreme form of circling is "waltzing," in which the head is held low and the body is nearly horizontal, as the outer wing is lowered to the point that the primaries touch the ground, and the inner wing is nearly concealed by the flank feathers (see fig. 21). Between bouts of waltzing the bird may stand erect and utter a long call, sounding like *errrrrrrr* or *errrk.* The female usually responds to these displays simply by continuing her normal activities, such as foraging, preening, or dusting.

As the male loses his aggressive tendencies, perhaps by recognition of the nonaggressive female-like responses of the other bird, he may move off some distance and begin pecking at various edible or nonedible objects. This tidbitting display is performed in association with a special call, sounding like a rapid *tu-tu-tu-tu-tu,* becoming progressively more rapid and higher in pitch. A second call, sounding like *pitoo,* may also be uttered while tidbitting. If the female is sexually active, she may then run to the male and begin pecking in the same area. The male then moves off in a stiff-legged "high-stepping" posture, gradually working toward the rear of the female and again performing tidbitting. This behavior may lead to copulation, which begins with the female facing away from the male and crouching. The male stands erect briefly, often from three to ten feet away, then utters a precopulatory "rattle" note, *uh-uh-uh-uh,* and approaches in the high-stepping posture. As he mounts the female he stops calling and grasps her nape, and copulation then occurs. No calls are uttered during copulation, and afterward the male may move away in a high-stepping posture while the female vigorously shakes her feathers.

A second important element of sexual behavior between a pair is the "nest ceremony." In this display the male enters a clump of vegetation, crouches, raises and spreads his tail, and turns while performing nest-scraping motions. He also utters a special call, a soft, continuous *churrr,* and may vibrate his wings and tail. Females may perform the same ceremony, particularly when the mate is nearby, and Stokes suggests that the display performs an important role in keeping the male closely associated with the female during the nesting period or for attaching the male to a clutch of eggs that he might take over for incubation.

Eggs are deposited in the nest by the female at the rate of about 1 to 1.9 days per egg, with the longer intervals typical earlier in the season and shorter extremes late in the season. Clutches range from about 10 to more than 20 eggs, with the average of four nests being 15.5 eggs (Mackie and Buechner, 1963). An incubation period of 24 days is typical. There is some uncertainty as to the role of the male in incubation and brood care. Some authorities (e.g., Galbreath and Moreland, 1953; Alcorn and Richardson, 1951; Mackie and Buechner, 1953) believe that the pair bond may normally last until early in the incubation period, after which the males may desert and gather together in groups. However, other observations (Goodwin, 1953; Stokes, 1961) suggest that the male may not only help raise the brood but may sometimes take over the first clutch, freeing the female to lay a second one. Mackie and Buechner (1963) noted that males were present in about 10 percent of 103 brood observations, but in many cases of two birds tending broods both appeared to be females. Christensen (1970) could find no definite case of a male chukar incubating under noncaptive conditions. There is little question that renesting by unsuccessful females does occur, but the incidence of such renesting has not yet been established. Mackie and Buechner doubt that renesting is likely after the final stages of incubation or after hatching, but they did find a nesting period extending for about five months from early March until mid-August.

Following hatching the young leave the nest with one or both parents and within a few weeks are likely to become mixed with members of other broods. Christensen (1970) reported seeing thirty to fifty chicks with from one to three adults and sometimes seeing coveys of more than one hundred chicks associated with up to ten adults. Perhaps the association of broods at watering places facilitates such interbrood transfers in this species, and thus brood-size data are of somewhat limited value. In Nevada, yearly state-wide averages of brood sizes have ranged, between 1960 and 1969, from 8.5 to 12.5 chicks, but it would seem that fall age-ratio data might provide a better index of reproductive success. Christensen noted that during 1968 and 1969 adult-to-young ratios of 1:4.14 (79.5 percent immature) and 1:5.05 (83.4 percent immature), respectively, existed. This ratio is close to those typical of bobwhites and suggestive of a high annual mortality rate. However, state-wide age ratios based on summer field surveys in Nevada between 1951 and 1969 have varied enormously, from 1:0.42 to 1:8.76, and would indicate remarkable yearly variation in productivity. Very low adult-to-young ratios were associated with drought years, such as 1953 and 1959, while high adult-to-young ratios were associated with years of favorable precipitation.

Vocal Signals

The studies of Stokes (1961, 1963) on the chukar and Goodwin (1953) on a related species of *Alectoris* provide the basis for the terminology of vocalizations in this genus. Several of these calls were mentioned in the preceding section, and need not be reviewed here. Alarm calls noted by Stokes (1961) include a ground alarm note, *whitoo*, which is also used when birds are flushed or are held in the hand. A short, gutteral *kerrr* note serves as an aerial predator note, which may be repeated as a continuing alarm or "on-guard" call while the bird soars overhead. An "all's-well" note, a soft, plaintive *coo-oor*, may be uttered when the source of alarm is gone or by loafing or feeding birds. Foraging birds also utter a food call, a slow *took* note or a rapidly repeated *tu-tu-tu-tu* series of notes, depending on the degree of excitement.

Several calls are present that may serve dual sexual and agonistic functions and are characteristic of the breeding season but not entirely limited to it. The best known of these is the rally call. This consists of a series of repeated *chuck* notes, which at progressively more intense stages sound like *per-chuck!* and *chuckara.* A single series of these calls may last up to twenty seconds, and as many as three series may be uttered in a minute. This call serves several different functions. It functions in both sexes as a scatter call to reassemble broken coveys throughout the nonbreeding period. Second, it may serve in unmated males as an advertising call that may attract available females. Third, during the breeding season it has aggressive characteristics and may serve to repel other males. To what extent this latter function might serve to space breeding pairs is still uncertain, but if it is a significant spacing mechanism for paired birds this would set the chukar's rally call apart functionally from the advertising calls of male New World quail, which are characteristic primarily of unpaired males and are only infrequently utilized after pair formation has occurred.

Besides the rally call, males in breeding condition may utter a harsh, repeated *chak* note reminiscent of an old steam engine, thus the name "steam-engine call." This call is evidently indicative of a conflict between attack and escape, especially when in the presence of a more dominant bird. Dominant males often alternate between the rally call and an excited squeaking series of notes, called by Stokes the *squee* call, apparently reflecting a stronger attack than escape tendency. A bird being attacked may also utter a raspy squealing note, lasting a second or more, indicative of extreme submission.

Finally, a series of strictly sexual notes are present, which are limited to the breeding season and characteristic of behavior associated with copulation and nesting. These include the copulation-intention note, the tidbitting and *pitoo* calls, and the nest-ceremony calls already mentioned earlier.

EVOLUTIONARY RELATIONSHIPS

As mentioned in the gray partridge account, there is little purpose in discussing the evolutionary relationships of these introduced species, since their nearest living relatives are beyond the limits established in this book. The reader is referred to Watson's discussion (1962a, b) of the problems of speciation in the *Alectoris* partridges.

Keys to Identification

THE three following keys can be used to identify unfamiliar species of North American grouse or quails that may be examined in the hand. Unless one is certain that the bird represents either a grouse or a quail, he should begin with the first key. The procedure, as in the use of all such keys, is to choose which of the two initial alternative descriptive couplets (A and AA) best fits the unknown bird. Having chosen one of these, proceed to the choice of couplets (B and BB) occurring immediately below the couplet chosen, without further regard for descriptions listed below the rejected alternative. After making a varying number of such choices in the first key, the reader will have identified his bird as to its major taxonomic group (family, subfamily, or tribe). If it belongs to either of the subfamilies of grouse or quails, the following two keys may be used to identify the specimen as to its species. These two keys operate in the same fashion as does the first, by the reader starting again with the choice of couplets A and AA and proceeding until the bird has been identified as to species. Measurements, where they are given, refer to adult birds, but in general the keys have been devised in a manner that will allow for identification regardless of the specimen's sex or, within limits, its age.

Key 1: The Major Groups of North American Galliformes

A. Hallux not elevated and more than half the length of lateral toes . . . family Cracidae

AA. Hallux elevated and less than half the length of lateral toes . . . family Phasianidae

- B. Head and upper neck naked, larger birds (folded wing over 300 mm, weight over 3,000 grams) . . . subfamily Meleagrinae (two species, excluded from present work)
- BB. Head and upper neck feathered, smaller birds (folded wing under 300 mm, weight under 3,000 grams)
    - C. Tarsus largely or entirely feathered, nostrils feathered, toes feathered and/or with comb-like (pectinate) margins . . . subfamily Tetraonidae, see key 2 for species identification
    - CC. Tarsus and nostrils unfeathered, toes never feathered or pectinate
        - D. Cutting edge of lower mandible usually with one or more slight indentations, tarsus never with sharp spur, 10–14 rectrices . . . subfamily Odontophorinae, see key 3 for species identification
        - DD. Cutting edge of lower mandible not indented, with or without spur on tarsus, 14 or more rectrices in introduced species
            - E. Larger, tarsus usually spurred, rectrices (12–32) longer than folded wing in most species . . . tribe Phasianini (one introduced species *Phasianus colchicus* with 18 rectrices)
            - EE. Smaller, tarsus little if at all spurred, rectrices (8–22) shorter than folded wing . . . tribe Perdicini (two introduced species, *Alectoris chukar* with 14 rectrices and *Perdix perdix* with 16–18 rectrices).

Key 2: Adults of North American Grouse Species (Tetraoninae)

A. Rectrices (tail feathers) all sharply pointed, larger birds (folded wing over 250 mm, weight of adults usually over 1,000 grams) . . . *Centrocercus urophasianus* (sage grouse)

AA. Rectrices not sharply pointed, usually with squarish tips, smaller birds (folded wing under 250 mm, weight usually under 1,200 grams)

- B. Lower half of tarsus unfeathered, sides of neck with broad, ornamental "ruff" feathers . . . *Bonasa umbellus* (ruffed grouse)
- BB. Lower half of tarsus feathered to base of toes or beyond, neck feathers not as described above
  - C. Outermost rectrices under 4/5 length of central ones, outer webs of primaries regularly patterned with white or buff spots . . . genus *Tympanuchus*
    - D. Central pair of rectrices considerably longer and different in color from others . . . *T. phasianellus* (sharp-tailed grouse)
    - DD. Central pair of rectrices not markedly different from others, neck with tapered, erectile pinnae . . . *T. cupido* (pinnated grouse)
  - CC. Outermost rectrices over 4/5 length of central ones, outer webs of primaries irregularly mottled or uniformly colored
    - D. Upper tail coverts not extending to tip of tail . . . genus *Dendragapus*
      - E. Rectrices 16 (rarely 18), underparts heavily barred . . . *D. canadensis* (spruce grouse)
      - EE. Rectrices 18–20 (rarely 16), underparts mostly grayish . . . *D. obscurus* (blue grouse)
    - DD. Upper tail coverts extending to tip of tail, normally with 16 rectrices . . . genus *Lagopus*
      - E. Lateral rectrices white . . . *L. leucurus* (white-tailed ptarmigan)
      - EE. Lateral rectrices dark brown or black
        - F. Bill black and heavier (usually over 9.5 mm high at base), folded wing over 195 mm . . . *L. lagopus* (willow ptarmigan)
        - FF. Bill slighter (usually under 8.5 mm high) and grayish at base, folded wing under 195 mm . . . *L. mutus* (rock ptarmigan).

In part after Ridgway and Friedmann, 1946.

Key 3: Adults of North American Quail Species (Odontophorinae)

A. Tail long (over 105 mm), considerable bare red skin present behind eye . . . genus *Dendrortyx*
  - B. Chin and throat gray, tail under 120 mm . . . *D. barbatus* (bearded tree quail)
  - BB. Chin and throat not gray, tail over 120 mm
    - C. Chin and throat white . . . *D. leucophrys* (buffy-crowned tree quail)
    - CC. Chin and throat black . . . *D. macroura* (long-tailed tree quail)

AA. Tail under 95 mm, little or no bare skin evident behind eye
  - B. Tail less than half the length of the folded wing, a bushy crest of soft, broad feathers usually present at nape, tips of extended feet reach beyond tail
    - C. Claws elongated, tips of lateral claws extend beyond base of middle claw
      - D. Tail feathers soft, narrowing toward tips, crest located at nape . . . genus *Cyrtonyx* (two closely related species, see p. 462)
      - DD. Tail feathers firm, broad, and with rounded tips, crest located at top of head . . . *Dactylortyx thoracicus* (singing quail)
    - CC. Claws not elongated, tips of lateral claws not reaching base of middle claw
      - D. Tail over 60 mm long, 12 rectrices present . . . genus *Odontophorus* (*O. guttatus*, spotted wood quail, only North American species)
      - DD. Tail under 50 mm long, with 10 rectrices . . . genus *Rhynchortyx* (one extralimital species)
  - BB. Tail longer than half the length of the folded wing, virtually crestless or bearing a distinct crest near front of head, tips of extended feet not reaching the end of tail
    - C. Scapulars and tertials spotted, tail under 70 mm long
      - D. With an erect, barred crest; sides and flanks vertically barred . . . *Philortyx fasciatus* (barred quail)
      - DD. Virtually crestless (except in extralimital species not covered by key), not vertically barred on flanks . . . genus *Colinus*

In part after Ridgway and Friedmann, 1946.

E. Chin and throat black or mostly black (males only)
  F. Breast and abdominal feathers with black edges, producing a scalloped appearance . . . *C. nigrogularis* (black-throated bobwhite)
  FF. Breast and abdominal feathers not as above . . . *C. virginianus* (bobwhite)
EE. Chin and throat not blackish (males or females)
  F. Throat white, limited ventrally by a black band . . . *C. virginianus* (bobwhite)
  FF. Throat buff-colored, little or no black present on neck or head (females)
    G. Crown and nape feathers edged with gray or grayish white; breast feathers with two terminal white spots . . . *C. nigrogularis* (black-throated bobwhite)
    GG. Crown and nape feathers edged with pale brown to buffy white; breast feathers lacking terminal white spots . . . *C. virginianus* (bobwhite)

CC. Scapulars and tertials unspotted, tail over 70 mm long
  D. Crest of two narrow, black plumes; folded wing over 120 mm . . . *Oreortyx pictus* (mountain quail)
  DD. Crest not as above, wing under 120 mm . . . genus *Callipepla*
    E. Crest of brown or black feathers that curve forward and are enlarged toward the tips
      F. Abdomen feathers edged with darker color in a scalloped pattern, flanks marked with olive brown . . . *C. californica* (California quail)
      FF. Abdomen feathers extensively blackish, or buffy with mottling or streaking, flanks marked with chestnut . . . *C. gambelii* (Gambel quail)
    EE. Crest feathers neither recurved nor enlarged toward tips
      F. Crest bushy and buff-colored, body feathers marked with dark scallops . . . *C. squamata* (scaled quail)
      FF. Crest pointed and brownish or cinnamon-colored; pale, rounded spots present on sides and abdomen . . . *C. douglasii* (elegant quail)

Name Derivations

Alectoris—from Latin *alector* (Greek *alectryōn*): cock
graeca—from Latin *Graecus*: of Greece
chukar—apparently onomatopoetic
Bonasa—from Latin *bonasum*: a bison (the drumming of the male resembling the bellowing of a bull); perhaps from Latin *bonus:* good, and *assum*: roast
umbellus—from Latin, meaning an umbrella (referring to the neck-tufts)
Callipepla—from Greek *kalos*: beautiful, and *peplos*: coat
californica—of California
douglasii—after David Douglas, Scottish explorer
gambelii—after William Gambel, American ornithologist
squamata—from Latin, meaning scale-like
Canachites—from Greek *kanacheō*: to make a noise, with formative suffix *-ites*
Centrocercus—from Greek *kentron*: a spine, and *kerkos*: tail
urophasianus—from Greek *oura*: tail, and *phasianos*: pheasant
Colinus—Latinized from the Spanish word *colín*, derived from Aztecan language Nahuatl, in which *zolin* means quail
virginianus—of Virginia
cristatus—from Latin, meaning crested
leucopogon—from Greek *leukos*: white, and *pōgōn*: beard

nigrogularis—from Latin *niger*: dark or black, *gula*: throat, and suffix *-aris*

Cyrtonyx—from Greek *kyrtos*: bent or curved, and *onyx*: nail or claw (referring to the long and curved claws)

montezumae—derived from the name of the Aztec warrior and emperor, Montezuma II

ocellatus—from Latin, meaning spotted as with little eyes

Dactylortyx—from Greek *daktylos*: a finger or toe, and *ortyx*: a quail (referring to the long toes)

thoracicus—from Greek *thōrax*: chest, with suffix *-icus*

Dendragapus—from Greek *dendron*: a tree, and *agapē*: love

obscurus—from Latin, meaning dusky

fuliginosus—from Latin, meaning sooty

canadensis—of Canada

Dendrortyx—from Greek *dendron*: a tree, and *ortyx*: a quail

barbatus—from Latin, meaning bearded

leucophrys—from Greek *leukos*: white, and *ophrys*: brow

macroura—from Greek *makros*: long or large, and *oura*: tail

Galliformes—from Latin *gallus*: a cock, and *forma*: appearance

gallinaceous—from Latin *gallina*: a hen, and adjective suffix *-aceus*, meaning pertaining to

grouse—probably from French *greoche, greiche* and *griais*, meaning spotted bird, and used in England as "grous" for the red grouse before being applied in North America for grouse in general

Lagopus—from Greek *lagōs*: a hare, and *pous*: a foot, meaning hare-footed (referring to the similarity between the feathered toes and the densely haired feet of rabbits)

leucurus—from Greek *leukos*: white, and *oura*: tail

mutus—from Latin *mutatus*: change or alteration (referring to the various plumages)

Lophortyx—from Greek *lophos*: a crest, and *ortyx*: a quail

Odontophorinae—from Greek *odous, odontos*: tooth, *phoros*: bearing, and subfamilial suffix *-inae*

Odontophorus—from Greek *odous, odontos*: tooth, and *phoros*: bearing (referring to the serrated maxilla)

guttatus—from Latin, meaning spotted or speckled

Oreortyx—from Greek *oros*: mountain, and *ortyx*: quail

partridge—from Middle English *pertriche*, and applied originally to the gray partridge. Also Scottish variants of *patrick, paitrick*, and *pertrick*

Pedioecetes—from Greek *pedion*: a plain, and *oiketes*: an inhabitant

phasianellus—diminutive of Latin *phasianus*: a pheasant

Perdix—from Latin, meaning partridge or quail

Phasianidae—from Latin *phasianus* (Greek *phasianos*): pheasant, and familial suffix *-idae*

pheasant—from Middle English *fesaunt* or *fesaun*, originally from Latin *phasianus*. Applied to the common pheasant, introduced into Europe from the area of the River Phasis (now Rioni) in Colchis, which flows into the Black Sea.

Philortyx—from Greek *philos*: loving, and *ortyx*: a quail

ptarmigan—from Gaelic *tārmachan;* applied originally to the rock ptarmigan

quail—from Old French *quaille* (modern French caille), and originally appled to the migratory quail (*Coturnix*)

Rhynchortyx—from Greek *rhynchos*: beak, and *ortyx*: a quail (referring to the large and strong beak)

cinctus—from Latin, meaning banded

Tetraoninae—from Greek *tetraōn*: a pheasant, and subfamilial suffix *-inae*

turkey—of uncertain origin, but first applied to peafowl (*Pavo*), guineafowl (*Numida*), or capercaillie (*Tetrao*) in England before being used for the New World turkey

Tympanuchus—from Greek *tympanon*: a drum, and *echō*: to have or hold

cupido—from Latin *Cupid* (referring to the "Cupid's wings" on the neck)

pallidicinctus—from Latin *pallidus*: pale, and *cinctus*: banded

pinnatus—from Latin, meaning plumed

Sources

Adams, J. L. 1956. A comparison of different methods of progesterone administration to the fowl in affecting egg production and molt. *Poultry Science* 35:323–26.

Agee, C. P. 1957. The fall shuffle in central Missouri bobwhites. *Journal of Wildlife Management* 21:324–35.

Aiken, C. E. H. 1930. A bobwhite × California quail hybrid. *Auk* 47:80–81.

Alcorn, J. R., and Richardson, F. 1951. The chukar partridge in Nevada. *Journal of Wildlife Management* 15:265–75.

Alden, P. 1969. *Finding the birds in western Mexico.* Tucson: University of Arizona Press.

Aldous, S. E. 1943. Sharp-tailed grouse in the sand dune country of north-central North Dakota. *Journal of Wildlife Management* 7:23–31.

Aldrich, J. W. 1946. The United States races of the bobwhite. *Auk* 63:493–508.

———. 1963. Geographic orientation of American Tetraonidae. *Journal of Wildlife Management* 27:529–45.

Aldrich, J. W., and Duvall, A. J. 1955. *Distribution of American gallinaceous game birds.* U.S. Department of Interior, Fish and Wildlife Service circular 34.

Aldrich, J. W., and Friedmann, H. 1943. A revision of the ruffed grouse. *Condor* 45:85–103.

Ali, S., and Ripley, S. D. 1969. *Handbook of the birds of India and Pakistan.* Vol. 2. London: Oxford University Press.

Allen, G. A., III. 1968. Keeping and raising blue grouse. *Game Bird Breeders, Aviculturalists and Conservationists' Gazette* 16(10–11):6–11.

Alvarez del Toro, M. 1952. *Los animales silvestres de Chiapas.* Tuxtla Gutiérrez, Chiapas: Ediciones del Gobierno del Estado.

———. 1964. *Lista de las aves de Chiapas: endémicas, emigrantes y de paso.* Tuxtla Gutiérrez, Chiapas: Instituto de Ciencias y Artes de Chiapas.

Amadon, D. 1966. The superspecies concept. *Systematic Zoology* 15:245.

American Ornithologists Union. 1957. *Check-list of North American Birds.* 5th ed. Baltimore: Lord Baltimore Press.

Ammann, G. A. 1944. Determining the age of pinnated and sharp-tailed grouse. *Journal of Wildlife Management* 8:170–71.

———. 1957. *The prairie grouse of Michigan.* Michigan Department of Conservation technical bulletin.

———. 1963a. Status and management of sharp-tailed grouse in Michigan. *Journal of Wildlife Management* 27:802–9.

———. 1963b. Status of spruce grouse in Michigan. *Journal of Wildlife Management* 27:591–93.

Anderson, R. K. 1969. Prairie chicken responses to changing booming-ground cover type and height. *Journal of Wildlife Management* 33:636–43.

Andrle, R. F. 1967. Birds of the Sierra de Tuxtla in Veracruz, Mexico. *Wilson Bulletin* 79:163–87.

Anthony, R. 1970. Ecology and reproduction of California quail in southeastern Washington. *Condor* 72:276–87.

Arnheim, N., Jr., and Wilson, A. C. 1967. Quantitative immunological comparison of bird lysozymes. *Journal of Biological Chemistry* 242: 3951–56.

Aubin, A. E. 1970. Territory and territorial behavior of male ruffed grouse in southwestern Alberta. Master's thesis, University of Alberta.

Bachant, J. P. 1969. Management implications of recent population trends of the Hungarian partridge (*Perdix perdix*) in Ohio. Abstract of paper presented at 78th Annual Meeting, Ohio Academy of Science, April 29, 1969.

Baepler, D. H. 1962. The avifauna of the Soloma region in Huehuetanango, Guatemala. *Condor* 64:140–53.

Bailey, A. M., and Niedrach, R. J. 1967. *A pictorial checklist of Colorado birds, with brief notes on the status of each species in neighboring states of Nebraska, Kansas, New Mexico, Utah and Wyoming.* Denver: Denver Museum of Natural History.

Bailey, F. M. 1928. *Birds of New Mexico.* Santa Fe: New Mexico Department of Fish and Game.

Bailey, V. 1928. A hybrid scaled × Gambel's quail from New Mexico. *Auk* 45:210.

Baker, M. F. 1952. Population changes of the greater prairie chicken in Kansas. *Transactions 17th North American Wildlife Conference,* pp. 259–366.

______. 1953. *Prairie chickens of Kansas.* University of Kansas Museum of Natural History and State Biological Survey miscellaneous publication no. 5.

Baldini, J. T.; Roberts, R. E.; and Kirkpatrick, C. M. 1952. Studies of the reproductive cycle of the bobwhite quail. *Journal of Wildlife Management* 16:91–93.

Banks, R. C., and Walker, L. W. 1964. A hybrid scaled × Douglas quail. *Wilson Bulletin* 76:378–80.

Bannerman, D. A. 1963. *The birds of the British Isles.* Vol. 12. Edinburgh: Oliver and Boyd.

Barth, E. K. 1953. Calculation of egg volume based on loss of weight during incubation. *Auk* 70:151–59.

Bartholomew, G. A., and MacMillen, R. E. 1961. Water economy of the California quail and its use of sea water. *Auk* 78:505–14.

Bartholomew, G. A., and Dawson, W. R. 1958. Body temperatures in California and Gambel's quail. *Auk* 75:150–56.

Bateman, A. 1968. Raising mountain quail. *Game Bird Breeders, Aviculturalists and Conservationists' Gazette* 16(2):11–13.

Batterson, W. M., and Morse, W. B. 1948. *Oregon sage grouse.* Oregon Game Commission fauna series no. 1.

Beebe, W. 1926. *Pheasants, their lives and homes.* 2 vols. Garden City, N.Y.: Doubleday, Page & Co.

Beer, J. 1943. Food habits of the blue grouse. *Journal of Wildlife Management* 7:32–44.

Bendell, J. F. 1955a. Disease as a control of a population of blue grouse, *Dendragapus obscurus fuliginosus* (Ridgway). *Canadian Journal of Zoology* 33:195–223.

______. 1955b. Age, molt and weight characteristics of blue grouse. *Condor* 57:354–61.

______. 1955c. Age, breeding behavior and migration of sooty grouse, *Dendragapus obscurus fuliginosus* (Ridgway). *Transactions 20th North American Wildlife Conference,* pp. 367–81.

Bendell, J. F., and Elliott, P. W. 1966. Habitat selection in blue grouse. *Condor* 68:431–46.

———. 1967. *Behavior and the regulation of numbers in blue grouse.* Canadian Wildlife Serice report series no. 4.

Bendire, C. 1892. *Life histories of North American birds.* U.S. National Museum special bulletin, vol. 1. no. 1.

Bennitt, R. 1951. *Some aspects of Missouri quail and quail hunting 1938–1948.* Missouri Conservation Commission technical bulletin 2.

Bent, A. C. 1932. *Life histories of North American gallinaceous birds.* U.S. National Museum bulletin 162.

Berger, D. D.; Hamerstrom, F.; and Hamerstrom, F. N., Jr. 1963. The effect of raptors on prairie chickens on booming grounds. *Journal of Wildlife Management* 27:778–91.

Bergerud, A. T. 1970a. Vulnerability of willow ptarmigan to hunting. *Journal of Wildlife Management* 34:282–85.

———. 1970b. Population dynamics of the willow ptarmigan *Lagopus lagopus alleni* L. in Newfoundland 1955 to 1965. *Oikos* 21:299–325.

Bergerud, A. T.; Peters, S. S.; and McGrath, R. 1963. Determining sex and age of willow ptarmigan in Newfoundland. *Journal of Wildlife Management* 27:700–711.

Berndt, R., and Meise, W. 1962. *Naturgeschichte der Vögel.* Vol. 2. Stuttgart: Franckh.

Berrett, D. G. 1963. The birds of the Mexican state of Tabasco. Ph.D. dissertation, Louisiana State University.

Binford, L. C. 1968. A preliminary survey of the avifauna of the Mexican state of Oaxaca. Ph.D. dissertation, Louisiana State University.

Bishop, R. A. 1964. The Mearns quail (*Cyrtonyx montezumae mearnsi*) in southern Arizona. Master's thesis, University of Arizona.

Bishop, R. A., and Hungerford, C. R. 1965. Seasonal food selection of Arizona Mearns quail. *Journal of Wildlife Management* 29:813–19.

Blackford, J. L. 1958. Territoriality and breeding behavior of a population of blue grouse in Montana. *Condor* 60:145–58.

———. 1963. Further observations on the breeding behavior of a blue grouse population in Montana. *Condor* 60:485–513.

Blake, E. R. 1950. Report on a collection of birds from Guerrero, Mexico. *Fieldiana, Zoology* 32:417–74.

———. 1958. Birds of Volcan de Chirique, Panama. *Fieldiana, Zoology* 36:498–577.

Blake, E. R., and Hanson, H. C. 1942. Notes on a collection of birds from Michoacan, Mexico. *Field Museum of Natural History, Zoological Series* 22:513–51.

Blank, T. H., and Ash, J. S. 1956. The concept of territory in the partridge, *Perdix p. perdix. Ibis* 98:379–89.

Blank, T. H.; Southwood, T. R. E.; and Cross, D. J. 1967. The ecology of the partridge: I. Outline of population processes with a particular reference to the chick mortality and nest density. *Journal of Animal Ecology* 36:549–56.

Boag, D. A. 1965. Indicators of sex, age, and breeding phenology in blue grouse. *Journal of Wildlife Management* 29:103–8.

______. 1966. Population attributes of blue grouse in southwestern Alberta. *Canadian Journal of Zoology* 44:799–814.

Boag, D. A., and Sumanik, K. M. 1969. Characteristics of drumming sites selected by ruffed grouse in Alberta. *Journal of Wildlife Management* 33:621–28.

Bohl, W. H. 1957. *Chukars in New Mexico 1931–1957.* New Mexico Department of Game and Fish bulletin no. 6.

Boucard, A. 1883. On a collection of birds from Yucatan. *Proceedings of the Zoological Society of London*, pp. 434–61.

Bradbury, W. C. 1915. Notes on the nesting of the white-tailed ptarmigan in Colorado. *Condor* 17:214–22.

Brander, R. B. 1967. Movements of female ruffed grouse during the mating season. *Wilson Bulletin* 79:28–36.

Braun, C. E. 1969. Population dynamics, habitat, and movements of white-tailed ptarmigan in Colorado. Ph.D. dissertation, Colorado State University.

______. 1970. Distribution and habitat of white-tailed ptarmigan in Colorado and New Mexico. Abstract of paper presented at 46th Annual Meeting, Southwestern and Rocky Mountain Division, A.A.A.S., April 22–25, 1970, Las Vegas, New Mexico.

Braun, C. E., and Pattie, D. L. 1969. A comparison of alpine habitats and white-tailed ptarmigan occurrence in Colorado and northern Wyoming. Abstract of paper presented at 45th Annual Meeting, Southwestern and Rocky Mountain Division, A.A.A.S., May 8, 1969, Colorado Springs, Colorado.

Braun, C. E., and Rogers, G. E. 1967a. *Determination of age and sex of the southern white-tailed ptarmigan.* Colorado Game, Fish and Parks Department game information leaflet no. 54.

______. 1967b. Habitat and seasonal movements of white-tailed ptarmigan in Colorado. Abstract of paper presented at meeting of Colorado-Wyoming Academy of Science, April 28, 1967, Boulder, Colorado.

Braun, C. E., and Willers, W. B. 1967. The helminth and protozoan parasites of North American grouse (family: Tetraonidae): A checklist. *Avian Diseases* 11:170–87.

Bremer, P. E. 1967. *Sharp-tailed grouse in Minnesota.* Minnesota Department of Conservation informational leaflet no. 15.

Brodkorb, P. 1939. New subspecies of birds from the District of Soconusco, Chiapas. University of Michigan Museum of Zoology occasional paper no. 401.

———. 1964. Catalog of fossil birds: part 2 (Anseriformes through Galliformes). *Bulletin of the Florida State Museum, Biological Sciences* 8: 195–335.

Brooks, A. 1907. A hybrid grouse, Richardson's × sharp-tail. *Auk* 24:167–69.

———. 1926. The display of Richardson's grouse, with some notes on the species and subspecies of the genus *Dendragapus. Auk* 43:281–87.

———. 1930. The specialized feathers of the sage hen. *Condor* 32:205–7.

Brown, C. P. 1946. Food of Maine ruffed grouse by seasons and cover types. *Journal of Wildlife Management* 10:17–28.

———. 1954. Distribution of the Hungarian partridge in New York. *New York Fish and Game Journal* 1:119–29.

———. 1956. Distribution of bobwhite quail in New York. *New York Fish and Game Journal* 2:191–99.

Brown, R. L. 1969a. Ecological study of Mearn's quail. Job Progress Report, in *Wildlife Research in Arizona*, 1968, pp. 35–47.

———. 1969b. Effect of hunting on Mearn's quail. Job Progress Report, in *Wildlife Research in Arizona*, 1968, pp. 49–52.

———. 1970. Mearn's quail management information. Mimeographed. Arizona Game and Fish Department job completion report, project W-53-20.

Buckley, J. L. 1954. Animal population fluctuations in Alaska—a history. *Transactions 19th North American Wildlife Conference*, pp. 338–54.

Bump, G. 1951. The chukar partridge (*Alectoris graeca*) in the Middle East with observations on its adaptability to conditions in the southwestern United States. Mimeographed. U. S. Fish and Wildlife Service.

Bump, G., and Bohl, W. H. 1964. Summary of foreign game bird propagations and liberations 1960 to 1963. U.S. Fish and Wildlife Service special scientific report (Wildlife) no. 80.

Bump, G.; Darrow, R.; Edminster, F.; and Crissey, W. 1947. *The ruffed grouse: life history, propagation, management.* Albany: New York State Conservation Department.

Campbell, H. 1957. Fall foods of Gambel's quail *(Lophortyx gambelii)* in New Mexico. *Southwestern Naturalist* 2:122–28.

———. 1968. Seasonal precipitation and scaled quail in eastern New Mexico. *Journal of Wildlife Management* 32:641–44.

Campbell, H., and Harris, B. K. 1965. Mass population dispersal and long-distance movements in scaled quail. *Journal of Wildlife Management* 29:801–5.

Campbell, H., and Lee, L. 1953. Studies on quail malaria in New Mexico and notes on other aspects of quail populations. Santa Fe: New Mexico Department of Game and Fish.

Carr, R. 1969. Raising ptarmigan and spruce grouse. *Game Bird Breeders, Aviculturalists and Conservationists' Gazette* 17(1):6–9.

Casey, W. H. 1965. Some speculations on the minimum habitat requirements of bobwhite quail. *Proceedings 19th Annual Conference S.E. Association of Game and Fish Commissioners*, pp. 30–39.

Chambers, R. E., and Sharp, W. M. 1958. Movement and dispersal within a population of ruffed grouse. *Journal of Wildlife Management* 22:231–39.

Chapman, F. M. 1896. Notes on birds observed in Yucatan. *Bulletin of the American Museum of Natural History* 8:271–90.

______. 1929. *My tropical air castle.* New York: Appleton and Co.

Choate, T. S. 1960. Observations on the reproductive activities of the white-tailed ptarmigan *(Lagopus leucurus)* in Glacier Park, Montana. Master's thesis, Montana State University.

______. 1963. Habitat and population dynamics of white-tailed ptarmigan in Montana. *Journal of Wildlife Management* 27:684–99.

Christensen, G. C. 1954. *The chukar partridge in Nevada.* Nevada Fish and Game Commission biological bulletin no. 1.

______. 1970. *The chukar partridge: its introduction, life history, and management.* Nevada Department of Fish and Game biological bulletin no. 4.

Christisen, D. M. 1967. A vignette of Missouri's native prairie. *Missouri Historical Review* 61:166–86.

______. 1969. National status and management of the greater prairie chicken. *Transactions 34th North American Wildlife and Natural Resources Conference*, pp. 207–17.

Cink, C. L. 1971. Comparative behavior and vocalizations of three *Colinus* species and their hybrids. Master's thesis, University of Nebraska.

Clark, H. K. 1899. The feather-tracts of North American grouse and quail. *Proceedings U.S. National Museum* 21:641–53.

Clarke, C. H. D. 1954. *The bobwhite quail in Ontario.* Ontario Department of Lands and Forests technical bulletin, fish and wildlife, series 2.

Coats, J. 1955. Raising lesser prairie chickens in captivity. *Kansas Fish and Game* 13(2):16–20.

Cockrum, E. L. 1952. A check-list and bibliography of hybrid birds in North America north of Mexico. *Wilson Bulletin* 64:140–59.

Conover, H. B. 1926. Game birds of the Hooper Bay region, Alaska. *Auk* 43:162–80, 303–18.
Cooke, C. H. 1958. Behaviour and display of the common partridge. *Countryside* 18:241–44.
Copelin, F. F. 1963. *The lesser prairie chicken in Oklahoma.* Oklahoma Wildlife Conservation Department technical bulletin no. 6.
Cottam, C.; Nelson, A. L.; and Saylor, L. W. 1940. *The chukar and Hungarian partridges in America.* U.S. Department of Interior Biological Survey Wildlife Leaflet 159.
Cracraft, J. 1968. Reallocation of the Eocene fossil *Palaeophasianus meleagroides* Shuffeldt. *Wilson Bulletin* 80:281–85.
Crawford, J. E. 1960. The movements, productivity and management of sage grouse in Clark and Fremont counties, Idaho. Master's thesis, University of Idaho.
Crichton, V. 1963. Autumn and winter foods of the spruce grouse in central Ontario. *Journal of Wildlife Management* 27:597.
Crispens, C. G., Jr. 1960a. *Quails and partridges of North America: a bibliography.* Seattle: University of Washington Press.
———. 1960b. Food habits of California quail in eastern Washington. *Condor* 72:473–77.
Crunden, C. W. 1963. Age and sex of sage grouse from wings. *Journal of Wildlife Management* 27:846–50.
———, ed. 1959. The western states sage grouse questionaire. Nevada Fish and Game Department.
Dalke, P. D.; Pyrah, D. B.; Stanton, D. C.; Crawford, J. E., and Schlatterer, E. F. 1960. Seasonal movements and breeding behavior of sage grouse in Idaho. *Transactions 25th North American Wildlife Conference*, p. 396–407.
———. 1963. Ecology, productivity, and management of sage grouse in Idaho. *Journal of Wildlife Management* 27:811–41.
Damon, D. 1949. Winter foods of quail in Nebraska. Nebraska Game, Forestation and Parks Commission *Wildlife Management Notes* 1:23–26.
Davis, D. E., and Domm, L. V. 1942. The sexual behavior of hormonally treated domestic fowl. *Proceedings of the Society for Experimental Biology and Medicine*, 48:667–69.
Davis, J. A. 1968. The postjuvenal wing and tail molt of the ruffed grouse *(Bonasa umbellus monticola)* in Ohio. *Ohio Journal of Science* 68:305–12.
———. 1969a. Aging and sexing criteria for Ohio ruffed grouse. *Journal of Wildlife Management* 33:628–36.
———. 1969b. Relative abundance and distribution of ruffed grouse in Ohio, past, present and future. Mimeographed. Ohio Department of

Natural Resources, Division of Wildlife inservice document no. 62.
Davis, W. B. 1944. Notes on summer birds of Guerrero. *Condor* 46:9–14.
Davison, V. E. 1940. An 8-year census of lesser prairie chickens. *Journal of Wildlife Management* 4:55–62.
Dawson, W. L. 1923. *The birds of California.* Vol. 3. San Francisco: South Moulton Co.
Delacour, J. 1951. *The pheasants of the world.* London: Country Life.
______. 1961–1962. The American quails (tribe Odontophorini). *Avicultural Magazine* 67:12–18; 68:15–19.
Dellinger, J. O. 1967. My experiences breeding and raising Mearns quail. *Game Bird Breeders, Aviculturalists and Conservationists' Gazette,* 16(4):9–10.
Dement'ev, G. P., and Gladkov, N. A. eds. 1967. *Birds of the Soviet Union.* Vol. 4. Jerusalem: Israel Program for Scientific Translations.
Dickey, D. R., and van Rossem, A. J. 1938. *The birds of El Salvador.* Zoological Series, no. 23. Chicago: Field Museum of Natural History.
Dixon, J. 1927. Contribution to the life history of the Alaska willow ptarmigan. *Condor* 29:213–23.
Dixon, K. L. 1959. Ecological and distributional relations of desert scrub birds in western Texas. *Condor* 61:397–409.
Domm, L. V. 1927. New experiments on ovariotomy and the problem of sex inversion in the fowl. *Journal of Experimental Zoölogy* 48:31–196.
Dorney, R. S. 1959. *Relationship of ruffed grouse to forest cover types in Wisconsin.* Wisconsin Conservation Department technical bulletin no. 18.
______. 1963. Sex and age structure of Wisconsin ruffed grouse populations. *Journal of Wildlife Management* 27:599–603.
Dorney, R. S., and Holzer, F. V. 1957. Spring aging methods for ruffed grouse cocks. *Journal of Wildlife Management* 21:268–74.
Duck, L. G. 1943. Seasonal movements of bobwhite quail in northwestern Oklahoma. *Journal of Wildlife Management* 7:365–68.
Duncan, D. A. 1968. Food of California quail on burned and unburned central California foothill rangeland. *California Fish and Game* 54: 123–27.
Dwight, J., Jr. 1900. The moult of the North American Tetraonidae (quails, partridge and grouse). *Auk* 17:34–51, 143–66.
Edminster, F. C. 1947. *The ruffed grouse: its life story, ecology and management.* New York: Macmillan.
______. 1954. *American game birds of field and forest.* New York: Charles Scribner's Sons.
Edwards, E. P. 1968. *Finding birds in Mexico.* 2d ed. Sweet Briar, Virginia: Author.

Edwards, E. P., and Lea, R. B. 1955. Birds of the Montserrate area, Chiapas, Mexico. *Condor* 57:31–54.

Edwards, E. P., and Martin, P. S. 1955. Further notes on birds of the Lake Patzcuaro region, Mexico. *Auk* 72:174–78.

Ellis, C. R., Jr., and Stokes, A. W. 1966. Vocalizations and behavior in captive Gambel quail. *Condor* 68:72–80.

Ellison, L. 1966. Seasonal foods and chemical analysis of winter diet of Alaskan spruce grouse. *Journal of Wildlife Management* 30:729-35.

———. 1968a. Sexing and aging Alaskan spruce grouse by plumage. *Journal of Wildlife Management* 32:12–16.

———. 1968b. Movements and behavior of Alaskan spruce grouse during the breeding season. Mimeographed. Transactions of the meeting of the California-Nevada Section of The Wildlife Society, 1968.

Emlen, J. T., Jr. 1939. Seasonal movements of a low-density valley quail population. *Journal of Wildlife Management* 3:118–30.

———. 1940. Sex and age ratios in survival of the California quail. *Journal of Wildlife Management* 4:92–99.

Emlen, J. T., Jr., and Glading, B. 1945. *Increasing valley quail in California.* University of California Agricultural Experiment Station bulletin 695.

Eng, R. L. 1955. A method for obtaining sage grouse age and sex ratios from wings. *Journal of Wildlife Management* 19:267–72.

———. 1959. A study of the ecology of male ruffed grouse (*Bonasa umbellus* L.) on the Cloquet Forest Research Center, Minnesota. Ph.D. dissertation, University of Minnesota.

———. 1963. Observations on the breeding biology of male sage grouse. *Journal of Wildlife Management* 27:841–46.

Evans, K. E., and Gilbert, D. L. 1969. A method for evaluating greater prairie chickens habitat in Colorado. *Journal of Wildlife Management* 33:643–49.

Evans, R. M. 1961. Courtship and mating behavior of sharptailed grouse *Pedioecetes phasianellus jamesi.* Master's thesis, University of Alberta.

———. 1969. Territorial stability in sharp-tailed grouse. *Wilson Bulletin* 81:75–78.

Ezra, A. 1938. A successful breeding of the mountain quail *(Oreortyx picta)* by Alfred Ezra. *Avicultural Magazine* 3:275–76.

Farner, D. S. 1955. Birdbanding in the study of population dynamics. In *Recent Studies in Avian Biology.* Urbana: University of Illinois Press.

Fatora, J. R.; Provost, E. E.; and Jenkins, J. H. 1967. Preliminary report on the breeding periodicity and brood mortality in bobwhite quail on the AEC Savannah River Power Plant. *Proceedings 20th Annual Conference, S.E. Association of Game and Fish Commissioners,* pp. 146–54.

Fay, L. D. 1963. Recent success in raising ruffed grouse in captivity. *Journal of Wildlife Management* 27:642–47.

Figge, H. 1946. Scaled quail management in Colorado. *Proceedings of 26th Meeting of the Western Association of Game and Fish Commissioners*, pp. 161–67.

Flieg, G. M. 1970. Observations on the first North American breeding of the spot-winged wood quail *(Odontophorus capueira). Avicultural Magazine* 76:1–4.

Fowle, C. D. 1960. A study of the blue grouse (*Dendragapus obscurus* [Say]) on Vancouver Island, British Columbia. *Canadian Journal of Zoology* 38:701–13.

Francis, W. J. 1965. Double broods in California quail. *Condor* 67:541–42.

Friedmann, H.; Griscom, L.; and Moore, R. T. 1950. *Distributional check-list of the birds of Mexico.* Cooper Ornithological Club Pacific Coast Avifauna no. 29.

Fuertes, L. A. 1903. With the Mearns quail in southwestern Texas. *Condor* 5:113–16.

Gabrielson, I. N., and Jewett, S. G. 1940. *Birds of Oregon.* Corvallis: Oregon State University.

Galbreath, D. S., and Moreland, R. 1953. *The chukar partridge in Washington.* Washington Department of Game biological bulletin 11.

Game bird geography. 1961. *Utah Fish and Game* 17(11):4–7.

Genelly, R. E. 1955. Annual cycle in a population of California quail. *Condor* 57:263–85.

Gilfillan, M. C., and Bezdek, H. 1944. Winter foods of the ruffed grouse in Ohio. *Journal of Wildlife Management* 8:208–10.

Gill, R. B. 1966. *A literature review on the sage grouse.* Colorado Department of Game, Fish and Parks and Colorado Cooperative Wildlife Research Unit special report no. 6.

Girard, G. L. 1937. Life history, habits and food of the sage grouse. *University of Wyoming Publication* 3:1–56.

Glading, B. 1938a. A male California quail hatches a brood. *Condor* 40:261.

———. 1938b. Studies on the nesting cycle of the California valley quail in 1937. *California Fish and Game* 24:318–40.

———. 1941. Valley quail census methods and populations at the San Joaquin Experimental Range. *California Fish and Game* 27:33–38.

Glading, B.; Selleck, D. M.; and Ross, F. T. 1945. Valley quail under private management at the Dun Lakes Club. *California Fish and Game* 31:166–83.

Godfrey, W. E. 1966. *The birds of Canada.* Ottawa: Queens Printer.

Goodnight, C. J., and Goodnight, M. L. 1956. Some observations on a tropical rain forest in Chiapas, Mexico. *Ecology* 37:139–50.

Goodwin, D. 1953. Observations on voice and behaviour of the red-legged partridge *Alectoris rufa. Ibis* 95:581–614.

______. 1958. Further notes on pairing and submissive behavior of the red-legged partridge *Alectoris rufa. Ibis* 100:59–66.

Gorsuch, D. M. 1934. *Life history of the Gambel quail in Arizona.* University of Arizona biological science bulletin 2.

Gould, J. 1850. *A monograph of the Odontophorinae or partridges of North America.* London: Author.

Gower, W. C. 1939. The use of the bursa of Fabricius as an indication of age in game birds. *Transactions of the 4th North American Wildlife Conference*, pp. 426–30.

Grange, W. W. 1948. *Wisconsin grouse problems.* Wisconsin Conservation Department publication no. 338.

Gray, A. P. 1958. *Bird hybrids.* Farnham Royal: Commonwealth Agricultural Bureau.

Green, W. E., and Hendrickson, G. O. 1938. The European partridge in north central Iowa. *Iowa Bird Life* 8:18–22.

Greenberg, D. B. 1949. *Raising gamebirds in captivity.* Princeton: Van Nostrand.

Greenewalt, C. H. 1968. *Bird song: acoustics and physiology.* Washington: Smithsonian Institution Press.

Griner, L. A. 1939. A study of the sage grouse, *Centrocercus urophasianus*, with special reference to life history, habitat requirements and numbers and distribution. Master's thesis, Utah State Agricultural College.

Grinnell, J.; Bryant, H. C.; and Storer, T. I. 1918. *The game birds of California.* Berkeley: University of California Press.

Grinnell, J., and Miller, A. H. 1944. *The distribution of the birds of California.* Cooper Ornithological Club Pacific Coast avifauna no. 27.

Grinnell, J., and Storer, T. I. 1924. *Animal life in the Yosemite.* Berkeley: University of California Press.

Griscom, L. 1932. The distribution of bird-life in Guatemala: a contribution to the study of the origin of Central American bird-life. *Bulletin of the American Museum of Natural History* 64:1–439.

Gromme, O. J. 1963. *Birds of Wisconsin.* Madison: University of Wisconsin Press.

Gross, A. O. 1928. The heath hen. *Memoirs Boston Society of Natural History* 6:491–588.

Gross, W. B. 1964. Voice production by the chicken. *Poultry Science* 43: 1005–8.

______. 1968. Voice production by the turkey. *Poultry Science* 47:1101–5.

Guiguet, C. J. 1955. *The birds of British Columbia:* (4) *upland game birds.*

British Columbia Provincial Museum handbook no. 10.
Gullion, G.W., and Gullion, A.M. 1961. Weight variations of captive Gambel quail in the breeding season. *Condor* 63:95-97.
Gullion, G. W. 1956a. Evidence of double-brooding in Gambel quail. *Condor* 58:232–34.
———. 1956b. *Let's go desert quail hunting.* Nevada Fish and Game Commission biological bulletin no. 2.
———. 1960. The ecology of Gambel's quail in Nevada and the arid southwest. *Ecology* 41:518–36.
———. 1962. Organization and movements of coveys of a Gambel quail population. *Condor* 64:402–15.
———. 1966. A viewpoint concerning the significance of studies of game bird food habits. *Condor* 68:372–76.
———. 1967a. Selection and use of drumming sites by male ruffed grouse. *Auk* 84:87–112.
———. 1967b. The ruffed grouse in northern Minnesota. University of Minnesota Forest Wildlife Relations Project, Cloquet Forest Research Center, Cloquet, Minnesota.
———. 1969. Aspen–ruffed grouse relationships. Abstract of paper presented at 31st Midwest Wildlife Conference, December 8, 1969, St. Paul, Minnesota.
Gullion, G. W., and Christensen, G. C. 1967. A review of the distribution of gallinaceous game birds in Nevada. *Condor* 59:128–38.
Gullion, G. W., and Marshall, W. H. 1968. Survival of ruffed grouse in a boreal forest. *The Living Bird* 7:117–67.
Hale, J. B., and Dorney, R. S. 1963. Seasonal movements of ruffed grouse in Wisconsin. *Journal of Wildlife Management* 27:648–56.
Hale, J. B.; Wendt, R. F.; and Halazon, G. C. 1954. *Sex and age criteria for Wisconsin ruffed grouse.* Wisconsin Conservation Department technical wildlife bulletin no. 9.
Hamerstrom, F.; Berger, D. D.; and Hamerstrom, F. N., Jr. 1965. The effect of mammals on prairie chickens on booming grounds. *Journal of Wildlife Management* 29:536–42.
Hamerstrom, F. N., Jr. 1939. A study of Wisconsin prairie chicken and sharp-tailed grouse. *Wilson Bulletin* 51:105–20.
———. 1963. Sharptail brood habitat in Wisconsin's northern pine barrens. *Journal of Wildlife Management* 27:793–802.
Hamerstrom, F. N., Jr., and Hamerstrom, F. 1949. Daily and seasonal movements of Wisconsin prairie chickens. *Auk* 66:313–37.
———. 1951. Mobility of the sharptailed grouse in relation to its ecology and distribution. *American Midland Naturalist* 46:174–226.

———. 1960. Comparability of some social displays of grouse. *Proceedings XII International Ornithological Congress,* 1958, pp. 274–93.

———. 1961. Status and problems of North American grouse. *Wilson Bulletin* 73:284–94.

Hamerstrom, F. N., Jr.; Mattson, O. E.; and Hamerstrom, F. 1957. *A guide to prairie chicken management.* Wisconsin Conservation Department technical wildlife bulletin no. 15.

Hamilton, T. H. 1962. The habitats of the avifauna of the mesquite plains of Texas. *American Midland Naturalist* 76:85–105.

Hammond, M. C. 1941. Fall and winter mortality among Hungarian partridges in Bottineau and McHenry counties, North Dakota. *Journal of Wildlife Management* 4:375–82.

Harper, F. 1953. Birds of the Nueltin Lake Expedition, Keewatin, 1947. *American Midland Naturalist* 49:1–116.

———. 1958. *Birds of the Ungava Peninsula.* University of Kansas Museum of Natural History miscellaneous publication no. 17.

Harper, H. T. 1963. The red-legged partridge in California. *Proceedings of the 43rd Annual Conference of the Western Association of State Game and Fish Commissioners,* pp. 193–95.

Harper, H. T.; Harry, B. H.; and Bailey, W. D. 1958. The chukar partridge in California. *California Fish and Game* 44:5–50.

Harris, C. L.; Gross, W. B.; and Robeson, A. 1968. Vocal acoustics of the chicken. *Poultry Science* 47:107–12.

Harrison, C. J. O. 1965. Plumage patterns and behaviour in the painted quail. *Avicultural Magazine* 71:176–84.

Hart, C. M.; Lee, O. S.; and Low, J. B. 1952. *The sharp-tailed grouse in Utah, its life history, status, and management.* Utah Department of Fish and Game publication no. 3.

Hart, F. E. 1943. Hungarian partridges in Ohio. *Ohio Conservation Bulletin* 7:4–5, 26.

Hartman, F. A. 1955. Heart weight in birds. *Condor* 57:221–38.

Haugen, A. O. 1957. Distinguishing juvenile from adult bobwhite quail. *Journal of Wildlife Management* 21:29–32.

Hellmayr, C. E., and Conover, B. 1942. *Catalogue of birds of the Americas and adjacent islands.* Zoological Series, no. 13. Chicago: Field Museum of Natural History.

Henderson, C. W. 1971. Comparative temperature and moisture responses in Gambel and scaled quail. *Condor* 71:430–36.

Henderson, F. R. 1964. Grouse and grass, twin crops. *South Dakota Conservation Digest* 31(1):16–19.

Henderson, F. R.; Brooks, F. W.; Wood, R. E.; and Dahlgren, R. B. 1967.

Sexing of prairie grouse by crown feather patterns. *Journal of Wildlife Management* 31:764–69.
Hensley, M. M. 1954. Ecological relations of the breeding bird population of the desert biome in Arizona. *Ecological Monographs* 24:185–207.
Hickey, J. J. 1955. Some American population research on gallinaceous birds. *Recent Studies in Avian Biology*. Urbana: University of Illinois Press.
Hjorth, I. 1967. Fortplantningsbeteende inom Hönsfågelfamiljen Tetraonidae [Reproductive behavior in male grouse]. *Vår Fågelvärld* 26:193–243.
______. 1970. Reproductive behaviour in Tetraonidae, with special reference to males. *Viltrevy* 7:183–596.
Hobmaier, A. 1932. The life history and the control of the cropworm, *Capillaria contorta*, in quail. *California Fish and Game* 18:290–96.
Hoffman, D. M. 1963. The lesser prairie chicken in Colorado. *Journal of Wildlife Management* 27:726–32.
______. 1965. *The scaled quail in Colorado*. Colorado Department of Game, Fish and Parks technical publication no. 18.
Hoffmann, R. 1927. *Birds of the Pacific states*. Boston: Houghton Mifflin.
Hoffmann, R. S. 1956. Observations on a sooty grouse population at Sage Hen Creek, California. *Condor* 58:321–36.
______. 1961. The quality of the winter food of blue grouse. *Journal of Wildlife Management* 25:209–10.
Höhn, E. O. 1957. Observations on display and other forms of behavior of certain Arctic birds. *Auk* 74:203–14.
______. 1969. *Die Schneehuhner (Gattung Lagopus)*. Leipzig: Die Neue Brehm-Bücherei.
Holman, J. A. 1961. Osteology of living and fossil New World quails (Aves, Galliformes). *Bulletin of the Florida State Museum (Biological Sciences)* 6:131–233.
______. 1964. Osteology of gallinaceous birds. *Quarterly Journal of the Florida Academy of Science* 27:230–52.
Höst, P. 1942. Effect of light on the moults and sequences of plumage in the willow ptarmigan. *Auk* 59:388–403.
Howard, H. 1966. Two fossil birds from the Lower Miocene of South Dakota. *Los Angeles County Museum, Contributions in Science*, No. 107, pp. 1–8.
Howard, H., and Miller, A. H. 1933. Bird remains from cave deposits in New Mexico. *Condor* 35:15–18.
Howard, W. E., and Emlen, J. T., Jr. 1942. Intercovey social relationships in the valley quail. *Wilson Bulletin* 54:162–70.
Howes, J. R. 1968. Raising game birds: principles of good management.

Game Bird Breeders, Aviculturalists and Conservationists' Gazette, 16(2):8–11.

Hubbard, J. P. 1966. A possible back-cross hybrid involving scaled and Gambel's quail. *Auk* 83:136–37.

Hudson, G. E. 1955. An apparent hybrid between the ring-necked pheasant and the blue grouse. *Condor* 57:304.

Hudson, G. E.; Lanzillotti, P. J.; and Edward, G. D. 1959. Muscles of the pelvic limb in galliform birds. *American Midland Naturalist* 61:1–67.

Hudson, G. E.; Parker, R. A.; Berge, J. V.; and Lanzillotti, P. J. 1966. A numerical analysis of the modifications of the appendicular muscles in various genera of gallinaceous birds. *American Midland Naturalist* 76:1–73.

Humphrey, P. S., and Parkes, K. C. 1959. An approach to the study of molts and plumages. *Auk* 76:1–31.

Hungerford, C. R. 1960. Water requirements of Gambel's quail. *Transactions 25th North American Wildlife Conference,* pp. 231–40.

———. 1962. Adaptations shown in selection of food by Gambel quail. *Condor* 64:213–19.

———. 1964. Vitamin A and productivity in Gambel's quail. *Journal of Wildlife Management* 28:141–47.

Huxley, J. S., and Bond, F. W. 1942. The display of Rheinart's pheasant *(Rheinardia ocellata). Proceedings of the Zoological Society of London* 111a:277–8.

Irving, L. 1960. Birds of Anaktuvuk Pass, Kobuk, and Old Crow: a study in Arctic adaptation. *U.S. National Museum Bulletin* 217:1–409.

Irving, L.; West, C. G.; Peyton, L. J.; and Paneak, S. 1967. Migration of willow ptarmigan in Arctic Alaska. *Arctic* 20:77–85.

Jackson, A. S., and DeArment, R. 1963. The lesser prairie chicken in the Texas panhandle. *Journal of Wildlife Management* 27:733–37.

Jaeger, E. C. 1957. *The North American deserts.* Stanford: Stanford Press.

Janson, R. 1953. Prairie chickens in South Dakota. *Conservation Digest* 20(2):11, 15–16.

Janson, V. 1969. *Bobwhite quail management in Michigan.* Michigan Department of Natural Resources.

Jehl, J. R. 1969. Fossil grouse of the genus *Dendragapus. Transactions of the San Diego Society of Natural History* 15:165–74.

Jenkins, D. 1961. Social behaviour in the partridge *Perdix perdix. Ibis* 103a:157–88.

Jenkins, D.; Watson, A.; and Miller, G. R. 1963. Population studies on red grouse, *Lagopus lagopus scoticus* (Lath.), in north-east Scotland. *Journal of Animal Ecology* 32:317–76.

Jensen, R. A. C. 1967. A comparative chromosome study of certain galliform and other birds. Ph.D. dissertation, University of California at Davis.

Jewett, S. G.; Taylor, W. P.; Shaw, W. T.; and Aldrich, J. W. 1953. *Birds of Washington State.* Seattle: University of Washington Press.

Johansen, H. 1956. Revision und Entstehung der arktischen Vögelfauna. *Acta Arctica* 8:1–98.

Johnsgard, P. A. 1970. A summary of intergeneric New World quail hybrids, and a new intergeneric hybrid combination. *Condor* 72:85–88.

______. 1971. Experimental hybridization in the New World quail (Odontophorinae). *Auk* 88:264–75.

Johnsgard, P. A., and Wood, R. W. 1968. Distributional changes and interactions between prairie chickens and sharp-tailed grouse in the midwest. *Wilson Bulletin* 80:173–88.

Johnson, D. A. 1960. Chukars, 'the exotic partridge,' *Naturalist* 11(2):29–32.

Johnson, M. D. 1964. *Feathers from the prairie: a short history of upland game birds.* North Dakota Game and Fish Department project report W-67-R-5.

Johnson, R. E., and Lockner, J. R. 1968. Heart size and altitude in ptarmigan. *Condor* 70:185.

Johnston, D. W. 1963. Heart weights of some Alaska birds. *Wilson Bulletin* 75:435–46.

Johnston, R. F. 1964. The breeding birds of Kansas. *University of Kansas, Publications of the Museum of Natural History* 12:575–655.

Jollie, M. 1955. A hybrid between the spruce grouse and the blue grouse. *Condor* 57:213–15.

Jones, Robert. 1963. Identification and analysis of lesser and greater prairie chicken habitat. *Journal of Wildlife Management* 27:757–78.

______. 1964a. The specific distinctness of the greater and lesser prairie chickens. *Auk* 81:65–73.

______. 1964b. Habitat used by lesser prairie chickens for feeding related to seasonal behavior of plants in Beaver County, Oklahoma. *Southwestern Naturalist* 9:111–17.

______. 1966. Spring, summer, and fall foods of the Columbian sharp-tailed grouse in eastern Washington. *Condor* 68:536–40.

Jones, Richard. 1969a. Epidermal hyperplasia in the incubation patch of the California quail, *Lophortyx californicus,* in relation to pituitary prolactin content. *General and Comparative Endocrinology* 12:498–502.

______. 1969b. Hormonal control of incubation patch development in the

California quail, *Lophortyx californicus. General and Comparative Endocrinology* 13:1–14.

Jonkel, C. J., and Greer, K. R. 1963. Fall food habits of spruce grouse in northwest Montana. *Journal of Wildlife Management* 27:593–96.

Judd, S. 1905a. The bob-white and other quails of the United States in their economic relations. *U.S. Biological Survey Bulletin* 21:1–66.

———. 1905b. The grouse and wild turkeys of the United States, and their economic value. *U.S. Biological Survey Bulletin* 24:1–55.

Juhn, M., and Harris, P. C. 1955. Local effects on the feather papilla of thyroxine and of progesterone. *Proceedings of the Society for Experimental Biology and Medicine* 90:202–4.

———. 1968. Molt of capon feathering with prolactin. *Proceedings of the Society for Experimental Biology and Medicine* 98:669–72.

June, J. 1967. A comprehensive report on the Wyoming grouse family. *Wyoming Wildlife* 10(1):19–24.

Kabat, C., and Thompson, D. R. 1963. *Wisconsin quail, 1834–1962: population dynamics and habitat management.* Wisconsin Conservation Department technical bulletin no. 30.

Kealy, R. D. 1970. Storage and incubation of game bird eggs. *Modern Game Breeding* 6(4):19–21.

Keith, L. B. 1963. *Wildlife's ten-year cycle.* Madison: University of Wisconsin Press.

Keller, R. J.; Shepherd, H. R.; and Randall, R. H. 1941. *Survey of 1941: North Park, Jackson County, Moffat County, including comparative data of previous seasons.* Colorado Game and Fish Commission sage grouse survey 3.

Kellogg, F. E.; Doster, G. L.; and Williamson, L. L. 1970. A bobwhite density greater than one bird per acre. *Journal of Wildlife Management* 34:464–66.

Kelso, L. H. 1937. *Food of the scaled quail.* U.S. Department of Agriculture, Biological Survey wildlife leaflet 84.

Kessel, B., and Schaller, G. B. 1960. Birds of the upper Sheenjek Valley, northeastern Alaska. *Biological Papers, University of Alaska,* 4:1–59.

Kirkpatrick, C. M. 1955. Factors in photoperiodism of bobwhite quail. *Physiological Zoölogy* 28:255–64.

Kirkpatrick, C. M., and Leopold, A. C. 1952. The role of darkness in sexual activity of the quail. *Science* 116:280–81.

Klaas, E. E. 1968. Summer birds from the Yucatan Peninsula, Mexico. *University of Kansas, Publications of Museum of Natural History* 17: 581–611.

Klebenow, D. A. 1969. Sage grouse nesting and brood habitat in Idaho.

Journal of Wildlife Management 33:649–62.
Klebenow, D. A., and Gray, G. M. 1968. Food habits of juvenile sage grouse. *Journal of Range Management* 21:80–83.
Klimstra, W. D. 1950a. Notes on bobwhite nesting behavior. *Iowa Bird Life* 20:2–7.
______. 1950b. Bobwhite quail nesting and production in southeastern Iowa. *Iowa State Journal of Science* 24:385–95.
Klimstra, W. D., and Scott, T. G. 1957. Progress report on bobwhite nesting in southern Illinois. *Proceedings 10th Annual Conference, S.E. Association of Game and Fish Commissioners,* p. 351–55.
Klimstra, W. D., and Ziccardi, V. C. 1963. Night-roosting habitat of bobwhites. *Journal of Wildlife Management* 27:202–14.
Klonglan, E. D., and Hlavka, G. 1969. Iowa's first ruffed grouse hunting season in 45 years. *Proceedings Iowa Academy of Science* 76:226–30.
Kobayashi, H. 1958. On the induction of molt in birds by 17 α oxyprogesterone-17 capronate. *Endocrinology* 63:420–30.
Kobriger, G. D. 1965. Status, movements, habitats, and foods of prairie grouse on a sandhills refuge. *Journal of Wildlife Management* 29:788–800.
Koivisto, I. 1965. Behavior of the black grouse, *Lyrurus tetrix* (L.), during the spring display. *Finnish Game Research* 26:1–60.
Korschgen, L. J. 1948. Late fall and early-winter food habits of bobwhite quail in Missouri. *Journal of Wildlife Management* 12:46–57.
______. 1962. Food habits of greater prairie chickens in Missouri. *American Midland Naturalist* 68:307–18.
______. 1966. Foods and nutrition of ruffed grouse in Missouri. *Journal of Wildlife Management* 30:86–100.
Kruijt, J. P. 1962. On the evolutionary derivation of wing display in Burmese red junglefowl and other gallinaceous birds. *Symposia of the Zoological Society of London* 8:25–35.
Kruijt, J. P., and Hogan, H. A. 1967. Social behavior on the lek in the black grouse, *Lyrurus tetrix tetrix* (L.). *Ardea* 55:203–40.
Kulenkamp, A. W.; Coleman, T. H.; and Ernst, R. A. 1967. Artificial insemination of bobwhite quail. *British Poultry Science* 8:177–82.
Kulenkamp, A. W., and Coleman, T. H. 1968. Egg production in bobwhite quail. *Poultry Science* 47:1687–88.
Lack, D. 1947. The significance of clutch-size in the partridge *(Perdix perdix). Journal of Animal Ecology* 16:19–23.
______. 1966. *Population studies of birds.* Oxford: Oxford University Press.
______. 1968. *Ecological adaptations for breeding in birds.* London: Methuen & Co.
Lahnum, W. H. 1944. A study of the mountain quail with suggestions for

management in Oregon. Ph.D. dissertation, Oregon State College.

Lance, A. N. 1970. Movements of blue grouse on the summer range. *Condor* 72:437–44.

Larimer, E. J. 1960. *Winter foods of the bobwhite in southern Illinois.* Illinois Natural History Survey biological notes no. 42.

Larrison, E. J., and Sonnenberg, K. G. 1968. *Washington birds: their location and identification.* Seattle: Seattle Audubon Society.

Lauckhart, J. B. 1957. Animal cycles and food. *Journal of Wildlife Management* 21:230–33.

Leach, H. R., and Browning, B. M. 1958. A note on the food of sage grouse in the Madeline Plains area of California. *California Fish and Game* 44:73–76.

Leach, H. R., and Hensley, A. L. 1954. The sage grouse in California, with special reference to food habits. *California Fish and Game* 40:385–94.

Lee, L. 1950. Kill analysis for the lesser prairie chicken in New Mexico, 1949. *Journal of Wildlife Management* 14:475–77.

Lee, R. O. 1966. Through Quintana Roo, 1965. *Explorer's Journal* 44:83–88.

LeFebvre, E. A., and LeFebvre, J. H. 1958. Notes on the ecology of *Dactylortyx thoracicus. Wilson Bulletin* 70:372–77.

Lehmann, V. W. 1941. *Attwater's prairie chicken: its life history and management.* U.S. Department of the Interior, Fish and Wildlife Service North American fauna no. 57.

———. 1946. Bobwhite quail reproduction in southwestern Texas. *Journal of Wildlife Management* 10:111–23.

———. 1968. The Attwater prairie chicken, current status and restoration opportunities. *Transactions 33rd North American Wildlife Conference,* p. 398–407.

Lehmann, V. W., and Mauermann, R. G. 1963. Status of Attwater's prairie chicken. *Journal of Wildlife Management* 27:713–25.

Lehmann, V. W., and Ward, H. 1941. Some plants valuable to quail in southwestern Texas. *Journal of Wildlife Management* 5:131–35.

Lemburg, W. W. 1962. Rearing sharp-tailed grouse. *Game Bird Breeders, Pheasant Fanciers and Aviculturalists' Gazette* 13(9):10–11.

Leopold, A. 1933. *Game management.* New York: Charles Scribner's Sons.

———. 1949. *A Sand County Almanac.* New York: Oxford University Press.

Leopold, A. S. 1939. Age determination in quail. *Journal of Wildlife Management* 3:261–65.

———. 1953. Intestinal morphology of gallinaceous birds in relation to food habits. *Journal of Wildlife Management* 17:197-203.

______. 1959. *Wildlife of Mexico: The game birds and mammals.* Berkeley: University of California Press.

Leopold, A. S., and McCabe, R. A. 1957. Natural history of the Montezuma quail in Mexico. *Condor* 59:3–26.

Levy, S. H.; Levy, J. J.; and Bishop, R. A. 1966. Use of tape-recorded female quail calls during the breeding season. *Journal of Wildlife Management* 30:426–28.

Lewin, V. 1963. Reproduction and development of young in a population of California quail. *Condor* 65:249–78.

______. 1965. The introduction and present status of California quail in the Okanagan Valley of British Columbia. *Condor* 67:61–66.

Lewis, J. B.; McGowan, J. D.; and Baskett, T. S. 1968. Evaluating ruffed grouse reintroduction in Missouri. *Journal of Wildlife Management* 32:17–28.

Ligon, J. S. 1952. The vanishing masked bobwhite. *Condor* 54:48–50.

______. 1961. *New Mexico birds and where to find them.* Albuquerque: University of New Mexico Press.

Lint, K. C. 1965. Courtship and display—great argus pheasant. *Zoonooz* 38(8):5–7.

Lofts, B., and Murton, R. K. 1968. Photoperiodic and physiological adaptations regulating avian breeding cycles and their ecological significance. *Journal of Zoology* 155:327–94.

Longley, W. H. 1951. Bobwhite quail. *Conservation Volunteer* 14(1):27–31; 14(2):29–31.

Loveless, C. M. 1958. *The mobility and composition of bobwhite quail populations in South Florida: with notes on the post-nuptial and post-juvenal molts.* Florida Game and Fresh Water Fish Commission technical bulletin no. 4.

Lowery, G. H., Jr., and Newman, R. J. 1951. Notes on the ornithology of southeastern San Luis Potosi. *Wilson Bulletin* 63:315–22.

Lowery, J., and Dalquest, W. W. 1951. Birds from the state of Veracruz, Mexico. *University of Kansas, Publications of the Museum of Natural History* 3:531–649.

Lumsden, H. G. 1961a. Displays of the spruce grouse. *Canadian Field-Naturalist* 75:152–60.

______. 1961b. The display of the capercaillie. *British Birds* 54:257–72.

______. 1965. *Displays of the sharptail grouse.* Ontario Department of Lands and Forests technical series research report no. 66.

______. 1968. *The displays of the sage grouse.* Ontario Department of Lands and Forests research report (wildlife) no. 83.

———. 1969. A hybrid grouse, *Lagopus* × *Canachites*, from northern Ontario. *Canadian Field-Naturalist* 83:23–30.

———. 1970. The shoulder-spot display of grouse. *The Living Bird* 9:65–74.

Lumsden, H. G., and Weeden, R. B. 1963. Notes on the harvest of spruce grouse. *Journal of Wildlife Management* 27:587–91.

Lundell, C. L. 1937. *The vegetation of Petén.* Carnegie Institute publication no. 478.

MacDonald, S. D. 1968. The courtship and territorial behavior of Franklin's race of the spruce grouse. *The Living Bird* 7:4–25.

———. 1970. The breeding behavior of the rock ptarmigan. *The Living Bird* 9:195–238.

Mackie, R. J., and Buechner, H. K. 1963. The reproductive cycle of the chukar. *Journal of Wildlife Management* 27:246–60.

Maher, W. J. 1959. Habitat distribution of birds breeding along the upper Laolak River, northern Alaska. *Condor* 61:351–68.

Marsden, H. M., and Baskett, T. S. 1958. Annual mortality in a banded bobwhite population. *Journal of Wildlife Management* 22:414–19.

Marshall, W. H. 1946. Cover preferences, seasonal movements, and food habits of Richardson's grouse and ruffed grouse in southern Idaho. *Wilson Bulletin* 58:42–52.

———. 1965. Ruffed grouse behavior. *Bio-Science* 15:92–94.

Marshall, W. H., and Jensen, M. S. 1937. Winter and spring studies of the sharp-tailed grouse in Utah. *Journal of Wildlife Management* 1:87–99.

Martin, A. C.; Zim, H. S.; and Nelson, A. L. 1951. *American wildlife and plants.* New York: McGraw Hill Book Co.

Martin, N. S. 1970. Sagebrush control related to habitat and sage grouse occurrence. *Journal of Wildlife Management* 34:313–20.

Martin, P. S. 1955. Zonal distribution of vertebrates in a Mexican cloud forest. *American Naturalist* 89:347–61.

Masson, W. V., and Mace, R. U. 1962. *Upland game birds.* Oregon State Game Commission wildlife bulletin no. 5.

May, T. A. 1970. Seasonal foods of white-tailed ptarmigan in Colorado. Master's thesis, Colorado State University.

May, T. A., and Braun, C. E. 1969. Observations on winter foods of Colorado white-tailed ptarmigan. Abstract of paper presented at meeting of the Southwestern and Rocky Mountain Division, A.A.A.S., May 8, 1969, Colorado Springs, Colorado.

———. 1970. Seasonal trends in the selection of foods by Colorado white-tailed ptarmigan. Abstract of paper presented at meeting of the Central Mountain and Plains Section, The Wildlife Society, April 18, 1970, Pingree Park, Colorado.

Mayr, E., and Amadon, D. 1951. A classification of recent birds. *American Museum Novitates* 1496:1–42.

Mayr, E., and Short, L. L., Jr. 1970. *Species taxa of North American birds: a contribution to comparative systematics.* Nuttall Ornithological Club publication no. 9.

McCabe, R. A. 1954. Hybridization between the bobwhite and scaled quail. *Auk* 71:293–97.

McCabe, R. A., and Hawkins, A. S. 1946. The Hungarian partridge in Wisconsin. *American Midland Naturalist* 36:1–75.

McColm, M. 1970. Ruffed grouse. *Nevada Outdoors* 4(3):27.

McEwen, L. C.; Knapp, D. B.; and Hilliard, E. A. 1969. Propagation of prairie grouse in captivity. *Journal of Wildlife Management* 33:276–83.

McLean, D. D. 1930. The quail of California. California Division of Fish and Game bulletin no. 2.

McMillan, I. I. 1964. Annual population changes in California quail. *Journal of Wildlife Management* 28:702–11.

McNabb, F. M. A. 1969. A comparative study of water balance in three species of quail. *Comparative Biochemistry and Physiology* 28:1045–74.

Mercer, E., and McGrath, R. 1963. *A study of a high ptarmigan population on Brunette Island, Newfoundland, in 1962.* Newfoundland Wildlife Division, Department of Mines and Resources.

Meyer de Schauensee, R. 1966. *The species of birds of South America and their distribution.* Philadelphia: Academy of Natural Science of Philadelphia.

Miller, A. H., and Stebbins, R. C. 1964. *The lives of desert animals in Joshua Tree National Monument.* Berkeley: University of California Press.

Miller, L. 1943. Notes on the Mearns quail. *Condor* 45:104–9.

Miller, W. De Witt. 1905. List of birds collected in southern Sinaloa, Mexico, by J. H. Batty, during 1903–1904. *Bulletin of the American Museum of Natural History* 21:339–69.

Moffitt, J. 1938. The downy young of *Dendragapus. Auk* 55:589–95.

Mohler, L. L. 1944. Distribution of upland game birds in Nebraska. *Nebraska Bird Review* 12:1–6.

______. 1963. Winter surveys of Nebraska prairie chickens and management implications. *Journal of Wildlife Management* 27:737–38.

Monroe, B. L. J. 1968. *A distributional survey of the birds of Honduras.* American Ornithologists Union ornithological monograph no. 7.

Monson, G., and Phillips, A. R. 1964. The species of birds in Arizona. In *The vertebrates of Arizona.* Tucson: University of Arizona Press.

Moreland, R. 1950. Success of chukar partridge in the state of Washington. *Transactions 15th North American Wildlife Conference,* pp. 399–409.

Mosby, H. S., ed. 1963. *Wildlife investigational techniques.* 2d ed. Blacksburg, Virginia: The Wildlife Society.

Moss, R. 1968. Food selection and nutrition in ptarmigan *(Lagopus mutus). Symposia of the Zoological Society of London* 21:207–16.

———. 1969. Rearing red grouse and ptarmigan in captivity. *Avicultural Magazine* 75:256–61.

Murphy, D. A., and Baskett, T. S. 1952. Bobwhite mobility in central Missouri. *Journal of Wildlife Management* 16:498–510.

Mussehl, T. W. 1960. Blue grouse production, movements, and populations in the Bridger Mountains, Montana. *Journal of Wildlife Management* 24:60–68.

———. 1963. Blue grouse brood cover selection and land-use implications. *Journal of Wildlife Management* 27:547–55.

Mussehl, T. W., and Leik, T. H. 1963. Sexing wings of adult blue grouse. *Journal of Wildlife Management* 27:102–6.

Mussehl, T. W., and Schladweiler, P. 1969. *Forest grouse and experimental spruce budworm insecticide studies.* Montana Fish and Game department technical bulletin no. 4.

Myers, C. W. 1969. The ecological geography of cloud forest in Panama. *American Museum Novitates* 2396:1–52.

Myers, J. A. 1917. Studies on the syrinx of *Gallus domesticus. Journal of Morphology* 29:165–214.

National survey of fishing and hunting. 1965. U.S. Bureau of Sport Fisheries and Wildlife Resource publication 27.

Nelson, A. L., and Martin, A. C. 1953. Gamebird weights. *Journal of Wildlife Management* 17:36–42.

Nelson, O. C. 1955. A field study of the sage grouse in southeastern Oregon with special reference to reproduction and survival. Master's thesis, Oregon State College.

Ohmart, R. B. 1967. Comparative molt and pterylography in the quail genera *Callipepla* and *Lophortyx. Condor* 69:535–48.

Ormiston, J. H. 1966. The food habits, habitat and movements of mountain quail in Idaho. Master's thesis, University of Idaho.

Orr, R. T., and Webster, J. D. 1968. New subspecies of birds from Oaxaca (Aves: Phasianidae, Turdidae, Parulidae). *Proceedings of the Biological Society of Washington* 81:37–40.

Our feathered friends. n. d. Wyoming Game and Fish Commission.

Palmer, R. S. 1949. Maine birds. *Harvard Museum of Comparative Zoology Bulletin* 102:1–656.

Palmer, W. L. 1954. Unusual ruffed grouse density in Benzie County, Michigan. *Journal of Wildlife Management* 18:542–43.

______. 1963. Ruffed grouse drumming sites in northern Michigan. *Journal of Wildlife Management* 27:656–63.
Parmelee, D. F.; Stephens, H. A.; and Schmidt, R. H. 1967. The birds of southeastern Victoria Island and adjacent small islands. *National Museum of Canada Bulletin* 222:1–229.
Patterson, R. L. 1949. Sage grouse along the Oregon Trail. *Wyoming Wild Life* 13(8):1–16, 34–37.
______. 1952. *The sage grouse in Wyoming.* Denver: Sage Books.
Paynter, R. J., Jr. 1955. The ornithogeography of the Yucatán Peninsula. Yale University, *Bulletin of the Peabody Museum* No. 9, pp. 1–347.
______, ed. 1957. Biological investigations in the Selva Lacandona, Chiapas, Mexico. *Harvard Museum of Comparative Zoology Bulletin* 116:193–298.
Peck, M. E. 1911. A hybrid quail. *Condor* 3:149–51.
Pendergast, B. A., and Boag, D. A. 1970. Seasonal changes in diet of spruce grouse in central Alberta. *Journal of Wildlife Management* 34:605–11.
______. 1971. Maintenance and breeding of spruce grouse in captivity. *Journal of Wildlife Management* 35:177–79.
Peterle, T. J. 1951. Intergeneric galliform hybrids: a review. *Wilson Bulletin* 63:219–24.
Peters, J. L. 1934. *Check-list of birds of the world.* Vol. 2. Cambridge: Harvard University Press.
Peters, S. S. 1958. Food habits of the Newfoundland willow ptarmigan. *Journal of Wildlife Management* 22:384–94.
Peterson, J. G. 1970. The food habits and summer distribution of juvenile sage grouse in central Montana. *Journal of Wildlife Management* 34: 147–55.
Peterson, R. T. 1960. *A field guide to the birds of Texas.* Boston: Houghton Mifflin.
Petrides, G. A. 1942. Age determination in American gallinaceous game birds. *Transactions 7th North American Wildlife Conference,* p. 308–28.
______. 1949. Viewpoint on the analysis of open season sex and age ratios. *Transactions 14th North American Wildlife Conference,* p. 391–410.
______. 1951. Notes on age determination in juvenal European quail. *Journal of Wildlife Management* 15:116–17.
Petrides, G. A., and Nestler, R. B. 1943. Age determination in juvenal bobwhite quail. *American Midland Naturalist* 30:774–82.
Phillips, A. R. 1959. Las subespecies de la codornis de Gambel y el problema de los cambios climaticos en Sonora. *Anales del Instituto de Biología, Mexico Universidad Nacional* 29:361–74.
______. 1966. Further systematic notes on Mexican birds. *Bulletin of the British Ornithologists' Club* 86:86–94.

Phillips, A. R.; Marshall, J.; and Monson, G. 1964. *Birds of Arizona.* Tucson: University of Arizona Press.

Phillips, R. L. 1967. Fall and winter food habits of ruffed grouse in northern Utah. *Journal of Wildlife Management* 31:827–29.

Pocock, R. I. 1911. The display of the peacock-pheasant *Polyplectron chinguis. Avicultural Magazine* 2:229–37.

Porter, R. D. 1955. The Hungarian partridge in Utah. *Journal of Wildlife Management* 19:93–109.

Price, J. B. 1938. An incubating male California quail. *Condor* 40:87.

Principal game birds and mammals of Texas. 1945. Austin: Texas Game, Fish, and Oyster Commission.

Prososki, A. E. 1970. Social behavior and adult vocalizations of some *Colinus* and *Lophortyx* hybrids. Master's thesis, University of Nebraska.

Pynnönen, A. 1954. Beiträge zur Kenntnis der Lebensweise des Haselhuhns *Tetrastes bonasia* (L.). *Papers on Game Research* 12:1–90.

Pyrah, D. B. 1954. A preliminary study toward sage grouse management in Clark and Fremont counties based on seasonal movement. Master's thesis, University of Idaho.

———. 1963. *Sage grouse investigations.* Idaho Fish and Game Department, Wildlife Restoration Division job completion report, project W 125-R-2.

———. 1964. Sage chickens in captivity. *Game Bird Breeders, Pheasant Fanciers and Aviculturalists' Gazette* 13(9):10–11.

Quick, H. F. 1947. Winter food of white-tailed ptarmigan in Colorado. *Condor* 49:233–35.

Raitt, R. J., Jr. 1960. Breeding behavior in a population of California quail. *Condor* 62:284–92.

———. 1961. Plumage development and molts of California quail. *Condor* 63:294–303.

Raitt, R. J., and Genelly, R. E. 1964. Dynamics of a population of California quail. *Journal of Wildlife Management* 28:127–41.

Raitt, R. J., and Ohmart, R. D. 1966. Annual cycle of reproduction and molt in Gambel quail of the Rio Grande Valley, southern New Mexico. *Condor* 68:541–61.

———. 1968. Sex and age ratios in Gambel quail of the Rio Grande Valley, southern New Mexico. *Southwestern Naturalist* 13:27–33.

Rare and endangered fish and wildlife of the United States. 1968. U.S. Bureau of Sport Fisheries and Wildlife Resource publication 34.

Rasmussen, D. I., and Griner, L. A. 1938. Life history and management studies of the sage grouse in Utah, with special reference to nesting and feeding habits. *Transactions 3rd North American Wildlife Conference,* pp. 852–64.

Rawley, E. V., and Bailey, W. J. 1964. *Utah upland game birds.* Utah State Department of Fish and Game publication 63–12.

Resadny, C. D. 1965. Huns on the move. *Wisconsin Conservation Bulletin* 30(6):21–23.

Richardson, F. 1941. Results of the southern California quail banding program. *California Fish and Game* 27:234–49.

Ricklefs, R. E. 1969. An analysis of nesting mortality in birds. *Smithsonian Contributions to Zoology* no. 9.

Ridgway, R., and Friedmann, H. 1946. *The birds of North and Middle America: part X (Galliformes).* Washington: Smithsonian Institution.

Ripley, T. H. 1957. *The bobwhite in Massachusetts.* Massachusetts Division of Fish and Game bulletin no. 15.

Rippin, A. P. 1970. Social organization and recruitment on the arena of sharp-tailed grouse. Master's thesis, University of Alberta.

Robel, R. J. 1965. Quantitative indices to activity and territoriality of booming *Tympanuchus cupido pinnatus* in Kansas. *Transactions Kansas Academy of Science* 67:702–12.

______. 1966. Booming territory size and mating success of the greater prairie chicken *(Tympanuchus cupido pinnatus). Animal Behaviour* 14:328–31.

______. 1967. Significance of booming grounds of greater prairie chickens. *Proceedings American Philosophical Society* 111:109–14.

______. 1970. Possible role of behaviour in regulating greater prairie chicken populations. *Journal of Wildlife Management* 34:306–12.

Robel, R. J.; Briggs, J. N.; Cebula, J. J.; Silvey, N. J.; Viers, C. E.; and Watt, P. G. 1970. Greater prairie chicken ranges, movements, and habitat usage in Kansas. *Journal of Wildlife Management* 34:286–306.

Roberts, H. S. 1963. Aspects of the life history and food habits of rock and willow ptarmigan. Master's thesis, University of Alaska.

Roberts, T. S. 1932. *The birds of Minnesota.* Vol. 1. Minneapolis: University of Minnesota Press.

Robeson, S. B. 1963. Some notes on the summer whistling of male bobwhite quail. *New York Fish and Game Journal* 10:228.

Robinson, T. S. 1957. *The ecology of bobwhites in south-central Kansas.* University of Kansas Museum of Natural History and State Biological Survey miscellaneous publication no. 15.

Robinson, W. L. 1969. Habitat selection by spruce grouse in northern Michigan. *Journal of Wildlife Management* 33:113–20.

Robinson, W. L., and Maxwell, D. E. 1968. Ecological study of the spruce grouse on the Yellow Dog Plains. *Jack-Pine Warbler* 46:75–83.

Rogers, G. B. 1964. *Sage grouse investigations in Colorado.* Colorado

Game, Fish and Parks Department, Game Research Division technical publication no. 16.
———. 1968. *The blue grouse in Colorado.* Colorado Game, Fish and Parks Department, Game Research Division technical publication no. 21.
———. 1969. *The sharp-tailed grouse in Colorado.* Colorado Game, Fish and Parks Department, Game Research Division technical publication no. 23.
Rogers, G. E., and Braun, C. E. 1967. Ptarmigan. *Colorado Outdoors* 16(4):22–28.
———. 1968. Ptarmigan management in Colorado. Mimeographed. Proceedings 48th Annual Conference of the Western Association of Game and Fish Commissioners, July 9, 1968, Reno, Nevada.
Romanoff, A. L.; Bump, G.; and Holm, E. 1938. *Artificial incubation of some upland game bird eggs.* New York Conservation Department bulletin no. 2.
Rosene, W. 1969. *The bobwhite quail: its life and management.* New Brunswick, N.J.: Rutgers University Press.
Rowan, W. 1952. The Hungarian partridge *(Perdix perdix)* in Canada. *Transactions Royal Society of Canada, Section III* 46:161–62.
Rowley, J. S. 1966. Breeding records of birds of the Sierra Madre del Sur, Oaxaca, Mexico. *Proceedings of the Western Foundation of Vertebrate Zoology* 1:107–204.
Russell, P. 1932. The scaled quail of New Mexico. Master's thesis, University of New Mexico.
Russell, S. M. 1964. *A distributional study of the birds of British Honduras.* American Ornithologists' Union ornithological monograph no. 1.
Salomonsen, F. 1939. Moults and sequence of plumages in the rock ptarmigan *(Lagopus mutus* [Montin]*). Dansk Ornithologisk Forenings Tidsskrift* 103:1–491.
Sandfort, W. W. 1954. Evaluation of chukar partridge range in Colorado. *Proceedings 34th Annual Conference of the Western Association of Game and Fish Commissioners,* p. 244–50.
Sands, J. L. 1968. Status of the lesser prairie chicken. *Audubon Field Notes* 22:454–56.
Saunders, G. B.; Halloway, A. D.; and Handley, C. O., Jr. 1950. *A fish and wildlife survey of Guatemala.* U.S. Fish and Wildlife Service special scientific report (wildlife) no. 5.
Schaldach, W. J., Jr. 1963. The avifauna of Colima and adjacent Jalisco, Mexico. *Proceedings of the Western Foundation of Vertebrate Zoology* 1:1–100.
Schemnitz, S. D. 1959. Past and present distribution of scaled quail *(Calli-*

pepla squamata) in Oklahoma. *Southwestern Naturalist* 43:148–52.
———. 1961. Ecology of the scaled quail in the Oklahoma panhandle. *Wildlife Monographs* 8:1–47.
———. 1964. Comparative ecology of bobwhite and scaled quail in the Oklahoma panhandle. *American Midland Naturalist* 71:429–33.
Schenkel, R. 1956–1958. Zur Deutung der Balzleistungen einiger Phasianiden und Tetraoniden. *Ornithologische Beobachter* 53:182–201; 55:65–95.
Schlatterer, E. F. 1960. Productivity and movements of a population of sage grouse in southeastern Idaho. Master's thesis, University of Idaho.
Schlotthauer, P. H. 1967. All about quail and grouse. *Game Bird Breeders, Conservationists, and Aviculturalists' Gazette* 16(3):9–11.
Schmidt, R. K., Jr. 1969. Behavior of white-tailed ptarmigan in Colorado. Master's thesis, Colorado State University.
Schneegas, E. R. 1967. Sage grouse and sagebrush control. *Transactions 32nd North American Wildlife Conference* 32:270–74.
Schwartz, C. W. 1945. The ecology of the prairie chicken in Missouri. *University of Missouri Studies* 20:1–99.
Scott, J. W. 1942. Mating behavior of the sage grouse. *Auk* 59:472–98.
———.1950. A study of the phylogenetic or comparative behavior of three species of grouse. *Annals of the New York Academy of Science* 51:477–98.
Scott, W. E. 1943. The Canada spruce grouse in Wisconsin. *Passenger Pigeon* 5:61–72.
———. 1947. The Canada spruce grouse *(Canachites canadensis canace). Wisconsin Conservation Bulletin* 12(3):27–30.
Seth-Smith, D. 1914. Notes from the Zoological Gardens. *Wild Life* 4:54.
———. 1925. The argus pheasant and its display. *Avicultural Magazine,* Series 4, 3:175–79.
———. 1929a. American quails or colins. *Avicultural Magazine,* Series 4, 7:64–67.
———. 1929b. Grouse (Tetraonidae). *Avicultural Magazine,* Series 4, 7:96–98.
Shaffner, C. S. 1955. Progesterone induced molt. *Poultry Science* 34:840–42.
Sharp, W. M. 1963. The effects of habitat manipulation and forest succession on ruffed grouse. *Journal of Wildlife Management* 27:664–71.
Sharpe, R. S. 1968. The evolutionary relationships and comparative behavior of prairie chickens. Ph.D. dissertation, University of Nebraska.
Shields, P. W., and Duncan, D. A. 1966. Fall and winter food of California quail in dry years. *California Fish and Game* 52:275–82.
Shoemaker, H. H. 1961. Rearing of young prairie chickens in captivity. *Illinois Wildlife* 16(4):1–4.
———. 1964. Report on studies of captive prairie chickens. *Illinois Wildlife* 19(3):6–8.

Short, L. L., Jr. 1967. A review of the genera of grouse (Aves, Tetraoninae). *American Museum Novitates* 2289:1–39.

Shrader, T. A. 1944. Ruffed and spruce grouse. *Conservation Volunteer* 7(40):36–40.

Sibley, C. G. 1957. The evolutionary and taxonomic significance of sexual dimorphism and hybridization in birds. *Condor* 59:166–91.

———. 1960. The electrophoretic patterns of avian egg-white proteins as taxonomic characters. *Ibis* 102:215–84.

Simpson, G. 1935. Breeding blue grouse in captivity. *Transactions American Game Conference* 21:218–19.

Sisson, L. H. 1970. Distribution and selection of sharptailed grouse dancing grounds in the Nebraska sand hills. *Proceedings of the 8th Conference of the Prairie Grouse Technical Council,* 1969.

Skutch, A. F. 1947. Life history of the marbled wood-quail. *Condor* 49: 217–32.

Slud, P. 1964. The birds of Costa Rica: distribution and ecology. *Bulletin of the American Museum of Natural History* 128:1–430.

Smith, N. D., and Buss, I. O. 1963. Age determination and plumage observations of blue grouse. *Journal of Wildlife Management* 27:566–78.

Smith, R. H., 1961. Age classification of the chukar partridge. *Journal of Wildlife Management* 25:84–86.

Smithe, F. B. 1966. *The birds of Tikal.* New York: Natural History Press.

Snyder, L. L. 1935. *A study of the sharp-tailed grouse.* University of Toronto Studies, Biological Series no. 40.

———. 1957. *Arctic birds of Canada.* Toronto: University of Toronto Press.

Southwood, T. R. E. 1967. The ecology of the partridge: II. the role of pre-hatching influences. *Journal of Animal Ecology* 36:557–62.

Southwood, T. R. E., and Cross, D. J. 1969. The ecology of the partridge: III. breeding success and the abundance of insects in natural habitats. *Journal of Animal Ecology* 38:497–509.

Sowls, L. K. 1960. Results of a banding study of Gambel's quail in southern Arizona. *Journal of Wildlife Management* 24:185–90.

Stanford, J. A. 1953. Quail do have second broods. *Missouri Conservationist* 14(12):5–6, 12.

Stanton, D. C. 1958. A study of breeding and reproduction in a sage grouse population in southeastern Idaho. Master's thesis, University of Idaho.

Stempel, M. E., and Rogers, S., Jr. 1961. History of prairie chickens in Iowa. *Proceedings Iowa Academy of Science* 68:314–22.

Stenlund, M. H., and Magnus, L. T. 1951. The spruce hen comes back. *Conservation Volunteer* 14(84):20–24.

Steward, P. A. 1967. Hooting of Sitka blue grouse in relation to weather, season, and time of day. *Journal of Wildlife Management* 31:28–34.

Stirling, I. 1968. Aggressive behavior and the dispersion of female blue grouse. *Canadian Journal of Zoology* 46:405–8.

Stirling, I., and Bendell, J. F. 1966. Census of blue grouse with recorded calls of a female. *Journal of Wildlife Management* 30:184–87.

———. 1970. The reproductive behavior of blue grouse. *Syesis* 3:161–71.

Stoddard, H. L. 1931. *The bobwhite quail: its habits, preservation, and increase.* New York: Charles Scribner's Sons.

Stokes, A. W. 1961. Voice and social behavior of the chukar partridge. *Condor* 63:111–27.

———. 1963. Agonistic and sexual behavior in the chukar partridge, *Alectoris graeca. Animal Behaviour* 11:121–34.

———. 1967. Behavior of the bobwhite, *Colinus virginianus. Auk* 84:1–33.

Stokes, A. W., and Williams, H. W. 1968. Antiphonal calling in quail. *Auk* 85:83–89.

Stone, W. 1890. On birds collected in Yucatan and southern Mexico. *Proceedings Academy of Science of Philadelphia* 43:201–18.

Stoneberg, R. P. 1967. A preliminary study of the breeding biology of the spruce grouse in northwestern Montana. Master's thesis, University of Montana.

Stonehouse, B. 1966. Egg volume from linear dimensions. *Emu* 65:227–28.

Storer, R. W. 1961. *Two collections of birds from Campeche, Mexico.* University of Michigan Museum of Zoology occasional paper no. 621.

Stresemann, E. 1966. Die Mauser der Vögel, *Journal für Ornithologie,* Vol. 107 (special edition), pp. 1–448.

Stupka, A. 1963. *Notes on the birds of the Great Smoky Mountains National Park.* Knoxville: University of Tennessee Press.

Sturkie, P. D. 1965. *Avian physiology.* 2d. ed. Ithaca, N. Y.: Cornell University Press.

Sumner, E. L. 1935. A life history study of the California quail, with recommendations for its conservation and management. *California Fish and Game* 21:167–253, 275–342.

Sumner, L., and Dixon, J. S. 1953. *Birds and mammals of the Sierra Nevada.* Berkeley: University of California Press.

Suskin, P. P. 1927. Notes on *Alectoris chukar,* with descriptions of six new subspecies. *Bulletin British Ornithologists' Club* 48(317):22–27.

Sutherland, C. A., and McChesney, D. S. 1965. Sound production in two species of geese. *The Living Bird* 4:99–106.

Sutton, G. M. 1967. *Oklahoma birds.* Norman: University of Oklahoma Press.

———. 1968. The natal plumage of the lesser prairie chicken. *Auk* 85:679.

Sutton, G. M., and Parmelee, D. F. 1956. The rock ptarmigan in southern Baffin Island. *Wilson Bulletin* 68:52–62.

Swank, W. G., and Gallizioli, S. 1954. The influence of hunting and of rainfall upon Gambel's quail populations. *Transactions 19th North American Wildlife Conference*, p. 283–95.

Swarth, H. S. 1909. Distribution and molt of the Mearns quail. *Condor* 11:39–43.

Symington, D. F., and Harper, T. A. 1957. *Sharp-tailed grouse in Saskatchewan.* Saskatchewan Department of Natural Resources conservation bulletin no. 4.

Tait, H. D., coordinator. 1968. *Index to Federal aid Publications in sport fish and wildlife restoration and selected cooperative research project reports, to March 1968.* Washington: U.S. Department of Interior.

Thomas, K. P. 1969. Sex determination of bobwhites by wing criteria. *Journal of Wildlife Management* 33:215–16.

Thompson, D. R., and Kabat, C. 1950. The wing molt of the bobwhite. *Wilson Bulletin* 62:20–31.

Thurman, J. R. 1966. Ruffed grouse ecology in southeastern Monroe County, Indiana. Master's thesis, Purdue University.

Todd, W. E. C. 1920. A revision of the genus *Eupsychortyx*. *Auk* 37:189–220.

———. 1940. Eastern races of the ruffed grouse. *Auk* 57:390–97.

———. 1963. *Birds of the Labrador Peninsula and adjacent areas.* Toronto: University of Toronto Press.

Tordoff, H. B. 1951. A quail from the Oligocene of Colorado. *Condor* 53:203.

Trippensee, R. E. 1948. *Wildlife management: upland game and general principles.* New York: McGraw Hill Book Co.

Trueblood, R. W. 1954. The effect of grass reseeding in sagebrush lands on sage grouse populations. Master's thesis, Utah State Agricultural College.

Tsukamota, G. 1970. Poor man's grand slam. *Nevada Outdoors* 4(3):18–20.

Tuck, L. M. 1968. Recent Newfoundland bird records. *Auk* 85:304–11.

Tufts, R. W. 1961. *The birds of Nova Scotia.* Halifax: Nova Scotia Museum.

Upland game birds of Idaho. 1951. Idaho Fish and Game Commission.

van Rossem, A. J. 1925. Flight feathers as age indicators in *Dendragapus. Ibis*, Series 12, 1:417–22.

———. 1931. Report on a collection of birds from Sonora, Mexico. *Transactions San Diego Society of Natural History* 6:237–304.

———. 1945. *A distributional survey of the birds of Sonora, Mexico.* Louisiana State University Museum of Zoology occasional paper no. 21.

van Tyne, J. 1935. *The birds of northern Petén, Guatemala.* University of

Michigan Museum of Zoology miscellaneous publication no. 27.
Verheyen, W. 1962. The Congo peacock *Afropavo congensis* Chapin 1936 at Antwerp Zoo. *International Zoo Yearbook* 4:87–91.
Vorhies, C. T. 1929. Do southwestern quail require water? *Arizona Wildlife* 2:154–58.
Wallmo, O. C. 1951. *Fort Huachuca wildlife area surveys, 1950–1951.* Tucson: Arizona Game and Fish Commission.
______. 1954. Nesting of Mearns quail in southeastern Arizona. *Condor* 56:125–28.
______. 1956a. Determination of sex and age of scaled quail. *Journal of Wildlife Management* 20:154–58.
______. 1956b. *Ecology of scaled quail in west Texas.* Austin: Texas Game and Fish Commission.
______. 1957. A study of blues. *Texas Game and Fish* 15(8):4–7.
Warner, D. W. 1959. The song, nest, eggs, and young of the long-tailed partridge. *Wilson Bulletin* 71:307–12.
Warner, D. W., and Harrell, B. E. 1957. The systematics and biology of the singing quail, *Dactylortyx thoracicus. Wilson Bulletin* 69:123–48.
Warner, N. L., and Szenberg, A. 1964. The immunological function of the bursa of Fabricius in the chicken. *Annual Review of Microbiology* 18:253–68.
Watson, A. 1965. A population study of ptarmigan *(Lagopus mutus)* in Scotland. *Journal of Animal Ecology* 34:135–72.
Watson, A., and Jenkins, D. 1964. Notes on the behaviour of the red grouse. *British Birds* 57:137–70.
Watson, A.; Parr, R.; and Lumsden, H. G. 1969. Differences in the downy young of red and willow grouse and ptarmigan. *British Birds* 62:150–53.
Watson, G. E. 1962a. Three sibling species of *Alectoris* partridge. *Ibis* 104:353–67.
______. 1962b. Sympatry in Palearctic *Alectoris* partridges. *Evolution* 16: 11–19.
______. 1962c. Molt, age determination, and annual cycle in the Cuban bobwhite. *Wilson Bulletin* 74:28–42.
______. 1963. Incomplete first prebasic molt in the chukar partridge. *Auk* 80:80–81.
Wayre, P. 1964. Display of the common koklass. *Annual Report of the Ornamental Pheasant Trust,* p. 13.
Weaver, H. R., and Haskell, W. L. 1968. Age and sex determination of the chukar partridge. *Journal of Wildlife Management* 32:46–50.
Weeden, R. B. 1959. Ptarmigan research project, final report. Mimeographed. *Arctic Institute of North America.*

———. 1961. Outer primaries as indicators of age among rock ptarmigan. *Journal of Wildlife Management* 25:337–39.

———. 1963. Management of ptarmigan in North America. *Journal of Wildlife Management* 27:673–83.

———. 1964. Spatial separation of sexes in rock and willow ptarmigan in Winter. *Auk* 81:534–41.

———. 1965a. Breeding density, reproductive success, and mortality of rock ptarmigan at Eagle Creek, Alaska, from 1960–1964. *Transactions 30th North American Wildlife Conference*, p. 336–48.

———. 1965b. Grouse and ptarmigan in Alaska: their ecology and management. Mimeographed. Alaska Department of Fish and Game.

———. 1967. Seasonal and geographic variation in the foods of adult white-tailed ptarmigan. *Condor* 69:303–9.

Weeden, R. B., and Watson, A. 1967. Determining the age of rock ptarmigan in Alaska and Scotland. *Journal of Wildlife Management* 31:825–26.

West, G. C., and Meng, M. S. 1966. Nutrition of willow ptarmigan in northern Alaska. *Auk* 83:603–15.

Westerskov, K. 1956. Age determination and dating nesting events in the willow ptarmigan. *Journal of Wildlife Management* 20:274–79.

———. 1965. Winter ecology of the partridge *(Perdix perdix)* in the Canadian prairie. *Proceedings New Zealand Ecological Society* 12:23–30.

———. 1966. Winter food and feeding habits of the partridge *(Perdix perdix)* in the Canadian prairie. *Canadian Journal of Zoology* 44:303–22.

Wetmore, A. 1941. Notes on birds of the Guatemala highlands. *Proceedings of the U.S. National Museum* 89:523–81.

———. 1960. A classification for the birds of the world. *Smithsonian Miscellaneous Collections* 139(11):1–37.

———. 1965. *The birds of the Republic of Panamá: pt. 1. Tinamidae (tinamous) to Rhynchopidae (skimmers).* Smithsonian Miscellaneous Collections no. 150.

White, C. L., and Foscue, E. J. 1939. Henequen: the green gold of Yucatán. *Journal of Geography* 38:151–55.

Williams, G. R. 1967. The breeding biology of California quail in New Zealand. *New Zealand Ecological Society Proceedings* 14:88–99.

Williams, H. W. 1969. Vocal behavior of adult California quail. *Auk* 84: 631–59.

Williams, H. W., and Stokes, A. W. 1965. Factors affecting the incidence of rally calling in the chukar partridge. *Condor* 67:31–43.

Wing, L. 1946. Drumming flight in the blue grouse and courtship characters in the Tetraonidae. *Condor* 48:154–57.

Wing, L.; Beer, J.; and Tidyman, W. 1944. The brood habits and growth of "blue grouse." *Auk* 61:426–40.

Wood-Gush, D. G. M. 1954. The courtship of the brown leghorn cock. *British Journal of Animal Behaviour* 2:95–102.

———. 1956. The agonistic and courtship behavior of the brown leghorn cock. *British Journal of Animal Behaviour* 4:133–42.

Wright, V. L. 1966. Status of the gray partridge in Indiana. Master's thesis, Purdue University.

Yeatter, R. E. 1935. *The Hungarian partridge in the Great Lakes Region.* University of Michigan School of Forestry and Conservation Bulletin no. 5.

———. 1943. The prairie chicken in Illinois. *Illinois Natural History Bulletin* 22:377–416.

———. 1963. Population responses of prairie chickens to land-use changes in Illinois. *Journal of Wildlife Management* 27:739–57.

Yocom, C. F. 1943. The Hungarian partridge, *Perdix perdix* (L.), in the Palouse region, Washington. *Ecological Monographs* 13:167–201.

Yocom, C. F., and Harris, S. W. 1952. Food habits of mountain quail *(Oreortyx picta)* in eastern Washington. *Journal of Wildlife Management* 17:204–7.

Zimmerman, D. A., and Harry, G. B. 1951. Summer birds of Autlan, Jalisco. *Wilson Bulletin* 63:302–14.

Zwickel, F. C. 1966. Early mortality and the numbers of blue grouse. Ph.D. dissertation, University of British Columbia.

———. 1967. Early behaviour in young blue grouse. *Murrelet* 48:2–7.

Zwickel, F. C., and Bendell, J. F. 1967. Early mortality and the regulation of numbers in blue grouse. *Canadian Journal of Zoology* 45:817–51.

Zwickel, F. C., Brigham, J. H., and Buss, I. O. 1966. Autumn weights of blue grouse in north-central Washington, 1954 to 1963. *Condor* 68:488–96.

Zwickel, F. C., Buss, I. O., and Brigham, J. H. 1968. Autumn movements of blue grouse and their relevance to populations and management. *Journal of Wildlife Management* 32:456–68.

Zwickel, F. C., and Lance, A. N. 1965. Renesting in blue grouse. *Journal of Wildlife Management* 29:202–4.

———. 1966. Determining the age of young blue grouse. *Journal of Wildlife Management* 30:712–17.

Zwickel, F. C., and Martinsen, C. F. 1967. Determining the age and sex of Franklin spruce grouse by tails alone. *Journal of Wildlife Management* 31:760–63.

Index

Vernacular Names

ENGLISH vernacular names indexed here are for the most part those used in this book for species or larger groupings. Vernacular names for subspecies as well as alternative vernacular names for species are included only if they are in general usage or have been referred to in the text discussions. Plates and figures are identified by number, and pages containing major discussions of each species are indicated by boldface.

Scientific Names

Names indexed here are restricted to those of subspecies, species, or larger groupings of gallinaceous birds mentioned in the text. Technical names of other animal groups and plants are not indexed. Entries shown here are for the major page references; the index to vernacular names should be consulted for secondary references and references to illustrations.